Cyber Security and Critical Infrastructures

Cyber Security and Critical Infrastructures

Editors

**Leandros Maglaras
Helge Janicke
Mohamed Amine Ferrag**

MDPI • Basel • Beijing • Wuhan • Barcelona • Belgrade • Manchester • Tokyo • Cluj • Tianjin

Editors

Leandros Maglaras
School of Computer Science
and Informatics, De Montfort
University, Leicester LE1
9BH, UK

Helge Janicke
Cyber Security Cooperative
Research Centre, Building 15
Level 2/270 Joondalup Dr,
Joondalup, WA 6027,
Australia

Mohamed Amine Ferrag
Department of Computer
Science, Guelma University,
Guelma, Algeria

Editorial Office
MDPI
St. Alban-Anlage 66
4052 Basel, Switzerland

This is a reprint of articles from the Topic published online in the open access journals *Sensors* (ISSN 1424-8220), *Applied Sciences* (ISSN 2076-3417), *Electronics* (ISSN 2079-9292), and *Future Internet* (ISSN 1999-5903) (available at: https://www.mdpi.com/topics/Cyber_Security_Critical_Infrastructures).

For citation purposes, cite each article independently as indicated on the article page online and as indicated below:

LastName, A.A.; LastName, B.B.; LastName, C.C. Article Title. *Journal Name* **Year**, *Volume Number*, Page Range.

ISBN 978-3-0365-4845-6 (Hbk)
ISBN 978-3-0365-4846-3 (PDF)

© 2022 by the authors. Articles in this book are Open Access and distributed under the Creative Commons Attribution (CC BY) license, which allows users to download, copy and build upon published articles, as long as the author and publisher are properly credited, which ensures maximum dissemination and a wider impact of our publications.

The book as a whole is distributed by MDPI under the terms and conditions of the Creative Commons license CC BY-NC-ND.

Contents

About the Editors vii

Preface to "Cyber Security and Critical Infrastructures" ix

Leandros Maglaras, Helge Janicke, Mohamed Amine Ferrag
Cybersecurity of Critical Infrastructures: Challenges and Solutions
Reprinted from: *Sensors* **2022**, *22*, 5105, doi:10.3390/s22145105 1

Aleksandr Ometov, Oliver Liombe Molua, Mikhail Komarov and Jari Nurmi
A Survey of Security in Cloud, Edge, and Fog Computing
Reprinted from: *Sensors* **2022**, *22*, 927, doi:10.3390/s22030927 5

Zixiang Bi, Guoai Xu, Guosheng Xu, Chenyu Wang and Sutao Zhang
Bit-Level Automotive Controller Area Network Message Reverse Framework Based on Linear Regression
Reprinted from: *Sensors* **2022**, *22*, 981, doi:10.3390/s22030981 33

Mingrui Ma, Lansheng Han and Yekui Qian
CVDF DYNAMIC—A Dynamic Fuzzy Testing Sample Generation Framework Based on BI-LSTM and Genetic Algorithm
Reprinted from: *Sensors* **2022**, *22*, 1265, doi:10.3390/s22031265 63

Najla Al-Taleb and Nazar Abbas Saqib
Towards a Hybrid Machine Learning Model for Intelligent Cyber Threat Identification in Smart City Environments
Reprinted from: *Appl. Sci.* **2022**, *12*, 1863, doi:10.3390/app12041863 85

Auwal Sani Iliyasu, Usman Alhaji Abdurrahman and Lirong Zheng
Few-Shot Network Intrusion Detection Using Discriminative Representation Learning with Supervised Autoencoder
Reprinted from: *Appl. Sci.* **2022**, *12*, 2351, doi:10.3390/app12052351 101

Chih-Chieh Hung, Chuang-Chieh Lin, Hsien-Chu Wu, Chia-Wei Lin
A Study on Reversible Data Hiding Technique Based on Three-Dimensional Prediction-Error Histogram Modification and a Multilayer Perceptron
Reprinted from: *Appl. Sci.* **2022**, *12*, 2502, doi:10.3390/app12052502 119

Ángel Longueira-Romero, Rosa Iglesias, Jose Luis Flores and Iñaki Garitano
A Novel Model for Vulnerability Analysis through Enhanced Directed Graphs and Quantitative Metrics
Reprinted from: *Sensors* **2022**, *22*, 2126, doi:10.3390/s22062126 143

Hasan Alkahtani and Theyazn H. H. Aldhyani
Artificial Intelligence Algorithms for Malware Detection in Android-Operated Mobile Devices
Reprinted from: *Sensors* **2022**, *22*, 2268, doi:10.3390/s22062268 171

Murilo Góes de Almeida and Edna Dias Canedo
Authentication and Authorization in Microservices Architecture: A Systematic Literature Review
Reprinted from: *Appl. Sci.* **2022**, *12*, 3023, doi:10.3390/app12063023 197

Chen Hajaj, Nitay Hason and Amit Dvir
Less Is More: Robust and Novel Features for Malicious Domain Detection
Reprinted from: *Electronics* **2022**, *11*, 969, doi:10.3390/electronics11060969 217

Joakim Kävrestad, Allex Hagberg, Marcus Nohlberg, Jana Rambusch, Robert Roos and Steven Furnell
Evaluation of Contextual and Game-Based Training for Phishing Detection
Reprinted from: *Future Internet* **2022**, *14*, 104, doi:10.3390/fi14040104 . . . . **237**

João Vitorino, Nuno Oliveira and Isabel Praça
Adaptative Perturbation Patterns: Realistic Adversarial Learning for Robust Intrusion Detection
Reprinted from: *Future Internet* **2022**, *14*, 108, doi:10.3390/fi14040108 . . . . **253**

Hua Shen, Xinyue Liu and Xianchao Zhang
A Detection Method for Social Network Images with Spam, Based on Deep Neural Network and Frequency Domain Pre-Processing
Reprinted from: *Electronics* **2022**, *11*, 1081, doi:10.3390/electronics11071081 . . . . **271**

Simona Ramanauskaitė, Anatoly Shein, Antanas Čenys and Justinas Rastenis
Security Ontology Structure for Formalization of Security Document Knowledge
Reprinted from: *Electronics* **2022**, *11*, 1103, doi:10.3390/electronics11071103 . . . . **281**

Mahmoud Al-Dwairi, Ahmed S. Shatnawi, Osama Al-Khaleel and Basheer Al-Duwairi
Ransomware-Resilient Self-Healing XML Documents
Reprinted from: *Future Internet* **2022**, *14*, 115, doi:10.3390/fi14040115 . . . . **295**

About the Editors

Leandros Maglaras

Leandros Maglaras is a Professor of cybersecurity at the School of Computer Science and Informatics, De Montfort University, UK. From September 2017 to November 2019, he was the Director of the National Cyber Security Authority of Greece. Contact him at leandros.maglaras@dmu.ac.uk.

Helge Janicke

Helge Janicke is the Research Director of the Cyber Security Cooperative Research Centre, Australia, Professor at Edith Cowan University and Visiting Professor at De Montfort University, U.K. Contact him at helge.janicke@cybersecuritycrc.org.au.

Mohamed Amine Ferrag

Mohamed Amine Ferrag is a Senior Lecturer at the Department of Computer Science, Guelma University, Algeria and Visiting Senior Researcher at the NAU-Lincoln Joint Research Center of Intelligent Engineering, Nanjing Agricultural University, China. Contact him at ferrag.mohamedamine@univ-guelma.dz.

Preface to "Cyber Security and Critical Infrastructures"

Critical infrastructures are essential for national public security, economic well-being and national security. Critical vulnerabilities in such infrastructures are increasing with the proliferation of information technology. As critical infrastructures become more vulnerable to cyber-attacks, protecting them has become an important issue for any organization or country. Due to the apparent impact of such conditions, the risks to ongoing operations, such as failure to upgrade legacy infrastructure or failure to comply with required regulatory regimes, are high.

With the rapid proliferation of complex cyber threats targeting critical infrastructures, which have a significant disruptive impact, cybersecurity for critical infrastructures is an important issue for academics, professionals and policy-makers. The effective cybersecurity management of critical infrastructures requires a comprehensive overview of the technical, political, human and behavioral aspects. Furthermore, the coronavirus pandemic poses new challenges for companies adapting to business models in which working from home has become "the new normal". Businesses are accelerating digital transformation and cybersecurity is now a major concern.

This book presents manuscripts that were accepted after a careful peer-review process for publication in the topic "Cyber Security and Critical Infrastructures" by the Applied Sciences, Electronics, Future Internet, Sensors and Smart Cities MDPI journals. The book includes sixteen articles: an editorial, fifteen original research papers describing current challenges, innovative solutions and real-world experiences involving critical infrastructures, and one review paper focusing on the security and privacy challenges in Cloud, Edge, and Fog computing.

Leandros Maglaras, Helge Janicke, and Mohamed Amine Ferrag
Editors

Editorial

Cybersecurity of Critical Infrastructures: Challenges and Solutions

Leandros Maglaras [1,*], Helge Janicke [2] and Mohamed Amine Ferrag [3]

1. School of Computer Science and Informatics, De Montfort University, Leicester LE1 9BH, UK
2. Cyber Security Cooperative Research Centre, Edith Cowan University, Perth 6027, Australia; helge.janicke@cybersecuritycrc.org.au
3. Department of Computer Science, Guelma University, Guelma 24000, Algeria; ferrag.mohamedamine@univ-guelma.dz
* Correspondence: leandros.maglaras@dmu.ac.uk

Citation: Maglaras, L.; Janicke, H.; Ferrag, M.A. Cybersecurity of Critical Infrastructures: Challenges and Solutions. *Sensors* **2022**, *22*, 5105. https://doi.org/10.3390/s22145105

Received: 21 June 2022
Accepted: 6 July 2022
Published: 7 July 2022

Publisher's Note: MDPI stays neutral with regard to jurisdictional claims in published maps and institutional affiliations.

Copyright: © 2022 by the authors. Licensee MDPI, Basel, Switzerland. This article is an open access article distributed under the terms and conditions of the Creative Commons Attribution (CC BY) license (https://creativecommons.org/licenses/by/4.0/).

People's lives are becoming more and more dependent on information and computer technology. This is accomplished by the enormous benefits that the ICT offers for everyday life. Digital technology creates an avenue for communication and networking, which is characterized by the exchange of data, some of which are considered sensitive or private. There have been many reports recently of data being hijacked or leaked, often for malicious purposes. Maintaining security and privacy of information and systems has become a herculean task. It is therefore imperative to understand how an individual's or organization's personal data can be protected. Moreover, critical infrastructures are vital resources for the public safety, economic well-being and national security.

The major target of cyber attacks can be a country's Critical National Infrastructures (CNIs) like ports, hospitals, water, gas or electricity producers, that use and rely on Industrial Control Systems but are affected by threats to any part of the supply chain. Cyber attacks are increasing at rate and pace, forming a major trend. The widespread use of computers and the Internet, coupled with the threat of activities of cyber criminals, has made it necessary to pay more attention to the detection or improve the technologies behind information security. The rapid reliance on cloud-based data storage and third-party technologies makes it difficult for industries to provide security for their data systems. Cyber attacks against critical systems are now common and recognized as one of the greatest risks facing today's world [1].

This editorial presents the manuscripts accepted, after a careful peer-review process, for publication in the topic "Cyber Security and Critical Infrastructures" of the MDPI journals Applied Sciences, Electronics, Future Internet, Sensors and Smart Cities. The first volume includes sixteen articles: one editorial article, fifteen original research papers describing current challenges, innovative solutions, and real-world experiences involving critical infrastructures and one review paper focusing on the security and privacy challenges on Cloud, Edge, and Fog computing.

Many companies have recently decided to use cloud, edge and fog computing in order to achieve high storage capacity and efficient scalability. The work presented in [2] mainly focuses on how security in Cloud, Edge, and Fog Computing systems is achieved and how users' privacy can be protected from attackers. The authors mention that there is a huge potential for vulnerabilities in security and privacy of such system. One good way of screening systems for possible vulnerabilities is by performing auditing of the systems based on security standards.

The recent EU Directive on security of network and information systems (the NIS Directive) has identified transport as one of the critical sectors that need to be secured in a European level. Smart cars is changing the transport landscape by introducing new capabilities along with new threats. Focusing on vehicle security, the authors in [3] examine the bit-level CAN bus reverse framework using a multiple linear regression model. The increasingly diverse features in today's vehicles offer drivers and passengers a more relaxed driving experience and greater convenience along with new security threats. The reverse capability of the proposed system can help automotive security researchers to describe vehicle behavior using CAN messages when DBC files are not available.

Vulnerabilities in computer programs have always been a serious threat to software security, which may cause denial of service, information leakage and other attacks. The authors in [4] propose a new framework of fuzzy testing sample generation called CVDF DYNAMIC. which consists of three parts: Sample generation based on a genetic algorithm, sample generation based on a bi-LSTM neural network and sample reduction based on a heuristic genetic algorithm.

The transformation of cities into smart cities is on the rise. Through the use of innovative technologies such as the Internet of Things (IoT) and cyber–physical systems (CPS) that are connected through networks, smart cities offer better services to the citizens. The authors in propose a novel machine learning solution for threat detection in a smart city [5].The proposed hybrid Deep learning model that consists of QRNN and CNN improves cyber threat analysis accuracy, loweres False Postitive rate, and provides real-time analysis. The authors evaluated the proposed model on two datasets that were simulated to represent a realistic IoT environment and proved its superiority.

The next article in this collection [6] proposes a novel framework for few-shot network intrusion detection. Based on the fact that DL methods have been widely successful as network-based IDSs but require sizeable volumes of datasets which are not always feasible, the authors focus on few-shot solutions. Their proposed method is suitable for detecting specific classes of attacks. This model could be very helpful for deploying novel IDSs for Industrial Control Systems, which are the core of Critical Infrastructures, where there is a general lack of datasets.

In [7] the authors propose a novel reversible data hiding (RDH) scheme that can be applied to either remote medical diagnosis or even military secret transmission. The authors utilize a trained multi-layer perception neural network in order to be able to predict pixel values and then combining those with prediction error expansion techniques (PEE) to achieve (RDH). The proposed method although efficient is very time consuming and the authors propose in the future to implement novel solution to improve this aspect.

Focusing on Industrial components that are the main parts of critical infrastructures the authors in [8] propose a model for vulnerability analysis through the their entire life-cycle. The model can Identify the root causes and nature of vulnerabilities for the industrial components. This information is useful extracting new requirements and test cases, support the prioritization of patching and track vulnerabilities during the whole life-cycle of industrial components. The proposed model is applicable to existing systems and can be a good source of information for defining patching, training and security needs.

Android mobile devices are becoming the targets of several attacks nowadays since they support many of the everyday digital needs of the users. Since many sensitive applications are offered in these smart devices, like e-banking, adversaries have launched a number of new attacks. IoT enhances the power of malicious entities or people to perform attacks on critical systems or services. A lot of connected devices additionally mean a bigger attack surface for attacks and greater risk. Hackers using infected devices can generate many frequent, organized and complex malicious attacks. The authors in [9] propose novel IDS for malware in android devices combining several machine learning techniques. The proposed classifiers achieved good accuracy outperforming existing state-of-the-art models.

Having identified a lack of studies related to security in microservices architecture and especially for for authentication and authorization to such systems, the authors in [10] perform an analysis about this open issue. Microservices can increase scalability, availability and reliability of the system but come with an increase in the attack surface and new threats in the communication between them. Since microservices can become an integral part of critical systems, a thorough research on the attacks and defence against them is crucial. The article concludes that several existing solutions can be applied to make the systems robust but also novel methods need to be proposed that are tailored to the new architectures.

In another article that deals with machine learning as a defence mechanism for smart systems, the authors in [11] focus on the correct feature selection. Feature selection is the process of correctly identifying those features that help the machine learning algorithm be robust against an adversary. The article proposes a smart feature selection process and a novel feature engineering process which are proven to be more precise in terms of manipulated data while maintaining good results on clean data. The proposed solutions can be easily adopted in real environments in order to deal with sophisticated attacks against critical infrastructures.

Information Security Awareness Training is used to raise awareness of the users against cyber attacks and help them build a responsible behavior. In [12] the authors try to answer the question whether game-based training and Context-Based Micro-Training (CBMT) can help users correctly identify phishing against legitimate emails. IN order to answer this question the authors conducted a simulated experiment with 41 participants and the results showed that both methods managed to improve user behavior in relation to phishing emails. The paper concludes that training is a strong tool against cyber attacks but must be combined with other security solutions.

A vital challenge faced nowadays by federal and business decision-makers for choosing cost-efficient mitigations to scale back risks from supply chain attacks, particularly those from adversarial attacks that are complex, hard to detect and can lead to severe consequences. Focusing on adversarial attacks and how these can alter the performance of AI based detection systems, the authors in [13] propose a novel robust solution. Their proposed model was evaluated in both Enterprise and Internet of Things (IoT) networks and is proven to be efficient against adversarial classification attacks and adversarial training attacks.

There are many reasons why it's vital to know what users can perceive as believable. It is crucial for service suppliers to grasp their vulnerabilities so as to assess their exposure to risks and also the associated problems. moreover, recognizing what the vulnerabilities are interprets into knowing from wherever the attacks are likely to come which leads for appropriate technical security measures to be deployed to protect against attacks. In [14] the authors present a solution that combines deep neural network and frequency domain pre-processing in order to detect images with embedded spam in social networks. The proposed method is proven to be superior against state-of-the-art detection models in terms of detection accuracy and efficiency. One of the major contributions of the authors is the creation of a novel dataset that contains images with embedded spam, which will be expanded in the near future.

Finding the correct sources that include vital information about securing critical systems is very important. Unfortunately, the lack of a fully functioning semantic web or text-based solutions to formalize security data sources limits the exploitation of existing cyber intelligence data sources. In [15] the authors aim to empower ontology-based cyber intelligence solutions by presenting a security ontology framework for storing data in an ontology from various textual data sources, supporting knowledge traceability and evaluating relationships between different security documents.

Ransomware has become one of the major threats against critical systems the latest years. The recent report from ENISA has ranked ransomware attacks first in terms of severity and frequency. Current solutions against ransomware do not cover all possible risks of data loss. In this article [16], the authors try to address this aspect and provide an effective solution that ensures efficient recovery of XML documents after ransomware attacks.

Author Contributions: All the authors contributed equally to this editorial. All authors have read and agreed to the published version of the manuscript.

Funding: This research received no external funding.

Institutional Review Board Statement: Not applicable.

Informed Consent Statement: Not applicable.

Data Availability Statement: Not applicable.

Conflicts of Interest: The authors declare no conflict of interest.

References

1. Maglaras, L.; Ferrag, M.A.; Derhab, A.; Mukherjee, M.; Janicke, H. Cyber security: From regulations and policies to practice. In *Strategic Innovative Marketing and Tourism*; Springer: Berlin/Heidelberg, Germany, 2019; pp. 763–770.
2. Ometov, A.; Molua, O.L.; Komarov, M.; Nurmi, J. A Survey of Security in Cloud, Edge, and Fog Computing. *Sensors* **2022**, *22*, 927. [CrossRef] [PubMed]
3. Bi, Z.; Xu, G.; Xu, G.; Wang, C.; Zhang, S. Bit-Level Automotive Controller Area Network Message Reverse Framework Based on Linear Regression. *Sensors* **2022**, *22*, 981. [CrossRef] [PubMed]
4. Ma, M.; Han, L.; Qian, Y. CVDF DYNAMIC—A Dynamic Fuzzy Testing Sample Generation Framework Based on BI-LSTM and Genetic Algorithm. *Sensors* **2022**, *22*, 1265. [CrossRef] [PubMed]
5. Al-Taleb, N.; Saqib, N.A. Towards a Hybrid Machine Learning Model for Intelligent Cyber Threat Identification in Smart City Environments. *Appl. Sci.* **2022**, *12*, 1863. [CrossRef]
6. Iliyasu, A.S.; Abdurrahman, U.A.; Zheng, L. Few-shot network intrusion detection using discriminative representation learning with supervised autoencoder. *Appl. Sci.* **2022**, *12*, 2351. [CrossRef]
7. Hung, C.C.; Lin, C.C.; Wu, H.C.; Lin, C.W. A Study on Reversible Data Hiding Technique Based on Three-Dimensional Prediction-Error Histogram Modification and a Multilayer Perceptron. *Appl. Sci.* **2022**, *12*, 2502. [CrossRef]
8. Longueira-Romero, Á.; Iglesias, R.; Flores, J.L.; Garitano, I. A Novel Model for Vulnerability Analysis through Enhanced Directed Graphs and Quantitative Metrics. *Sensors* **2022**, *22*, 2126. [CrossRef] [PubMed]
9. Alkahtani, H.; Aldhyani, T.H. Artificial Intelligence Algorithms for Malware Detection in Android-Operated Mobile Devices. *Sensors* **2022**, *22*, 2268. [CrossRef] [PubMed]
10. de Almeida, M.G.; Canedo, E.D. Authentication and Authorization in Microservices Architecture: A Systematic Literature Review. *Appl. Sci.* **2022**, *12*, 3023. [CrossRef]
11. Hajaj, C.; Hason, N.; Dvir, A. Less is more: Robust and novel features for malicious domain detection. *Electronics* **2022**, *11*, 969. [CrossRef]
12. Kävrestad, J.; Hagberg, A.; Nohlberg, M.; Rambusch, J.; Roos, R.; Furnell, S. Evaluation of Contextual and Game-Based Training for Phishing Detection. *Future Internet* **2022**, *14*, 104. [CrossRef]
13. Vitorino, J.; Oliveira, N.; Praça, I. Adaptative Perturbation Patterns: Realistic Adversarial Learning for Robust Intrusion Detection. *Future Internet* **2022**, *14*, 108. [CrossRef]
14. Shen, H.; Liu, X.; Zhang, X. A Detection Method for Social Network Images with Spam, Based on Deep Neural Network and Frequency Domain Pre-Processing. *Electronics* **2022**, *11*, 1081. [CrossRef]
15. Ramanauskaitė, S.; Shein, A.; Čenys, A.; Rastenis, J. Security Ontology Structure for Formalization of Security Document Knowledge. *Electronics* **2022**, *11*, 1103. [CrossRef]
16. Al-Dwairi, M.; Shatnawi, A.S.; Al-Khaleel, O.; Al-Duwairi, B. Ransomware-Resilient Self-Healing XML Documents. *Future Internet* **2022**, *14*, 115. [CrossRef]

Review

A Survey of Security in Cloud, Edge, and Fog Computing

Aleksandr Ometov [1,2], Oliver Liombe Molua [1], Mikhail Komarov [3] and Jari Nurmi [1,*]

1. Electrical Engineering Unit, Faculty of Information Technology and Communication Sciences, Tampere University, 33720 Tampere, Finland; aleksandr.ometov@tuni.fi (A.O.); oliverliombe.molua@tuni.fi (O.L.M.)
2. Laboratory of Cryptographic Methods of Information Security, Faculty of Secure Information Technologies, ITMO University, 191002 St. Petersburg, Russia
3. Graduate School of Business, National Research University—Higher School of Economics, 101000 Moscow, Russia; mkomarov@hse.ru
* Correspondence: jari.nurmi@tuni.fi

Abstract: The field of information security and privacy is currently attracting a lot of research interest. Simultaneously, different computing paradigms from Cloud computing to Edge computing are already forming a unique ecosystem with different architectures, storage, and processing capabilities. The heterogeneity of this ecosystem comes with certain limitations, particularly security and privacy challenges. This systematic literature review aims to identify similarities, differences, main attacks, and countermeasures in the various paradigms mentioned. The main determining outcome points out the essential security and privacy threats. The presented results also outline important similarities and differences in Cloud, Edge, and Fog computing paradigms. Finally, the work identified that the heterogeneity of such an ecosystem does have issues and poses a great setback in the deployment of security and privacy mechanisms to counter security attacks and privacy leakages. Different deployment techniques were found in the review studies as ways to mitigate and enhance security and privacy shortcomings.

Keywords: computing; survey; security; privacy; distributed systems; computational offloading

1. Introduction

The goal of having a huge capacity for storage with efficient scalability has recently been the driving force for different enterprises, organizations, and small companies when switching to Cloud, Edge, and Fog paradigms from standalone execution [1]. Significantly, this shift brings numerous challenges along the way. This work mainly focuses on how security in Cloud, Edge, and Fog Computing systems is achieved and users' privacy protected from attackers. Essentially, the vision is a holistic management style for personal data at the global centers hosting Edge, Fog, and Cloud.

As of today, security and privacy issues have become a major concern when Cloud providers holding large amounts of data and essential applications share them with customers. As a result of these concerns, related topics present major problems in the computing paradigms research field [2]. Currently, the most attention in each computing model is on protecting users' privacy from unauthorized groups or individuals gaining access and hindering attacks. Moreover, keeping data integrity intact and also maintaining it is a very vital aspect. This research takes an approach to review the security and privacy aspects in Cloud, Edge, and Fog paradigms [3–5].

The rapid and ever-increasing need for novel computational offloading strategies is a great challenge when it comes to protecting personal information and other important data [6]. Historically, Cloud customers possess legitimate access to their individual information and data (in other words, users should have the right as to how, when, and to what extent other people can gain access to their personal information) [7]. Importantly, five

different features relating to security and privacy aspects are raised in any order: integrity, accountability, confidentiality, availability, and the preservation of privacy [7–9].

Recently, there has been a sharp, universal shift from traditional operations in organizations to embracing innovations such as Cloud Computing and other paradigms. These different paradigms have been the subject of many academic studies and reviews from students and researchers. It is both difficult and very challenging for different Information and Communication Technology (ICT) engineers, researchers, and students to generally keep up with the ever-growing pace of new journals, literature, and article reviews. One important area concerning the various paradigms is the security and privacy aspect, which we shall systematically review based on PRISMA guidelines [10].

The rest of the paper is organized as follows. First, Section 2 briefly outlines the explanation of different computing paradigms. Next, Section 3 provides an outlook on the specifics of security and privacy for each paradigm and their similarities. Furthermore, Section 4 provides the major identified challenges and vulnerabilities. Section 5 concludes the discussion.

2. Background on Computing Paradigms

Before diving deeper into the main sections of the paper, a general overview of the different mentioned paradigms needs to be provided. For clarity and consistency, each paradigm is carefully discussed concisely. The reason for discussing each of these paradigms is to have an overview that will guide the understanding of the research goal for this paper, which is primarily the information security and privacy aspects for each paradigm.

2.1. Cloud-Related Aspects

Historically, the growth and expansion of the infrastructures of many companies have come from evolving technologies and innovations. Cloud computing is seen as a unique solution to provide applications for enterprises [11]. It uses different components such as hardware and software to render services, especially over the Internet. The possibility of accessing various data and applications provided was originally made straightforward by Cloud computing.

Several industrial giants and standardization bodies attempted to define Cloud computing in their understandings and views. The National Institute of Standards and Technology (NIST) is widely considered to provide the most reliable and precise definition for Cloud computing as "a model for enabling ubiquitous, convenient, on-demand network access to a shared pool of configurable computing resources (e.g., networks, servers, storage, applications, and services) that can be rapidly provisioned and released with minimal management effort or service provider interaction" [12].

Five different models particularly characterize Cloud computing: on-demand self-service, broad network access, multi-tenancy and resource pooling, rapid elasticity, and scalability. Generally, more Cloud computing resources can be provided as required by manufacturers and different enterprises while avoiding interactions with humans involving service providers, e.g., database instances, storage space, virtual machines, and many others. Having access to corporate Cloud accounts is essential as it helps corporations to virtualize the various services, Cloud usage, and supply of services as demanded [13].

Simultaneously, there is a need for broad network access, i.e., accessing capabilities via established channels across the network advance the use of heterogeneous thick and thin customer devices such as workstations, tablets, laptops, and mobile phones [14]. This access leads to the resource pooling aspect, i.e., computing resources from the provider are grouped using a particular multi-tenant model used in serving various clients. The unseen and non-virtual resources are carefully allocated and reallocated according to the customer's needs. Usually, customers do not understand or access the spot-on position or area provided. However, location specification can be established at an advanced state of situation or abstraction followed by various examples of resources such as network bandwidth, processing, memory, and storage [15].

Such a massive heterogeneous environment leads to the scalability aspect [16]. The growth of a client marketplace or business is made possible due to the tremendous ability to create specific Cloud resources, enabling improvement or reducing costs. Sometimes, changes might occur on the user's need for Cloud computing, which will be immediately responded to by the platform or system.

Finally, the resource use is keenly observed, regulated, and feedback is given to established billing based on usage (e.g., accounts of frequent customers, bandwidth, processing, and storage). The proper reporting of essential services used can be done transparently if the used resources are adequately looked into, controlled and account is given [12].

From the architectural perspective, big, medium, and small enterprises use Cloud computing technology to save or store vital data in the Cloud, enabling them to access this stored information from any part of the world via connecting to the Internet. Service-oriented and event-driven architectures are the main combination that makes up the Cloud computing architecture. The two important parts dividing the Cloud computing architecture are naturally Front End (FE) and Back End (BE) [17].

As seen in Figure 1, various components are involved in the computing architecture [6]. Furthermore, we take a brief look at each architecture's different features. Furthermore, we can see that a network connects both front and back ends via the wired or wireless medium.

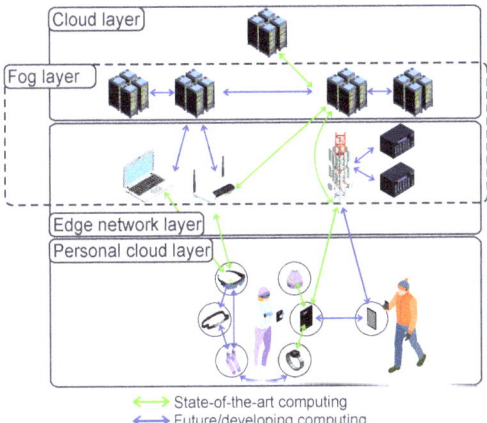

Figure 1. Most common task offloading models.

2.2. Edge-Related Aspects

As a new generation of computational offloading, Edge arrived to allocate the resources at the network edge, i.e., closer to various office and home appliances such as mobile devices, Internet of Things (IoT) devices, clients, and client's sensors. In recent years, there has been fast growth in industrial and research investment in Edge computing. The pivot for Edge computing is the physical availability and closeness, of which end-to-end latency is influenced by this essential point of Cloudlets, with bandwidth achievable economically, trust creation, and ability to survive [18].

Communication overheads between a customer and a server site are reduced due to a decrease in actual transmission distances (in terms of geography and number of hops) brought about by the Edge computing in the network. As one of the definitions, "Edge computing is a networking philosophy focused on bringing computing as close to the source of data as possible to reduce latency and bandwidth use. In simpler terms, Edge computing means running fewer processes in the Cloud and moving those processes to local places, such as on a user's computer, an IoT device, or an Edge server" [19]. Some other definitions of Edge computing are "a physical compute infrastructure positioned on the spectrum between the device and the hyper-scale Cloud, supporting various applications. Edge computing brings processing capabilities closer to the end-user/device/source of data which eliminates the journey to the Cloud data center and reduces latency" [20]. There are several cases in which architectural designs are specifically intended, considering their work plan and setting up the infrastructure is based on its need.

Considered a state-of-the-art paradigm, Edge computing takes services and applications from the Cloud known to be centralized to the nearest sites to the main source and offers computational power to process data. It also provides added links for connecting the Cloud and the end-user devices. One of the best ways to solve or reduce Cloud computing issues is to make sure there is an increase in Edge nodes in a particular location, which will also help in decreasing the number of devices attributed to a sole Cloud [21].

Overall, the main Edge service consumers are resource-constrained devices, e.g., wearables, tracker bands for fitness and medical uses, or smartphones [22]. Fog devices, in turn, subdues the shortcomings of Cloud by transferring some of the core functions of Cloud towards the network Edge while keeping the Cloud-like operation possible [23], e.g., Edge and Fog nodes may act as interfaces attaching these devices to the Cloud [24].

A typical Edge computing architecture comprises three important nodes (see Figure 1): the Cloud, local Edge, and the Edge Device. Notably, Local Edge involves a well-defined structure with several sublayers of different Edge servers with a bottom-up power flow in computation. Both Access Points (APs) and Base Stations (BSs) are Edge servers situated at the sublayer considered to be the lowest together with proximity-based communications [25]. These are particularly installed to obtain data during communication from various Edge devices, returning a control flow using several wireless interfaces.

Cellular BSs transmit the data to the Edge servers found in the (upper) sublayer after receiving data from Edge devices. Here, the upper sublayer is particularly concerned with operating computation work. Very fundamental analysis and computation are done after data are forwarded from BSs. At a recent Edge server, the computational restriction is placed such that if the difficulty in a given work surpasses it, the work is offloaded and sent to the upper sublayers with adequate computation abilities. A chain of flow control is then concluded by these servers with passing back to the access points, and finally, in the end, send them to Edge devices [26].

The Edge architecture allowed to switch more delay intolerant applications closer to the computation demanders, e.g., Augmented/Virtual/Mixed Reality (AR/VR/MR) gaming, cellular offloading, etc., all together following the proximity-driven nature of the paradigm [27]. Generally, there are two approaches to the proximity between the Edge and user's equipment: physical and logical proximity.

Physical proximity refers to the exact distance between the top segment of data computation and user equipment. Logical proximity refers to the count of hops between the Edge computing segment and the users' equipment. There are potential occurrences of congestion because of the lengthy route caused by multiple hops, leading to increased latency issues. To avoid queuing that can result in delays, logical proximity needs to limit such events at the back-haul of the computing network systems.

Despite the shortcomings of the normal Cloud paradigm innovations to match up with great demands, given lower energy level, real-time, and in particular security and privacy aspects, the Edge paradigm is not considered a substitute for the Cloud paradigm. Edge and Cloud paradigms are known to assist each other in a cordial manner in several situations. The Cloud and Edge paradigms cooperate in some network areas, including autonomous cars, industrial Internet, as well as smart cities, offices and homes. Importantly, Edge and Cloud paradigm collaboration offers many chances for reduced latency in robust software such as autonomous cars, network assets of companies, and information analysis on the IoT [28].

Nevertheless, Edge operation is executed through supported capabilities from several actors. Cellular LTE, short-range Bluetooth Low Energy (BLE), Zigbee, and Wi-Fi are various technologies that create connectivity by linking endpoint equipment and nodes of the Edge computing layer. There is great importance for access modalities as it establishes the endpoint equipment bandwidth availability, the connection scope, and the various device type assistance rendered [29].

2.3. Fog-Related Aspects

Access gateways or set-top-boxes are end devices that can accommodate Fog computing services. The new paradigm infrastructure permits applications to operate nearby to observe activities easily and handle huge data originating from individuals, processes, or items. The creation of automated feedback is a driving value for the Fog computing concept [30]. Customers benefit from Fog and Cloud services, such as storage, computation, application services, and data provision. In general, it is possible to separate Cloud from Fog, which is closer to clients in terms of proximity, mobile assistance for mobility, and dense location sharing [31], while keeping the Cloud functionality in a distributed and transparent for the user manner.

According to NIST, "Fog computing is a layered model for enabling ubiquitous access to a shared continuum of scalable computing resources. The model facilitates the deployment of distributed, latency-aware applications and services, and consists of fog nodes (physical or virtual), residing between smart end-devices and centralized (cloud) services. The fog nodes are context aware and support common data management and communication system. They can be organized in clusters – either vertically (to support isolation), horizontally (to support federation), or relative to fog nodes' latency-distance to the smart end-devices" [32]. Generally, Fog computing is considered to be an extension or advancement of Cloud computing, as the latter one ideally focuses mostly on a central system for computing, and it occurs on the upper section of the layers, and Fog is responsible for reducing the load at the Edge layer, particularly at the entrance points and for resource-constrained devices [33].

The use of the term "Fog Computing" and " Edge Computing" refers to the hosting and performing duties from the network end by Fog devices instead of having a centralized Cloud platform. This means putting certain processes, intelligence, and resources to the Cloud's Edge rather than deriving use and storage in the Cloud. Fog computing is rated as the future huge player when it comes to the Internet of Everything (IoE) [34], and its subgroup of the Internet of Wearable Things (IoWT) [35].

Communication, storage, control, decision-making, and computing close to the Edge of the network are specially chosen by Fog architecture. Here, the executions and data storage are executed to solve the shortcomings of the current infrastructure to access critical missions and use cases, e.g., the data density. OpenFog consortium defines Fog computing as "a horizontal, system-level architecture that distributes computing, storage, control, and networking functions closer to the users along a Cloud-to-thing continuum" [36]. Another definition explains Fog as "an alternative to Cloud computing that puts a substantial amount of storage, communication, control, configuration, measurement, and management at the Edge of a network, rather than establishing channels for the centralized Cloud storage and use, which extends the traditional Cloud computing paradigm to the network Edge" [37].

The deployment of Fog computing systems is somewhat similar to Edge but dedicated to applications that require higher processing power while still being closer to the user. This explains why devices belonging to the Fog are heterogeneous, raising the question of the ability of Fog computing to overcome the newly created adversaries of managing resources and problem-solving in this heterogeneous setup. Therefore, investigation of related areas such as simulations, resource management, deployment matters, services, and fault tolerance are very simple requirements [38].

As of today, Fog computing architecture lacks standardization, and until recently, there is no definite architecture with given criteria. Despite so, many research articles and journals have managed to develop their versions of Fog computing architecture. In this section, an attempted explanation is detailed in an understanding manner, which describes the different components which make up the general architecture [38].

Generally, most of the research projects performed on Fog computing have mostly been represented as a three-layer model in its architecture [39], see Figure 2. Moreover, there is a detailed N-layer reference architecture [40], established by the OpenFog Consortium, being regarded as an improvement to the three-layer model. However, we will be looking at a three-layer architecture.

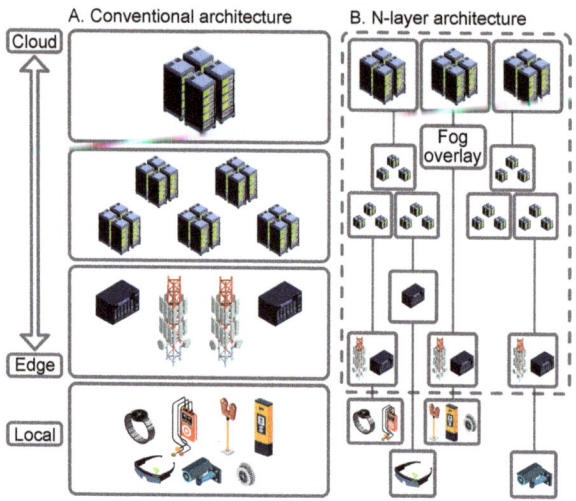

Figure 2. Most commonly analyzed computing architectures.

Fog computing is considered to be non-trivial addition regarding Cloud computing based on Cloud-to-Things setup. In fact, it displays a middle layer (also known as the Fog layer), closing the gap between the local end devices and Cloud infrastructure [41].

Notably, and as in the Cloud, the Fog layer also uses local virtualization technologies. On the other hand, taking into consideration the available resources, it will be more adequate to implement virtualization with container-based solutions [38]. It should also be remembered that Fog nodes found in this layer are large in number. Based on OpenFog Consortium, Fog node is referred to as "the physical and logical network element that implements Fog computing services" [42]. Fog nodes have the capability of performing computation, transmission, and also storing data temporarily and are located in between the Cloud and end-user devices [43].

The essential pushes for the eminent migration from Cloud computing to Fog computing are caused by load from computations and bringing Cloud computing close to Edge. Several characteristics define Fog computing by the tremendous variety of applications and IoT design services [44]. The major one corresponds to the extreme heterogeneity of the ecosystem, which provides services between centralized Cloud and different devices found at the Edge, such as end-user applications via Fog. The heterogeneity of Fog computing servers comprises shared locations with hierarchically structured blocks.

At the same time, the entire system is highly distributed geographically. Fog computing models consist of extensively shared deployments in actuality to offer a Quality of Service (QoS) regarding mobile and non-mobile user appliances [45]. The nodes and sensors of the Fog computing are geographically shared in the case of various stage environments, for instance, monitoring different aspects such as chemical vats, healthcare systems, sensors, and the climate.

The ability to effectively react to the primary goal and objective can be called cognition. Customers' requirements are better alerted by analytics in a Fog-focused data gateway, which helps give a good position to understand where to make a transmission, storage possibilities, and the control operations along the whole process from Cloud to the Internet of Things continuum. Customers enjoy the best experience due to applications' closeness to user devices and creating a better precision and reactiveness concerning the clients' needs [46].

2.4. Differences and Similarities of Paradigms

The main goal of Fog and Edge paradigms are similar in some areas, unlike the Cloud. Both of those bring the capabilities of the Cloud closer to the users and offer customers with lower latency services while making sure, on the one hand, that highly delay-tolerant applications would achieve the required QoS, and, on the other hand, lowering the overall network load [47]. It is not straightforward to differentiate and compare Cloud, Edge, and Fog Computing. This subsection attempts to discern and look into similar features between the computing paradigms [48]. The differences and similarities of the various paradigms are summarized in Table 1.

Table 1. Comparison on different computing paradigms.

Attributes	Cloud Computing	Edge Computing	Fog Computing
Architecture	Centralized	Distributed	Distributed
Expected Task Execution Time [1]	High	High-Medium	Low
Provided Services	Universal services	Often uses mobile networks	Vital for a particular domain and distributed
Security	Centralized (guaranteed by the Cloud provider)	Centralized (guaranteed by the Cellular operator)	Mixed (depending on the implementation)
Energy Consumption	High	Low	Varying but higher than for Edge
Identifying location	No	Yes	Yes
Main Providers	Amazon and Google	Cellular network providers	Proprietary
Mobility	Inadequate	Offered with limited support	Supported
Interaction in Real-Time	Available	Available	Available
Latency	High	Low	Varying but higher than for Edge
Bandwidth Cost	High	Low	Low
Storage capacity and Computation	High	Very limited	Varying
Scalability	Average	High	High
Overall usage	Computation distribution for huge data (Google MapReduce), Apps virtualization, Storage of data scalability	Control of traffic, data caching, wearable applications	CCTV surveillance, imaging of subsurface in real-time, IoT, Smart city, Vehicle-to-Vehicle (V2X)

[1] Importantly, Edge may provide higher results but only for computationally simple tasks (benefiting in terms of communication latency), while Fog would provide higher computational speed maintaining the latency (for, e.g., AR/VR applications). Executions in the Cloud would always provide the worst results as the computational unit is geographically distant from the user, which would naturally require tremendous communication overheads compared to geographically closer locations.

Nonetheless, it is essential to overview each of these indicated paradigms to address security and privacy aspects in Cloud, Edge, and Fog paradigms. This subsection described some fundamental features that constitute each of the said paradigms, making them unique in their ways. We looked into the different architectures, how these paradigms are characterized and how beneficial they are to the industries, and addressed some scenarios in which they are applied.

Cloud being a centralized architecture and an IoT promoter has several shortcomings such as high latency, location sensibility, and computation time, just to name a few. Researchers then suggested upgraded technologies known as Edge and Fog paradigms to lessen the burden on Cloud systems and resolve the issues indicated. Ultimately, we see that those two paradigms have helped decrease the large quantity of data sent to the Cloud.

Finally, the Edge paradigm is advantageous over the Cloud paradigm, especially regarding security and privacy. However, the Fog paradigm consisting of Fog nodes is regarded as an outstanding architecture uniquely created so that IoT appliances render improved services and support. Next, we shall present some security and privacy analyses relating to Cloud, Edge, and Fog paradigms, respectively.

3. Security and Privacy of Computing Paradigms

Security and privacy have a symbiotic relationship and are closely related. Many academics and organizations see the two terms closely related to the ICT domain. The influence of digitalization has tremendously shaped our daily activities [30]. Industrial giants currently deal with various computing paradigms involving huge computation and processing of Big Data. Thus, transmitting these data from one source to another makes it vulnerable and requires protection. In this section, we will define security, privacy, threats, countermeasures, and security mechanisms, and we will see some differences and possible similarities between security and privacy [49].

3.1. Cloud-Related Aspects

The majority of today's networks and the idea of storing data remotely is greatly inclined to technologies relating to Cloud computing. One of the exceptional demands is for the Cloud to see that services are always made available consistently, the reliability is maintained, and data are supplied as demanded. As mentioned earlier, one of the prime reasons organizations or individuals are reluctant to embrace the quick movement to the Cloud model is the huge concern for information security and privacy. Some acknowledged issues tied to security and privacy in Cloud computing include confidentiality, data security, phishing, and multi-tenancy [50]. This section looks into the various threats aligned with security and privacy within the Cloud computing system and suggests some modalities for threat mitigation.

Cloud computing users adopt different distributed Cloud models based on their specific needs, and because of this, the Cloud security and privacy threats differ according to the infrastructure hosted in the Cloud. According to the Cloud Security Alliance (CSA), major regular threats are information leakages, Denial of Service (DoS) Attack, and Advanced Persistent Threats (APT) [51].

Adequate Cloud infrastructural security largely depends on the established protective technologies with many layers. This brings about the importance of adapting an Intrusion Detection System (IDS) specifically to trace suspected threats intelligently and intercept potential attacks over a network. Furthermore, the various events witnessed can be separated to carry out network status analysis. Resources and services of Cloud CIA are said to encounter different types of threats originating from either inside or outside intruders [52].

3.1.1. Cloud Data Security

Data security is an essential aspect that plays a significant role in handling Cloud devices and keeps them running. This may involve protection and restoration guides for data and centers for Cloud services, and data involved in transmissions or transfers must always be protected.

Generally, there is a need for simple yet robust mechanisms that offer a smooth method of learning about Cloud service capabilities before deployment and those that align with Cloud security features during the establishing stage. The presence of Cloud service providers and Cloud customers also plays a role in the deployment plan since both parties must meet certain data security requirements [53]. Here, issues such as service level negotiation, information traffic, and especially data security will arise [54]. It is important for Cloud service suppliers to properly protect customers' data stored in the Cloud to reduce or eliminate security shortcomings. Techniques used in encrypting data must be very strong to guarantee better data security and implement authentication mechanisms that monitor other information access. Access control through data encryption should be established so that only the rightfully selected employees can reach the data.

3.1.2. Cloud Data Privacy

The public Cloud faces more privacy threats, although these threats are very different based on their Cloud model variants. Some of the concerns of the danger here are the proliferation of information, malicious usage by an unauthorized person, and incapability to control by clients [55]. Clients' sensitive documents stored in the Cloud can be reached by attackers using the file's hash codes, with the help of a mechanism used in duplicating information [56]. Risks about privacy are regarded from several angles, such as access control, Cloud systems, customers, and stored information [57]. Knowing data privacy and other relating privacy principles will enormously assist in dealing with the known threat concerns. One vital setback holding some organizations from moving to the Cloud is the fear of losing classified data through information leakage [58].

Most often, people's privacy is breached either knowingly or unknowingly. Accessing a person's private data without their knowledge or authorization is strongly considered an invasion of privacy. Different trends can occur, such as open disclosure, privacy attack, data violation, and other means of attacks. Privacy leakage can be very damaging, but privacy issues can be better managed with the points mentioned below:

- Trust: Disclosing data of an individual or organization is considered a breach of privacy. Trust plays a very pivotal role in decreasing or eliminating fear [59]. There are various trust standards every customer can agree to, but in general, their concern is to see minimal or zero breaches of privacy at a reasonable scale [60].
- Access Control: Cloud systems present massive issues, such that an unauthorized person or group of individuals can obtain access if not properly addressed. An effective way of handling this is by answering the questions [61]:
    - Who? The privileged persons to access certain data and who not to.
    - What? Some detailed data are not made accessible to every worker. So what specific files are permitted for whom?
    - When? Some data are needed for a period of time, and that period must strictly be controlled when that information has been accessed.

  These can be made functional by establishing management policies, checks on multi-domain, and providing strong management keys.
- Encryption of data needs to be sufficiently strong to protect the privacy of the client's files. Weak encryption of data poses a serious challenge to Cloud privacy [61].

3.2. Edge-Related Aspects

Since Cloud computing's performance dropped greatly caused by various factors, including the growing number of nodes, Edge computing has provided a significant paradigm shift. Edge Computing is observed as an innovation because it can carry applications with its new technological capabilities in shared computing while also performing information processing right at the point of need, without transporting the data to the Cloud. Users overall have a better feeling when data are processed close to them, improving their response time. This is made possible thanks to the computation that is directly carried out at the nodes of distributed equipment [62].

Fifth Generation (5G) networks are taking over many areas and operations of our daily activities [63]. Edge computing is undeniably the pivot of all these changes being a part of

5G network, making it vital in terms of smaller resource-constrained devices and how they interact. Edge Computing shows a relationship with heterogeneous equipment and several cross-connected networks. The inter-connectivity of these Edge supporting technologies exposes it to the most concerning aspect of any device, technology, network, and above all, organizations, which is safety. The threats involved here cannot be taken for granted, and this now led us to the subject matter, security, and privacy in Edge Computing. With computation at the node of Edge devices, other security circumstances will show up and still require continuous research work for improvements [64].

In Edge, the chances for imminent threats and attacks are very likely because of the decentralized design of the Edge computing system, even though the processing of information at the nodes offers some security and privacy protection. Smart devices also expose security issues and dangerous malware to Edge computing. The structure of Edge computing cannot adequately support the mechanisms for securing and protecting information. This, therefore, implies that the complexity of this Edge node at the network leaves the data very exposed and hard to secure.

Despite the growing nature of Edge computing technologies, its security and privacy development remain a continuous process and tells why there exist not so many research findings. Researchers and other academics globally have been putting every effort in performing relevant research work to develop countermeasures to improve the security and privacy of Edge systems. Different simple mobile Edge computing methods were used for carrying out security checks, presentation of an overall security and protection scheme with proposals from the research work done. The Edge security findings do present a relevant citation from a theoretical approach. As mentioned previously, the existing known issues in this work relating to Edge computing information security and privacy are partitioned into four separate parts [65]: Access Control, Identity Authentication, Information Security, and Privacy Protection. Based on the focused theme of this work, "Security and Privacy Aspects", we shall be looking more into only Information security and data protection.

3.2.1. Edge Data Security

Data integrity, confidentiality, and attack detection are the common goal and reasons for data security. It assists in designing an Edge-computing system that is secured. Issues such as information breach and information loss are resolved by outsourcing information under control, non-fixed storage, and sharing responsibility. Data duties are allowed to be carried out securely by customers. Presently, it is still challenging to identify works on Edge Computing security, and privacy since many academics do mostly focus on Cloud paradigms [66], or perhaps Fog paradigm [67]. The major aim of information security in Edge systems is to securely move data and ease the heavy load by creating a shared model with a smoothly operating system. As a result, very acceptable shared information security and lightweight designs are developed for both end-users and remote nodes.

A key responsibility in safeguarding customers' secrets and upholding the confidence involved, especially at the Edge network, should be rendered, e.g., a digitalized building constructed with many IoT devices, which can be a prime target due to its huge quantity of personal data produced. Therefore, a more regarded approach to protect the privacy of customers and gain their confidence is to make sure that data processing occurs at the Edge network or node of the house [68].

In addition to aspects detected earlier, the following notable Edge-specific elements should be considered. Note, cloud challenges also generally apply to Edge operation scenarios:

- Confidentiality, in the case of mobile clients intending to use the services of mobile applications, is always taken seriously, and for this reason, some clients find it difficult to decide whether to use it [69]. The authors of [70] list some shortcomings relating to Edge computing confidentiality, showing a very high risk posed by the providers of services gaining unpermitted passage to classified information. This occurs during data transmission in a distributed or unsecured network later stored and processed in the Edge distributed network. Data security has constantly been breached. Good enough, restricting access today to project confidentiality is achievable due to some newly created mechanisms [71].

- Detecting Attacks: Edge systems can operate smoothly with the assistance of Edge nodes where the Edge applications are located to offer maximum standard services. This ensures that the entire Edge system is free from abnormalities or threats. The Edge node consists of harsh surroundings with an inadequate security guarantee, exposing the Edge nodes to threats. The performance of an Edge system can massively be hindered when the threats from one Edge node are mismanaged and might subsequently extend to another Edge node. Thus, finding a quick solution can be hard because of the weight of the threat that spreads across the Edge nodes. Furthermore, added costs would be incurred to find the baseline reason for the problem, and even recovery might take a while [72]. Therefore, regular checks must be performed to detect any previous potential or imminent attacks.

3.2.2. Edge Data Privacy

In Edge computing, accessing the system does not reflect trust. Averagely accepted systems store important data, resulting in critical privacy leakage. Examples of clients' data stored are personal information, location, and identity. The focus areas to be discussed herein any order include privacy, identity, and location privacy safeguarding [73].

Edge computing always raises much concern in stark contrast to other existing computing models protecting information. This is because the challenges, e.g., leakages relating to Edge data privacy, are daunting. An Edge information center, services, infrastructure suppliers, and even certain clients are the potential weak link or at least establishments you cannot fully trust with such interwoven computing/cellular networks. With regard to this, the act of keeping safe the private information of clients is an obligation that requires very close attention [74]:

- Protection of Data Privacy: At the Edge nodes, huge amounts of data belonging to clients are retrieved from applications and other users' pieces of equipment. This collected information is then processed and analyzed. Despite the trustworthiness of the Edge computing nodes, they can still display some level of vulnerability. Classified information such as an individual's medical data must be top secret. Therefore, information privacy protection is very important to avoid leakage at the nodes of Edge computing [75].
- Identity Privacy: Compared to the Cloud systems, especially Mobile Cloud, Edge models still lack adequate research attention in protecting the identity of customers well. Identity privacy protection is a major concern for several organizations and even individual customers. The third-party identity-designed model is said to still pose vulnerability [76].
- Location Privacy: Several software and services from Worldwide Web render functional capabilities based on location. For a client to gain access when they want to use the services in Edge computing, that client must deliver their location as required by the service provider [77,78]. One of the particularly concerning fears is breaching data location through possible leaks. Different researchers gave some solution schemes on how to deal with issues on data leakage. A dynamic distribution in location privacy protection was presented in a mobile model of social internet platforms. This model can sort out visitors with low trust levels within a certain range of social interactions. It performs this by dividing customers' data location (unidentifiable) and personalities in individual storage systems. This separation enables the service provider to hide customers' location data safely. The importance of this model is that even if an attacker manages to breach one of the storage facilities, for example, data location, it will not pose a major threat since the identity of the client is not leaked or exposed [79].

3.3. Fog-Related Aspects

Many businesses have transformed massively, especially with the fast growth in large data usage, due to Cloud computing [80]. Meanwhile, the quest for private services also began to grow hugely. A great number of well-centralized systems is offered by Cloud computing platforms [81,82], although with some shortcomings. Clouds and their endpoints show certain unwanted long and irregular delays and time-conscious services to some [83]. There is a pertinent high risk in a situation whereby there is a breakdown in the information building and between network interconnected systems. One potential breach here is possible privacy exposure. To mitigate this challenge, the Fog computing [84] model was introduced, and it assisted Cloud-Edge in improving computation, security, and privacy, which is now the leading and most recommended computing service.

Fog devices are considered to be separate and distributed pieces of equipment ranging from gateways, routers, switches, or professional installation of traditional servers [85]. Furthermore, with the current demand for huge emission reduction, Fog computing is highly viewed as a smart green platform with sustainability and great security benefits. Many fog Nodes (FNs) are seen as renewable constitute the Fog computing system. The geographical placing of FNs can be spread throughout several locations. A great level of pressure exerted in the information center during computation is vastly decreased due to the different FNs working independently but together through a well-calculated formula. Fog can separate or sifter the processing at the central layer found at the middle of the endpoint and Cloud [86], which may significantly enhance the QoS and brings down expenses [87]. Fog computing was highly considered in great demand to deal with the ever-growing IoT issues, as we shall see in the next sub-Section [88].

Fog computing was established as the most viable approach because of its ability to cross-connect every digital equipment, wireless endpoint, and local device. This interconnectivity is vulnerable to vital security and privacy violations such as disclosing clients' data location, leaking classified documents, and stealing private accounts. First considered by Cisco, Fog computing was brought to expand the Cloud activities to the system's Edge. The consideration of Fog computing surfaces as an option to local Cloud offering huge assistance in terms of QoS, latency, and location distribution [45]. Services such as networking, storage, and most importantly, computing between the customer and information center are rendered by Fog computing hugely considered a virtualized system [89], carrying the related vulnerabilities along the way.

According to the Edge system, every single unit in the Edge computing functions independently to see that information is not forwarded to the Cloud, and instead, it is locally handled. On the other hand, transferring to Cloud or processing the data from various information origins is always a decision made by Fog computing nodes, taking into account its assets. Fog computing can expand some Cloud services that are not assisted in Edge structure, such as Infrastructure as a service (IaaS), software as a service (SaaS), and platform as a service (PaaS). Fog computing is completely Edge inclined but can be supported by Fog computing while at the Edge of the network, expansion of communication assets and computation are performed [90].

3.3.1. Fog Data Security

Some attacks usually threaten private and government entities since they function in Cloud, Edge, and Fog computing. To offer a level of protection to the architecture, a Threat Intelligence Platform (TIP) is important to be developed [91]. Data security is the most prioritized aspect in the industrial sector, especially as information must be safeguarded. Intelligent equipment and sensor devices are deployed to reduce threats and security attacks extensively. The feature about heterogeneity and geographical sharing impacts the implementation of Cloud security frameworks into Fog computing systems [5]. Some of the considered security challenges are confidentiality, authentication, availability, and information privacy. These mentioned frameworks assist in creating and monitoring accesses to persons and organizations.

Considering the medical field, we see that patients' health history involves classified information and the Fog architecture has several nodes that might present some vulnerabilities. These vulnerabilities can be unpermitted access to information when stored or at the time of transfer, untrustworthy insiders, and during system distribution of information. Fog system by means of cable or wireless network consistently receives information transferred from sensors of medical devices. Tampering with patients' personal data, integrity, and device availability is obvious and can occur when communication systems and sensors are targeted. Some through channels as Denial of Service (DoS) can easily be perpetrated due to the vulnerabilities found in wireless networks. On the other hand, the absence of proper frameworks to control access to the Fog nodes that process important information can compromise information through leakage because of account theft, unpermitted access, and possibly some unsafe passage. The mentioned problems can be mitigated through thorough analysis and stringent rules and regulations to establish standard control mechanisms such as personal systems, selective (limited) encryption, and reciprocated authentication [92].

Overall, Fog provides Edge-like challenges while bridging those even more towards the decentralized and distributed environment.

3.3.2. Fog Data Privacy

Protecting the privacy of individuals and enterprises is often a primary concern encountered by the Fog paradigm, especially with the Fog nodes positioned near the individuals and facilitates the gathering of vital information sometimes relating to geographical location, identity, social security numbers, and many. One great challenge is that it is quite hard to keep centralized monitoring due to the distributed nature of Fog nodes.

During transmission, attackers can easily gain access to steal essential information when the Fog nodes are not well secured. More practical studies are needed to understand privacy problems better and innovate current solutions to preserve data privacy [93]. Privacy leakage often happens, even though end-users are never in accordance to release their personal information. There are some main areas of clients' privacy: data privacy, location privacy, identity privacy, and usage privacy [94].

4. Main Security and Privacy Challenges

This section briefly describes the major challenges per paradigm and provides a concise table highlighting the essential ones and the proposed countermeasures identified in the literature.

4.1. Cloud Paradigm Challenges

Data loss, privacy leakage, multi-tenancy, unpermitted access to management platforms, Internet protocol, injection attacks are some of the main challenges faced in Cloud [95,96]. Such challenges turn to make room for potential attacks, letting access control to cybercriminals, granting access to unauthorized services, therefore disclosing several classified data, if not all.

Cloud computing faces enormous threats when involved with these vulnerabilities and thus affects business too, either directly or indirectly. One of the most reliable ways to repel threats and attacks is to identify any found and analyze the behavior properly. This section explains the different Cloud computing issues [97].

- Multi-tenancy is used in providing services to different customers and organizations with a particular software operating on the SaaS provider's servers within the architectural design. Every user company can use an application that is virtually designed in dividing data and configuring it virtually with the help of specially designed software. In this SaaS model, there is a high risk of vulnerability because clients turn to work with applications of multi-tenancy manufactured by Cloud Service Providers (CSP). The maximum-security of customer's data is the direct responsibility of the Cloud provider since sensitive information such as financial and individual data are hosted in their Cloud system [55].

Managing resources and scheduling work are some methods used by certain Cloud providers [98], but hardware potential is fully attained through virtualization by CSPs providers. Sandboxed setups refer to Virtual Machines (VM)being completely separate. Hardware sharing with the clients is considered safe according to this mindset. On the other hand, cybercriminals can gain access to the host when the sandboxed system has security setbacks [99]. The virtualization software is strongly recommended since it is capable of showing recent vulnerabilities in Cloud security, such as retrieving data by targeting a VM on one machine through attacks through cross-Virtual Machine side channel [100].

- Data Integrity: Security attention is greatly put on data integrity in the Cloud, which means any reply to a data request sent must be from someone with an access privilege. Establishing a general basic data integrity standard is important, though it is not still in place [101]. Trust is one of those many values that clients are expected to demonstrate in the computing facet. Today, a lot of companies or institutions encounter the issue of trust, and this hugely impacts the handling of their data [102].
- Unauthorized Access: One of the most vulnerable aspects of Cloud computing is giving unauthorized access to management platforms and resources. Users are exposed to this due to the shared technologies often involved in Cloud services. An acceptable way of mitigating the security solution of such a scenario is by introducing access control, and this helps in securing the client's personal information and its domain for privacy [103]. It is worth noting that cybercriminals can simply have unauthorized access to Cloud service systems because of a single-style authentication model and not very strong authentication mechanisms being used [104].
- Data loss and Leakage: The low cost of Cloud services is one reason customers turn to migrate to the Cloud, and it is warned that customers should pay attention to their important information since various diverse aspects can easily breach their data security. There is an increased chance of data leakage or loss due to high traffic and usage of the Cloud. The vulnerabilities and threats in Cloud service are undeniable, posing a great security threat to businesses and institutions. Significantly, it can be frustrating when you cannot retrieve and restore data after accidentally deleting files from the Cloud due to a lack of a backup system [105].
- Malicious Insider: Every organization has different rules and regulations regarding recruitment policies and employee information. However, some employees have higher status, which guarantees them the privilege of accessing certain essential data within the company. Based on CSA, they proposed the implementation of transparency in the general data security and management activities standard, outlining notification procedures during security failures, while using Service Level Agreement (SLA) as a demand for human resource, and finally establishing and exercising strict rules in the management of supply chain [105].

  It may be far easier for a person with malicious ideas to work for a CSP since no one is seen as a suspect [106]. This individual can quickly be involved in malicious events, especially if they have unhindered access to sensitive information, especially if the CSP cannot strictly monitor its workers.
- Identity Theft: Victims or organizations can suffer heavy impact due to weak passwords due to phishing attacks by some attackers who turn to disguise as authentic persons to steal the different important data of their victims. The sole reason for identity theft is to gain access to sensitive digital resources of individuals and companies by any malicious means. Every protected communication within the Cloud system happens with access control, and this is made possible using an encryption key [107].
- Man-in-the-Middle Attack: During the flow of data from one end to another or between different systems, cybercriminals can easily take advantage and gain access, therefore having control of classified data. This can easily occur when the secure socket layer (SSL) is insecure due to inadequate configuration. Specifically, in Cloud systems, hackers can attack the communication within the information centers. Effi-

cient SSL configuration and data analysis among accepted entities can go a long way to significantly lower the threat posed by a middle-man attacker [108].
- The DoS attack aims to limit or stop the execution of service and from accessing needed data. This creates a scenario where actual users partially or fully lack service availability. Whenever the right person uses the Cloud services to reach the data server to access information, access is denied. This happens because the attacker uses a method in which he constantly congests the server of a precise resource through request flooding, and the targeted server will then be unable to reply to a legitimate access request. There exist several ways this attack can be performed, for example, by way of SQL injection attack, bandwidth wastage, and also by way of incorrectly using model resources [109].
- Phishing Attack is one of the most common attacks in which the criminal turns to impersonate and deceive their victims by leading them to malicious links. The presence of the Cloud makes it flexible for hackers to hide their Cloud hosting of numerous accounts of different clients that uses Cloud services using phishing activities. There are two kinds of threat divisions in which phishing can be grouped. Primary, irresponsible attitude whereby a cybercriminal can also make full use of Cloud services to simply host a site for a phishing attack. Secondary, Cloud computing services and their many accounts can be hijacked [110].

4.2. Edge Paradigm Challenges

The Edge paradigm is considered to offer huge benefits to Edge customers such as storage, data processing, just to name a few. However, unlike the Cloud paradigm, Edge computing still faces big security and privacy challenges, which we will explore despite these many gains in this subsection.

- Data Injection: When a machine is vulnerable, an attacker can push harmful information to share negative information. The act of injecting dangerous data by a malicious attacker into a device is known as poisoning. Data can be faked, then used to create fraudulent messages to render the nodes of the target compromised, and it is called an external forgery, for example, in a modern digital industrial production line where the adversary happens to give false machine readings, therefore causing severe functional changes with the bad aim to harm the devices [65].
- Eavesdropping: In this scenario, an attacker can mask itself and observe network traffic during transmission and capture data illegally. It is quite hard to point out this type of attack because the attacker happens to hide inside the platform [111].
- Privacy Leakage: The absence of strict access control to the node of Edge can easily lead to data privacy being tampered with. However, the attack strength is very low. The information generated from devices situated at Edge proximity is stored and processed in the Edge data building. Customers classified these Edge data buildings can leak information since the content is known [112].
- Distributed DoS: Attackers usually take advantage of network protocol vulnerabilities to launch attacks on Edge nodes, causing network damage and restricting resource access and provision of services. Attackers carry out these attacks by loading the server with many data packets to shut down the channel by jamming the server's bandwidth. Another option is where the Cloud data server or the Edge systems are being flooded with data packets to massively take out resources [65].
- Permission and Access Control: Unauthorized access is a major challenge in the Edge paradigm. It is important to know an individual or employee before authorizing them to access any sensitive information in the system. It can be achieved by establishing access control protocols. Connectivity between several pieces of equipment and other services can be considered secured when access control measures and permission are implemented [113].

4.3. Fog Paradigm Challenges

The Cloud paradigm has countermeasures for its security and privacy threats. Nevertheless, these countermeasures may not apply to the Fog paradigm due to the active presence at the network Edge of Fog entities. The immediate vicinity where Fog entities operate will confront various threats which may not constitute a good functioning Cloud. The security solutions in the Fog paradigm are improving and increasing as well. However, most of the published literature on Fog computing security and privacy does not provide insights with an extensive assessment of the various issues. Importantly, we elaborate on some security and privacy challenges encountered in the Fog paradigm.

- Trust Issue: Fog systems face trust design challenges due to the reciprocal demand for trust and the distributed nature of their network. Cloud computing platforms are different since they already consist of pre-designed security models that match the industrial security requirements, granting customers and enterprises some trust measures within the Cloud system. However, this is not so with Fog computing networks which are more exposed and liable to security and privacy attacks. Even though the same security mechanism can be deployed to every Fog node that makes up the Fog computing network, the distributed design also makes it quite challenging to resolve the trust problem [24].
- Malware Attacks: Infecting the Fog computing system with a malware attack is a very high-level challenge in the network. It is carried out to steal sensitive data, breach confidential information, and even refuse service with the help of a virus, spyware, Trojan horse, or Ransomware. To assist Fog computing applications in mitigating these malicious attacks, authentic defense mechanisms for virus or worm detection and advanced anti-malware must be introduced [114].
- Computation—Data Processing: Fog nodes often receive data collected from end-user equipment, processed, sent to the Cloud system, or end-user pieces of equipment are forwarded information transmitted from the Cloud. After the various processes, the data sent from end-users to Cloud systems and the data sent from Fog nodes to the Cloud are different in size and nature. Another challenge here is that several providers have these Fog nodes, making them hard to be trusted due to the many security and privacy shortcomings arising after the processing of data [115].
- Node Attack: Here, the attacker engages physically by targeting to capture the vulnerable nodes. There are moments when the attacker can decide to alter the whole node, cause defects to the hardware, or steal sensitive information from the Fog nodes by digitally sending messages and causing sensor nodes distortion of classified data. Such attacks can have damaging effects on the nodes of the Fog network, and observing these node sensors will help identify issues and deploy some node capturing defense of algorithmic cryptography [114].
- Privacy Preservation: There is a huge concern as customers using CSP, IoT, and wireless systems face data leaks of personal information. It is not easy to preserve this privacy in the Fog network due to the closeness of Fog nodes to the customers' environment, and it can also facilitate gathering plenty of vital information such as identity, location, and utility usages. Privacy leakage can also occur when communication between Fog nodes becomes more frequent [94].

4.4. Major Attacks and Countermeasures

It is essential to note that vulnerabilities, threats, or security attacks can appear differently in different paradigms, and there exists no specific way of solving the various security issues. Thus, several designed models must be considered to safeguard a Cloud, Edge, or Fog computing system. This will help create a joint force of many reliable layer defense models [116].

Table 2 presents a detailed comparison of Cloud, Edge, and Fog paradigms based on a designated OSI model layer. Different attack examples were common to the three involved paradigms associated with the various layers. These identified security attacks and privacy leakages are matched to a specific proposed countermeasure. In some situations, the same countermeasure of a particular paradigm can be applied to the other ones. However, due to the complexity of these paradigms or their ecosystem, this deployment of a single countermeasure is challenging.

Table 2. Attack specifics of paradigms and suggested countermeasures.

Layer	Brief Description	Attack	Specifics of Paradigm/Main Proposed Countermeasures		
			Cloud	Edge	Fog
Application	Data inclined applications faces attacks and if breached, unpermitted access on websites is reached. Malware is of different forms, e.g., Trojan horses and viruses. An illegal software used to access legitimate information. Attacks HTTP [117].	HTTP Flood	Application monitoring is highly recommended. Web Application Firewalls (WAF), Anti-virus, privacy protection management [118].	Filtering mechanisms and intrusion detection systems [26].	HTTP-Redirect scheme [119].
		SQL Injection	SQL injection detection using adaptive deep learning [120].	Modifying circuits to minimize information leakage by adding random noise or delay, implementing a constant execution path code and balancing Hamming weights [121].	SQL injection detection using Elastic-pooling [122].
		Malwares	Use of Antivirus Softwares [118].	Signature-based and behavior-based detection [123].	Mirai botnet detector [119].
Session/Presentation	"It is defined as a pool of virtualized computer resources." Virtualization offers better usage of hardware assets with an opportunity for additional services avoiding extra costs for infrastructures. Customers are provided with virtual storage [124].	Hyper-visor	Strong configurations, up-to-date Operating System (OS).	Computational Auditing	Robust Authentication scheme.
		Data leakage	Encrypt stored data/use secured transmission medium, e.g., SSL/TLS, Virtual Firewall [125]	Homomorphic Encryption [126].	Isolation of user's data, Access control strictly based on positions [114].
		VM-Based	Anti-viruses, anti-spyware to monitor illegal events in guest OS [127].	Identity and Authentication scheme such as Identity-Based Encryption (IBE) [126].	Intrusion detection and prevention mechanism use for anomaly detection, behavioral assessment, and machine learning approach in classifying attacks [119].
Transport	"Provides a total end-to-end solution for reliable communications". The two main protocols are TCP and UDP. The smooth performance in communication strongly depends on TCP/IP between user and server [128].	TCP Flood	Firewalls, SYN Cache [129].	SYN cookies [130].	Integrated Firewalls [131].
		UDP Flood	Graphene design for secure communication [132].	Response rate for UDP packets should be reduced [131].	Response rate for UDP packets same as in Edge, should be reduced [131].
		Session hijacking	AES-GCM symmetric encryption [132].	User light-weight authentication algorithm [130].	Encrypting communication using two-ways or multi-purpose authentication [92].

Table 2. Cont.

Layer	Brief Description	Attack	Specifics of Paradigm/Main Proposed Countermeasures		
			Cloud	Edge	Fog
Network	The routing of data packets across different networks from a source to an end node, is performed by the network layer [133].	DoS attack	Intrusion Detection System (IDS) [134], Access Security	Network Authentication mechanisms	Deploy routing security and observing the behaviour of nodes [135].
		MITM	Data Encryption [118].	Time stamps, encryption algorithm [121].	Use of Authentication schemes [114].
		Spoofing attacks	Identity Authentication [118].	Secure trust schemes [39].	Secured identification and Strong authentication [39].
PHY/MAC	The manner how types of equipment are physically hooked up to a wired or wireless network system and can be sorted for physical addressing with the help of a designated MAC address [136].	Eaves-dropping	Encryption, Cryptography [137]	Data Encryption using asymmetric AES scheme [121].	Protection of identity by use of IBC [138].
		Tampe-ring	Detection of behavioural pattern	Observe manner of behaviour [137].	Multicast authentication as PKI [67].
		Replay attack	Dynamic identity-based authentication model [139].	Authentication mechanisms [140].	Key generation approach [140].

As of now, end devices do not involve any established security measures. For this reason, during data transmission, security vulnerabilities are likely to be present. Some vulnerability research is underway to understand the different ways an end device or layer can face an attack. It is of significance that vulnerability research projects must be carried out extensively and in-depth when studying attacks and their aspects [141]. At each layer, we can deduce that security vulnerabilities are safeguarded differently. This attains the basic security demands such as confidentiality, authenticity, integrity, and not the least, availability. Cryptography is suggested for data confidentiality in stopping data leakages to illegitimate persons. Although cryptography turns out to offer better data confidentiality, it does need additional computation power, therefore causing latency. Users and end-devices have proximity to each other. For example, FNs pose some level of reach to individuals' data, especially where the information is generated. Data processed in FNs are significant security-wise due to their sensitivity more than data being processed in Cloud servers, thus requiring enhanced protection.

Overall, Cloud, Edge, and Fog paradigms consist of applications, resources, and a massive quantity of end-devices within a given centralized or decentralized area, existing together and inter-communicating. Therefore, the huge potential for vulnerabilities in security and privacy does exist. One good way of screening systems for possible vulnerabilities is by auditing security standards.

Vulnerabilities in any system might expressly grant attackers partial or full access to cause severe harm. If data are breached, it can expose critical information of individuals or organizations, and an attack can cause serious malfunctioning of an entire network and create disruptions. We found that the main target of gaining access to sensitive data is threats, seizures, or vulnerabilities of the examined paradigms, whether joint or apart.

Importantly, we found that these vulnerabilities can be properly discovered with the right tools and approaches. Despite the constant search for vulnerabilities in systems by attackers (hackers/cybercriminals), there are up-to-date, sophisticated countermeasures to mitigate such threats, internal or external. Most essentially, each vulnerability has a specific mechanism to counter its threats and attacks. Moreover, another important aspect is that the vulnerabilities turn to undermine the security and privacy of the related paradigms, exposing them (data) to potential security attacks and privacy leakages.

5. Discussion and Conclusions

The essential aim of this work was to execute a comprehensive article review on Cloud, Edge, and Fog paradigms, respectively, with a special focus on identifying similarities, differences, attacks, and countermeasures based on security and privacy aspects.

Cloud, Edge, and Fog paradigms create a substantial heterogeneous quantity of data capable of being managed over a centralized or distributed system. Looking at the discussions presented in this work, we deduced that the security and privacy issues on the heterogeneity of this ecosystem are a significant challenge. Data transfer from one end to another opens a way for many security and privacy vulnerabilities, even though some of these weaknesses can be detected and eliminated quickly. Solutions cannot be swiftly deployed to user devices simply because of the complexity of the ecosystem. However, IDS mechanisms are largely significant for different paradigms, as some are considered effective in countering DoS/DDoS attacks (Zero-day-attack). In certain scenarios, IDS mechanisms introduce gateway devices to provide higher processing power if needed.

Security and privacy are considered primary drawbacks, limiting several institutions and organizations to adopt computational offloading technology. As mentioned earlier, these paradigms face different security and privacy threats, but the most outstanding are DoS/DDoS attacks. For instance, Cloud customers can suffer heavily if Cloud services and resources are breached for a moment by attackers. Cloud systems encounter high latency and high costs in communication and data storage. These issues are present because of the centralized nature of the Cloud and its geographical distance from end-devices that produce data. To resolve these shortcomings in the Cloud, Edge Computing was introduced as a Cloud Computing extension.

As identified during the review, Edge provides much less latency than Cloud platform to end-devices; thus, there is a rapid drop in security when migrating from the Cloud platform to the Edge platform due to the Edge network being decentralized (distributed) in nature. Furthermore, observing the migration of data to end-devices from Cloud platform via Edge network, the storage capacity sharply reduces. There is also a rapid decrease in real-time operations as data moves from end-devices via the Edge platform to the Cloud platform. For longer storage needs, a Cloud platform is used. Storage or processing of data from the end-devices occurs in the Edge platform. Despite the emerging of Edge Computing, vulnerabilities and threats still exist, and this, therefore, calls for strict measures with enhanced security and privacy techniques. Fog paradigm was considered to ameliorate Cloud and Edge paradigms.

As with the Edge paradigm, Fog is rendering services (computation, networking, data storage, etc.) closer to the end-devices rather than moving data to the Cloud platform but in a distributed manner. However, the introduction of the Fog paradigm is seen to improve the infrastructural network to match the demands of large data quantity while enhancing the processing strength efficiently. Fog paradigm can improve mobility, complexity in a distribution environment, location identity, real-time response, as well as security and privacy. The fog paradigm does not depend on the Cloud data center but instead relies on end-devices to store and process its data. Broader availability of node access gives some level of flexibility to the applications. Like the Fog paradigm, the Edge paradigm also permits computation handling at the network edge, near where data are generated. What makes the Fog paradigm different from the Edge paradigm is its ability for Fog nodes to interconnect, while the Edge paradigm operates with separate Edge nodes.

Confidentiality, integrity, and availability are information systems' most significant security and privacy properties. The transfer and storage of data must be confidential, with integrity, and made available. Confidentiality grants data access only to individuals and organizations that own these data. During the transfer of data within the different user layers, the main network, storing and processing data in Cloud, Edge, or Fog paradigm, its access is strongly restricted. Encrypting data is a way of achieving confidentiality. Data correctness and consistency is a model of integrity which avoids information being tampered with or modified. Some mechanisms can be used for verifying sent and received data integrity. Only authorized persons are granted access to available data. Thus, availability determines that data must be available anywhere based on established policies. To attain these expectations, various instruments, patterns, methodologies, and mechanisms such as cryptography, encryption, authentication, and others are deployed to the multiple platforms (layers) when data are being transferred and stored.

Overall, Cloud, Edge, and Fog paradigms exhibit the same view of providing QoS to customers, but they all have a separate set of features that makes them differ from one another, as we have explained in this work. Notably, the Fog paradigm is designated the most effective and reliable system to better handle the security and privacy challenges encountered.

To summarize, even though the Fog paradigm can offer better security and privacy services to end-devices in general, some features of the Fog paradigm, such as decentralization, constraints of resources, homogeneity, and virtualized systems, are vulnerable to security and privacy challenges in comparison to the Cloud paradigm, which is centralized. Due to the absence of standardization regarding countermeasures deployment, highly effective security and privacy mitigation in the Cloud paradigm cannot be implemented straight to the Fog paradigm because of the named features above. Therefore, Fog systems do need innovative countermeasures to address these challenges. Future research should also address new techniques and mechanisms that fit Fog paradigm features and possibly cross-platform countermeasure tools. Hence, they should be suggestions for effective and efficient solutions.

Review Methodology: The systematic literature review is based on PRISMA guidelines [10]. The publication date range was set from 2017 to 2021. We used the most popular ICT sector databases for research works, such as IEEE, Web of Science, Science Direct, Springer, and Scopus, while not considering pre-prints, duplicates, and gray literature. Later on, we analyzed the titles, abstracts, and keywords of the various academic publications to figure out specific journal articles and other important papers related to security and privacy in Cloud, Edge, and Fog paradigms. The following search query was formulated for reproducibility:

> TITLE (((cloud OR Edge OR fog) AND computing) AND (security OR privacy)) AND
> (LIMIT-TO(PUBYEAR, 2021) AND LIMIT-FROM (PUBYEAR, 2017)) AND
> (LIMIT-TO(SUBJAREA, "COMP") OR LIMIT-TO (SUBJAREA, "ENGI")) AND
> (LIMIT-TO(LANGUAGE, "English")) AND (LIMIT-TO (PUBSTAGE, "final"))

Some exclusion criteria were set to narrow the search outcomes during the first screening stage from the paper's titles and abstracts:

- Not related to security and privacy in Cloud, Edge, and Fog computing;
- Not in English;
- Works with no technical content;
- Purely review papers;
- Full text not available.

After applying the exclusion criteria, the selected number of publications was lowered from 1390 to 447. Sixty-one duplicates were found and were taken off the list. The headings of the various articles, their abstracts, and important words of the retained 386 papers were screened, and 187 papers were dismissed since they did not match the exclusion criteria. The number of papers left was 199, and their whole content were thoroughly analyzed. After the additional screening, 122 papers were still rejected since they were unrelated to the topic.

Author Contributions: Conceptualization, A.O., J.N.; methodology, A.O.; validation, J.N.; formal analysis, J.N., M.K.; investigation, O.L.M., A.O.; writing, original draft preparation, O.L.M., A.O.; writing, review and editing, A.O., M.K., J.N.; visualization, A.O.; supervision, A.O., J.N.; project administration, A.O., J.N.; funding acquisition, J.N. All authors have read and agreed to the published version of the manuscript.

Funding: This project has received financial support from the Priority 2030 Federal Academic Leadership Program.

Acknowledgments: The work was executed as part of the second author's Master's thesis work titled "Security and Privacy Aspects of Cloud, Edge, and Fog Paradigms: A Systematic Review".

Conflicts of Interest: The authors declare no conflict of interest.

Abbreviations

The following abbreviations are used in this manuscript:

5G	5th Generation Networks
AES	Advanced Encryption Standard
AP	Access Point
APT	Advanced Persistent Threats
AR	Augmented Reality
BE	Back End
BLE	BLuetooth Low Energy
BS	Base station
CCTV	Closed-circuit television
CSA	Cloud Security Alliance
CSP	Cloud service providers
DDoS	Distributed Denial of Service
DoS	Denial of Service
FE	Front End
FN	Fog Nodes
GCM	Galois/Counter Mode
HTTP	Hypertext Transfer Protocol
LTE	Long Term Evolution
IaaS	Infrastructure as a service
IBC	Identity Based Cryptography
IBE	Identity-Based Encryption
ICT	Information and Communication Technology
IDS	Intrusion Detection System
IoE	Internet of Everything
IoWT	Internet of Wearable Things
MAC	Mediul Access Control
MITM	Man-in-the-Middle Attack
MR	Mixed Reality
NIST	National Institute of 66 Standards and Technology
OS	Operating System
OSI	Open Systems Interconnection model
PaaS	Platform as a Service
PKI	Public Key Infrastructure
PRISMA	Preferred Reporting Items for Systematic Reviews and Meta-Analyses
QoS	Quality of Service
SaaS	Software as a Service
SLA	Service Level Agreement
SQL	Structured Query Language
SSL	Secure Socket Layer

SYN	SYNchronize message
TCP	Transmission Control Protocol
TIP	threat intelligence Platform
TLS	Transport Layer Security
UDP	User Datagram Protocol
V2X	Vehicle-to-Vehicle
VM	Virtual Machines
VR	Virtual Reality
WAF	Web Application Firewalls
Wi-Fi	Wireless Fidelity

References

1. Chalapathi, G.S.S.; Chamola, V.; Vaish, A.; Buyya, R. Industrial Internet of Things (IIoT) Applications of Edge and Fog Computing: A Review and Future Directions. In *Fog/Edge Computing For Security, Privacy, and Applications*; Springer: Cham, Switzerland, 2021; pp. 293–325.
2. Ranaweera, P.; Jurcut, A.D.; Liyanage, M. Survey on Multi-Access Edge Computing Security and Privacy. *IEEE Commun. Surv. Tutor.* **2021**, *23*, 1078–1124. [CrossRef]
3. Alhroob, A.; Samawi, V.W. Privacy in Cloud Computing: Intelligent Approach. In Proceedings of the International Conference on High Performance Computing Simulation (HPCS), Orléans, France, 16–20 July 2018; pp. 1063–1065.
4. Parikh, S.; Dave, D.; Patel, R.; Doshi, N. Security and Privacy Issues in Cloud, Fog and Edge Computing. *Procedia Comput. Sci.* **2019**, *160*, 734–739. [CrossRef]
5. Aljumah, A.; Ahanger, T.A. Fog Computing and Security Issues: A Review. In Proceedings of the 7th International Conference on Computers Communications and Control (ICCCC), Oradea, Romania, 8–12 May 2018; pp. 237–239.
6. Ometov, A.; Chukhno, O.; Chukhno, N.; Nurmi, J.; Lohan, E.S. When Wearable Technology Meets Computing in Future Networks: A Road Ahead. In Proceedings of the 18th ACM International Conference on Computing Frontiers, Virtual Event, Italy 11–13 May 2021; pp. 185–190.
7. Guilloteau, S.; Venkatesen, M. *Privacy in Cloud Computing-ITU-T Technology Watch Teport March 2012*; International Telecommunication Union: Geneva, Switzerland, 2013.
8. Cook, A.; Robinson, M.; Ferrag, M.A.; Maglaras, L.A.; He, Y.; Jones, K.; Janicke, H. Internet of Cloud: Security and Privacy Issues. In *Cloud Computing for Optimization: Foundations, Applications, and Challenges*; Springer: Berlin/Heidelberg, Germany, 2018; pp. 271–301.
9. Xiao, Z.; Xiao, Y. Security and Privacy in Cloud Computing. *IEEE Commun. Surv. Tutor.* **2013**, *15*, 843–859. [CrossRef]
10. PRISMA Guidelines. Available online: http://www.prisma-statement.org/ (accessed on 21 December 2021).
11. Nieuwenhuis, L.J.; Ehrenhard, M.L.; Prause, L. The Shift to Cloud Computing: The Impact of Disruptive Technology on the Enterprise Software Business Ecosystem. *Technol. Forecast. Soc. Chang.* **2018**, *129*, 308–313. [CrossRef]
12. NIST Special Publication 800-145: Definition of Cloud Computing Recommendations of the National Institute of Standards and Technology. Available online: https://nvlpubs.nist.gov/nistpubs/Legacy/SP/nistspecialpublication800-145.pdf (accessed on 21 December 2021).
13. Five Characteristics of Cloud Computing. Available online: https://www.controleng.com/articles/five-characteristics-of-cloud-computing/ (accessed on 21 December 2021).
14. Application Management in the Cloud. Available online: http://www.sciencedirect.com/science/article/pii/B9780128040188000048 (accessed on 21 December 2021).
15. Cloud Computing. Available online: https://masterworkshop.skillport.com/skillportfe/main.action?assetid=47045 (accessed on 21 December 2021).
16. Spatharakis, D.; Dimolitsas, I.; Dechouniotis, D.; Papathanail, G.; Fotoglou, I.; Papadimitriou, P.; Papavassiliou, S. A Scalable Edge Computing Architecture Enabling Smart Offloading for Location Based Services. *Pervasive Mob. Comput.* **2020**, *67*, 101217. [CrossRef]
17. Jadeja, Y.; Modi, K. Cloud Computing—Concepts, Architecture and Challenges. In Proceedings of the International Conference on Computing, Electronics and Electrical Technologies (ICCEET), Nagercoil, India, 21–22 March 2012; pp. 877–880.
18. Satyanarayanan, M. Edge Computing. *Computer* **2017**, *50*, 36–38. [CrossRef]
19. Edge Computing Learning Objectives. Available online: https://www.cloudflare.com/en-gb/learning/serverless/glossary/what-is-edge-computing/ (accessed on 21 December 2021).
20. Edge Computing—What Is Edge Computing? Available online: https://stlpartners.com/edge-computing/what-is-edge-computing/ (accessed on 21 December 2021).
21. Gezer, V.; Um, J.; Ruskowski, M. An Extensible Edge Computing Architecture: Definition, Requirements and Enablers. In Proceedings of the UBICOMM, Barcelona, Spain, 12–16 November 2017.
22. Mäkitalo, N.; Flores-Martin, D.; Berrocal, J.; Garcia-Alonso, J.; Ihantola, P.; Ometov, A.; Murillo, J.M.; Mikkonen, T. The Internet of Bodies Needs a Human Data Model. *IEEE Internet Comput.* **2020**, *24*, 28–37. [CrossRef]

23. Sarkar, S.; Misra, S. Theoretical Modelling of Fog Computing: A Green Computing Paradigm to Support IoT Applications. *IET Netw.* **2016**, *5*, 23–29. [CrossRef]
24. Mukherjee, M.; Matam, R.; Shu, L.; Maglaras, L.; Ferrag, M.A.; Choudhury, N.; Kumar, V. Security and Privacy in Fog Computing: Challenges. *IEEE Access* **2017**, *5*, 19293–19304. [CrossRef]
25. Ometov, A.; Olshannikova, E.; Masek, P.; Olsson, T.; Hosek, J.; Andreev, S.; Koucheryavy, Y. Dynamic Trust Associations over Socially-Aware D2D Technology: A Practical Implementation Perspective. *IEEE Access* **2016**, *4*, 7692–7702. [CrossRef]
26. Xiao, Y.; Jia, Y.; Liu, C.; Cheng, X.; Yu, J.; Lv, W. Edge Computing Security: State of the Art and Challenges. *Proc. IEEE* **2019**, *107*, 1608–1631. [CrossRef]
27. Kozyrev, D.; Ometov, A.; Moltchanov, D.; Rykov, V.; Efrosinin, D.; Milovanova, T.; Andreev, S.; Koucheryavy, Y. Mobility-Centric Analysis of Communication Offloading for Heterogeneous Internet of Things Devices. *Wirel. Commun. Mob. Comput.* **2018**, *2018*, 3761075. [CrossRef]
28. Jiang, C.; Cheng, X.; Gao, H.; Zhou, X.; Wan, J. Toward Computation Offloading in Edge Computing: A Survey. *IEEE Access* **2019**, *7*, 131543–131558. [CrossRef]
29. Dolui, K.; Datta, S.K. Comparison of Edge Computing Implementations: Fog Computing, Cloudlet and Mobile Edge Computing. In Proceedings of the Global Internet of Things Summit (GIoTS), Geneva, Switzerland, 6–9 June 2017; pp. 1–6.
30. Mäkitalo, N.; Aaltonen, T.; Raatikainen, M.; Ometov, A.; Andreev, S.; Koucheryavy, Y.; Mikkonen, T. Action-Oriented Programming Model: Collective Executions and Interactions in the Fog. *J. Syst. Softw.* **2019**, *157*, 110391. [CrossRef]
31. Stojmenovic, I.; Wen, S.; Huang, X.; Luan, H. An Overview of Fog Computing and Its Security Issues. *Concurr. Comput. Pract. Exp.* **2016**, *28*, 2991–3005. [CrossRef]
32. NIST Special Publication 500-325: Fog Computing Conceptual Model Recommendations of the National Institute of Standards and Technology. Available online: https://nvlpubs.nist.gov/nistpubs/SpecialPublications/NIST.SP.500-325.pdf (accessed on 21 December 2021).
33. Ometov, A.; Shubina, V.; Klus, L.; Skibińska, J.; Saafi, S.; Pascacio, P.; Flueratoru, L.; Gaibor, D.Q.; Chukhno, N.; Chukhno, O.; et al. A Survey on Wearable Technology: History, State-of-the-Art and Current Challenges. *Comput. Netw.* **2021**, *193*, 108074. [CrossRef]
34. Mahmood, Z.; Ramachandran, M. Fog Computing: Concepts, Principles and Related Paradigms. In *Fog Computing*; Springer International Publishing: Cham, Switzerland, 2018; pp. 3–21.
35. Qaim, W.B.; Ometov, A.; Molinaro, A.; Lener, I.; Campolo, C.; Lohan, E.S.; Nurmi, J. Towards Energy Efficiency in the Internet of Wearable Things: A Systematic Review. *IEEE Access* **2020**, *8*, 175412–175435. [CrossRef]
36. *IEEE Std 1934-2018; IEEE Standard for Adoption of OpenFog Reference Architecture for Fog Computing*; IEEE: New York, NY, USA, 2018; pp. 1–176.
37. Peng, M.; Yan, S.; Zhang, K.; Wang, C. Fog-Computing-based Radio Access Networks: Issues and Challenges. *IEEE Netw.* **2016**, *30*, 46–53. [CrossRef]
38. Naha, R.K.; Garg, S.; Georgakopoulos, D.; Jayaraman, P.P.; Gao, L.; Xiang, Y.; Ranjan, R. Fog Computing: Survey of Trends, Architectures, Requirements, and Research Directions. *IEEE Access* **2018**, *6*, 47980–48009. [CrossRef]
39. Lin, J.; Yu, W.; Zhang, N.; Yang, X.; Zhang, H.; Zhao, W. A Survey on Internet of Things: Architecture, Enabling Technologies, Security and Privacy, and Applications. *IEEE Internet Things J.* **2017**, *4*, 1125–1142. [CrossRef]
40. OpenFog Consortium. *OpenFog Reference Architecture for Fog Computing*; OpenFog Consortium: Fremont, CA, USA, 2017; pp. 1–162.
41. Hu, P.; Dhelim, S.; Ning, H.; Qiu, T. Survey on Fog Computing: Architecture, Key Technologies, Applications and Open Issues. *J. Netw. Comput. Appl.* **2017**, *98*, 27–42. [CrossRef]
42. OpenFog Consortium Architecture Working Group. OpenFog Architecture Overview. *White Pap. OPFWP001* **2016**, *216*, 35.
43. De Donno, M.; Tange, K.; Dragoni, N. Foundations and Evolution of Modern Computing Paradigms: Cloud, IoT, Edge, and Fog. *IEEE Access* **2019**, *7*, 150936–150948. [CrossRef]
44. Fog Computing: An Overview of Big IoT Data Analytics. Available online: https://www.hindawi.com/journals/wcmc/2018/7157192/#references (accessed on 21 December 2021).
45. Bonomi, F.; Milito, R.; Zhu, J.; Addepalli, S. Fog Computing and Its Role in the Internet of Things. In Proceedings of the MCC workshop on Mobile Cloud Computing, Helsinki, Finland, 17 August 2012; pp. 13–16.
46. Chiang, M.; Zhang, T. Fog and IoT: An Overview of Research Opportunities. *IEEE Internet Things J.* **2016**, *3*, 854–864. [CrossRef]
47. Khan, S.U. The Curious Case of Distributed Systems and Continuous Computing. *IT Prof.* **2016**, *18*, 4–7. [CrossRef]
48. Anawar, M.R.; Wang, S.; Azam Zia, M.; Jadoon, A.K.; Akram, U.; Raza, S. Fog Computing: An overview of big IoT data analytics. *Wirel. Commun. Mob. Comput.* **2018**, *2018*, 7157192. [CrossRef]
49. Zar, J. Privacy and Security As Assets: Beyond Risk Thinking to Profitable Payback. In Proceedings of the IEEE Global Telecommunications Conference, New Orleans, LA, USA, 30 November–4 December 2008; pp. 1–6.
50. Lee, K. Security Threats in Cloud Computing Environments. *Int. J. Secur. Appl.* **2012**, *6*, 25–32.
51. Cloud Security Alliance. Cloud Security Alliance Releases 'The Treacherous Twelve' Cloud Computing Top Threats, 2016. Available online: https://cloudsecurityalliance.org/press-releases/2016/02/29/cloud-security-alliance-releases-the-treacherous-twelve-cloud-computing-top-threats-in-2016/ (accessed on 21 December 2021).
52. Modi, C.; Patel, D.; Borisaniya, B.; Patel, H.; Patel, A.; Rajarajan, M. A Survey of Intrusion Detection Techniques in Cloud. *J. Netw. Comput. Appl.* **2013**, *36*, 42–57. [CrossRef]

53. Chang, V.; Ramachandran, M. Towards Achieving Data Security with the Cloud Computing Adoption Framework. *IEEE Trans. Serv. Comput.* **2016**, *9*, 138–151. [CrossRef]
54. Fox, A.; Griffith, R.; Joseph, A.; Katz, R.; Konwinski, A.; Lee, G.; Patterson, D.; Rabkin, A.; Stoica, I.; Zaharia, M. *Above the Clouds: A Berkeley View of Cloud Computing*; Technical Report UCB/EECS-2009-28; EECS Department, University of California: Berkeley, CA, USA, 2009.
55. Pearson, S.; Benameur, A. Privacy, Security and Trust Issues Arising from Cloud Computing. In Proceedings of the IEEE Second International Conference on Cloud Computing Technology and Science, Indianapolis, IN, USA, 30 November–3 December 2010; pp. 693–702.
56. Mulazzani, M.; Schrittwieser, S.; Leithner, M.; Huber, M.; Weippl, E. Dark Clouds on the Horizon: Using Cloud Storage as Attack Vector and Online Slack Space. In Proceedings of the 20th USENIX conference on SecurityAugust, San Francisco, CA, USA, 8–12 August 2011; pp. 1–11.
57. Di Vimercati, S.D.C.; Foresti, S.; Jajodia, S.; Paraboschi, S.; Samarati, P. Over-Encryption: Management of Access Control Evolution on Outsourced Data. In Proceedings of the 33rd International Conference on Very Large Data Bases, Vienna, Austria, 23–27 September 2007; pp. 123–134.
58. Mogull, R.; Arlen, J.; Gilbert, F.; Lane, A.; Mortman, D.; Peterson, G.; Rothman, M.; Moltz, J.; Moren, D.; Scoboria, E. Security Guidance for Critical Areas of Focus in Cloud Computing v4.0. Cloud Security Alliance. 2017. Available online: https://downloads.cloudsecurityalliance.org/assets/research/security-guidance/security-guidance-v4-FINAL.pdf (accessed on 21 December 2021).
59. Tyagi, A.K.; Niladhuri, S.; Priya, R. Never Trust Anyone: Trust-Privacy Trade-Offs in Vehicular Ad-hoc Networks. *J. Adv. Math. Comput. Sci.* **2016**, *19*, 1–23. [CrossRef]
60. Rusk, J.D. Trust and Decision Making in the Privacy Paradox? In Proceedings of the Southern Association for Information Systems Conference, Macon, GA, USA, 21–22 March 2014.
61. Sun, P.J. Privacy Protection and Data Security in Cloud Computing: A Survey, Challenges, and Solutions. *IEEE Access* **2019**, *7*, 147420–147452. [CrossRef]
62. Ai, Y.; Peng, M.; Zhang, K. Edge Computing Technologies for Internet of Things: A Primer. *Digit. Commun. Netw.* **2018**, *4*, 77–86. [CrossRef]
63. Moltchanov, D.; Ometov, A.; Andreev, S.; Koucheryavy, Y. Upper Bound on Capacity of 5G mmWave Cellular with Multi-Connectivity Capabilities. *Electron. Lett.* **2018**, *54*, 724–726. [CrossRef]
64. França, R.P.; Iano, Y.; Monteiro, A.C.B.; Arthur, R. Lower Memory Consumption for Data Transmission in Smart Cloud Environments with CBEDE Methodology. In *Smart Systems Design, Applications, and Challenges*; IGI Global: Hershey, PA, USA, 2020; pp. 216–237.
65. Roman, R.; Lopez, J.; Mambo, M. Mobile Edge Computing, Fog et al.: A Survey and Analysis of Security Threats and Challenges. *Future Gener. Comput. Syst.* **2018**, *78*, 680–698. [CrossRef]
66. Zissis, D.; Lekkas, D. Addressing Cloud Computing Security Issues. *Future Gener. Comput. Syst.* **2012**, *28*, 583–592. [CrossRef]
67. Stojmenovic, I.; Wen, S. The Fog Computing Paradigm: Scenarios and Security Issues. In Proceedings of the Federated Conference on Computer Science and Information Systems, Warsaw, Poland, 7–10 September 2014; pp. 1–8.
68. Bhat, S.A.; Sofi, I.B.; Chi, C.Y. Edge Computing and Its Convergence With Blockchain in 5G and Beyond: Security, Challenges, and Opportunities. *IEEE Access* **2020**, *8*, 205340–205373. [CrossRef]
69. Khan, A.N.; Ali, M.; Khan, A.R.; Khan, F.G.; Khan, I.A.; Jadoon, W.; Shamshirband, S.; Chronopoulos, A.T. A Comparative Study and Workload Distribution Model for Re-encryption Schemes in a Mobile Cloud Computing Environment. *Int. J. Commun. Syst.* **2017**, *30*, e3308. [CrossRef]
70. Du, M.; Wang, K.; Chen, Y.; Wang, X.; Sun, Y. Big Data Privacy Preserving in Multi-Access Edge Computing for Heterogeneous Internet of Things. *IEEE Commun. Mag.* **2018**, *56*, 62–67. [CrossRef]
71. Hou, Y.; Garg, S.; Hui, L.; Jayakody, D.N.K.; Jin, R.; Hossain, M.S. A Data Security Enhanced Access Control Mechanism in Mobile Edge Computing. *IEEE Access* **2020**, *8*, 136119–136130. [CrossRef]
72. Zeyu, H.; Geming, X.; Zhaohang, W.; Sen, Y. Survey on Edge Computing Security. In Proceedings of the International Conference on Big Data, Artificial Intelligence and Internet of Things Engineering (ICBAIE), Fuzhou, China, 12–14 June 2020; pp. 96–105.
73. Cao, K.; Liu, Y.; Meng, G.; Sun, Q. An Overview on Edge Computing Research. *IEEE Access* **2020**, *8*, 85714–85728. [CrossRef]
74. Zhang, J.; Chen, B.; Zhao, Y.; Cheng, X.; Hu, F. Data Security and Privacy-Preserving in Edge Computing Paradigm: Survey and Open Issues. *IEEE Access* **2018**, *6*, 18209–18237. [CrossRef]
75. Liu, D.; Yan, Z.; Ding, W.; Atiquzzaman, M. A Survey on Secure Data Analytics in Edge Computing. *IEEE Internet Things J.* **2019**, *6*, 4946–4967. [CrossRef]
76. Khalil, I.; Khreishah, A.; Azeem, M. Consolidated Identity Management System for Secure Mobile Cloud Computing. *Comput. Netw.* **2014**, *65*, 99–110. [CrossRef]
77. Flueratoru, L.; Shubina, V.; Niculescu, D.; Lohan, E.S. On the High Fluctuations of Received Signal Strength Measurements with BLE Signals for Contact Tracing and Proximity Detection. *IEEE Sens. J.* **2021**. [CrossRef]
78. Shubina, V.; Ometov, A.; Andreev, S.; Niculescu, D.; Lohan, E.S. Privacy versus Location Accuracy in Opportunistic Wearable Networks. In Proceedings of the International Conference on Localization and GNSS (ICL-GNSS), Tampere, Finland, 2–4 June 2020; pp. 1–6.

79. Wei, W.; Xu, F.; Li, Q. MobiShare: Flexible Privacy-Preserving Location Sharing in Mobile Online Social Networks. In Proceedings of the IEEE INFOCOM, Orlando, FL, USA, 25–30 March 2012; pp. 2616–2620.
80. Li, R.; Liu, A.X.; Wang, A.L.; Bruhadeshwar, B. Fast and Scalable Range Query Processing with Strong Privacy Protection for Cloud Computing. *IEEE/ACM Trans. Netw.* **2015**, *24*, 2305–2318. [CrossRef]
81. Wang, K.; Du, M.; Yang, D.; Zhu, C.; Shen, J.; Zhang, Y. Game-Theory-based Active Defense for Intrusion Detection in Cyber-Physical Embedded Systems. *ACM Trans. Embedded Comput. Syst.* **2017**, *16*, 1–21. [CrossRef]
82. Shi, W.; Zhang, L.; Wu, C.; Li, Z.; Lau, F.C. An Online Auction Framework for Dynamic Resource Provisioning in Cloud Computing. *ACM SIGMETRICS Perform. Eval. Rev.* **2014**, *42*, 71–83. [CrossRef]
83. Ma, F.; Luo, X.; Litvinov, E. Cloud Computing for Power System Simulations at ISO New England—Experiences and Challenges. *IEEE Trans. Smart Grid* **2016**, *7*, 2596–2603. [CrossRef]
84. Chen, X.; Jiao, L.; Li, W.; Fu, X. Efficient Multi-User Computation Offloading for Mobile-Edge Cloud Computing. *IEEE/ACM Trans. Netw.* **2015**, *24*, 2795–2808. [CrossRef]
85. Chen, S.; Irving, S.; Peng, L. Operational Cost Optimization for Cloud Computing Data Centers Using Renewable Energy. *IEEE Syst. J.* **2015**, *10*, 1447–1458. [CrossRef]
86. Zeng, D.; Gu, L.; Guo, S.; Cheng, Z.; Yu, S. Joint Optimization of Task Scheduling and Image Placement in Fog Computing Supported Software-Defined Embedded System. *IEEE Trans. Comput.* **2016**, *65*, 3702–3712. [CrossRef]
87. Wang, K.; Yuan, L.; Miyazaki, T.; Zeng, D.; Guo, S.; Sun, Y. Strategic Antieavesdropping Game for Physical Layer Security in Wireless Cooperative Networks. *IEEE Trans. Veh. Technol.* **2017**, *66*, 9448–9457. [CrossRef]
88. Rimal, B.P.; Maier, M. Workflow Scheduling in Multi-Tenant Cloud Computing Environments. *IEEE Trans. Parallel Distrib. Syst.* **2016**, *28*, 290–304. [CrossRef]
89. Aazam, M.; Huh, E.N. Fog Computing: The Cloud-IoT\IoE Middleware Paradigm. *IEEE Potentials* **2016**, *35*, 40–44. [CrossRef]
90. Mahmud, R.; Kotagiri, R.; Buyya, R. Fog Computing: A Taxonomy, Survey and Future Directions. In *Internet of Everything*; Springer: Berlin/Heidelberg, Germany, 2018; pp. 103–130.
91. El-Sayed, H.; Sankar, S.; Prasad, M.; Puthal, D.; Gupta, A.; Mohanty, M.; Lin, C.T. Edge of Things: The Big Picture on the Integration of Edge, IoT and the Cloud in a Distributed Computing Environment. *IEEE Access* **2017**, *6*, 1706–1717. [CrossRef]
92. Khan, S.; Parkinson, S.; Qin, Y. Fog Computing Security: A Review of Current Applications and Security Solutions. *J. Cloud Comput.* **2017**, *6*, 1–22. [CrossRef]
93. Atlam, H.F.; Walters, R.J.; Wills, G.B. Fog Computing and the Internet of Things: A Review. *Big Data Cogn. Comput.* **2018**, *2*, 10. [CrossRef]
94. Ni, J.; Zhang, K.; Lin, X.; Shen, X. Securing Fog Computing for Internet of Things Applications: Challenges and Solutions. *IEEE Commun. Surv. Tutor.* **2018**, *20*, 601–628. [CrossRef]
95. Modi, C.; Patel, D.; Borisaniya, B.; Patel, A.; Rajarajan, M. A Survey on Security Issues and Solutions at Different Layers of Cloud Computing. *J. Supercomput.* **2013**, *63*, 561–592. [CrossRef]
96. Khorshed, M.T.; Ali, A.S.; Wasimi, S.A. A Survey on Gaps, Threat Remediation Challenges and Some Thoughts for Proactive Attack Detection in Cloud Computing. *Future Gener. Comput. Syst.* **2012**, *28*, 833–851. [CrossRef]
97. Nenvani, G.; Gupta, H. A Survey on Attack Detection on Cloud Using Supervised Learning Techniques. In Proceedings of the Symposium on Colossal Data Analysis and Networking (CDAN), Indore, India, 18–19 March 2016; pp. 1–5.
98. Ciurana, E. *Developing with Google App Engine*; Springer: New York, NY, USA, 2009.
99. Kortchinsky, K. CloudBurst: A VMware Guest to Host Escape Story. BlackHat USA. 2009. Available online: https://docplayer.net/42925918-Cloudburst-a-vmware-guest-to-host-escape-story.html (accessed on 21 December 2021).
100. Ristenpart, T.; Tromer, E.; Shacham, H.; Savage, S. Hey, You, Get Off of My Cloud: Exploring Information Leakage in Third-Party Compute Clouds. In Proceedings of the 16th ACM Conference on Computer and Communications Security, Chicago IL, USA, 9–13 November 2009; pp. 199–212.
101. Naccache, D.; Stern, J. A New Public Key Cryptosystem Based on Higher Residues. In Proceedings of the 5th ACM Conference on Computer and Communications Security, San Francisco, CA, USA, 2–5 November 1998; pp. 59–66.
102. Hay, B.; Nance, K.; Bishop, M. Storm Clouds Rising: Security Challenges for IaaS Cloud Computing. In Proceedings of the 44th Hawaii International Conference on System Sciences, Kauai, HI, USA, 4–7 January 2011; pp. 1–7.
103. Almtrf, A.; Alagrash, Y.; Zohdy, M. Framework Modeling for User Privacy in Cloud Computing. In Proceedings of the 9th Annual Computing and Communication Workshop and Conference (CCWC), Las Vegas, NV, USA, 7–9 January 2019; pp. 819–826.
104. Patel, A.; Shah, N.; Ramoliya, D.; Nayak, A. A Detailed Review of Cloud Security: Issues, Threats Attacks. In Proceedings of the 4th International Conference on Electronics, Communication and Aerospace Technology (ICECA), Coimbatore, India, 5–7 November 2020; pp. 758–764.
105. Archer, J.; Boehme, A.; Cullinane, D.; Kurtz, P.; Puhlmann, N.; Reavis, J. Top Threats to Cloud Computing v1.0. Cloud Security Alliance. 2010; pp. 1–14. Available online: https://ioactive.com/wp-content/uploads/2018/05/csathreats.v1.0-1.pdf (accessed on 21 December 2021).
106. Wrenn, B.; ISSEP CISSP. When Security and Compliance Are Essential, Trust Unisys. 2010. Available online: https://www.unisys.com/solutions/cloud-and-infrastructure-solutions/cloud-security-solutions/ (accessed on 21 December 2021).
107. Grabosky, P. Organized Cybercrime and National Security. In *Cybercrime Risks and Responses*; Springer: Berlin/Heidelberg, Germany, 2015; pp. 67–80.

108. Freier, A.; Karlton, P.; Kocher, P. *RC 6101: The Secure Sockets Layer (SSL) Protocol Version 3.0*; Internet Engineering Task Force (IETF): Fremont, CA, USA, 2011.
109. Chonka, A.; Xiang, Y.; Zhou, W.; Bonti, A. Cloud Security Defence to Protect Cloud Computing Against HTTP-DoS and XML-DoS Attacks. *J. Netw. Comput. Appl.* **2011**, *34*, 1097–1107. [CrossRef]
110. Amara, N.; Zhiqui, H.; Ali, A. Cloud Computing Security Threats and Attacks with Their Mitigation Techniques. In Proceedings of the International Conference on Cyber-Enabled Distributed Computing and Knowledge Discovery (CyberC), Nanjing, China, 12–14 October 2017; pp. 244–251.
111. He, D.; Chan, S.; Guizani, M. Security in the Internet of Things Supported by Mobile Edge Computing. *IEEE Commun. Mag.* **2018**, *56*, 56–61. [CrossRef]
112. Yi, S.; Qin, Z.; Li, Q. Security and Privacy Issues of Fog Computing: A survey. In Proceedings of the International Conference on Wireless Algorithms, Systems, and Applications, Qufu, China, 10–12 August 2015; Springer: Berlin/Heidelberg, Germany, 2015; pp. 685–695.
113. Abomhara, M.; Køien, G.M. Security and privacy in the Internet of Things: Current status and open issues. In Proceedings of the International Conference on Privacy and Security in Mobile Systems (PRISMS), Aalborg, Denmark, 11–14 May 2014; pp. 1–8.
114. Veerraju, T.; Kumar, K.K. A Survey on Fog Computing: Research Challenges in Security and Privacy Issues. *Int. J. Eng. Technol.* **2018**, *7*, 335–340. [CrossRef]
115. Guan, Y.; Shao, J.; Wei, G.; Xie, M. Data Security and Privacy in Fog Computing. *IEEE Netw.* **2018**, *32*, 106–111. [CrossRef]
116. Alkadi, O.; Moustafa, N.; Turnbull, B. A Review of Intrusion Detection and Blockchain Applications in the Cloud: Approaches, Challenges and Solutions. *IEEE Access* **2020**, *8*, 104893–104917. [CrossRef]
117. Li, Y.; Li, D.; Cui, W.; Zhang, R. Research based on OSI model. In Proceedings of the 3rd International Conference on Communication Software and Networks, Xi'an, China, 27–29 May 2011; pp. 554–557.
118. Ara, A.; Al-Rodhaan, M.; Tian, Y.; Al-Dhelaan, A. A Secure Service Provisioning Framework for Cyber Physical Cloud Computing Systems. *arXiv* **2015**, arXiv:1611.00374.
119. Krishnan, P.; Duttagupta, S.; Achuthan, K. SDN/NFV Security Framework for Fog-to-Things Computing Infrastructure. *Softw. Pract. Exp.* **2020**, *50*, 757–800. [CrossRef]
120. Li, Q.; Li, W.; Wang, J.; Cheng, M. A SQL Injection Detection Method Based on Adaptive Deep Forest. *IEEE Access* **2019**, *7*, 145385–145394. [CrossRef]
121. Alwarafy, A.; Al-Thelaya, K.A.; Abdallah, M.; Schneider, J.; Hamdi, M. A Survey on Security and Privacy Issues in Edge-Computing-Assisted Internet of Things. *IEEE Internet Things J.* **2021**, *8*, 4004–4022. [CrossRef]
122. Xie, X.; Ren, C.; Fu, Y.; Xu, J.; Guo, J. SQL Injection Detection for Web Applications Based on Elastic-Pooling CNN. *IEEE Access* **2019**, *7*, 151475–151481. [CrossRef]
123. Soni, N.; Malekian, R.; Thakur, A. Edge Computing in Transportation: Security Issues and Challenges. *arXiv* **2020**, arXiv:2012.11206.
124. Turel, Y.; Kotowski, R. Cloud Computing Virtualization and Cyber Attacks: Evidence Centralization. 2015. Available online: https://www.researchgate.net/publication/275021701 (accessed on 21 December 2021).
125. Almutairy, N.M.; Al-Shqeerat, K.H. A Survey on Security Challenges of Virtualization Technology in Cloud Computing. *Int. J. Comput. Sci. Inf. Technol. (IJCSIT)* **2019**, *11*. [CrossRef]
126. Tao, Z.; Xia, Q.; Hao, Z.; Li, C.; Ma, L.; Yi, S.; Li, Q. A Survey of Virtual Machine Management in Edge Computing. *Proc. IEEE* **2019**, *107*, 1482–1499. [CrossRef]
127. Kazim, M.; Zhu, S.Y. Virtualization Security in Cloud Computing. In *Guide to Security Assurance for Cloud Computing*; Springer: Berlin/Heidelberg, Germany, 2015; pp. 51–63.
128. Alotaibi, A.M.; Alrashidi, B.F.; Naz, S.; Parveen, Z. Security issues in Protocols of TCP/IP Model at Layers Level. *Int. J. Comput. Networks Commun. Secur.* **2017**, *5*, 96–104.
129. Kumarasamy, S.; Gowrishankar, A. An Active Defense Mechanism for TCP SYN Flooding Attacks. *arXiv* **2012**, arXiv:1201.2103.
130. Butun, I.; Österberg, P.; Song, H. Security of the Internet of Things: Vulnerabilities, Attacks, and Countermeasures. *IEEE Commun. Surv. Tutor.* **2020**, *22*, 616–644. [CrossRef]
131. Sinha, P.; Jha, V.K.; Rai, A.K.; Bhushan, B. Security Vulnerabilities, Attacks and Countermeasures in Wireless Sensor Networks at Various Layers of OSI Reference Model: A Survey. In Proceedings of the International Conference on Signal Processing and Communication (ICSPC), Coimbatore, India, 28–29 July 2017; pp. 288–293.
132. Faisal, A.; Zulkernine, M. A Secure Architecture for TCP/UDP-based Cloud Communications. *Int. J. Inf. Secur.* **2021**, *20*, 161–179. [CrossRef]
133. Radhakrishnan, R.; Edmonson, W.W.; Afghah, F.; Rodriguez-Osorio, R.M.; Pinto, F.; Burleigh, S.C. Survey of Inter-Satellite Communication for Small Satellite Systems: Physical Layer to Network Layer View. *IEEE Commun. Surv. Tutor.* **2016**, *18*, 2442–2473. [CrossRef]
134. Younis, O.H.; Essa, S.E.; Ayman, E.S. A Survey on Security Attacks/Defenses in Mobile Ad-Hoc Networks. *Commun. Appl. Electron.* **2017**, *6*, 1–9.
135. Le, A.; Loo, J.; Lasebae, A.; Vinel, A.; Chen, Y.; Chai, M. The Impact of Rank Attack on Network Topology of Routing Protocol for Low-Power and Lossy Networks. *IEEE Sens. J.* **2013**, *13*, 3685–3692. [CrossRef]

136. Dimic, G.; Sidiropoulos, N.; Zhang, R. Medium Access Control—Physical Cross-Layer Design. *IEEE Signal Process. Mag.* **2004**, *21*, 40–50. [CrossRef]
137. Pan, F.; Pang, Z.; Luvisotto, M.; Xiao, M.; Wen, H. Physical-Layer Security for Industrial Wireless Control Systems: Basics and Future Directions. *IEEE Ind. Electron. Mag.* **2018**, *12*, 18–27. [CrossRef]
138. Echeverría, S.; Klinedinst, D.; Williams, K.; Lewis, G.A. Establishing Trusted Identities in Disconnected Edge Environments. In Proceedings of the IEEE/ACM Symposium on Edge Computing (SEC), Washington, DC, USA, 27–28 October 2016; pp. 51–63.
139. Li, C.T.; Lee, C.C.; Weng, C.Y. A Dynamic Identity-Based User Authentication Scheme for Remote Login Systems. *Sec. Commun. Netw.* **2015**, *8*, 3372–3382. [CrossRef]
140. Wang, D.; Bai, B.; Lei, K.; Zhao, W.; Yang, Y.; Han, Z. Enhancing Information Security via Physical Layer Approaches in Heterogeneous IoT with Multiple Access Mobile Edge Computing in Smart City. *IEEE Access* **2019**, *7*, 54508–54521. [CrossRef]
141. Davis, B.D.; Mason, J.C.; Anwar, M. Vulnerability Studies and Security Postures of IoT Devices: A Smart Home Case Study. *IEEE Internet Things J.* **2020**, *7*, 10102–10110. [CrossRef]

Article

Bit-Level Automotive Controller Area Network Message Reverse Framework Based on Linear Regression

Zixiang Bi, Guoai Xu *, Guosheng Xu, Chenyu Wang and Sutao Zhang

School of Cyberspace Security, Beijing University of Posts and Telecommunications, Beijing 100876, China; bzx@bupt.edu.cn (Z.B.); guoshengxu@bupt.edu.cn (G.X.); wangchenyu@bupt.edu.cn (C.W.); lunter-zst@bupt.edu.cn (S.Z.)
* Correspondence: xga@bupt.edu.cn

Citation: Bi, Z.; Xu, G.; Xu, G.; Wang, C.; Zhang, S. Bit-Level Automotive Controller Area Network Message Reverse Framework Based on Linear Regression. *Sensors* **2022**, *22*, 981. https://doi.org/10.3390/s22030981

Academic Editors: Leandros Maglaras, Helge Janicke and Mohamed Amine Ferrag

Received: 9 December 2021
Accepted: 24 January 2022
Published: 27 January 2022

Publisher's Note: MDPI stays neutral with regard to jurisdictional claims in published maps and institutional affiliations.

Copyright: © 2022 by the authors. Licensee MDPI, Basel, Switzerland. This article is an open access article distributed under the terms and conditions of the Creative Commons Attribution (CC BY) license (https://creativecommons.org/licenses/by/4.0/).

Abstract: Modern intelligent and networked vehicles are increasingly equipped with electronic control units (ECUs) with increased computing power. These electronic devices form an in-vehicle network via the Controller Area Network (CAN) bus, the de facto standard for modern vehicles. Although many ECUs provide convenience to drivers and passengers, they also increase the potential for cyber security threats in motor vehicles. Numerous attacks on vehicles have been reported, and the commonality among these attacks is that they inject malicious messages into the CAN network. To close the security holes of CAN, original equipment manufacturers (OEMs) keep the Database CAN (DBC) file describing the content of CAN messages, confidential. This policy is ineffective against cyberattacks but limits in-depth investigation of CAN messages and hinders the development of in-vehicle intrusion detection systems (IDS) and CAN fuzz testing. Current research reverses CAN messages through tokenization, machine learning, and diagnostic information matching to obtain details of CAN messages. However, the results of these algorithms yield only a fraction of the information specified in the DBC file regarding CAN messages, such as field boundaries and message IDs associated with specific functions. In this study, we propose multiple linear regression-based frameworks for bit-level inversion of CAN messages that can approximate the inversion of DBC files. The framework builds a multiple linear regression model for vehicle behavior and CAN traffic, filters the candidate messages based on the decision coefficients, and finally locates the bits describing the vehicle behavior to obtain the data length and alignment based on the model parameters. Moreover, this work shows that the system has high reversion accuracy and outperforms existing systems in boundary delineation and filtering relevant messages in actual vehicles.

Keywords: Controller Area Network; electronic control units; database CAN; reverse; multiple linear regression; bit-level; vehicle behavior

1. Introduction

The increasingly diverse features in today's vehicles offer drivers and passengers a more relaxed driving experience and greater convenience. Vehicle connectivity provides real-time information and a variety of entertainment options. In addition, vehicle support features such as advanced driver assistance systems (ADAS), reduce driving stress and make driving safer. These capabilities have multiplied due to the increasing number of electronic control units (ECUs) and higher computing power. Current vehicles are equipped with up to 150 ECUs [1], that need to communicate in a unified network that requires the vehicles to provide sophisticated real-time performance, sufficient data transmission volume, and adequate reliability. Control Area Network (CAN), a technology that meets these requirements, became the international standard for intra-vehicle network communication in 1993 [2]. However, since CAN uses broadcast communication and lacks security mechanisms such as encryption and authentication, it increases the probability that the vehicle will be attacked [3–6].

Many examples of attacks on vehicles have confirmed that it is possible to attack the vehicle and perform negative control [7–9]. The most typical attack case is the attack by Miller et al. on a Jeep Cherokee that was driving on the highway and used a CAN bus-connected entertainment system and ECU firmware, that resulted in acceleration and brake failures [10]. More recently, Keen Labs in China exploited vulnerabilities in Tesla's assisted driving system to drive the vehicle into the reverse lane and even remotely control the vehicle's steering with a gamepad [11]. Regardless of the type of vulnerability, the common denominator of the attack is the need to inject information into the CAN bus to cause the vehicle to behave dangerously [12]. To prevent the CAN bus from being infiltrated with targeted attacks, original equipment manufacturers (OEMs) privatize the database CAN (DBC) file. The DBC file defines the structure, content, and meaning of each message in the CAN network [13,14]. Even the DBC file is different for different models of the same brand. It is very time-consuming for an attacker to work reverse before implementing CAN bus attacks. For security researchers, private DBC files are a massive obstacle to CAN security research. The most affected area is the automotive intrusion detection system (IDS), a crucial research element in automotive security. CAN intrusion detection systems have been proposed to detect anomalies by analyzing CAN traffic [15–23], but these studies are based on message transmission characteristics that are practically irrelevant to the behavior and status of the vehicle. Therefore, the existing IDSs for the CAN are not very powerful. Another hindered study is the fuzzy test on the CAN bus, which is often used to automatically test and discover unknown vulnerabilities in ECUs [24–28]. Since the DBC files are hidden, which causes the fuzzy test intelligence to construct data blindly, brute force and random data make the test inefficient. In addition, the lack of DBC files with detailed descriptions of CAN messages hinders automotive aftermarket development. Without effective access to vehicle status, automotive driver assistance systems and status display tools become meaningless.

The detailed specification of CAN messages is crucial for CAN network intrusion detection, fuzz testing, and automotive aftermarket products. To obtain the CAN message description in the DBC document, the security research field has proposed CAN bus reversion methods such as CAN message tokenization algorithm, machine learning-based inversion method, and onboard diagnostics II (OBD-II) diagnostic information matching. The earliest CAN message tokenization algorithm was the FBCA algorithm proposed by Markovitz et al. in 2017 [29], followed by the READ algorithm proposed by Marchetii and Stabili in 2018 [30]. The automatic CAN message translator LibreCAN was proposed by Pesé et al. in 2019 [31]. Recently, the ReCAN [32] dataset was published by Zago et al. in 2020 using a similar approach to READ. However, they are limited to classifying and subdividing data changes, such as constants, multiple values, counters, sensors. These cannot obtain specific information, such as the meaning and alignment of each tagged data. It is of minimal help for IDS research and aftermarket. The most typical of the machine learning-based CAN message reversal methods are Jaynes et al. proposed a method for efficient identification of sending ECUs, which identifies CAN frame by analyzing a similarity construction model describing uniform vehicle state information [33]. A data-driven CAN bus reversion method proposed by Buscemi et al. used already available open-source DBC files to train a machine learning model to identify unknown CAN message contents [34], a scheme similar to the unsupervised machine learning-based scheme proposed by Ezeobi et al. [35]. The accuracy of this type of solution depends entirely on the coverage of the training set. Since each vehicle is configured with a unique DBC file, it is almost impossible for the training set of such algorithms to cover all vehicle models. These approaches have been validated only on simulated data and are practically infeasible. Methods based on matching OBD-II diagnostic information describe the vehicle status in CAN information by comparing and matching OBD-II responses. Song and Kim et al. first proposed to create windows before and after the OBD-II response information to find candidate information that exactly duplicates the response data and repeat it several times to determine the information describing the response [36]. Blaauwendraad proposed a matching method using correlation coefficients based on Song's method [37]. While these methods can yield some inversion results, they can only identify specific vehicle behavior in CAN messages. The insufficient number of supported vehicle behaviors for per vehicle diagnostics limits the application of this scheme. Additionally, the CANHUNTER [38] proposed by Wen et al. in 2020 reverses the CAN message by disassembling the control APP that interacts with the car. Although this is a novel idea, this method can only obtain what is specified in the APP, and the scheme will be completely invalid once the APP commands are escaped at the server-side. In addition, since such APPs are only valid for the specified car model, this scheme also receives the limitation of the car model. In summary, existing CAN message reversal techniques are limited in their implementation by the number of available DBC files and vehicle models, and their results are unsatisfactory. Solutions that are not limited by vehicle models and can achieve close to the DBC file reversal results are urgently needed.

The CAN frame data tags alone do not reveal any valuable information, and one needs to have DBC files to decode them. However, the DBC files are hidden and usually different for each model. Reverse engineering solutions for CAN information that are not constrained by the vehicle model and can access critical information in the DBC files are urgently needed. To achieve CAN message reversal close to the DBC file, this study innovatively proposes a multiple linear regression model after an in-depth analysis of the way the DBC file specifies the vehicle behavior. The model is built using each bit of the CAN message data field as the independent variable and the vehicle behavior data as the dependent variable. As the input of our framework additionally includes sensor data, our framework needs to be very useful. First, the framework uses the R^2 of the model to filter the candidate messages related to vehicle behavior, which has an excellent filtering result on related messages compared to existing schemes. In addition, the framework outperforms existing systems in terms of data boundary delineation by locating the bits describing the vehicle behavior and obtaining the details of field functions, starting bits, field lengths, and alignment formats in the DBC file based on the β value of each model. Finally, since commercially available vehicles must be configured with a standard CAN data interface and the vehicle behavior can be captured by commonly used sensors, the inverse framework proposed in this study is independent of the vehicle model and brand.

The structure of this study is as follows. Section 2 introduces the CAN bus, DBC file, multiple linear regression models preliminary introductions and describes the feasibility of the study's ideas. Section 3 describes the design and implementation ideas of the framework. Section 4 evaluates the performance of the CAN reverse framework in actual vehicles, the reverse accuracy, the time required, the advantages over existing solutions, and the applicability of the framework. Section 5 concludes the study.

2. Background and Feasibility

2.1. CAN Bus Overview

The CAN bus is a serial communication bus originally developed by Bosch [39]. Later the international standards organization (ISO) issued the international standard ISO11898 for CAN in 1993 [40]. CANs have become one of the most widely used fieldbuses globally due to their high transmission rate and high real-time characteristics.

The standard format of a CAN message is shown in Figure 1. It begins with the start of frame (SOF), followed by an 11-bit identifier (ID) and a remote transmission request (RTR). The ID defines the meaning and type of the message and is also used to filter irrelevant messages when the node receives the messages. The ID is also used for arbitration when multiple nodes send data simultaneously; the smaller the ID is, the higher the priority is. RTR is used to distinguish the type of message. A six-bit control field follows this: identifier extension (IDE) and r0 specify the length of the frame, and the data length code (DLC) specifies the number of bytes in the data field. The data field is the core of the CAN message and is 64 bits long. It contains the vehicle control commands, the status data, and any other data to be transmitted (e.g., counters, checksum values, etc.). This is followed by the Circular Check Code (CRC), the Acknowledgement Field (ACK), and the end of frame (EOF), respectively.

Figure 1. Standard CAN message frame.

For CAN message reversal work, the main targets of the reversal are the identifier (ID) and the data fields. When reversing CAN messages, the relevant message ID is usually locked first, and then the data fields are analyzed to obtain specific bit fields that characterize the vehicle behavior.

2.2. DBC File

The form and content of each type of CAN message are defined in the DBC file, so each OEM keeps it private to avoid leakage from the data source and prevent negative control and modification of the car. However, all CAN messages must be fully translated using the DBC file as a table, making sense for CAN reverse work. The contents defined in the DBC file are listed in Table 1. The Name, ID, Cycle Time, and Length describe the entire message. The Function specifies one or more vehicle behaviors in the message data fields. Byte Order, Start Byte, Start Bit, Bit Length, Units, Precision, and Offset specify how the message describes the specific behavior. Typically, the data fields of a message contain multiple functions.

Table 1. DBC file content definition

Field Name	Definition
Name	The overall function of this message (e.g., body, speed, etc.)
ID	The identifier of this message
Cycle time	The sending period of this message
Length	The length of this message
Function	The specific function contained in this message (e.g., angel change)
Byte order	The arrangement of the specific function
Start byte	The starting byte of the specific function
Start bit	The starting bit in first byte
Bit length	The length of the function
Unit	The unit of the function
Resolution	The resolution of the function
Offset	The offset of the function

The message with ID 0x198 is used to explain the correspondence between the DBC file and the CAN message content. As shown in Figure 2a, the DBC file defines the name of the message as angle, the message sending period is 10 ms, the message length is 64 bits, and it contains 3 vehicle behaviors: steering angle, brake pedal angle, and gas pedal angle. The steering angle is arranged in Motorola (LSB) form from bit 0 to bit 15 with a resolution of 0.01. Similarly, the gas pedal and brake pedal angles are arranged in bits 16–23 and 48–55 of the data field, respectively. The alignment is Intel (MSB) and Motorola. When capturing any message with ID 0x198, its data can be decoded according to the provisions of the DBC file. According to the definition of DBC, the message shown in Figure 2b describes the angle information of the vehicle at this moment, where the brake pedal angle is $(2^2 + 2^4 + 2^5 + 2^7) \times 0.1 = 19.1°$, the steering angle is $(2^0 + 2^4 + 2^5 + 2^7) \times 0.01 = 1.77°$, and the throttle angle is 0.

Name	ID	Cycle time	Length	Function	Byte order	Start byte	Start bit	Bit length	Unit	Resolution	Offset
Angel	0x198	10	64	Steer angle	Motorola	0	0	16	deg	0.01	0
				Brake angle	Intel	2	0	8	deg	0.1	0
				Acceleration angle	Motorola	6	0	8	deg	1	0

(a)

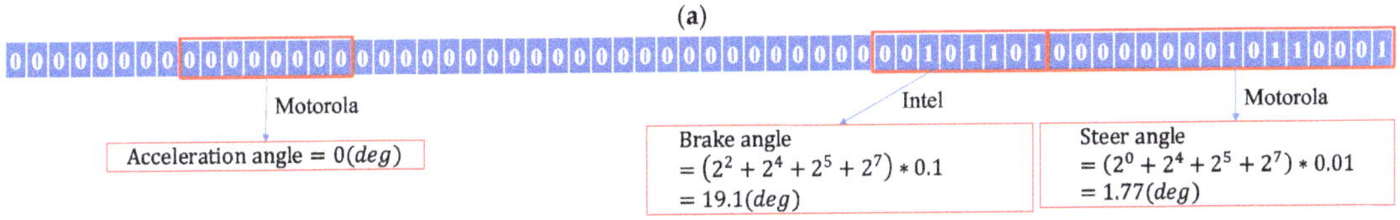

Motorola: Acceleration angle = 0(deg)

Intel: Brake angle = $(2^2 + 2^4 + 2^5 + 2^7) * 0.1 = 19.1(deg)$

Motorola: Steer angle = $(2^0 + 2^4 + 2^5 + 2^7) * 0.01 = 1.77(deg)$

(b)

Figure 2. Correspondence diagram between DBC file and CAN messages: (**a**) 0x198 Message definition in DBC; (**b**) Message data decoded according to DBC.

In summary, the DBC file is vital to study the CAN messages in-depth, which makes the DBC file a realistic target for reverse work.

2.3. Linear Regression Preliminary

In statistics, the multiple linear regression model describes the linear relationship [41,42] between the scalar dependent variable y and several explanatory variables defined as $X = (x_1, x_2, \ldots, x_k)$ and the model function is shown in Equation (1), where $\beta = (\beta_0, \ldots, \beta_k)$ is an unknown model parameter that can be estimated by giving sample set of y and X. The ordinary least squares method is the most commonly used method for parameter estimation. For a given sample set y_e (see Equation (2)) and X_e (see Equation (3)), the ordinary least squares method first creates a new matrix Ω, as shown in Equation (4).

$$y = \beta_0 + \beta_1 x_1 + \beta_2 x_2 + \ldots + \beta_k x_k \tag{1}$$

$$y_e = \begin{pmatrix} y_1 \\ y_2 \\ \vdots \\ y_m \end{pmatrix} \tag{2}$$

$$X_e = \begin{pmatrix} x_{11} & \cdots & x_{1k} \\ x_{21} & \cdots & x_{2k} \\ \vdots & \ddots & \vdots \\ x_{m1} & \cdots & x_{mk} \end{pmatrix} \tag{3}$$

$$\Omega = \begin{pmatrix} 1 & x_{11} & \cdots & x_{1k} \\ 1 & x_{21} & \cdots & x_{2k} \\ \vdots & \vdots & \ddots & \vdots \\ 1 & x_{m1} & \cdots & x_{mk} \end{pmatrix} \tag{4}$$

The estimation $\hat{\beta}$ can be obtained from Equation (5), where Ω^T is the transpose of Ω. The determination coefficient R^2 indicates how well the samples fit the linear model created with $\hat{\beta}$ and is calculated by Equation (6), where $\hat{y}_i = \hat{\beta}_0 + \hat{\beta}_1 x_{i1} + \ldots + \hat{\beta}_i x_{ik}$ is the y_i estimated with the linear model and $\overline{y}_i$ is the mean of y_e. The value of R^2 is in the range $[0, 1]$, and 1.0 is the best fit.

$$\hat{\beta} = \left(\Omega^T \Omega\right)^{-1} \Omega^T y_e \tag{5}$$

$$R^2 = 1 - \frac{\sum_{i=1}^{m}(y_i - \hat{y}_i)^2}{\sum_{i=1}^{m}(y_i - \overline{y_i})^2} \tag{6}$$

2.4. Feasibility

Based on the way the CAN messages are defined in the DBC file and the characteristics of the multiple linear regression model, this section presents the feasibility of a bit-level inverse CAN message.

According to the definition of the DBC file, the vehicle behavior in the CAN message is expressed as a binary serial number of bits, and there is also a resolution and an offset between the actual vehicle behavior data and this value. As shown in Figure 3, the relationship between the actual vehicle behavior and the corresponding bits in the CAN message is linear, and the adjacent linear coefficients satisfy the two-fold relationship. A multiple linear regression model of sensor data and each bit in the CAN message can be constructed when sensors are used to obtain vehicle behavior data. If the adjacent regression coefficients β satisfy the doubling relationship, the consecutive bits corresponding to the coefficients describe the vehicle behavior. In addition, the length, boundary, and alignment of the data can be determined based on the β that satisfies the condition.

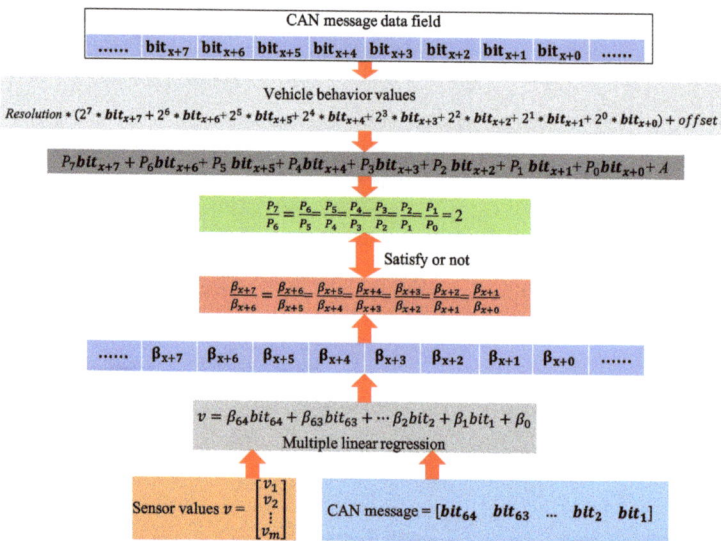

Figure 3. Reverse feasibility based on linear regression.

3. Framework Design

According to the previous description, the DBC file defines the detailed content and form of each message, which is critical for both the research and aftermarket communities. For the scientific field, obtaining the specific meaning of CAN messages facilitates the construction of better Intrusion Detection Prevention Systems (IDPS), instead of just finding anomalies based on data variation patterns. In addition, fuzzy testing can also improve efficiency by performing more targeted data injections based on the content of CAN messages. For aftermarket manufacturers, DBC files can help produce more driver assistance products, such as head-up displays and driver assistance devices. However, for confidentiality and security reasons, OEMs keep DBC files private. In addition, most of the existing CAN message reversal solutions are focused on sorting and ID filtering of data fields. The current CAN message reversal results are limited, obtaining the tags of the data types, data boundaries, and the message IDs associated with some car behaviors.

In this study, a bit-level automotive CAN message reverse framework is proposed by building a multiple linear regression model for CAN message data fields and actual physical measurements of the vehicle. Based on the optimal model parameters, the messages related to vehicle behavior are filtered. The data content, data boundary, encoding format, and linear relationship of CAN messages are extracted to maximize the recovery of the DBC file. Figure 4 provides an overview of the framework in three phases: data collection and processing, related message filtering, and bit-level message reverse. The variables used in each phase are defined below.

- X: the raw CAN dataset of the vehicle obtained from the OBD-II interface, containing the entire behavioral trajectory of the vehicle.
- Y: the sensor dataset, containing the complete set of measurable vehicle behavior measurements, collected simultaneously with X.
- Y_r: the raw set of measurements of a particular vehicle behavior collected using the sensor. r is the particular vehicle behavior that includes speed, acceleration, steering wheel steering angle, brake pedal angle, accelerator pedal angle, gear angle, and switches angle.
- Y_s: a more detailed vehicle behavior dataset obtained after processing Y_r, where s represents more detailed vehicle behavior.
- X_i: the dataset containing data fields of messages with ID i in X, and $i \in (id_0, id_1, \ldots, id_n)$.
- Y_{si}: the result of resampling of Y_s according to the frequency of X_i.
- R^2_{si}: the coefficient of determination of a multiple linear regression model between X_i and Y_{si}.
- β_{si}: the regression coefficient set of the multiple linear regression model between X_i and Y_{si}.
- Δ_s the threshold value used for the message filter.
- m_i the CAN message with ID i.
- T_β: the threshold used for filtering the β.

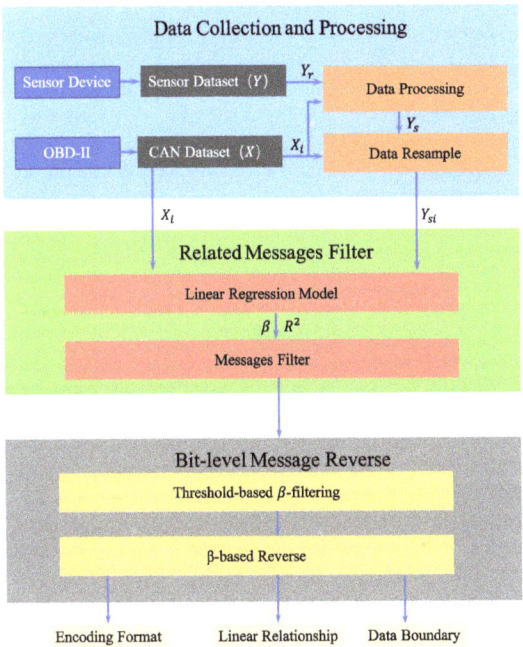

Figure 4. Overview of the framework.

3.1. Data Collection and Processing

This phase aims to acquire and process vehicle behavior measurements, as well as in-vehicle CAN traffic. The flowchart of this phase is shown in Figure 5, which is mainly divided into data acquisition, data processing, and data resampling.

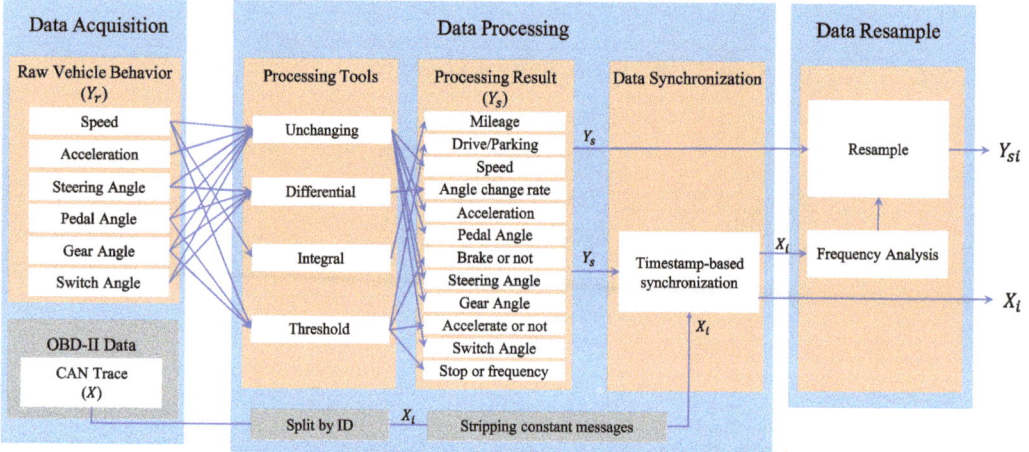

Figure 5. Data collection and processing flow.

3.1.1. Data Collection

The basic data needed to execute the message reverse framework are the in-vehicle CAN bus traces, X, and the raw physical measurements, Y_r. Where Y_r is the original sensor data for a particular behavior of the vehicle and X is the CAN trajectory obtained when the vehicle performs that behavior. The current phase requires the simultaneous acquisition of X and Y_r to reduce errors in linear regression modeling. Therefore, the data acquisition device shown in Figure 6 is used in this phase, using the same timestamp for synchronization. The CAN trace acquisition device is shown in Figure 6a. This device is a combined cable consisting of an OBD-II to DB9 diagnostic cable and a PCAN-USB FD adapter. The cable connects from the OBD-II port of the vehicle to the USB port on the side of the computer to allow the real-time collection of CAN traffic. The behavioral measurements of the vehicle are collected using the sensor device shown in Figure 6b. The device consists of a global positioning system (GPS) antenna, a universal serial bus (USB) interface, and a gyroscope angle sensor with a 0–200 Hz sampling frequency. Although the device is only $78.56 [43], it has a speed sampling accuracy of 0.001 km/h and an angle sampling accuracy of 0.1°. To reduce the error of the sensor sampling, the sampling device should be installed in such a way that the direction of sample change is consistent with the direction of either axis of the sensor. For example, the Y-axis of the sensor is aligned with the head direction when collecting vehicle speed, and the X-axis of the sensor is aligned with the angle change direction when collecting angle data. To represent the behavior and condition of the vehicle as completely as possible, the location of the sensor deployment and the collected data are listed in Table 2. The synchronous work of the above two devices provide the raw data for the reverse framework.

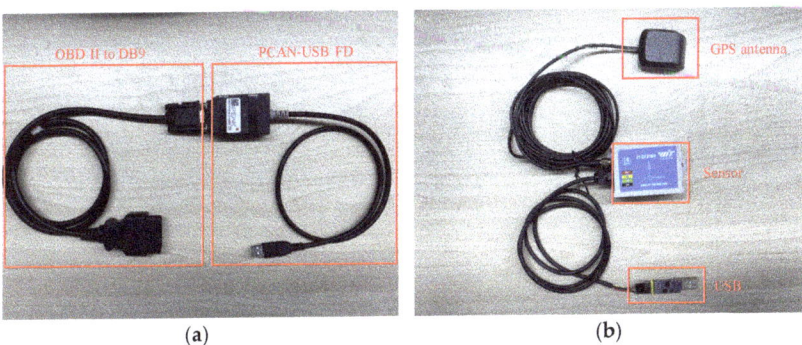

Figure 6. Data acquisition equipment: (**a**) OBD-II data collection equipment; (**b**) Vehicle behavior sensor.

Table 2. Sensor locations and associated physical value.

Location	Physical Characteristics
Bodywork	Speed, Acceleration
Steering wheel	Steering angle
Brake pedal	Pedal angle
Accelerator pedal	Pedal angle
Gear knob	Gear angle
Wiper switch	Switch angle

3.1.2. Data Processing and Resampling

Since the raw data collected by the sensors is limited and does not provide a good picture of the various vehicle states, the collected Y_r must be processed to reveal more vehicle-related state information. Integral, derivative, and discretization processes are performed on the obtained Y_r to get more information. Based on the vehicle behavior in each Y_r, the rate of behavior change is obtained by derivative, the total amount of change is obtained by integral, and the discrete behavioral states are obtained based on a threshold value. Take speed as an example, the acceleration of the vehicle could be obtained by calculating its derivative to time, and the mileage is obtained by calculating its integral for time. Based on the vehicle speed and the threshold of 1 km/h, the vehicle can be classified into two discrete states of stationary and driving. The data processing methods and results are shown in Table 3. After the extension, there are 13 types of vehicle behaviors. The output after data processing is Y_s, which contains more detailed vehicle states.

When processing the raw CAN data collected through the OBD-II port, this framework classifies the raw CAN messages based on the ID and removes the constant data field CAN messages. Since the ID identifies the type of the CAN message, X_i is first determined by grouping by the ID during processing to facilitate the subsequent modeling of the messages for each ID. Since the framework proposed in this study is based on vehicle behavior to reverse CAN messages, constant CAN messages during sensor acquisition of vehicle behavior do not describe any vehicle behavior and are therefore considered as noise. This noisy data is defined as constant data in READ and LibreCAN, CAN message with constant data fields. Noisy messages can be removed to reduce the number of resamples and subsequent modeling, thus reducing the overall time required.

Table 3. Methods and results of raw data processing.

Raw Data (r)	Operation	Detailed Vehicle Behavior (s)
Speed	- Integrals Judgment by threshold	Speed Mileage Drive/Parking
Brake Pedal Angle	- Differential Judgment by threshold	Brake pedal angle Angle change rate Brake or not
Accelerator Pedal Angle	- Differential Judgment by threshold	Accelerator pedal angle Angle change rate Accelerate or not
Gear Angle	- Judgment by threshold	Gear angle P/R/N/D
Wiper Switch Angle	- Judgment by threshold	Wiper switch angle Stop or frequency

The next step of data processing is to synchronize the CAN messages with the vehicle behavior. In this study, the CAN messages in X_i are selected synchronously with the time interval of the beginning and the end of the vehicle behavior described by Y_s. Synchronizing the data ensures that the CAN messages in X_i and the behavior described by Y_s have the same vehicle behavior and state during this time interval.

Finally, multiple linear regression described in Section 2.2 is a method for modeling the dependent and explanatory variables in the same dimension. However, since the messages for each ID appear at a different frequency than the sampling rate of the sensor device, Y_s, must be resampled based on the frequency of X_i to ensure that the two have the same dimensionality [44]. In the data resampling process, this study uses the resampling method of time series in Python to resample each vehicle state Y_s according to the frequency of each X_i to facilitate subsequent modeling. The resampled data is Y_{si} with the same dimensions as X_i. In this step, a separate resampling must be performed for each Y_s based on the frequency of each X_i to obtain $13 \times n$ Y_{si}.

3.2. Related Messages Filter

Based on the results of data processing and resampling, the purpose of this stage is to build a linear regression model with Y_{si} as the dependent variable and each bit of the data field in X_i as the independent variable. Based on the R^2 of the model, the messages that are most relevant to the dependent variable are filtered out.

To obtain the relationship between each bit of the data field and the vehicle behavior, this step starts by expanding the data field in X_i in bit form, which is an $l \times 64$ matrix, where l is the number of messages with ID i. The dependent variable Y_{si}, which is an $l \times 1$ matrix, is defined to represent the vehicle state data resampled according to the message dimension, where s represents the different vehicle states, $s \in (s_1, s_2, \ldots, s_{13})$. A threshold Δ_s is defined to filter out the best model. The outputs of this stage are messages and linear regression models that are highly correlated with the individual vehicle behavior data. The flow of this phase is shown in Figure 7. The detailed process is shown below.

- **Step 1:** After processing, select a resampled vehicle behavior data Y_{si} and a data set X_i with ID i in the CAN bus trajectory.
- **Step 2:** Build a multiple linear regression model with Y_{si} as the dependent variable and X_i as the independent variable and calculate the model parameters R^2 and β.
- **Step 3:** Select the R^2 obtained in step 2 corresponding to Δ_s, and keep only the R^2 greater than Δ_s.
- **Step 4:** Iterate through each X_i and repeat step 1 to step 3. According to the filtering result, obtain the most relevant messages and the corresponding models with the vehicle behavior s.

- **Step 5:** Execute step 1 to step 4 for all s to obtain the candidate messages and the corresponding models for each vehicle behavior.

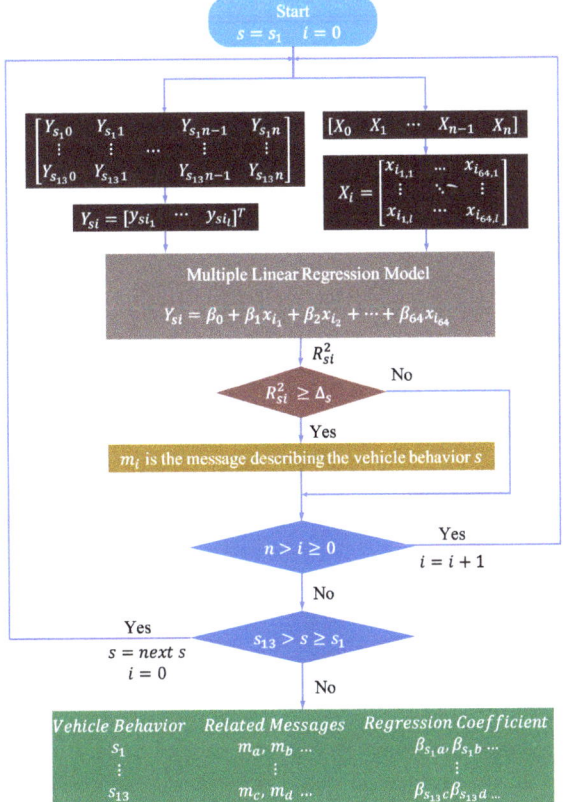

Figure 7. Message selection based on β.

3.3. Bit-Level Message Reverse

After the related message filtering phase, the most relevant candidate messages for the particular vehicle behavior and the corresponding linear regression models are determined. The linear regression models of Y_{si} and X_i are shown in Equation (7). This result clearly shows the relationship between the vehicle behavior and the data fields of m_i, where $\beta = (\beta_0, \beta_1, \ldots, \beta_{64})$ represents the linear relationship between this vehicle behavior data and each bit of the message.

$$Y_{si} = \beta_0 + \beta_1 x_{i1} + \beta_2 x_{i2} + \ldots + \beta_{64} x_{i64} \tag{7}$$

In this stage, the specific details of how the data fields of candidate CAN messages describe the behavior of the vehicle are determined by analyzing the regression coefficient β. As shown in Figure 8, the flow of the bit-level reverse for the candidate messages proceeds as follows.

- Iterate through each β_x in $\beta = (\beta_0, \beta_1, \ldots, \beta_{64})$, keeping only those β_x that are not less than the threshold value. If the value of β_x is less than the threshold, it means that the xth bit of the data field is not related to the specific vehicle behavior. Otherwise, this bit may represent how the behavior of the vehicle is recorded in the CAN messages. The result after threshold filtering is β'.

- If the filtered β' is discrete, the corresponding discrete bit likely represents the state of vehicle. If the filtered β' is continuous, then analyze whether Equation (8) or Equation (9) is satisfied between β'. If satisfied, the bits of the CAN message data field corresponding to the continuous β' describe the modeled vehicle behavior s. Moreover, the bits satisfying Equation (8) are in Motorola alignment, and those satisfying Equation (9) are in Intel alignment. When not satisfied, the CAN message has no relation to the vehicle's behavior.
- Analyzing the discrete β' values and the vehicle state data, the correspondence between the discrete bits and the vehicle state can be obtained reverse. For continuous β', the data length, the alignment form, and the linear relationship describing the vehicle behavior can be gained.

$$\beta_i = 2 \times \beta_{i+1} = 4 \times \beta_{i+2} = \ldots 2^n \times \beta_{i+n} \tag{8}$$

$$\beta_i = \frac{1}{2} \times \beta_{i+1} = \frac{1}{4} \times \beta_{i+2} = \ldots \frac{1}{2^n} \times \beta_{i+n} \tag{9}$$

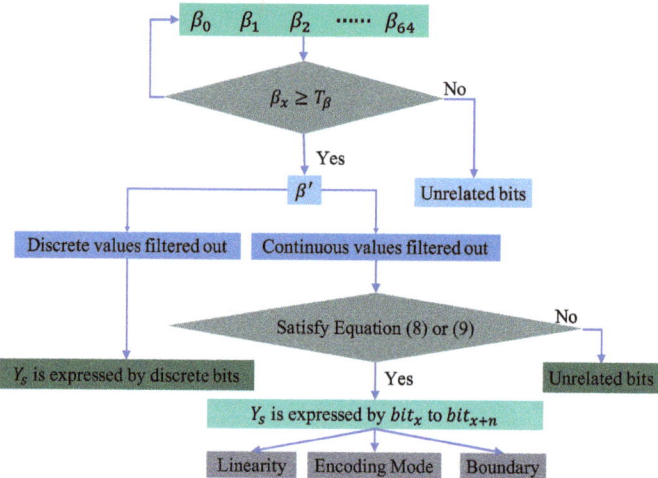

Figure 8. Diagram of bit-level reverse.

4. Performance Evaluation

To evaluate the proposed bit-level CAN bus reverse framework, this study implements it on an actual vehicle and obtains specific details of the vehicle CAN message data fields depicting the vehicle behavior for that vehicle. Using the reverse results, the accuracy of the algorithm is evaluated for practical applications based on the available DBC files [45]. In addition, this section evaluates the execution performance of the framework and compares the advantages of the algorithm over other reverse methods. Finally, the advantages of the algorithm in applications are discussed, and an example is given for reversing other vehicle messages when DBC files are not available.

4.1. Performance in Real Vehicle

4.1.1. Device Description and Data Processing

For the evaluation a 2017 Japanese B-Class sedan was used, whose internal network implements the standard CAN protocol and whose functionality is representative. A DBC file for this model has been obtained, which is used as ground truth for the reverse framework evaluation. To better represent the vehicle behavior, sensors are placed on the body, steering wheel, brake pedal, gas pedal, gear knob, and wiper switch to collect the behavioral data of the vehicle components, which are structured as shown in Figure 9.

The CAN data is collected through the OBD-II interface using the combination cable synchronously when collecting vehicle data. The collected CAN data is written to a log file using the upper computer program, containing the ID, type, length, data field, and timestamp of CAN messages. For accuracy evaluation, more than 3,661,000 consecutive CAN bus messages were collected, and more than 5,000,000 vehicle behavior sensor data were sequentially collected in the same period. The dataset (The dataset is partially open source and can be accessed at http://49.232.218.41:8000/data.zip accessed on 23 January 2022) is quantitatively described in Table 4, which describes the measurements and CAN data collected synchronously for each vehicle behavior.

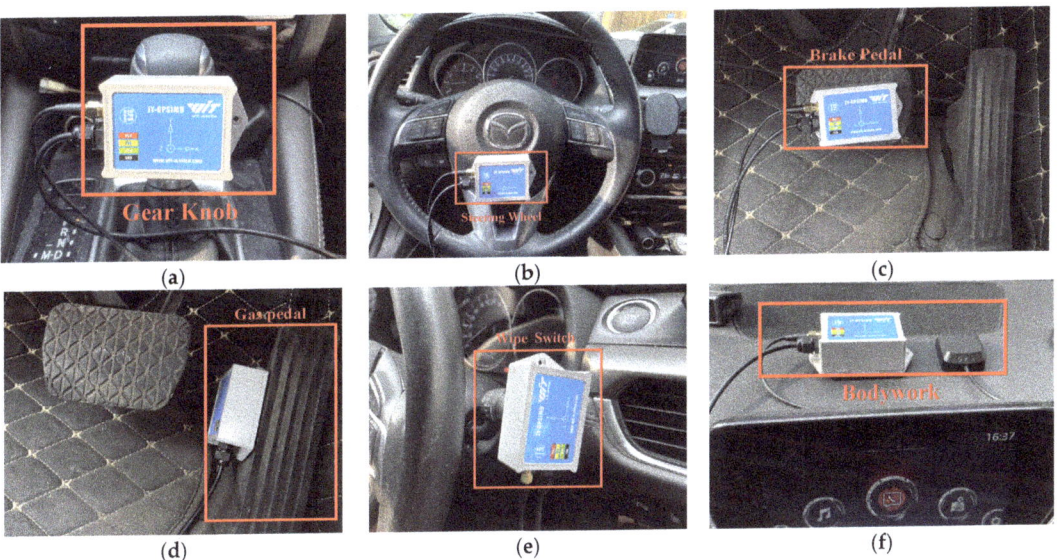

Figure 9. Sensor Acquisition Setup: (**a**) Gear angle; (**b**) Steering wheel angle; (**c**) Brake pedal angle; (**d**) Gas pedal angle; (**e**) Wiper switch angle; (**f**) Vehicle speed.

Table 4. Number of vehicle behaviors and CAN messages.

Vehicle Behavior	Number of Sensor Record	Number of CAN Messages
Bodywork	298,649	1,769,768
Steering Wheel	16,148	132,122
Brake Pedal	7961	57,399
Accelerator Pedal	6364	60,772
Gear Handle	13,105	113,001
Wiper Switch	12,876	118,095

By analyzing the collected CAN traces, the frequency distribution of the messages is shown in Figure 10. This result shows that the number of IDs collected from the test vehicle is 82, which means that there are 82 types of messages in the CAN network. For each type of CAN message, we analyze whether the data field of this CAN message changes and eliminate the messages with unchanged data fields. Based on the analysis and processing of CAN traces, the vehicle behavior data collected in Table 3 is resampled 82 times to obtain Y_{si}. A multiple linear regression model is built between Y_{si} and X_i according to the message filtering process.

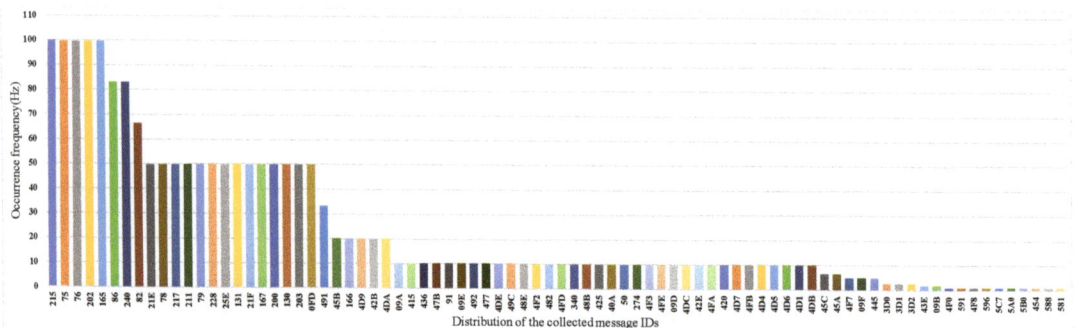

Figure 10. CAN message frequency distribution.

4.1.2. Message Filter Results

The results of the multivariate linear regression of the collected continuous vehicle behavior with each type of message are shown in Figure 11. The x-axis is the determination coefficient R^2 of the multiple linear regression model, and the y-axis is the effective ID distribution.

For the linear regression results of vehicle speed and CAN trace, according to the threshold value 0.6, three types of messages can be filtered out that directly record vehicle speed information with IDs 0x202, 0x215, and 0x217, as shown in Figure 11a. In addition, in this result, there are some R^2 values close to the threshold, such as 0x130, 0x165, 0x167 and 0x200. This is because they may describe information such as RPM, throttle, etc., that correlate with the vehicle speed, which explains their larger R^2. However, since these types of vehicle data cannot be collected by sensors, they cannot determine their exact meaning. As shown in Figure 11b, with a R^2 of 0.1 as the dividing line, the messages IDs strongly correlated with steering wheel angle are 0x086, 0x082, and 0x240. These messages may contain data describing steering wheel torque and steering rate in addition to the information directly representing steering angle. In the same way, messages related to the accelerator pedal are filtered out including messages with IDs 0x165, 0x167, 0x202, 0xFD, and 0x21F, with 0.2 as the divisor, as shown in Figure 11c. Messages with IDs 0x78, 0x202, and 0x165 are categorized as related to brake pedal angle with a threshold of 0.18 as shown in Figure 11d. The results of filtering information related to wiper switch and gear angle are shown in Figure 11e,f. With a threshold of 0.6, the message IDs related to the wiper are 0x9A, and the message IDs related to the gear are 0x165 and 0x228, respectively.

As can be seen from the results of the message filtering, the R^2 and threshold values for messages related to steering angle, acceleration, and brake pedal are generally small. This result is due to the slight variations in vehicle behavior when collecting these data. For example, the pedal is unlikely to be located at the lowest position when collecting the gas pedal angle while driving. In addition, the results for vehicle speed, gas pedal, and brake pedal show that a certain number of messages have an R^2 value that is below the threshold, but very close to it. Although these messages do not directly describe the state of the vehicle speed, gas pedal, and brake pedal, they do describe vehicle behavior correlated with the state. For example, the near-threshold telegrams in the throttle results describe the vehicle's speed, torque, and acceleration, among other things. However, since these messages do not directly describe the vehicle speed, they are classified as irrelevant messages by the threshold. Also, as shown in Figure 11e,f, the R^2 of the messages related to wiper and gears are clearly distinguished from others. Since the vehicle behavior (gear angle and wiper angle) data and the related CAN messages are all discrete, they can be clearly distinguished from the other messages when the linear regression modeling is performed.

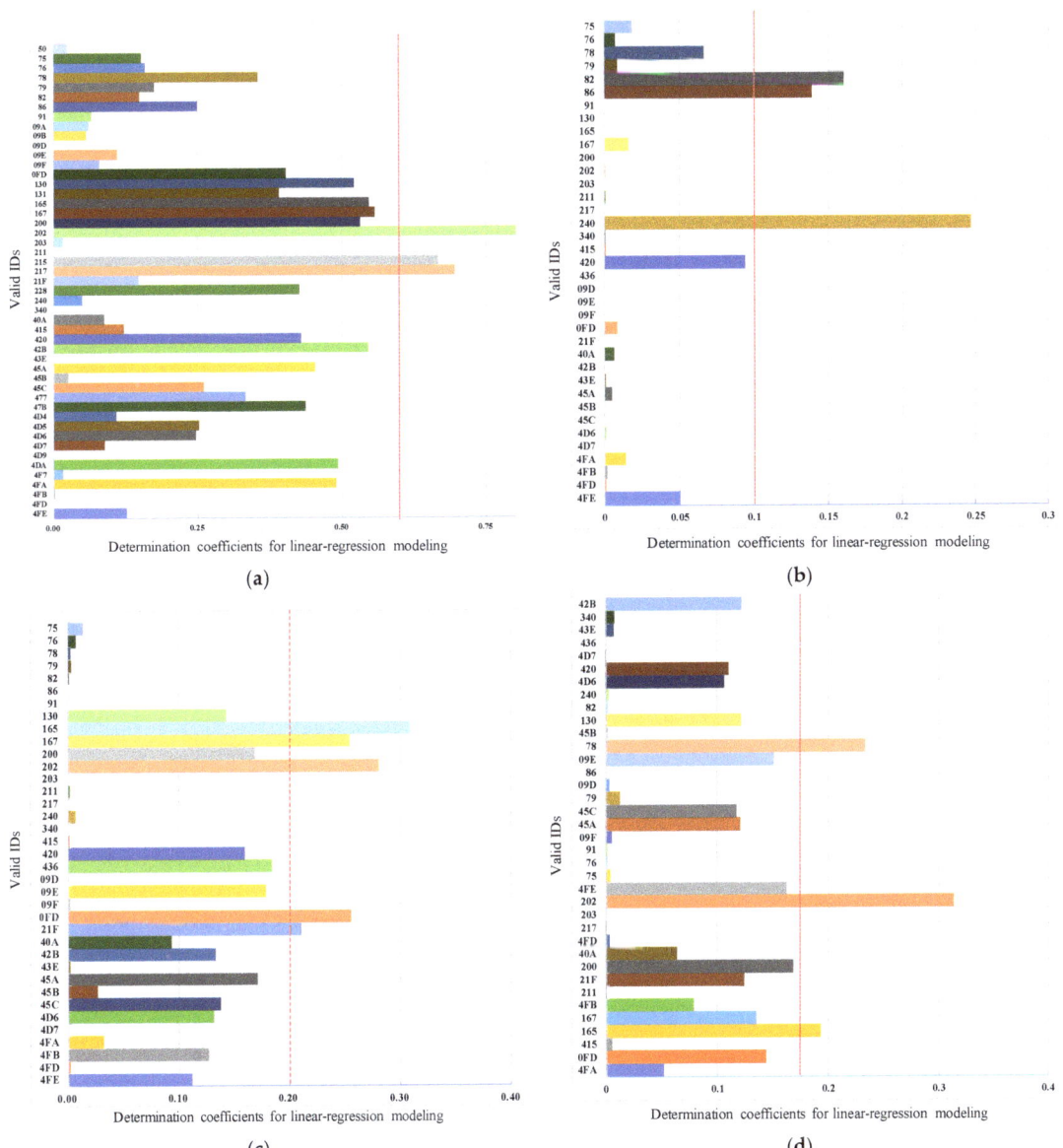

Figure 11. *Cont.*

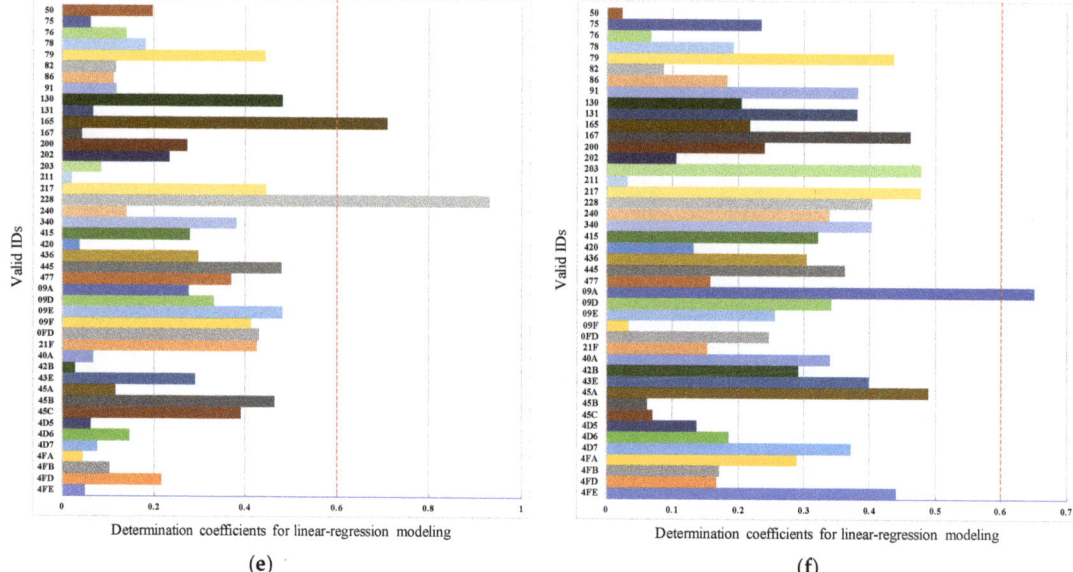

Figure 11. Real vehicle messages filter results: (**a**) speed-related messages; (**b**) steer angle-related messages; (**c**) gas pedal-related messages; (**d**) brake pedal-related messages; (**e**) gear-related messages; (**f**) wiper related-messages.

4.1.3. Bit-Level Reverse Results

By analyzing the linear regression result of the filtered messages, it is possible to reverse the portrayal of the vehicle behavior by the individual bits of the message.

The reverse result for the speed-related messages is shown in Figure 12. There is a two-fold relationship between the messages with IDs 0x202, 0x215, and 0x217 and the β of the vehicle speed. As shown in Figure 12a, bits 34 to 42 in the message with ID 0x202 indicate the vehicle's speed, arranged in the format of Motorola. For the message with ID 0x215, according to Figure 12b, bits 0 to 12, bits 16 to 28, bits 32 to 44, and bits 48 to 60 represent the vehicle speed information and the arrangement format is Motorola. The value for the β with ID 0x217 is shown in Figure 12c, and the bits describing the vehicle speed are 34 to 46, and the arrangement format is also Motorola.

The reverse results of the steering-related messages are shown in Figure 13. Bits 22 to 31 in the message with ID 0x82 describe the steering angle arranged in Motorola. In the corresponding message with 0x86, the steering angle is specified in bits 3 to 13 and 28 to 36, respectively. The message with ID 0x240 does not describe the steering angle directly, but because its R^2 is greater than the threshold, it is related to the change in steering.

Byte No. \ Bit No.	7	6	5	4	3	2	1	0
7	3357801971	−1628217732	−1.665294136	2.955850113	0.991322644	−0.289907624	−0.302687847	0.171739344
6	0.889935594	−0.412378188	0.151430476	0.126843613	0.01788121	−1655665397	1.686479037	1.5102
5	−519702221	4.603181641	4.782656932	0.928475752	1.453727623	−0.018194993	0.395604057	0.133305646
4	−0.004974229	0.019429916	0.534827322	0	0	0	0	0
3	3747208386	−666716471.7	−4093665735	18.80890454	7.041814705	33.087412692	15.94559698	7.694356925
2	3.82222913	2.052452633	0.733454925	0.39026707	0.18976073	0.095002291	−0.084735874	−0.09862176
1	0	0	0	1.52496	1.52496	0	0	1.52496
0	−0.229847605	0.392528104	0.10481122	−0.329716681	0.168024533	0.253546692	−0.028310573	0.053638706

(**a**)

Byte No. \ Bit No.	7	6	5	4	3	2	1	0
7	-1.9500×10^{8}	1.3835×10^{9}	-1.0858×10^{10}	496.0358	249.2333	126.1774	61.6175	29.6104
6	16.2949	7.1141	3.9536	1.7663	0.9877	0.4764	0.2377	0.1094
5	8.2500×10^{8}	3.3433×10^{8}	-7.5339×10^{10}	321.2619	159.5430	78.7844	39.3834	21.6369
4	10.6743	5.7481	2.3324	1.5307	0.8317	0.3773	0.1850	0.0872
3	-1.4109×10^{8}	2.8986×10^{8}	-7.5780×10^{10}	−368.1298	−182.5109	−93.1598	−44.0811	−22.5263
2	−12.1445	−5.9495	−2.6045	−1.4666	−0.8151	−0.3778	−0.2192	−0.1007
1	-1.0615×10^{8}	7.3628×10^{8}	-7.5471×10^{10}	−419.3960	−210.9166	−105.1524	−53.1068	−25.9040
0	−11.2769	−6.6105	−3.1541	−1.7671	−0.8195	−0.3136	−0.1652	−0.0913

(**b**)

Byte No. \ Bit No.	7	6	5	4	3	2	1	0
7	−0.2448	−0.1650	0.0102	−0.3607	0.5244	−0.5178	1.9200	0.2832
6	0.7781	−0.0524	0.1434	−0.3971	−0.3533	−0.1758	−0.1719	0.4453
5	0.2646	26.1434	13.2176	7.1642	3.2614	1.7673	0.8613	0.4206
4	0.2501	0.1133	0.0501	0.0265	0.0134	0.0887	1.2222	0.0000
3	0.0000	0.0000	−0.0076	−0.6139	−0.0034	−0.0945	0.7202	−0.0359
2	0.6625	−0.1119	−0.0314	0.1349	−0.4283	−0.0462	−1.8548	4.5949
1	0.0051	0.0000	0.0000	0.0162	0.0029	−0.0084	−0.0140	0.2394
0	−0.2080	−0.3410	0.0069	0.0527	0.0081	−0.0176	−0.0363	−0.0363

(**c**)

Figure 12. Speed-related messages reverse result: (**a**) ID 0x202 reverse result; (**b**) ID 0x215 reverse result; (**c**) ID 0x217 reverse result.

Byte No. \ Bit No.	7	6	5	4	3	2	1	0
7	0.0000	0.0000	0.0000	0.0000	0.0000	0.0000	0.0000	0.0000
6	0.0000	0.0000	0.0000	0.0000	0.0000	0.0000	0.0000	0.0000
5	−4.5691	3.2343	3.2343	3,2343	−39.6865	27.3546	11.5185	−12.8177
4	24.8477	30.8647	17.9640	6.4265	−2.3840	0.6244	−1,7468	0.0000
3	30.1037	14.1605	7.3889	3.1524	1.9637	1.2133	0.6756	0.3275
2	0.1524	0.0675	−0.4275	−1.1913	0.0000	−1.3348	0.0000	0.0000
1	0.0000	0.0000	0.0000	−85.8317	−4.8072	−0.3888	−5.5472	−6.9460
0	−7.0268	−20.8909	20.7759	3.0168	7.4394	−8.8125	12.4246	19.6513

(**a**)

Byte No. \ Bit No.	7	6	5	4	3	2	1	0
7	3.8521×10^{11}	1.1975×10^{13}	-4.36×10^{12}	3.5984×10^{12}	-4.012×10^{12}	9.4006	−1.0509	−4.42e12
6	2.0276×10^{13}	7.4398×10^{12}	8.416×10^{12}	7.5757	0.4792	−7.3607	3.2874	−3.7455
5	4.8201×10^{13}	4.6975×10^{12}	-4.93×10^{11}	7.6694×10^{11}	-1.402×10^{12}	26.4892	2.3737	−0.5498
4	5.2733	0.1092	−4.6697	−604.5745	−265.7115	−140.0021	−76.6727	−31.9156
3	−14.6939	−8.2973	−4.2185	−2.4103	−0.8919	0.5318	−1.1433	4.0898
2	−2.4496	0	0	0	4.7555×10^{13}	−4249609	4.6975×10^{12}	4.9275×10^{11}
1	7.6694×10^{11}	1.402×10^{12}	576.1385	276.3953	118.5238	47.5868	56.6046	29.7995
0	13.7582	8.9297	4.1675	2.3839	12.6086	−2.54e13	0	0

(**b**)

Byte No. \ Bit No.	7	6	5	4	3	2	1	0
7	3.4944×10^{12}	1.7257×10^{13}	1.7688×10^{12}	112.9218	11.9813	42.0565	25.1313	5.4629
6	1.1577e11	2.1990×10^{11}	5.8172×10^{10}	29.4187	27.3574	−15.4216	−15.2149	−18.2656
5	5.5942×10^{10}	6.1403×10^{10}	3.2494×10^{10}	1.2852×10^{10}	1.2993×10^{10}	3.6798×10^{10}	2.1176×10^{11}	3.844×10^{10}
4	5.4518×10^{11}	9.8716×10^{10}	5.5359×10^{10}	9.6373×10^{10}	2.4623×10^{11}	1.7064×10^{10}	6.5994×10^{11}	8.7764×10^{11}
3	7.7054×10^{13}	7.7054×10^{13}	−178.3829	−97.4542	−18.3842	−8.8223	6.6820	−2.7250
2	2.0554	4.7321×10^{9}	4.6150×10^{9}	4.6145×10^{10}	4.6140×10^{10}	6.1166×10^{10}	−82.1988	−32.3782
1	−2.9478	8.5155	0.0827	6.6903	−0.0349	4.3388	−1.0362	4.4696
0	−2.0336	2.4211	0.1935	0.8947	0.1155	0.0000	0.0000	0.0000

(**c**)

Figure 13. Steer-related messages reverse result: (**a**) ID 0x082 reverse result; (**b**) ID 0x086 reverse result; (**c**) ID 0x240 reverse result.

The results of the throttle-related message are shown in Figure 14. There is an approximate relationship of 2 times in the β corresponding to 0xFD, 0x167, and 0x202 in the results, so based on the β, we find that bits 49 to 55 in the message with 0xFD describe the gas pedal angle. As shown in Figure 14b, in the message whose ID is 0x167, bits 0 to 7 portray the angle of the gas pedal. The angle of the gas pedal in 0x202 is represented in bits 39 to 47. For the messages 0x165 and 0x21F, there is no 2x relationship in β. But the bits 40 to 43 of 0x21F indicate the rate of change of the gas pedal angle as shown in Figure 14d. For 0x165, the gas pedal angle is converted to a discrete state using a threshold: accelerated or not. The result of the discrete value is shown in Figure 14e, from which it can be seen that bit 29, and bits 22 to 26 of ID 0x165 describe whether the gas pedal is activated or not.

(a)

Bit No. / Byte No.	7	6	5	4	3	2	1	0
7	−3.52×10⁹	3.10×10¹¹	3.10×10¹¹	−6.98×10⁹	9.68×10⁹	1.01×10¹⁰	3.00×10¹¹	−1.73×10¹¹
6	9.02×10⁹	5.65×10⁹	2.98×10⁹	1.45×10⁹	8.38×10⁸	4.12×10⁸	2.01×10⁸	0.0
5	−1.07×10⁻²	−5.31×10⁻³	2.19×10⁻³	−2.40×10⁻³	0.0	0.0	0.0	0.0
4	0.0	0.0	0.0	0.0	0.0	0.0	0.0	0.0
3	0.0	0.0	0.0	−1.79×10¹¹	−1.79×10¹¹	−1.39×10¹¹	−1.39×10¹¹	0.0
2	0.0	0.0	0.0	0.0	0.0	0.0	9.67×10⁻¹	0.0
1	0.0	9.00×10⁻¹	8.27×10⁻¹	2.09×10⁻¹	−8.92×10⁻²	1.45×10⁻¹	1.01×10⁻¹	4.46×10⁻²
0	0.0	0.0	−1.79×10¹¹	−1.79×10¹¹	0.0	0.0	0.0	0.0

(b)

Bit No. / Byte No.	7	6	5	4	3	2	1	0
7	0.0000	0.3918	−0.0540	0.0061	0.4457	−0.3976	0.0773	0.1157
6	0.1405	0.0005	−8.1509	−8.1509	0.0000	0.0000	0.0000	0.0000
5	0.0000	0.0000	−8.1509	−8.1509	0.0000	−8.1509	0.0000	0.0000
4	0.0059	0.0037	0.0132	−0.0031	0.0000	0.0000	0.0000	0.0000
3	0.0000	−5.6113	−2.5396	−2.5396	−0.5518	−0.3847	−0.2282	0.1356
2	0.0401	−0.0890	−0.0185	−8.1509	−8.1509	0.0000	0.0000	0.0000
1	0.0000	0.0000	0.0000	0.0000	0.0000	0.0000	0.0000	0.0000
0	699,609.9052	349,804.9522	174,902.4813	87,451.2317	43,725.6307	21,862.8212	10,931.4122	5,465.7142

(c)

Bit No. / Byte No.	7	6	5	4	3	2	1	0
7	2.34×10⁻¹⁵	−6.80×10⁻¹⁵	3.47×10⁻¹	−2.64×10⁻¹	−1.80×10⁻¹	1.49×10⁻¹	−4.38×10⁻¹	1.33×10⁻¹
6	1.32×10⁻¹	−5.23×10⁻²	−5.22×10⁻²	−4.17×10⁻²	5.96×10⁻²	2.44×10⁻¹⁶	−1.18×10¹	−1.18×10¹
5	2.31×10⁻¹⁶	−1.21×10⁻¹⁶	−6.40×10⁻¹⁷	−3.80×10⁻¹⁷	−1.19×10⁻¹⁷	−6.76×10⁻¹⁸	−3.79×10⁻¹⁸	−1.53×10⁻¹⁸
4	−7.69×10⁻¹⁹	−1.19×10⁻¹⁷	−3.79×10⁻¹⁷	−4.72×10⁻¹⁸	0.0000	0.0000	0.0000	0.0000
3	0.0000	−3.63×10⁻²	−6.12×10⁻²	1.44×10⁻¹	4.41×10⁻¹	−2.80×10⁻¹	9.00×10⁻²	7.04×10⁻²
2	2.47×10⁻²	1.06×10⁻¹	0.0000	0.0000	0.0000	0.0000	0.0000	0.0000
1	0.0000	0.0000	0.0000	−1.18×10¹	−1.18×10¹	0.0000	0.0000	−1.18×10¹
0	−1.74×10⁻²	−2.84×10⁻¹	−1.16×10⁻¹	7.10×10⁻²	4.65×10⁻²	−6.69×10⁻²	−2.42×10⁻¹	−5.35×10⁻²

(d)

Bit No. / Byte No.	7	6	5	4	3	2	1	0
7	−1.6194×10⁻¹⁵	−1.6654×10⁻¹⁶	−2.4973×10⁻¹⁵	4.5258×10⁻¹⁵	6.9499×10⁻¹⁶	2.5087×10⁻¹⁵	−1.5002×10⁻¹⁵	−4.7190×10⁻¹⁶
6	−7.6787×10⁻¹⁶	0.0000	3.2174×10⁻¹⁶	0.0000	0.0000	−8.3207×10⁻¹	−1.2986×10⁻¹⁶	1.8555×10⁻¹⁶
5	−2.3070×10⁻¹⁶	7.5208×10⁻¹⁷	−5.5683×10⁻¹⁷	2.3650×10⁻¹⁷	5.5404×10⁻⁹	2.8454×10⁻⁹	1.5123×10⁻⁹	8.4469×10⁻¹⁰
4	−1.7759×10⁻¹⁷	0.0000	0.0000	0.0000	0.0000	0.0000	0.0000	0.0000
3	0.0000	0.0000	0.0000	0.0000	0.0000	0.0000	0.0000	−4.0665×10⁻¹
2	−2.9927×10⁻¹	−5.0412×10⁻¹	2.5011×10⁻¹	4.1907×10⁻¹	−3.0559×10⁻¹	1.0687×10⁻¹	3.9513×10⁻²	1.3640×10⁻²
1	6.2880×10⁻⁹	−8.9739×10⁻³	−6.2460×10⁻³	1.7465×10⁻²	7.4736×10⁻³	−5.6270×10⁻³	−9.8577×10⁻³	1.1375×10⁻³
0	−7.1771×10⁻³	7.8811×10⁻³	−2.7358×10⁻³	2.1376×10⁻²	3.8900×10⁻²	−7.0094×10⁻³	2.9474×10⁻³	8.3743×10⁻³

(e)

Bit No. / Byte No.	7	6	5	4	3	2	1	0
7	−30,722,903.3959	−780,259.2057	95,016.3169	7,016.7500	−103,882.7484	31,071.4599	−71,574.6187	39,885.1065
6	24,35,913.1645	447,637.7878	−416,934.3919	4,525,584.9459	1,243,563.1738	64,975,570.8269	108,015.1117	943,273.2207
5	−245,73,183.4337	0.0000	0.0000	812,245.6660	0.0000	0.6637	0.6834	0.1551
4	−0.0743	0.1130	0.1089	0.0979	0.0000	−0.0829	0.0000	0.0000
3	0.0000	0.0000	2,417,812,245.6660	0.0000	0.0000	2,417,812,245.6660	2,417,812,245.6660	2,417,812,245.6660
2	2,417,812,245.6660	2,417,812,245.6660	0.0000	0.0000	0.0000	0.0000	0.0000	0.0000
1	0.0000	0.0000	0.0000	0.0000	0.0000	0.0000	0.0000	0.6660
0	0.2348	−0.1436	0.1238	0.0247	−0.0284	−0.0201	0.0747	−0.0004

Figure 14. Gas-related messages reverse result: (**a**) ID 0x0FD reverse result; (**b**) ID 0x167 reverse result; (**c**) ID 0x202 reverse result; (**d**) ID 0x21F reverse result with gas angle change rate; (**e**) ID 0x165 reverse result with discrete state.

The results of the bit reverse for the brakes are shown in Figure 15. Based on the β of 0x78, the bits representing the brake pedal are bits 32 to 37, arranged as Motorola. Since there are no significant features in the β of 0x202 and 0x165, the linear regression β of these two types of IDs with discrete states of the brake pedal (braked or not) was calculated using the same method. The results show that in 0x165, bits 0, 1, 3, 7, and 8 indicate whether the vehicle's state is accelerated or not. For the message with 0x202 as ID, the results show that it does not describe the braking behavior but only the vehicle behavior with respect to braking.

(a)

Bit No. / Byte No.	7	6	5	4	3	2	1	0
7	1.0714×10^{10}	1.0769×10^{10}	-2.6298×10^{8}	1.0688×10^{10}	1.0708×10^{10}	1.0777×10^{10}	1.0731×10^{10}	-1.0664×10^{10}
6	-1.0501×10^{7}	3.5892×10^{8}	2.6142×10^{8}	1.3071×10^{8}	6.5355×10^{7}	3.2678×10^{7}	1.6339×10^{7}	-5.5794×10^{9}
5	-6.2889×10^{8}	-9.9866×10^{8}	-8.6633×10^{8}	-8.6227×10^{8}	-8.8981×10^{8}	-6.9424×10^{8}	-9.1693×10^{8}	1.3071×10^{8}
4	6.5355×10^{7}	3.2678×10^{7}	5.2284×10^{8}	2.6142×10^{8}	1.3071×10^{8}	6.5355×10^{7}	3.2678×10^{7}	1.6339×10^{7}
3	-1.0342×10^{10}	1.0457×10^{9}	1.6339×10^{7}	-1.4083×10^{8}	-2.0117×10^{9}	-3.2510×10^{2}	-2.2554×10^{-3}	0.0000
2	-1.0342×10^{10}	-1.0342×10^{10}	-1.0342×10^{10}	0.0000	0.0000	0.0000	-5.1423×10^{9}	-1.0342×10^{10}
1	1.3071×10^{8}	6.5355×10^{7}	3.2678×10^{7}	1.6339×10^{7}	0.0000	0.0000	0.0000	0.0000
0	-2.0914×10^{9}	1.0457×10^{9}	5.2284×10^{8}	2.6142×10^{8}	1.3071×10^{8}	6.5355×10^{7}	3.2678×10^{7}	1.6339×10^{7}

(b)

Bit No. / Byte No.	7	6	5	4	3	2	1	0
7	1.02	0.49	0.04	−0.30	−0.03	0.25	−0.18	−0.20
6	0.05	−0.20	0.16	−0.14	−0.02	−0.01	−0.03	0.69
5	−0.17	0.00	0.00	0.34	0.00	1.22	1.34	−0.16
4	0.28	0.22	0.19	−0.42	0.00	1.88	0.00	−0.45
3	−0.15	−0.06	−0.02	0.00	0.00	0.00	0.00	0.00
2	0.00	0.00	0.00	0.00	0.00	0.00	0.00	0.00
1	0.00	0.00	0.00	0.00	−0.02	0.06	0.00	35,984,984,372
0	35,984,984,372	0.49	−0.52	0.16	35,984,984,372	0.11	35,984,984,372	35,984,984,372

(c)

Bit No. / Byte No.	7	6	5	4	3	2	1	0
7	0.0000	0.0000	−1.5596	−1.0489	−0.4532	0.3652	−0.1251	0.0948
6	0.0823	0.0661	−0.0427	0.0620	0.0033	0.0000	−0.5051	−0.5051
5	0.0000	0.0000	0.0000	0.0000	0.0000	0.0000	0.0000	0.0000
4	0.0000	0.0000	0.0000	0.0000	0.0000	0.0000	0.0000	0.0000
3	0.0000	0.1442	−0.4417	−0.4461	−0.2856	0.1556	−0.3119	0.1625
2	−0.0977	−0.1675	0.0000	0.0000	0.0000	0.0000	0.0000	0.0000
1	0.0000	0.0000	0.0000	−0.5051	−0.5051	0.0000	0.0000	−0.5051
0	−0.5546	−0.2084	−0.4710	0.3023	−0.1523	−0.0358	0.0792	0.0284

Figure 15. Brake-related messages reverse result: (a) ID 0x078 reverse result; (b) ID 0x165 reverse result; (c) ID 0x202 reverse result.

The reverse results for the gears are shown in Figure 16. Since the gear behavior data is discrete, it is evident from the β that the message with 0x228 describes the gear information in bits 3, 5 to 7, 10 and 35 to 39, and 0x165 describes the gear in bits 51 to 54. The reverse result of the wipers is shown in Figure 17. The data describing the wiper speed in 0x9A are bits 37 to 38 and bit 50, And the specific reverse results are shown in Table 5.

(a)

Bit No. / Byte No.	7	6	5	4	3	2	1	0
7	-9.0171×10^{8}	1.1652×10^{9}	2.2511×10^{9}	3.4928	2.0663×10^{9}	-4.7794×10^{9}	-8.2490×10^{-02}	-4.7794×10^{9}
6	0.0000	0.0000	0.0000	0.0000	0.0000	0.0000	0.0000	3.1163
5	0.0000	0.0000	0.0000	0.0000	0.0000	0.0000	0.0000	0.0000
4	-1.1486×10^{11}	-1.1486×10^{11}	-1.1486×10^{11}	-1.1486×10^{11}	-1.1486×10^{11}	0.0000	0.0000	0.0000
3	0.0000	0.0000	0.0000	5.8351×10^{9}	5.8351×10^{9}	-1.6450×10^{10}	−1.1096	0.0000
2	0.0000	0.0000	0.0000	0.0000	0.0000	0.0000	0.0000	0.0000
1	0.0000	0.0000	0.0000	0.0000	-1.1486×10^{11}	0.0000	0.0000	0.0000
0	-1.1486×10^{11}	-1.1486×10^{11}	-1.1486×10^{11}	0.0000	-1.1486×10^{11}	0.0000	0.0000	0.0000

(b)

Bit No. / Byte No.	7	6	5	4	3	2	1	0
7	−1.11	2.09×10^{9}	4.86×10^{9}	-2.97×10^{10}	-1.78×10^{9}	-4.38×10^{9}	6.80×10^{9}	-3.47×10^{10}
6	-6.43×10^{10}	-1.12×10^{11}	-1.12×10^{11}	-1.12×10^{11}	-1.12×10^{11}	2.02×10^{9}	6.59×10^{9}	-1.09×10^{9}
5	-1.46×10^{9}	1.15×10^{9}	5.48×10^{9}	0.0000	-1.19×10^{9}	0.0000	7.30×10^{-1}	−1.45
4	-2.84×10^{-3}	2.39×10^{-1}	-5.64×10^{-1}	-1.74×10^{-1}	0.0000	0.0000	0.0000	0.0000
3	0.0000	0.0000	0.0000	0.0000	0.0000	0.0000	0.0000	0.0000
2	0.0000	0.0000	0.0000	0.0000	0.0000	0.0000	0.0000	0.0000
1	0.0000	0.0000	0.0000	0.0000	0.0000	1.10×10^{-2}	−9.09	1.55×10^{1}	1.16
0	1.60×10^{-1}	1.65	−4.77	−1.16	2.13×10^{-1}	-9.74×10^{-1}	2.18	−1.23

Figure 16. Gear-related messages reverse result: (a) ID 0x228 reverse result; (b) ID 0x165 reverse result.

Bit No. / Byte No.	7	6	5	4	3	2	1	0
7	-1.85×10^1	3.39	1.96×10^1	-4.83×10^{-1}	1.19×10^1	-5.14	-2.09×10^1	-2.07×10^1
6	1.92×10^1	-1.67×10^1	-7.31	2.27×10^1	9.22	3.53×10^{11}	1.49×10^1	2.51×10^1
5	1.94×10^1	-7.85	2.27×10^1	-2.20×10^1	5.86	1.74×10^1	-2.16×10^1	-7.46×10^{-1}
4	1.72×10^1	3.53×10^{11}	3.53×10^{11}	-1.01×10^1	5.38×10^{-1}	-2.25×10^1	-1.39×10^1	-8.58
3	2.41	3.38	-1.90×10^1	3.49	1.23×10^1	1.45	-2.20×10^1	5.49×10^{-1}
2	2.16×10^1	-5.38×10^{-1}	5.21	-6.12	-2.28×10^1	-2.45	1.98×10^1	-1.94×10^1
1	1.13×10^1	-1.66×10^1	1.73×10^1	8.06	-1.34×10^1	6.74	-2.33×10^1	1.03×10^1
0	-1.73×10^1	8.17	8.64	7.54	-1.64×10^1	-3.83	-2.16	7.45

Figure 17. Wiper-related messages reverse result.

Table 5. Results for gears and wipers of bit-level reverse.

	Status	ID 0x165		ID 0x228		
		Bits 54–51	Bits 39–35	Bit 10	Bits 7–5	Bit 3
Gear	P/N	0110	00010	1	110	0
	D	1100	10000	1	001	1
	R	1101	00010	1	010	1

	Status	ID 0x09A	
		Bit 50	Bits 38–37
Wiper	Auto	1	10
	Slow	0	10
	Fast	0	01

4.2. Framework Accuracy

The accuracy of the system proposed in this study is evaluated using the inverse results of the actual vehicles. The accuracy is evaluated using the DBC files of the test vehicle, which were determined to be the truth.

The accuracy of message filtering is shown in Table 6. All CAN traces are taken from the OBD-II interface, so the accuracy is expressed using the percentage of filtered quantities in the OBD-II. Among all the results, only the brake-related messages have an accuracy of 66.67%, while all other messages are filtered at 100%. The false-positive result for 0x202 for brakes is due to the fact the brakes are velocity-dependent to some extent. According to the DBC file, 0x202 does contain velocity information, which causes R^2 to be higher than the threshold. In addition, message 0x240 in the description of the DBC, describes the vehicle's torque information. Although it is a steering-related message, it cannot be inverted at the bit level because the torque measurement information is not directly available. It is also worth noting that the messages defined in the DBC file do not fully appear in the OBD-II interface. This phenomenon is due to a gateway in the vehicle CAN-bus network, which does not forward all bus traces to OBD-II, but only a portion of the traffic to the OBD-II interface. The rest of the CAN bus data, especially the traffic related to assisted driving and vehicle control, flows only within the vehicle and cannot be captured externally.

Table 6. Message filtering accuracy results for vehicle behavior.

Behavior	DBC Defined Messages	Messages Captured from OBD-II	Framework Filtering Results	Accuracy
Speed	0x25E, 0x217, 0x202, 0x215, 0x35F, 0x361	0x217, 0x202, 0x215	0x217, 0x202, 0x215	100%
Steer	0x86, 0x240, 0x243, 0x82	0x86, 0x240, 0x82	0x86, 0x240, 0x82	100%
Gas	0x202, 0x21C, 0xFD, 0x167, 0x165, 0x21F	0x202, 0xFD, 0x167, 0x165, 0x21F	0x202, 0xFD, 0x167, 0x165, 0x21F	100%
Brake	0x165, 0x78	0x165, 0x78	0x165, 0x78, 0x165	66.67%
Gear	0x228, 0x165	0x228, 0x165	0x228, 0x165	100%
Wiper	0x9A	0x9A	0x9A	100%

The bit-reverse accuracy is shown in Figure 18, which compares the bit reverse results of this framework with the vehicle behavior defined in the DBC file. Figure 18a shows the bit-inverse accuracy of the speed-dependent messages. It is observed that bits are written with speed in 0x202, 0x215, and 0x217 are partially reversed to obtain 9 bits for 16 bits in 0x202, 52 bits for 64 bits in 0x215, and 14 bits for 16 bits in 0x217. The bit reversal accuracy of the two steering-related messages, 0x082 and 0x086, is shown in Figure 18b. The proposed framework in this study correctly reverses 9 of the 16 bits in 0x082 and 18 of the 27 bits in 0x086. The accuracy of gas-related message reversal is shown in Figure 18c. 0x0FD gets 7 out of 8 bits, 0x167 completely reverses 8 bits, 0x202 gets 9 out of 16 bits, and both 0x21F and 0x165 have only one bit that is not reversed. Only bits 38 to 39 of 0x078 were not found in the brake-related messages' reverse results, as shown in Figure 18d. For the gear and wiper-related messages, the bits indicating the gear and wiper switches are both correctly reversed, which can be seen in Figure 18.

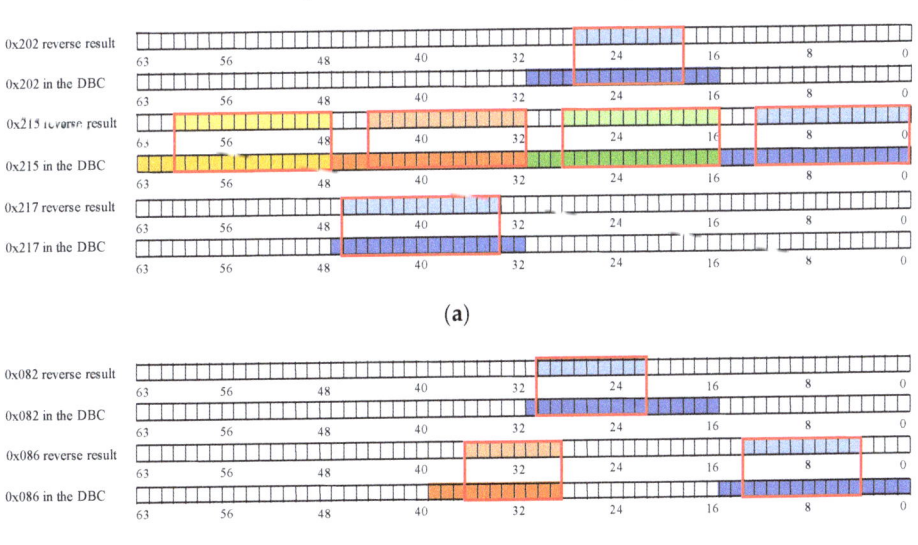

Figure 18. Cont.

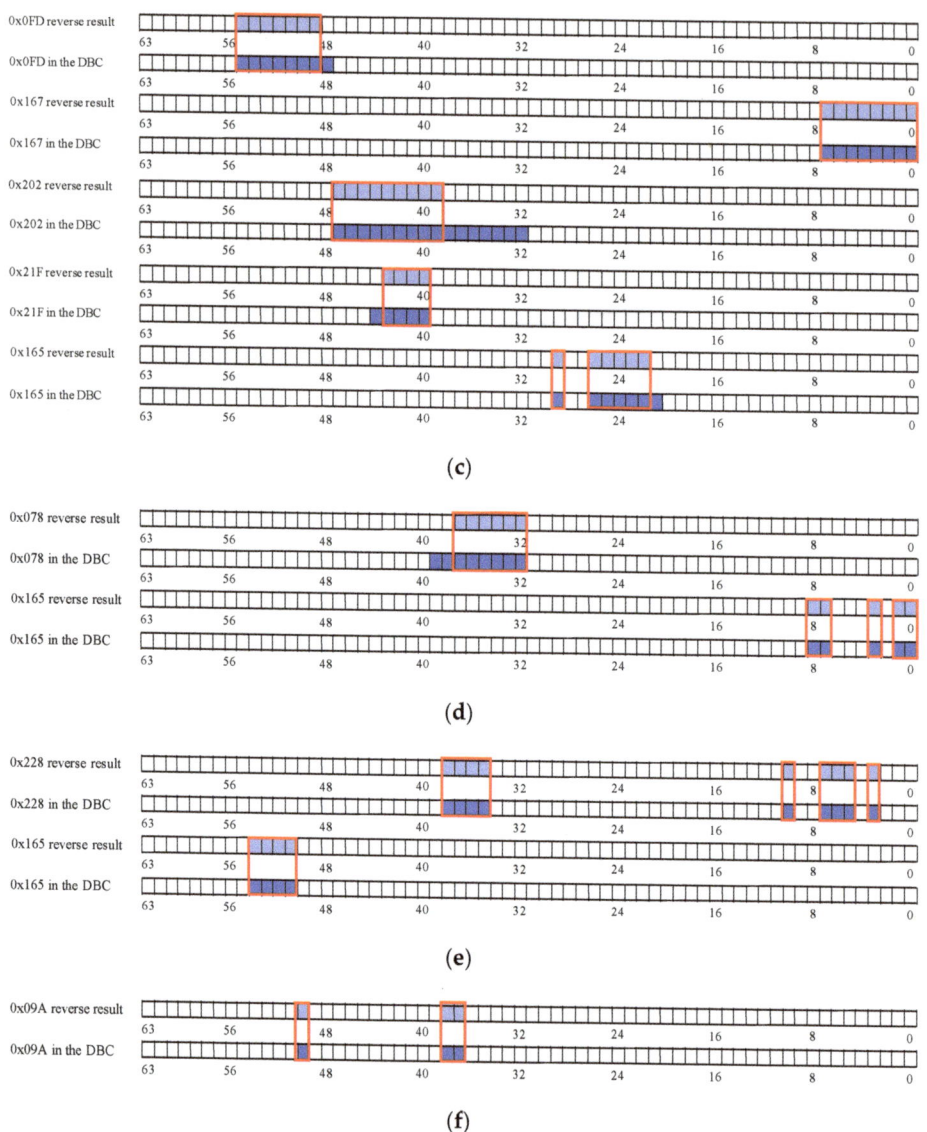

Figure 18. Bit reverse accuracy: (**a**) Speed reverse result; (**b**) Steer reverse result; (**c**) Gas reverse result; (**d**) Brake reverse result; (**e**) Gear reverse result; (**f**) Wiper reverse result.

The overall bit-reverse accuracy of the proposed framework for vehicle behavior is shown in Table 7. The overall reverse accuracy is over 76%, especially for gear, and wiper reversion can reach 100% because CAN messages and sensor data are discrete and not easily disturbed by other data. The reverse accuracy for vehicle speed, gas pedal, and accelerator pedal are all about 80% because these behaviors are difficult to reach the limit state during vehicle sampling, such as vehicle speed of 255 km/h, gas, and brake pedals kept at the maximum angle. Therefore, when reversing the messages related to these behaviors, their high values can barely be detected (i.e., the high value of β does not satisfy the two-fold relation), which results in poor accuracy. The steering-related information performs the worst, with only 65%. Due to the low degree of steering wheel variability in daily driving, the linear regression model is easily disturbed by irrelevant bits, resulting in poor accuracy of bit reversals.

Table 7. Bit reverse result with DBC file description.

Vehicle Behavior	Number of Relevant Bits in DBC	Reverse Results	Accuracy
Speed	96	74	77.1%
Steer	43	28	65.1%
Throttle	44	34	77.3%
Brake	13	11	84.6%
Gear	13	13	100%
Wiper	3	3	100%
Total	212	163	76.9%

4.3. Time Consumption

The framework's time performance analysis was performed on a CentOS server with an Intel® Xeon® Gold 6248 CPU @ 2.50 GHz and 8 GB of RAM using Python 3. The time is taken to compute the three critical stages of data resampling, multiple linear regression modeling, and a bitwise inversion was calculated separately during the evaluation. Table 8 shows the execution time results for each phase. The shortest time-consuming stage is the bit-inverse stage, which requires no more than 25 us in the longest case and can be completed within 7 us in the fastest case. The most time-consuming phase is the data resampling phase. The execution time of the data resampling phase varies from 1.15 s to 190.67 s, with an average time of 37.23 s, which is because this stage resamples the sensor data based on the number of IDs that occur. The essential linear regression phase does not take more than 0.84 s. Overall, the time required to reverse the content of a message correctly averages 37.41 s and does not exceed 191.5 s at most.

Table 8. Implementation time of each stage.

Step	Shortest (s)	Longest (s)	Average (s)
Resample	1.150728	190.674251	37.23192305
Linear regression model	0.007088	0.83345	0.179022554
Bit reverse	0.000007	0.000025	0.0000099
Total	1.157823	191.50772	37.4109555

4.4. Result of Comparison with Other Methods

This section presents the performance comparison results between the bit-level reverse framework proposed in this study and other CAN message reverse methods. Nowadays, the effective CAN message reversal algorithms are READ [30], LibreCAN [31], ReCAN, and Bram's proposed reversal algorithm based on the correlation coefficient [30]. Among them, READ, ReCAN [32], and LibreCAN algorithms use bit-flip rates to delimit CAN message data fields; LibreCAN and Bram's scheme [37] use correlation coefficients to find the message IDs describing specific vehicle behavior. The differences between the existing algorithms and the linear regression framework in reverse results are given in Table 9. Our proposed scheme is the only one that enables boundary delineation, correlated message identification, and bit reverse. READ and ReCAN only perform CAN message data boundary delineation, Bram's scheme only addresses correlated message screening, and LibreCAN achieves both results but cannot achieve bit-level inversion. Therefore, this section only compares the performance of this framework with existing algorithms in terms of boundary delineation, correlated message filtering, and execution complexity.

Table 9. Reverse function compared with existing algorithms.

Algorithm	Boundary Delineation	Related Message Filtering	Bit-Level Reverse
Bit-level reverse based on linear regression	√	√	√
READ	√	×	×
LibreCAN	√	√	×
ReCAN	√	×	×
Reverse engineering based on correlation coefficient	×	√	×

4.4.1. Boundary Delineation

In terms of boundary delineation, we compare the linear regression framework of this paper with the bit-flip rate algorithm used by READ, ReCAN, and LibreCAN. The performance of the methods in this study and the bit-flip rate method in delineating CAN messages with discrete states and continuous vehicle behavior is shown in Table 10. The framework in this paper can delineate the vehicle behavior within the corresponding range with 100% correctness, while the bit-flip-based rate is only 53.3% correct in delineating the boundaries. In particular, bit flipping has relatively good results in delineating CAN messages describing continuous behavior, but boundary delineation errors occur for fields corresponding to discrete vehicle behavior.

Table 10. Boundary Delineation Comparison.

Vehicle Behavior	ID	Linear Regression	Bit Flip (READ, ReCAN, LbreCAN)
Speed	202	√	√
	215	√	√
	271	√	√
Steer	082	√	√
	086	√	×
Throttle	0FD	√	×
	167	√	×
	202	√	√
	21F	√	√
	165	√	×
Brake	078	√	√
	165	√	√
Gear	228	√	×
	165	√	×
Wiper	09A	√	×
Total Accuracy		100%	53.33%

The reasons for the different performance of existing methods in delineating boundaries are explained in Figure 19 using 0x082 (for steering) and 0x228 (for gears) as examples. As shown in Figure 19a, this approach may not set the boundary for the boundary delineation of continuous values quite correctly, but the delineation is within the correct range. In contrast, the bit-flip rate approach is easily affected by bits with the exact change pattern or are completely changed when dividing the boundary, which leads to the boundary division outside the normal range. Figure 19b compares the delineation results of the two methods for discrete values. The bit-flip rate approach fails to delineate the boundary accurately because the flipped cases of individual bits are generalized to the same field as the adjacent invariant bits when delineating the boundary. Therefore, the framework proposed in this study gives better results for discrete values.

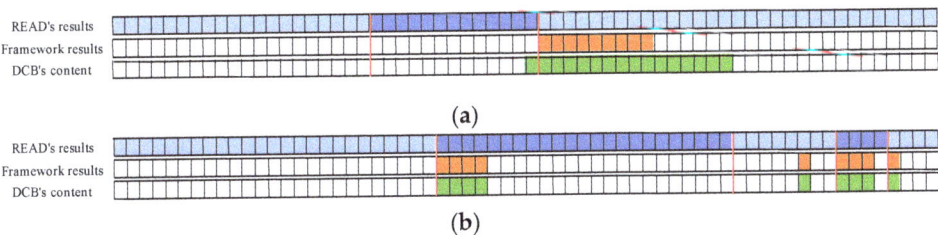

Figure 19. Boundary division results of bit-flip rate and proposed method: (**a**) Continuous value division result (0x082 for steering); (**b**) Discrete value division result (0x228 for gear).

4.4.2. Related Message Filtering

This section describes the outstanding performance of the framework in this paper compared to existing schemes in related message filtering, where existing schemes mainly use correlation coefficients (e.g., LibreCAN, Bram's method) to filter related messages. Figure 20 compares the performance between our proposed framework and the Pearson correlation coefficient for correlated message filtering. Regardless of the number of messages, the multiple linear regression method proposed in this study can filter messages related to vehicle behavior with 100% accuracy. When using the correlation coefficient to filter messages, although the accuracy of candidate message filtering increases as the number of messages rises, the accuracy still does not exceed 95%. When calculating the correlation between the two vectors, the results of the Pearson correlation coefficient are easily influenced by outliers in the two vectors, resulting in a reduced correlation coefficient that does not effectively filter out candidate messages [46]. In this paper, using multiple linear regression to model each bit of the data field as an independent variable, the effect of outliers is weakened, and the relevant messages are effectively filtered out. This result shows that the framework proposed in this study is more accurate than existing message filtering methods.

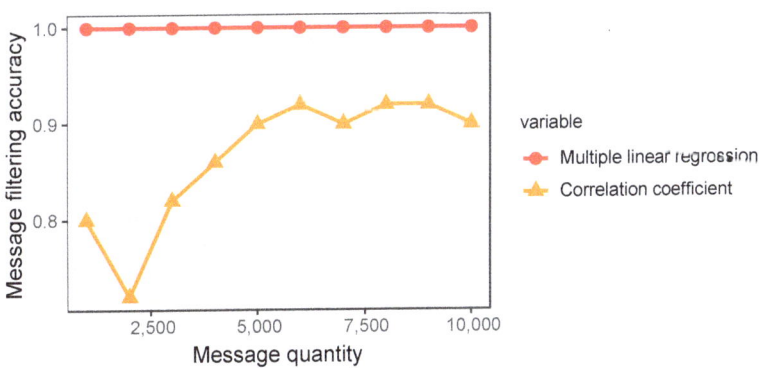

Figure 20. Comparison between correlation coefficient and multiple linear regression.

In addition, as shown in Table 11, the accuracy of the linear regression method is not affected by the number of messages, which remains 100%, while the correlation coefficient requires a higher number of messages to obtain a higher correct rate. This indicates that fewer messages are needed to locate messages related to vehicle behavior when using the linear regression method for CAN message screening, reducing data acquisition and computation time that speeds up the reverse work.

Table 11. The influence of different message counts on accuracy.

Methods	Number of Messages									
	1000	2000	3000	4000	5000	6000	7000	8000	9000	10,000
Linear regression	100%	100%	100%	100%	100%	100%	100%	100%	100%	100%
Correlation coefficients	80%	72%	82%	86%	90%	92%	90%	92%	92%	90%

4.4.3. Execution Complexity

We compare the algorithms in this section concerning the devices needed for their execution, the data requirement, the algorithm execution time, and the reverse results. As shown in Table 12, each algorithm relies on OBD-II data acquisition devices. Only the framework and LibreCAN require additional sensor devices and smartphones, respectively. In terms of data requirements, the READ and ReCAN require only CAN traffic, the linear regression method and LibreCAN require data from additional devices. However, the correlation coefficient method requires UDS data through interaction with the vehicle [47]. LibreCAN is the algorithm that takes the longest time to execute since some manual work is also required, and the fastest execution is the correlation coefficient method. The framework in this paper is close to the average time of the READ algorithm. However, in terms of reverse results, our scheme is the only one that can achieve bit-level reverse, outperforms the other algorithms in boundary delineation and message filtering, and does not require interaction with the vehicle. Although additional sensor devices are required, such sensors can be purchased very cheaply and used very simply in the market.

Table 12. Execution complexity comparison of different algorithms.

Algorithm	Devices Requirements	Data Requirements	Average Time	Reverse Results
Bit-level reverse based on linear regression	OBD-II data acquisition device, Behavior sensors	CAN traffic, Sensors data	37 s	Boundary Delineation, Related message filtering, Bit-level reverse
READ	OBD-II data acquisition device	CAN traffic	35.9 s	Boundary Delineation
ReCAN	OBD-II data acquisition device	CAN traffic	35.9 s	Boundary Delineation
LibreCAN	OBD-II data acquisition device, Smartphone	CAN traffic, Smartphone data	>60 s	Boundary Delineation, Related message filtering
Reverse engineering based on correlation coefficient	OBD-II data acquisition device	CAN traffic, UDS data	<20 s	Related message filtering

4.5. Application and Discussion

4.5.1. Application

The bit-level automotive CAN bus reverse framework proposed in this study can be used in almost all commercially available vehicles, independent of vehicle make and model. According to Table 12, the implementation of the framework requires OBD-II [48,49] data collection devices, sensors, and CAN traffic. In-vehicle CAN network traffic is typically collected using the OBD-II interface, a globally accepted automotive standard. It is required for almost all commercially available vehicles to be equipped with an OBD-II interface before they can be marketed [50–54]. Therefore, regardless of vehicle models on the market, the vehicle CAN traffic can be obtained after connecting OBD-II data collection devices. Therefore, regardless of vehicle models on the market, the vehicle CAN traffic can be obtained after connecting OBD-II data collection devices. For OBD-II data acquisition devices, such devices are readily available on the market today, with prices ranging from a few tens to a few hundred dollars. The sensor devices used in this framework are off-the-shelf motion sensors, which are inexpensive and easily placed in various vehicle parts to collect relevant data. Using CAN traffic and sensor data as input to our proposed framework, the algorithm proposed in this paper can obtain how CAN messages in any vehicle describe the vehicle state.

To verify the applicability of the framework, an electric car with completely a different power and brand was chosen to apply the framework. The reverse results are shown in Table A1 in Appendix A. In the absence of relevant DBC files, a script is provided in the appendix that can display CAN data changes in real-time to confirm the accuracy of each result. All filtered messages are consistent with the actual results in the actual results, and the reverse results of the bits remain consistent with the data bit changes. Overall, the method proposed in this study can be applied to most vehicle CAN message inversions and is not affected by vehicle changes.

4.5.2. Discussion

In this study, we propose an innovative bit-level reverse framework for automotive CAN messages. This framework builds a multiple linear regression model between CAN traces and sensor data, uses decision coefficients to filter candidate messages, and uses model parameters to determine how data fields represent vehicle behavior and maximally recover DBC files. In the test vehicle, this framework has high accuracy in both message screening and bit-inversion. However, the limitation of the test environment results in the unavailability of the extreme vehicle behavior data, leading to less than perfect results in bit-reversion. In addition, the framework reverses the candidate messages correctly in a short time, which improves the reversal efficiency. Our study proposes the only CAN message translator that can achieve bit-level reversal and has significant advantages over other existing methods for boundary delineation and message verification. Finally, the framework can be applied to any standard-compliant commercially available vehicle.

5. Conclusions

5.1. Implication

This study examines the bit-level CAN bus reverse framework using a multiple linear regression model. This framework is the only method that can achieve bit-level reversion. It uses sensor data as the dependent variable and each bit of the CAN message data field as the dependent variable to build a multiple linear regression model to obtain the carving of vehicle behavior for each bit based on the β. This study shows that the framework can accurately filter CAN messages related to vehicle behavior, reverse the way each bit represents vehicle behavior, and obtain the length, boundary, and alignment format of the signal. Compared to other methods, the framework can delineate the signal length and message filtering more accurately. In addition, the algorithm uses a globally available standard interface (OBD-II) and common motion sensors to capture CAN traffic and vehicle behavior data, which allows access to data that is not limited by model and make, making the algorithm more usable. The excellent reverse capability of the system can help automotive security researchers to quickly discover how CAN messages describe vehicle behavior when DBC files are not available. It is worth mentioning that attackers may also use our approach to find better attack approaches against cars. Although the framework makes DBC files less secret, it is more meaningful to study the automotive CAN detection and defense attack capabilities. In addition, a better attack prevention system could be developed based on the reverse results of this scheme.

5.2. Limitations and Future Work

The present study has three significant limitations that can be addressed in future studies.

First, the lack of extreme data affected the correctness of the experiment. When CAN traffic and vehicle behavior data were acquired, CAN data and sensor data could not cover extreme data, such as vehicle speed reaching 255 km/h, maximum steering wheel angle, and pedal reaching maximum angle. The lack of extreme data departs the highest position in the experimental results, resulting in unsatisfactory experimental results. Future research can obtain extreme data in closed scenarios to optimize the experimental results.

Second, insufficient DBC files. We use open-source DBC descriptions as truth when testing the results of validation experiments in vehicles. However, most of the current open-source DBC files are obtained by extracting the ECU firmware, resulting in a minimal number. This study can obtain the description of CAN messages without firmware, which provides a new idea to obtain DBC files for subsequent studies.

Finally, application limitations. Due to the limited number of test vehicles used, this framework validated its reverse effect in a subset of vehicles. According to the devices and data on which the framework relies, it can be applied to almost all vehicles. To address the difficulty of testing in actual vehicles, software and hardware simulations [55] of the internal networks of vehicles can be investigated in future research to address the application limitations.

Author Contributions: Formal analysis, G.X. (Guosheng Xu); Funding acquisition, G.X. (Guosheng Xu) and C.W.; Investigation, S.Z.; Methodology, Z.B. and G.X. (Guoai Xu); Project administration, G.X. (Guoai Xu); Resources, C.W.; Software, Z.B. and S.Z.; Supervision, G.X. (Guoai Xu); Validation, G.X. (Guosheng Xu); Writing—original draft, Z.B. and S.Z.; Writing—review & editing, C.W. All authors have read and agreed to the published version of the manuscript.

Funding: This research was funded by the National Natural Science Foundation of China under Grant No. 62102042, and the China Postdoctoral Science Foundation under Grant No. 2021T140074, and the Data Security Risk Monitoring Traceability & Integrated Management Platform project from the 2020 China Industrial Internet Innovation and Development Project.

Data Availability Statement: The data presented in this study are available in Section 4.4.1.

Acknowledgments: The authors would like to thank the editors and all the reviewers for their valuable comments.

Conflicts of Interest: The authors declare no conflict of interest.

Appendix A

Table A1 shows the framework's CAN message reverse for an electric vehicle manufactured in China. Although there is no DBC file to verify its correctness, we wrote a script (It can be found at http://49.232.218.41:8000/ accessed on 23 January 2022) that can display the data changes of the specified ID in real-time using the experimental equipment in this paper to verify the correctness of the results.

Table A1. Another vehicle reverse result.

Behavior	ID	Bits	Description
	0x212	48–56	real-time speed data
speed	0x23A	32–40, 56–64	real-time speed data
	0x21A	17–32	real-time speed data
mileage	0x21A	48–64	mileage per unit of time
steer	0x236	58–64	real-time steering data
brake pedal	0x668	0–16	brake pedal angle
	0x668	36	brake status
accelerate pedal	0x668	17–31	accelerate pedal angle
gear	0x235	39, 42, 44	D
		39, 42, 43	R

References

1. Number of Automotive Ecus Continues to Rise. Available online: https://www.eenewsautomotive.com/news/number-automotive-ecus-continues-rise (accessed on 15 May 2019).
2. CANbus—All You Need to Know. Available online: https://www.rs-online.com/designspark/canbus-all-you-need-to-know (accessed on 11 December 2020).
3. Bozdal, M.; Samie, M.; Aslam, S.; Jennions, I. Evaluation of can bus security challenges. *Sensors* **2020**, *20*, 2364. [CrossRef] [PubMed]

4. Farag, W.A. CANTrack: Enhancing automotive CAN bus security using intuitive encryption algorithms. In Proceedings of the 2017 7th International Conference on Modeling, Simulation, and Applied Optimization (ICMSAO), Sharjah, United Arab Emirates, 4–6 April 2017; IEEE: Piscataway, NJ, USA, 2017; pp. 1–5.
5. Van Herrewege, A.; Singelee, D.; Verbauwhede, I. CANAuth—A simple, backward compatible broadcast authentication protocol for CAN bus. In Proceedings of the ECRYPT Workshop on Lightweight Cryptography, Louvain-la-Neuve, Belgium, 28–29 November 2011; p. 20.
6. Bozdal, M.; Samie, M.; Jennions, I. A survey on can bus protocol: Attacks, challenges, and potential solutions. In Proceedings of the 2018 International Conference on Computing, Electronics & Communications Engineering (iCCECE), Southend, UK, 16–17 August 2018; IEEE: Piscataway, NJ, USA, 2018; pp. 201–205.
7. Miller, C.; Valasek, C. Remote exploitation of an unaltered passenger vehicle. In Proceedings of the Black Hat USA, Las Vegas, NV, USA, 1–8 August 2015.
8. TBONE—A Zero-Click Exploit for Tesla MCUs. Available online: https://kunnamon.io/tbone/tbone-v1.0-redacted.pdf (accessed on 28 April 2020).
9. Tesla Model S Can Be Hacked, and Fixed. Available online: https://www.npr.org/sections/alltechconsidered/2015/08/06/429907506/tesla-model-s-can-be-hacked-and-fixed-which-is-the-real-news (accessed on 6 August 2020).
10. Hackers Remotely Kill a Jeep on the Highway—With Me in It. Available online: https://www.wired.com/2015/07/hackers-remotely-kill-jeep-highway/ (accessed on 21 July 2015).
11. Tencent Keen Security Lab: Experimental Security Research of Tesla Autopilot. Available online: https://keenlab.tencent.com/en/2019/03/29/Tencent-Keen-Security-Lab-Experimental-Security-Research-of-Tesla-Autopilot/ (accessed on 29 March 2019).
12. Koscher, K.; Czeskis, A.; Roesner, F.; Patel, S.; Kohno, T.; Checkoway, S.; McCoy, D.; Kantor, B.; Anderson, D.; Shacham, H.; et al. Experimental security analysis of a modern automobile. In Proceedings of the 2010 IEEE Symposium on Security and Privacy, Berleley/Oakland, California, 16–19 May 2010; IEEE: Piscataway, NJ, USA, 2010; pp. 447–462.
13. CAN DBC File Explained—A Simple Intro. Available online: https://www.csselectronics.com/pages/can-dbc-file-database-intro (accessed on 10 May 2021).
14. DBC File Format Documentation. Available online: https://ishare.iask.sina.com.cn/f/3Yjd8GR3d.html (accessed on 20 September 2021).
15. Lee, H.; Jeong, S.H.; Kim, H.K. OTIDS: A novel intrusion detection system for in-vehicle network by using remote frame. In Proceedings of the 2017 15th Annual Conference on Privacy, Security and Trust (PST), Calgary, AB, Canada, 28–30 August 2017; IEEE: Piscataway, NJ, USA, 2017; pp. 57–5709.
16. Yu, K.S.; Kim, S.H.; Lim, D.W.; Kim, Y.S. A multiple Rényi entropy based intrusion detection system for connected vehicles. *Entropy* **2020**, *22*, 186. [CrossRef] [PubMed]
17. Song, H.M.; Woo, J.; Kim, H.K. In-vehicle network intrusion detection using deep convolutional neural network. *Veh. Commun.* **2020**, *21*, 100198. [CrossRef]
18. Cho, K.T.; Shin, K.G. Viden: Attacker identification on in-vehicle networks. In Proceedings of the 2017 ACM SIGSAC Conference on Computer and Communications Security, Dallas, TX, USA, 30 October 2017; pp. 1109–1123.
19. Marchetti, M.; Stabili, D. Anomaly detection of CAN bus messages through analysis of ID sequences. In Proceedings of the 2017 IEEE Intelligent Vehicles Symposium (IV), Los Angeles, CA, USA, 11–14 June 2017; IEEE: Piscataway, NJ, USA, 2017; pp. 1577–1583.
20. Tariq, S.; Lee, S.; Kim, H.K.; Woo, S.S. CAN-ADF: The controller area network attack detection framework. *Comput. Secur.* **2020**, *94*, 101857. [CrossRef]
21. Seo, E.; Song, H.M.; Kim, H.K. Gids: Gan based intrusion detection system for in-vehicle network. In Proceedings of the 2018 16th Annual Conference on Privacy, Security and Trust (PST), Belfast, UK, 28–30 August 2018; IEEE: Piscataway, NJ, USA, 2018; pp. 1–6.
22. Jin, S.; Chung, J.G.; Xu, Y. Signature-Based Intrusion Detection System (IDS) for In-Vehicle CAN Bus Network. In Proceedings of the 2021 IEEE International Symposium on Circuits and Systems (ISCAS), Daegu, Korea, 22–28 May 2021; IEEE: Piscataway, NJ, USA, 2021; pp. 1–5.
23. Lokman, S.F.; Othman, A.T.; Abu-Bakar, M.H. Intrusion detection system for automotive Controller Area Network (CAN) bus system: A review. *EURASIP J. Wirel. Commun. Netw.* **2019**, *2019*, 184. [CrossRef]
24. Fowler, D.S.; Bryans, J.; Shaikh, S.A.; Wooderson, P. Fuzz testing for automotive cyber-security. In Proceedings of the 2018 48th Annual IEEE/IFIP International Conference on Dependable Systems and Networks Workshops (DSN-W), Luxembourg, 25–28 June 2018; IEEE: Piscataway, NJ, USA, 2018; pp. 239–246.
25. Lee, H.; Choi, K.; Chung, K.; Kim, J.; Yim, K. Fuzzing can packets into automobiles. In Proceedings of the 2015 IEEE 29th International Conference on Advanced Information Networking and Applications, Gwangju, Korea, 24–27 March 2015; IEEE: Piscataway, NJ, USA, 2015; pp. 817–821.
26. McShane, J.; Kultinov, K. CAN Bus Fuzz Testing with Artificial Intelligence. *ATZelectronics Worldw.* **2021**, *16*, 62–64. [CrossRef]
27. Fowler, D.S.; Bryans, J.; Cheah, M.; Wooderson, P.; Shaikh, S.A. A method for constructing automotive cybersecurity tests, a CAN fuzz testing example. In Proceedings of the 2019 IEEE 19th International Conference on Software Quality, Reliability and Security Companion (QRS-C), Sofia, Bulgaria, 22–26 July 2019; IEEE: Piscataway, NJ, USA, 2019; pp. 1–8.

28. Fowler, D.S. A Fuzz Testing Methodology for Cyber-Security Assurance of the Automotive CAN Bus. Ph.D. Thesis, Coventry University, Coventry, UK, 2019.
29. Markovitz, M.; Wool, A. Field classification, modeling and anomaly detection in unknown CAN bus networks. *Veh. Commun.* **2017**, *9*, 43–52. [CrossRef]
30. Marchetti, M.; Stabili, D. READ: Reverse engineering of automotive data frames. *IEEE Trans. Inf. Forensics Secur.* **2018**, *14*, 1083–1097. [CrossRef]
31. Pesé, M.D.; Stacer, T.; Campos, C.A.; Newberry, E.; Chen, D.; Shin, K.G. LibreCAN: Automated CAN message translator. In Proceedings of the 2019 ACM SIGSAC Conference on Computer and Communications Security, London, UK, 11–15 November 2019; pp. 2283–2300.
32. Zago, M.; Longari, S.; Tricarico, A.; Carminati, M.; Pérez, M.G.; Pérez, G.M.; Zanero, S. ReCAN–Dataset for reverse engineering of Controller Area Networks. *Data Brief* **2020**, *29*, 105149. [CrossRef] [PubMed]
33. Jaynes, M.; Dantu, R.; Varriale, R.; Evans, N. Automating ECU identification for vehicle security. In Proceedings of the 2016 15th IEEE International Conference on Machine Learning and Applications (ICMLA), Anaheim, CA, USA, 18–20 December 2016; IEEE: Piscataway, NJ, USA, 2016; pp. 632–635.
34. Buscemi, A.; Castignani, G.; Engel, T.; Turcanu, I. A Data-Driven Minimal Approach for CAN Bus Reverse Engineering. In Proceedings of the 2020 IEEE 3rd Connected and Automated Vehicles Symposium (CAVS), Victoria, BC, Canada, 18 November–16 December 2020; IEEE: Piscataway, NJ, USA, 2020; pp. 1–5.
35. Ezeobi, U.; Olufowobi, H.; Young, C.; Zambreno, J.; Bloom, G. Reverse Engineering Controller Area Network Messages using Unsupervised Machine Learning. *IEEE Consum. Electron. Mag.* **2020**, *11*, 50–56. [CrossRef]
36. Song, H.M.; Kim, H.K. Discovering can specification using on-board diagnostics. *IEEE Des. Test* **2020**, *38*, 93–103. [CrossRef]
37. Blaauwendraad, B.; Kieberl, V. Automated reverse-engineering of CAN messages using OBD-II and correlation coefficients. Available online: https://www.os3.nl/_media/2019-2020/courses/rp2/p103_report.pdf (accessed on 23 January 2022).
38. Wen, H.; Zhao, Q.; Chen, Q.A.; Lin, Z. Automated cross-platform reverse engineering of CAN bus commands from mobile apps. In Proceedings of the 2020 Network and Distributed System Security Symposium (NDSS'20), San Diego, CA, USA, 23–26 February 2020.
39. CAN Specification. Available online: http://esd.cs.ucr.edu/webres/can20.pdf (accessed on 10 September 2021).
40. Texas Instruments. *Introduction to the Controller Area Network (CAN)*. Application Report SLOA101. 2002; pp. 1–17. Available online: https://www.rpi.edu/dept/ecse/mps/sloa101.pdf (accessed on 22 January 2022).
41. Uyanık, G.K.; Güler, N. A study on multiple linear regression analysis. *Procedia-Soc. Behav. Sci.* **2013**, *106*, 234–240. [CrossRef]
42. Tranmer, M.; Elliot, M. Multiple linear regression. *Cathie Marsh Cent. Census Surv. Res. (CCSR)* **2008**, *5*, 1–5.
43. Amazon-Acceleration Sensors. Available online: https://www.amazon.com/High-Stability-Inclinometer-High-Precision-Accelerometer-Navigation/dp/B072ZZ83JZ/ref=sr_1_3?crid=D2ETL9PC5TBQ&keywords=ten-axis+GPS+inertial+navigation+sensor&qid=1642921112&sprefix=ten-axis+gps+inertial+navigation+sensor%2Caps%2C855&sr=8-3 (accessed on 12 May 2021).
44. Chen, D.; Cho, K.-T.; Han, S.; Jin, Z.; Shin, K.G. Invisible sensing of vehicle steering with smartphones. In Proceedings of the 13th Annual International Conference on Mobile Systems, Applications, and Services, Florence, Italy, 18–22 May 2015; ACM: New York, NY, USA, 2015; pp. 1–13.
45. Opendbcfromcomma.ai. Available online: https://github.com/commaai/opendbc (accessed on 10 January 2021).
46. Benesty, J.; Chen, J.; Huang, Y.; Cohen, I. Pearson correlation coefficient. In *Noise Reduction in Speech Processing*; Springer: Berlin, German, 2009; pp. 1–4.
47. OBD-II PIDs. Available online: https://en.wikipedia.org/wiki/OBD-II_PIDs (accessed on 20 September 2021).
48. On-Board Diagnostics. Available online: https://en.wikipedia.org/wiki/On-_board_diagnostics#OBD-%20II_diagnostic_connector (accessed on 10 January 2021).
49. OBD II Generic PID Diagnosis. Available online: https://www.motor.com/magazinepdfs/092007_09.pdf (accessed on 10 August 2021).
50. Is Your Vehicle OBD II Compliant? Available online: https://www.plxdevices.com/obdii-compliant-vehicles-s/153.htm (accessed on 10 January 2022).
51. EPC (European Parliament and Council). Directive 98/69/EC of the European Parliament and of the Council of 13 October 1998 relating to measures to be taken against air pollution by emissions from motor vehicles and amending Council Directive 70/220/EEC. 1998. Available online: https://eur-lex.europa.eu/LexUriServ/LexUriServ.do?uri=CONSLEG:1998L0069:19981228:EN:PDF (accessed on 23 January 2022).
52. ISO 15765-4:2005 Road vehicles—Diagnostics on Controller Area Networks (CAN)—Part 4: Requirements for emissions-related systems. Available online: https://www.iso.org/standard/33619.html (accessed on 20 January 2022).
53. CAN Bus Explained—A Simple Intro. 2021. Available online: https://www.csselectronics.com/pages/can-bus-simple-intro-tutorial (accessed on 20 January 2022).
54. Limits and Measurement Methods for Emissions from Light-Duty Vehicles (CHINA 6). Available online: https://www.chinesestandard.net/PDF/BOOK.aspx/GB18352.6-2016 (accessed on 20 January 2022).
55. Mundhenk, P.; Mrowca, A.; Steinhorst, S.; Lukasiewycz, M.; Fahmy, S.A.; Chakraborty, S. Open source model and simulator for real-time performance analysis of automotive network security. *ACM Sigbed Rev.* **2016**, *13*, 8–13. [CrossRef]

Article

CVDF DYNAMIC—A Dynamic Fuzzy Testing Sample Generation Framework Based on BI-LSTM and Genetic Algorithm

Mingrui Ma [1], Lansheng Han [1,*] and Yekui Qian [2]

[1] School of Cyber Science and Engineering, Huazhong University of Science and Technology, Wuhan 430074, China; jkpathfinder@hust.edu.cn
[2] PLA Army Academy of Artillery and Air Defense, Zhengzhou 450052, China; scienceart2021@163.com
* Correspondence: 1998010309@hust.edu.cn

Citation: Ma, M.; Han, L.; Qian, Y. CVDF DYNAMIC—A Dynamic Fuzzy Testing Sample Generation Framework Based on BI-LSTM and Genetic Algorithm. *Sensors* **2022**, *22*, 1265. https://doi.org/10.3390/s22031265

Academic Editors: Athanasios V. Vasilakos and Vassilis S. Kodogiannis

Received: 22 December 2021
Accepted: 4 February 2022
Published: 7 February 2022

Publisher's Note: MDPI stays neutral with regard to jurisdictional claims in published maps and institutional affiliations.

Copyright: © 2022 by the authors. Licensee MDPI, Basel, Switzerland. This article is an open access article distributed under the terms and conditions of the Creative Commons Attribution (CC BY) license (https://creativecommons.org/licenses/by/4.0/).

Abstract: As one of the most effective methods of vulnerability mining, fuzzy testing has scalability and complex path detection ability. Fuzzy testing sample generation is the key step of fuzzy testing, and the quality of sample directly determines the vulnerability mining ability of fuzzy tester. At present, the known sample generation methods focus on code coverage or seed mutation under a critical execution path, so it is difficult to take both into account. Therefore, based on the idea of ensemble learning in artificial intelligence, we propose a fuzzy testing sample generation framework named CVDF DYNAMIC, which is based on genetic algorithm and BI-LSTM neural network. The main purpose of CVDF DYNAMIC is to generate fuzzy testing samples with both code coverage and path depth detection ability. CVDF DYNAMIC generates its own test case sets through BI-LSTM neural network and genetic algorithm. Then, we integrate the two sample sets through the idea of ensemble learning to obtain a sample set with both code coverage and vulnerability mining ability for a critical execution path of the program. In order to improve the efficiency of fuzzy testing, we use heuristic genetic algorithm to simplify the integrated sample set. We also innovatively put forward the evaluation index of path depth detection ability (pdda), which can effectively measure the vulnerability mining ability of the generated test case set under the critical execution path of the program. Finally, we compare CVDF DYNAMIC with some existing fuzzy testing tools and scientific research results and further propose the future improvement ideas of CVDF DYNAMIC.

Keywords: genetic algorithm; Bi-LSTM neural network; fuzzy testing sample generation; deep learning

1. Introduction and Background

Vulnerability in program has always been a serious threat to software security, which may cause denial of service, information leakage and other exceptions. Some typical cases of vulnerability exploitation, such as wannacry ransomware, have a disastrous impact on social economy and network security. Therefore, mining vulnerabilities scientifically and efficiently has been a hot topic.

At present, vulnerability mining technology can be divided into static vulnerability mining and dynamic testing (fuzzy testing) [1]. The former does not construct test cases nor run source code. By extracting the characteristics or key operations of the corresponding types of vulnerabilities, static code audit is carried out on the source code to detect the possibility of various vulnerabilities. The target source code of static vulnerability mining can be advanced language, assembly language generated by compiler, or binary file. The advantages of static vulnerability mining lie in fast mining speed, high efficiency, and good detection accuracy for vulnerabilities with obvious characteristics. However, static vulnerability mining often leads to high false positive rate and false negative rate for vulnerabilities with unclear features or diverse types and forms (such as null pointer reference vulnerability in C/C++). Dynamic fuzzy testing can solve this problem by constructing reasonable test examples. However, the efficiency of dynamic fuzzy testing is lower than that of static vulnerability mining because it needs to construct samples and run programs to determine whether there are vulnerabilities. Therefore, how to construct test cases with high pdda and code coverage is the key of fuzzy testing. In practical application, it is often necessary to combine static vulnerability mining with fuzzy testing to achieve better vulnerability detection performance. Existing mainstream fuzzy testing can be divided into the following three categories:

- Black box test (construct test cases to test without source code at all);
- White box test (analyze source code to generate corresponding test cases); e.g., [2];
- Grey box test (introduce lightweight program analysis technology to analyze program state), e.g., [3].

In black box test, the internal structure of the program is not understood at all, and the test cases are constructed blindly. Thus, its testing efficiency is very low. White box test uses program analysis methods [4] (such as path traversal and symbolic execution) to analyze the program source code and then constructs the corresponding test cases. The white box test can cover deeper test path, which causes a lot time cost and system resources with poor scalability. The grey box test [5] can achieve a good balance between the test efficiency and the coverage of test cases because of the introduction of lightweight program analysis technology. It is more effective than a black box test and more extensible than a white box test. At present, the grey box testing program is mainly guided by code coverage. The typical grey box fuzzers are AFL [6] and so on.

However, the problem of current grey box fuzzers is that they are designed to cover as many code execution paths as possible. In the regulation of seed energy, they usually use the idea of average distribution instead of regulating different energies for different test paths. Nevertheless, most of the source code vulnerabilities are concentrated on a small number of critical test paths in reality. Existing grey box fuzzers often spend a lot of time to detect the path whose vulnerability is not easy to be detected, thus reducing the efficiency of fuzzy testing.

Because the application of a single method in grey box fuzzy testing has its own limitations, more and more researchers have begun to integrate a variety of methods to achieve better fuzzy testing results, such as [7].

Based on existing research work [8], this paper proposes a new framework of fuzzy testing sample generation called CVDF DYNAMIC. It consists of three parts:

(1) The strategy of sample generation based on a genetic algorithm;
(2) The strategy of sample generation based on a bi-LSTM neural network;
(3) The strategy of sample reduction based on a heuristic genetic algorithm.

The genetic algorithm can improve the quality of test cases and expand the code coverage by simulating the natural process of gene recombination and evolution. The bi-LSTM time sequence can regulate different energy of the test path, which can make the seeds on the critical path iterate and mutate for many times, and enhance the path depth detection ability. The critical contribution of CVDF DYNAMIC is that it integrates the two methods of sample generation, and simplifies the sample set by using a heuristic genetic algorithm, which makes the test case set achieve a good balance in code coverage, path depth detection ability and sample set size. This paper also compares the proposed method with other fuzzy testing samples and further presents the improvement direction of that method.

2. Related Work

At present, researchers have applied fuzzy testing to different types of vulnerability mining. Lin et al. [9] proposed a priority-based path searching method (PBPS) to utilize the capability of concolic execution better. Peng et al. [10] proposed Angora, a new mutation-based fuzzer, and proved that Angora has better performance than other fuzzing tools. Wang et al. [8] used a neural network to guide the sample generation of fuzzy testing and proposed a solution called NeuFuzz. NeuFuzz has a very significant performance in the vulnerability mining of the critical execution path of the program. Zhang et al. [11] summarized existing fuzzy testing technologies and use case generation technologies of fuzzy testing. Zhang el al. [12] proposed an algorithm of sensitive region prediction based on a neural network and improved the detection efficiency and detection depth through the incremental learning method of sensitive areas. Combining fuzzy testing and symbolic execution, Xie [13] proposed a hybrid testing method based on a branch coverage called AFLeer. Xu et al. [14] applied a recurrent neural network to fuzzy testing sample generation. Luca et al. [2] designed a novel concolic executor to improve the efficiency of concolic execution and investigate whether techniques borrowed from the fuzzing domain can be used to solve the symbolic query problem. Stefan [15] proposed the notion of coverage-guided tracing to improve the efficiency of code coverage guided fuzzy testing. Yang et al. [16] proposed a novel programmable fuzzy testing framework. Developers only need to write a small number of fuzzy testing guidance programs to implement customized fuzzy testing. Patrice et al. [17] proposed learn&fuzz, which used a learned input probability distribution to intelligently guide fuzzing inputs. Li et al. [18] proposed symfuzz, which is a method combining directed fuzzy testing technology with selective symbolic execution technology and can realize vulnerability detection under complex path conditions. Liang et al. [19] proposed a machine-learning-based framework to improve the quality of seed inputs for fuzzing programs. Zou et al. [1] described the development from traditional automation to intelligent vulnerability mining in software vulnerability mining. This paper also pointed out that the application of traditional machine learning technology in the vulnerability mining field still has limitations. Ma et al. [20] proposed the optimization strategy of sample set reduction in the fuzzy process, including approximation algorithm. Cornelius [21] proposed IJON, an annotation mechanism that a human analyst can used to guide the fuzzer.

In the experimental part, this paper compares the simplification and efficiency of sample set between heuristic genetic algorithm and approximation algorithm. He et al. [22] proposed a tool called VCCFinder to find potential vulnerabilities. Nick et al. [23] used mined vulnerabilities by utilizing a code attribute graph for fuzzy testing. She et al. [24] proposed a novel program smoothing technique using a surrogate neural network models to achieve higher edge coverage and improve the ability of finding new bugs. Chen et al. [25] proposed POLYGLOT, a genetic fuzzing framework that generates high-quality test cases for exploring processors of different programing languages. Huang et al. [4] proposed PANGOLIN, an approach based on polyhedral path abstraction, which preserves the exploration state in the concolic execution stage and allows more effective mutation and constraint solving over existing techniques. Zhang et al. [26] proposed a novel incremental and stochastic rewriting technique STOCHFUZZ that piggy-backs on the fuzzing procedure. Liang et al. [3] presented DeepFuzzer, an enhanced greybox fuzzer with qualified seed generation, balanced seed selection and hybrid seed mutation. Chen et al. [7] proposed an ensemble fuzzing method, EnFuzz. Enfuzz contains many different heuristic genetic algorithms and achieves a better performance in terms of path coverage, branch coverage and bug discovery. The idea of ensemble is also similar to the CVDF DYNAMIC proposed in this paper. Yue et al. [27] presented a variant of the adversarial multi-armed bandit model for modeling AFL's power schedule process named EcoFuzz, which can effectively regulate seed energy in fuzzy testing. Zong et al. [28] proposed FuzzGuard, a deep-learning-based approach to predict the reachability of inputs and further improve the performance of DGF. Gan [5] proposed a data flow sensitive fuzzing solution GREYONE, which can further improve the performance of data flow analysis, and the experiments show that GREYONE has better performance than the existing fuzzy testing tools such as AFL. Sebastian et al. [29] proposed ParmeSan, a sanitizer guided fuzzing method to solve the low-bug coverage problem. Oleksii et al. [30] proposed specfuzz, which is a novel fuzzy testing method, which can be used to detect speculative execution vulnerabilities including spectre and out-of-order execution vulnerabilities. Compared with the traditional static analysis method, specfuzz has further improved the analysis accuracy. Lee et al. [31] proposed a constraint-guided directed greybox fuzzing method, which aims to satisfy a sequence of constraints rather than merely reaching a set of target sites. Christopher et al. [32] proposed a brand-new token-level fuzzing method. Different from the fuzzy method based on data flow or seed energy regulation, token-level fuzzing applies mutations at the token level instead of applying mutations either at the byte level or at the grammar level. The authors found many unknown bugs through the token-level fuzzing method on popular javascript engines. In recent years, the safety of deep learning technology has also attracted the attention of scholars. It is possible for attackers to deduce the sensitive training data of engineering through the unsafe deep learning model. Ximeng Liu et al. [33] briefly introduced four different types of attacks in deep learning, reviewed and summarized the security defense measures of deep learning attack methods and further discussed the remaining challenges and privacy issues of deep learning security. MB mollah et al. [34] proposed an efficient data-sharing scheme, which allows smart devices to share secure data with others at the edge of cloud-assisted IOT.

3. Algorithm Description

3.1. An Introduction of Existing Fuzzy Testing Sample Generation Methods

At present, the generation and variation methods of test cases are mainly described as follows:

The method based on symbolic execution [13].

The core idea of this method is to take the test case as the symbol value and search the core constraint information on the test path during the processing. A new test case is generated by constraint solving to cover different program execution paths. This method is suitable for testing programs with simple structure and less execution paths. However, the complexity of the program increases with the diversification of functions, resulting in the explosion of the number of paths. It is difficult for symbolic execution to be applied to constructing complex program test cases because of complex constraint solving problems.

The method based on taint analysis [10].

The core idea of this method is to mark the pollution source of the input data by using the dynamic taint analysis technology, focus on the spread process of the taint, extract the key taint information from it and use the taint information to guide the generation of seed variation and related test samples. It is an effective method to construct test samples for some key execution paths in programs and has good code coverage, such as Angora [10]. However, with the application of genetic algorithm and neural network in fuzzy testing, the disadvantage of low efficiency of taint analysis technology is gradually emerging.

The method based on evolutionary algorithm [35].

The evolutionary algorithm uses some core rules of biological evolution to guide the generation of fuzzy testing samples. At present, genetic algorithm is the most widely used evolutionary algorithm with the best performance. Its core idea is to carry out multiple rounds of iterative mutation on test cases, eliminate the test cases that do not meet the requirements according to some rules or select the samples with the best performance from them as the seeds of the next round of mutation. Genetic algorithm can be used not only to generate new test cases but also to simplify the sample set, so as to further improve the efficiency of fuzzy testing.

The method based on neural network [14].

As mentioned above, neural network has a very significant performance advantage in solving some nonlinear problems. The bi-LSTM neural network is used to mutate the seeds on a certain execution path to obtain a new test example. In the experiment, we prove that the bi-LSTM neural network has stronger path depth detection ability in specific key execution paths than that of the taint analysis. Moreover, Learn & Fuzz proposed by Patrice [17] et al. can improve the code coverage of fuzzy testing. Therefore, it can be predicted that the neural network will play a greater role in the future development of fuzzy testing.

3.2. Formal Definition

In order to facilitate the subsequent description of the algorithm, we give some related concepts and formal definitions of the evaluation index.

- Definition 1 PUT (input sample)

We define the program under test as PUT. For CVDF DYNAMIC, PUT is the corresponding binary executable program, and the corresponding test cases are mentioned in Section 4.1.

- Definition 2 Set Covering Problem (SCP)

A large number of facts show that there is an exponential proportional relationship between the growth number of execution paths of PUT and the growth number of its branch conditions, so the test cases cannot completely cover all execution paths. Therefore, in fuzzy testing, the problem of sample set coverage is transformed into the problem of minimum set coverage [36]. The minimum set covering problem is an NP hard problem [37]. The simplest algorithm idea is to use greedy algorithm to find the approximate optimal solution. The following formal definition is used to describe SCP problem:

For $A = [a_{ij}]$, it is a 0–1 matrix of m-row n-columns, where $C = C_j$ is an n-dimensional column vector. Let $p = [1,2,3\ldots\ldots m]$ and $q = [1,2,3\ldots\ldots n]$ be the row and column vectors of matrix A. Furthermore, let $C_j, j \in q$ represent the cost of a column. Without losing generality, we assume that $C_j > 0, j \in q$. It is specified here that if $a_{ij} = 1$, it means that column $j \in q$ at least covers one row $i \in p$. Therefore, the essence of the SCP problem is to find a minimum cost subset $S \subseteq q$. So, for every row $i \subseteq p$, it is covered by at least one column $j \subseteq S$. A natural mathematical model of SCP can be described as $v(SCP) = \min \sum_{j \in q} C_j x_j$, and it obeys $\sum_{j \in q} a_{ij} x_j \geq 1, i \in p, x_j \in (0,1)(j \in q)$. If $x_j = 1(j \in S)$, then $x_j = 0$.

- Definition 3 Path Depth Detection Ability

In fuzzy testing, there are many program-execution paths that may have vulnerabilities in PUT, so the generation of fuzzy testing samples should cover as many as possible for these program execution paths that may have vulnerabilities. For a program execution path, the number of detected vulnerabilities may be more than one, and different program execution paths can detect different numbers of vulnerabilities. We define the total number of vulnerabilities detected by the fuzzy testing sample under the current path as D_{NUM}, the total number of vulnerabilities contained in the current path as A_{NUM} and the weight of the total number of vulnerabilities contained in the current path as W. $DetectionCapability(DC)$ is a weighted result, and its operation method is shown in Equation (1):

$$DC = \frac{D_{NUM}}{A_{NUM}} \times W \qquad (1)$$

Among them, W increases with the number of vulnerabilities in the current path. This is because the number of vulnerabilities in different paths is different. For the variation method of the same fuzzy testing sample seed, if more vulnerabilities are contained in a path, the smaller the value of $\frac{D_{NUM}}{A_{NUM}}$ is. If the weight W is a constant, the DC value will decrease, and the path depth detection ability of a test case generation method cannot be objectively measured.

Suppose that a program under test has n execution paths, we define the average path detection ability as $WDC = \frac{\sum_{i=1}^{n} DC_i}{n}$: It can measure the ability of a fuzzy testing tool to detect the overall path depth

3.3. CVDF DYNAMIC Fuzzy Testing Sample Generation

The complete process of fuzzy testing sample generation of CVDF DYNAMIC is shown in Figure 1.

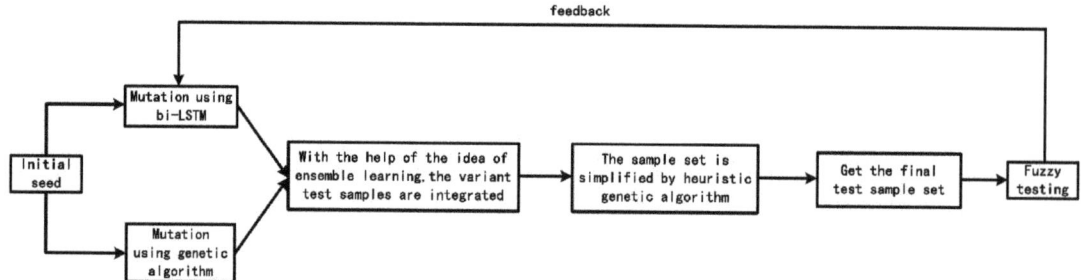

Figure 1. Complete Flow Chart of CVDF DYNAMIC Fuzzy Testing sample generation.

In the fuzzy testing part, we learn from the ensemble learning method in artificial intelligence. The seeds are mutated by genetic algorithm to generate a set of test cases, and then the seeds are mutated by the bi-LSTM neural network to generate another set of test cases. Finally, the two sets of test cases are integrated to obtain the final set of test cases.

Considering that the size of the sample set obtained by the integration of the two methods is too large, which reduces the efficiency of fuzzy testing, we use heuristic genetic algorithm to simplify the sample set. Finally, the reduced sample set is used for fuzzy testing, and the parameters in the bi-LSTM neural network are optimized according to the result feedback.

3.3.1. Theoretical Model and Training Process of BI-LSTM Neural Network

The BI-LSTM neural network training process of CVDF DYNAMIC is shown in the Figure 2.

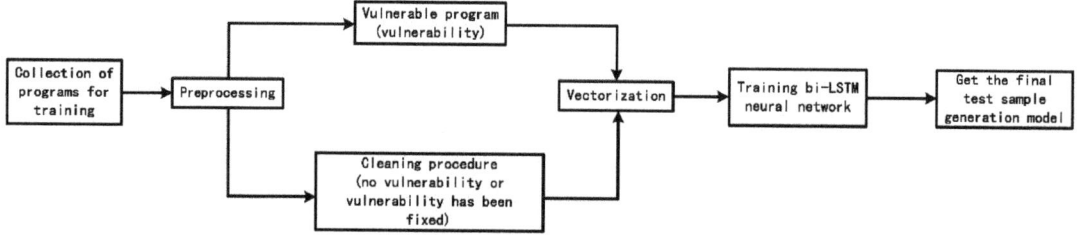

Figure 2. Training of neural network.

(a) Preprocessing and Vectorization

We preprocess the training dataset, including unifying the input format of the test cases and changing the format of some binary executable programs, so that they can adapt to the input of the neural network without changing the logic function of the original program.

Then, we use the PTFuzz tool, which is a tool to obtain the program execution path by using the Intel Processor Tracing module (IntelPT). PTFuzz makes a further improvement on the basis of AFL, which removes the dependence on the program instrument but uses PT to collect package information and filter package information, and finally obtains the execution path of the current seed according to the package information. In order to achieve this goal, our hardware environment should be based on Intel CPU platform and run under the appropriate version of Linux system. Since the PTFuzz tool stores the program execution path information in data packets in order to obtain the program execution path information that can be trained for neural networks, we need to decode the data packets in the corresponding memory and recover the complete program execution path according to the entry, exit and other relevant information of each data packet. The pseudocode of the Algorithm 1 Extracting program execution path is as follows:

Algorithm 1. Extracting program execution path

Start Func
Func ExtractPath(binary-source-code)
1: Start = LoadBinaryProgram(binary-source-code)
2: ProgStaddr = GetProgramEntry(Start)
3: ExecutionPath = []
4: **while** True:
5:   PackagePath = LoadCurrentPackage(ProgStaddr)
6:   ExecutionPath +|= PackagePath
7:   **If** ProgStaddr == JumpNextInstrument()
8:     ProgStaddr = GetNextInstruAddr()
9:   **If** ProgStaddr == EndOfMemSpace()
10:    break
11: **Return** ExecutionPath
End Func

In the pseudocode, JumpNextInstrument() and EndOfMemspace() are two judgment functions, which are used to judge whether to jump to the next instruction address and whether the end of the memory address of PTFuzz package has been reached, respectively. The ExecutionPath variable forms a complete program execution path by continuously connecting the PackagePath variable after decodeding. +|= is a concatenate operation.

After extracting the program execution path, we need to convert the program execution path containing instruction bytecodes into vector form and save the original semantic information of the original program execution path as much as possible.

We use the tool word2vec and regard a complete program execution path as a statement and an instruction as a word. Specifically, we regard the hexadecimal code of an instruction as a token, and then we use word2vec to train the corresponding bytecode sequence. In order to preserve as much context information as possible in the program execution path, we choose the Skip-Gram model in word2vec because it often has better performance in large corpus. The Skip-Gram model structure is shown in the Figure 3.

Finally, we need to transform the output of word2vec into an equal length coding input, which can be used as the input vector of the neural network. Let us set a maximum length, which is MaxLen. When the output length of word2vec is less than MaxLen, we use 0 to fill in the back end to make it MaxLen. When the output length of word2vec is larger than MaxLen, we truncate it from the front end and control the length to MaxLen.

(b)   BI-LSTM neural network structure and parameter optimization

The neural network structure we choose is bi-LSTM.

Bi-LSTM has excellent performance in dealing with long-term dependency problems, such as statement prediction and named entity recognition [38]. The statements associated with vulnerability characteristics may be far away in the whole program execution path, so we need the bi-LSTM neural network structure for the long-term memory of the information related to the vulnerability characteristics. In order to make the bi-LSTM neural network suitable for fuzzy testing, we modify the corresponding rules of the input gate, output gate and forgetting gate of the bi-LSTM. The specific structure of the single LSTM neuron and the specific rules of the input gate, output gate and forgetting gate are shown in Figure 4.

The number of hidden layers in the bi-LSTM neural network, epochs, batch size and other parameters will affect the final performance of the neural network. According to the experimental part in Section 4.2, we set the number of hidden layers to 5, the batch size to 64 and the drop rate to 0.4 and use a BPTT back-propagation algorithm to adjust the network weight, using random gradient descent (SGD) method to prevent the model from falling into the local optimal solution. For the hyper parameters in the bi-LSTM neural network, we choose to use dichotomy to accelerate the selection of corresponding values. Figure 5 shows the complete structure of the bi-LSTM neural network.

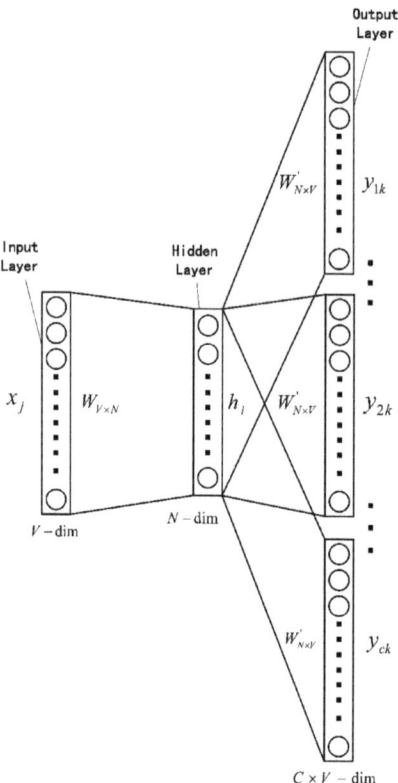

Figure 3. The Basic Structure Diagram Of Skip-Gram Model.

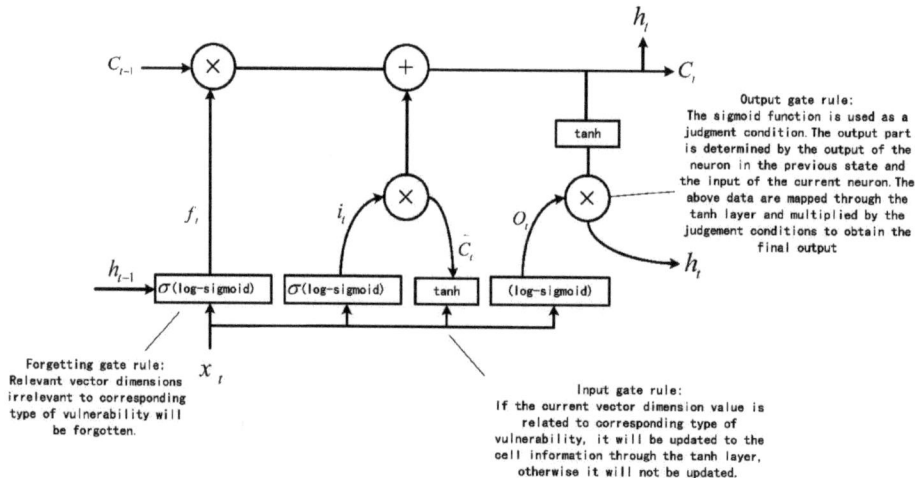

Figure 4. The Specific Structure of LSTM Neuron.

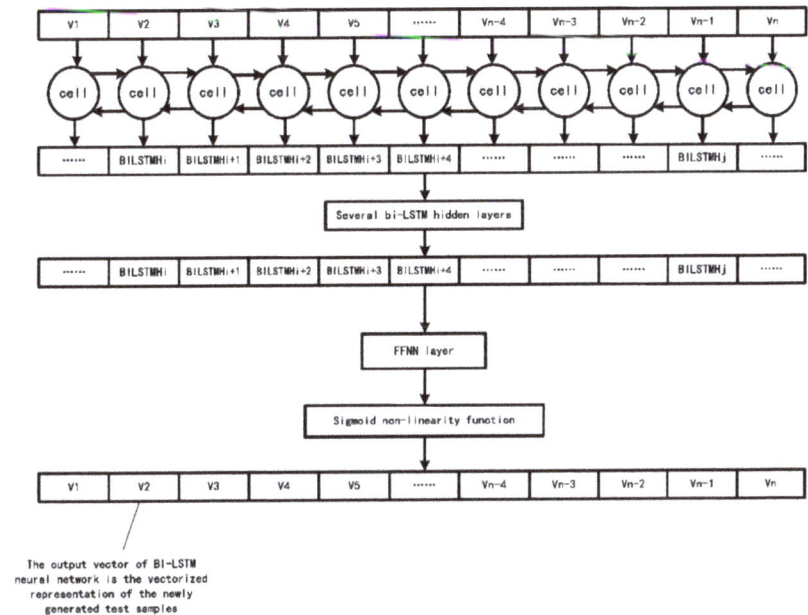

Figure 5. The Complete Structure of the bi-LSTM neural network.

From Figure 5, we make the coding input with length MaxLen pass through several bi-LSTM hidden layers to extract clearer context dependencies. We let the output of the last bi-LSTM hidden layer pass through a feed forward neural network layer and sigmoid activation function. The sigmoid activation function also normalizes the final output vector, which is the vector form of the fuzzy testing sample generated by the bi-LSTM neural network.

3.3.2. Genetic Algorithm for Constructing Test Cases

The core of the genetic algorithm used to construct samples can be divided into several parts, including population initialization, tracking and executing the tested program, fitness calculation and individual selection, crossover and mutation. The overall structure is shown in Figure 6.

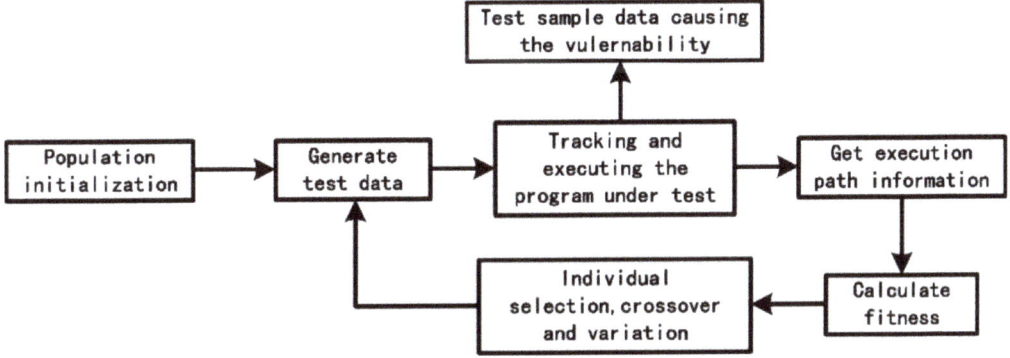

Figure 6. General Flow Chart of Generating Test Cases By Genetic Algorithm.

(a) Population initialization

In a genetic algorithm, the population is composed of several individuals. We abstract an individual as a chromosome. Let us set the length of the chromosome as D_{len}, which means the number of bytes of test data. Then, the ith individual in the population can be expressed as $X_i = (x_{i,1}, x_{i,2}, x_{i,3}, \ldots, x_{i,D_{len}})$. Population initialization is performed to assign a value to each gene $x_{i,k} (1 \leq k \leq D_{len}$ in $X_i)$. When there are initial test data, each byte of the initial test data is used to assign a value of $x_{i,k}$. Otherwise, the whole population can be initialized by randomized assignment.

(b) Tracking and executing the program under test

Tracking is divided into two aspects:

- Monitor whether the current test data will cause the tested program to crash;
- Record the execution path of the program

Because each program can be divided into many basic blocks during execution, the essence of the program execution is the process of execution and jump between basic blocks.

Each basic block has only one entry and exit. So, in a basic block, the program enters from the entry and exits from the exit. Therefore, we can use the entry address Inaddr of the basic block to represent each basic block. Then, the program execution process can be expressed as a sequence of basic blocks: $(Inaddr_1, Inaddr_2, \ldots, Inaddr_n)$ We define the jump of a basic block as $e = (Inaddr_k, Inaddr_{k+1})$, where $(1 \leq k \leq n-1)$.

Obviously, if every basic block is regarded as a point in a graph, then E is an edge in the graph. Since a basic block may be executed multiple times in the execution sequence, the graph is directed. In this case, the execution path of the program can be expressed as a sequence of edges $E_e = (e_1, e_2, e_3 \ldots e_{n-1})$.

Because some basic blocks may be repeated many times during program execution, some edges may appear many times. We combine the same edges to obtain a set of edges with the information of times of occurrence and analyze the frequency statistics of this set and further divide it into many groups according to the different times of occurrence 1, 2–3, 4–7, 8–15, 16–31, 32–63, 64–127 and 128.

It is easy to see that the significance of this classification is that it can use different bits of a byte to represent the times information, so it can improve the processing speed of the program. Finally, we will obtain a new set of occurrence information $F_e = (f_1, f_2, f_3 \ldots f_{n-1})$.

We use the above processing method for each basic block to get the final program execution path information.

(c) Fitness calculation

By tracking the program under test, we can see that an execution path information can be expressed as a sequence of edges. Therefore, in order to find a new execution path and improve the path coverage of CVDF DYNAMIC, we need to calculate the fitness. We define the sequence set of edges as $V = (V_1, V_2, \ldots, V_n)$, where each V_k $(1 \leq k \leq n)$ is equivalent to E_e. For any edge in E_e, let us assume that the final test data are X_i. We can obtain a binary set of edge information related to the test data, as shown in Equation (2):

$$Q_i = \{(e_{i,1}, X_{i,1}), (e_{i,2}, X_{i,2}) \ldots (e_{i,n}, X_{i,n})\} \quad (2)$$

It is not difficult to find that its essence is a weighted digraph, and the weight is the test data. We define that the fitness (adaptation) f of an individual consists of two functions, as shown in Equations (3) and (4).

Finding the number of new edges f_1 and the number of edges f_2 associated with them in Q_i:

$$f_1(X_i) = card(V_i - E_t) \quad (3)$$

$$f_2(X_i) = \sum_{q \in V_i}^{q} G(W_q, X_i) \quad (4)$$

$$G(X_1, X_2) = \begin{cases} 1(X1 = X2) \\ 0(X1 \neq X2) \end{cases} \tag{5}$$

Firstly, the fitness f_1 of each individual is calculated, and then the fitness f_2 of each individual is calculated after updating the set. The two sets used to calculate the fitness are updated after each round of testing. When comparing two individuals, first f_1 is compared; if f_1 cannot be distinguished, then compare f_2.

(d) Individual selection, crossover and variation

Our individual selection method uses elite selection to produce new individuals. It is a strategy of generating new individuals in genetic algorithm, which makes the individuals with high fitness enter the next generation. The method of crossover is 2-opt transformation. A number of random numbers are generated as the intersection points, and then the fragments of the intersection points in the chromosome are exchanged. Rather than using the random mutating method, this paper proposes a control mutation method to improve the effect of mutation. A motivating example of the Algorithm 2 Control Mutation is as follows:

Algorithm 2. Control Mutation

Start Func
Func ControlPROC(X,Y)
1: A = 1, B = 1
2: **IF** Y >= B **THEN**
3: **FORK1:** A = A × X, B = B + 1
4: **ELSE:**
5:    **IF** X >= A **THEN**
6:      FORK2: A = A + X, B = B − 1
7:    **ELSE:**
8:      FORK3: A = A − X, B = B/2
9: **RETURN** A
End Func

The input data format of the program is (X, Y) assuming that the template data are $(X = 1, Y = 1)$, and the variation factor is the operation of replacing 0. Therefore, two test data can be generated by mutation $(X = 1, Y = 0)$ and $(X = 0, Y = 1)$, which can cover FORK1 and FORK2. This form of testing could not achieve 100% branch coverage due to the failure to cover FORK3. For control variation, when the test data $(X = 1, Y = 0)$ generated by the variation make the program enter the new branch FORK2, the variation field of this time will be marked as an immutable field, and the variation will be carried out on the basis of the test data. In this example, the control variation marks $Y = 0$ as an immutable field and mutates the remaining fields, the X value, to 0, resulting in test data $(X = 0, Y = 0)$ that can be overridden by FORK3.

The control mutation strategy consists of the test data and control information that make the program enter the new branch. The control mutation process is as follows: Firstly, the control mutation strategy is taken out from the policy database, and the test data entering the new branch are taken as the mutation template. Secondly, check the stored control information and each byte in the template to confirm whether it is marked as control information; if so, check the next byte, if not, modify the byte in combination with random mutation strategy, generate test data and execute fuzzy testing, then continue to check the next byte. Finally, after all bytes are checked, we complete one time of mutation, and the above process is repeated.

After completing the above operations, we have completed a round of iteration of the genetic algorithm taking the newly generated chromosome data as the test data of the next round of mutation, that is, continuous iterative mutation.

3.3.3. Integrating New Test Data with Integration Idea

Firstly, through the above genetic algorithm, test cases with high path coverage are constructed from the original test case seeds. Then, for the test cases located on different execution paths, the bi-LSTM neural network is used to construct test cases with stronger path depth detection ability. Finally, we integrate the test case set constructed by the two methods to obtain the final test case set. Considering that the test case set generated by the above two methods may be too large and the efficiency of the fuzzy testing is reduced, this paper uses heuristic genetic algorithm to simplify the integrated test case set to ensure that the efficiency of fuzzy testing can be improved without losing the test performance.

3.3.4. Using Heuristic Genetic Algorithm to Reduce Sample Set

In order to reduce the sample set without losing the performance of fuzzy testing as much as possible, the screening principle of heuristic genetic algorithm in this paper is to give priority to the samples with stronger code coverage and Path Depth Detection Ability. Then, select the remaining test samples in the order of decreasing test performance, until the performance index basically covers the original fuzzy testing sample set (see the experiment in Section 4.4 for specific results). Here, our heuristic algorithm is a selection mutation algorithm for chromosomes.

(a) Using a compression matrix to represent chromosomes

At present, the common chromosome representation method is to use a 0–1 matrix [39]. The element of each row vector of the 0–1 matrix is 0 or 1. As mentioned earlier, we treat the basic block address as a collection of elements. Each basic block is equivalent to the gene in the genetic algorithm. Therefore, 1 in the 0–1 matrix indicates that a basic block exists in the sample, while 0 indicates that it does not exist. In this way, the sample set formed by all samples constitutes a 0–1 matrix, and the set of genes in each column is equivalent to a chromosome. Considering the complexity of the program execution path, the 0–1 matrix is a sparse matrix. If it is stored directly in the way of 0–1, the space efficiency will be significantly reduced. Therefore, this paper compresses the 0–1 matrix. Our storage method is a triple sequence $< Val, X_{cor}, Y_{cor} >$, where Val is the element with the storage value of 1, and X_{cor} and Y_{cor} are its X and Y coordinates in the original matrix, respectively. Since the value of Val is 1 by default, the value of this item can be omitted in the actual operation.

(b) Using heuristic genetic algorithm to improve chromosome

Each chromosome has its own independent gene sequence, but there will also be a large number of repeated and overlapping genes. Therefore, as mentioned above, we should solve the SCP when carrying out set coverage and reduce set redundancy as much as possible. Therefore, the heuristic function of the heuristic genetic algorithm is mainly reflected in eliminating the redundancy caused by gene duplication and screening better chromosomes through genetic iteration.

The specific algorithm is described as follows:

We deduce the chromosome from the position information in the compression matrix. For genes in the same column, if they contain more "1" values, it indicates that the performance priority of this column is relatively high, so we give priority to selection, mark the selected column and so on. Subsequently, we perform gene exchange on chromosomes. We assume that there are two different chromosomes Fa_1 and Fa_2 in the parent generation. After chromosome exchange, we can obtain the child's chromosomes Ch_1 and Ch_2. It is assumed that Ch_1 and Ch_2 can cover set S_1. We use sets T_1 and T_2 to store the line numbers not covered in the genes and use sets Cot_1 and Cot_2 to store the genes contained in Ch_1 and Ch_2. First, we calculate the performance priority of each gene in the parents Fa_1 and Fa_2, that is, count the number of "1" values in each column for screening. Then, we screen out the chromosomes with the highest performance priority in Fa_1 and Fa_2, copy them to Ch_1, count the genes contained in Ch_1 and delete the genes contained in Ch_1 from Cot_1. Then, we calculate the value of $Cot_1 - Ch_1$, which is the difference set, and store its line number in set T_1. Next, we continue to arrange the remaining genes of Fa_1 and Fa_2 using the same performance priority selection method, and then put them into Ch_1 again. The remaining genes will be put into Ch_2.

In the process of gene selection and gene exchange, there are some special cases with the same gene performance. At this time, we need to further screen them to obtain the optimal gene. Suppose that there are two genes, $Gene_1$ and $Gene_2$, with the same performance priority in Fa_1, and there is one gene $Gene_3$ in set Ch_1. At this time, we need to compare the results of $Gene_1 \cap Gene_3$ and $Gene_2 \cap Gene_3$ to screen out the larger results. Considering that there will be a corresponding mutation process in the genetic algorithm, the above calculation should be carried out before and after mutation to ensure that the optimal result is always selected.

From the above description, the heuristic genetic algorithm proposed in this paper uses the compression matrix on the basis of the original population and selects the optimal chromosome according to the way of gene selection and gene exchange. Therefore, this heuristic genetic algorithm essentially does not change the workflow of ordinary genetic algorithm, but through the optimization of search conditions, it simplifies the sample set and further improves the efficiency of fuzzy testing.

The specific process of the ordinary genetic algorithm has been described above. The heuristic genetic algorithm is different from ordinary genetic algorithm in the following aspects:

(c) Paternal selection

There are three common methods of paternal selection: random selection, tournament selection and roulette bet. Here, we use roulette method, the specific operation is as follows:

Step 1: The fitness of each individual in the population is calculated f_i ($i = 1,2,3, \ldots n$), where n is the population size.

Step 2: Calculate the probability $p_i = \frac{f_i}{\sum_1^n f_i}$ of each individual being inherited into the next generation population.

Step 3: Calculate the probability distribution of each individual:

$$q_i = \sum_{j=1}^{i} p(x_j). \tag{6}$$

Step 4: A pseudo-random number (rand) with uniform distribution is generated in the interval $(0, 1)$.

Step 5: When $rand < q_1$, q_1 is chosen; otherwise, if $q_{k-1} \leq rand \leq q_k$, individual K is chosen.

Step 6: Repeat step 4 and step 5 several times, and the number of repetitions depends on the size of the population.

(d) Cross rate selection

Crossover is the main way to produce new individuals. The crossover rate is the number of chromosomes in the crossover pool. A reasonable crossover rate can ensure that new individuals will be produced continuously in the crossover pool, but it will not produce too many new individuals, so as to prevent the genetic order from being destroyed. This paper adopts the most popular method of the adaptive crossover rate.

(e) Variation rate selection

The mutation rate is the proportion of the number of genes in a population based on the number of all genes. Because mutation is a way to produce new individuals, we can control the mutation by setting the number of genes or the rate of random mutation. Too low a mutation rate will lead to too few chromosomes involved in the mutation, which leads to the problem that the chromosome containing unique genes cannot be entered into the set. The high mutation rate will cause too many chromosomes involved in the mutation, which will generate some illegal data and increase the time cost. After the experiment and model tuning, the final mutation rate is 0.5.

(f) Elite ratio

The elite ratio means that the individuals with the highest fitness in the current population do not participate in crossover and mutation operations but replace the individuals with the lowest fitness in the current population after crossover and mutation operations. After the experiment and model optimization, the final elite ratio is 0.06.

(g) Stopping Criteria

The genetic algorithm has to go through several rounds of iterative evolution until it reaches the ideal result or reaches the threshold of the number of iterations. For the heuristic genetic algorithm, the threshold of iterations is 25.

4. Experiment and Evaluation

4.1. Data Sources

In the training part of the neural network, we need a large number of training samples to train our neural network so that the time series neural network can effectively capture the corresponding kinds of vulnerability characteristics from the training set. Therefore, we first collect a large number of vulnerability information from CVE and CNNVD national security vulnerability database, then screen out the vulnerability information, which is obviously suitable for neural network training. Then, we select the corresponding binary executable program and corresponding test cases from GitHub [40] and SARD [41] dataset and obtain a small number of training datasets from Symantec Security Company. The dataset we screened contains a variety of CWE vulnerability types, such as buffer overflow vulnerability (CWE-119, CWE-120, CWE-131), format string (CWE-134), etc. For the binary executable program corresponding to each vulnerability information, we filter out two versions, which are vulnerable version (no patch version) and clean version (with patch version). The purpose of using two different versions to train the neural network is to verify whether the corresponding test cases can trigger the vulnerability successfully. Second, we can further enhance the learning of the neural network for vulnerability features through this method of comparative training, so as to achieve a better training effect. The inspiration for the construction of this training dataset comes from the special training dataset constructed for generator G in GAN neural network, which contains labeled samples and unlabeled samples. Finally, all the datasets we get are shown in Table 1.

Table 1. Dataset Information of CVDF DYNAMIC.

Data Sources	Types of Vulnerabilities
SARD	(CWE-119, CWE-120, CWE-131, CWE-134 etc)
Security Focus	(CWE-119, CWE-120, CWE-189, CWE-369 etc)
Github	(CWE-415, CWE-476, CWE-119, CWE-763 etc)

We randomly select 80% of the data for the bi-LSTM neural network training set and the remaining 20% for CVDF DYNAMIC framework and subsequent experimental comparative analysis test set.

In the experiment, we mainly answer the following three questions:

Q1: Is the theoretical model of CVDF DYNAMIC valid?

Q2: Does CVDF DYNAMIC have a performance advantage in test case generation compared with the existing fuzzy testing tools?

Q3: What is the performance overhead of CVDF DYNAMIC? Does the reduction of sample sets improve the efficiency of CVDF DYNAMIC sample generation?

4.2. Evaluate the Validity of CVDF DYNAMIC's Theoretical Model

For Q1, our BI-LSTM neural network optimizes the parameters according to the method mentioned above, and after seven epochs training, the accuracy and loss performance of the model are shown in Figure 7.

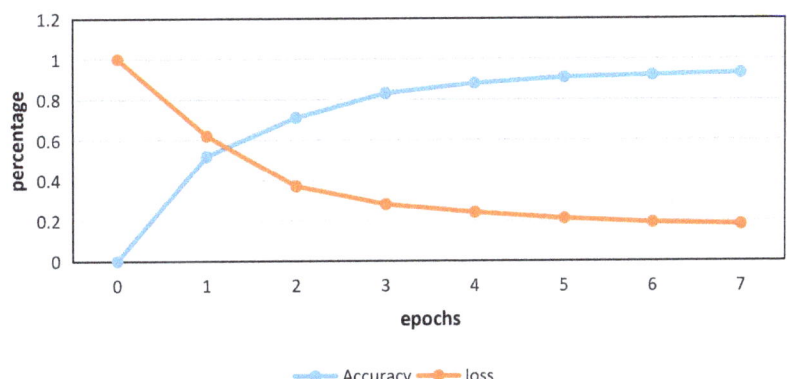

Figure 7. The Relationship Between the Accuracy and Loss Of bi-LSTM And Epochs.

It can be seen from Figure 7 that after seven training epochs, the accuracy of the BI-LSTM neural network is more than 90%, approaching 93% and stable, while the loss is less than 20% and tends to be stable.

Figure 8 shows a specific example of parameter optimization for the number of hidden layers of the bi-LSTM neural network. As can be seen from Figure 8, when the number of hidden layers is five, the performance of the bi-LSTM neural network on the three evaluation indices of precision, recall and accuracy is the best. Other parameters such as drop rate and batch size are optimized in a similar way.

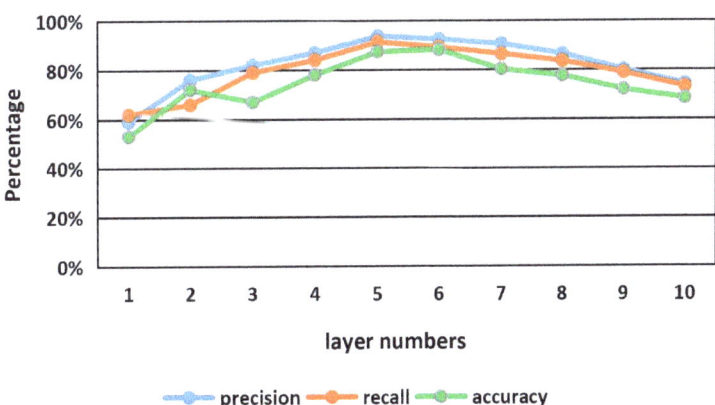

Figure 8. Relationship Between Evaluation Indices And Layer Numbers.

In the part of using the genetic algorithm to generate test cases, we compare the genetic algorithm with the existing fuzzy testing tool AFLFast under the two evaluation indices of code coverage and the number of generated edge sequences EdgeNum. The genetic algorithm has been generated through 25 rounds of iterations, and the test program uses the media processing program named FFmpeg [42] in the test set constructed above. The final experimental results are shown in Figure 9.

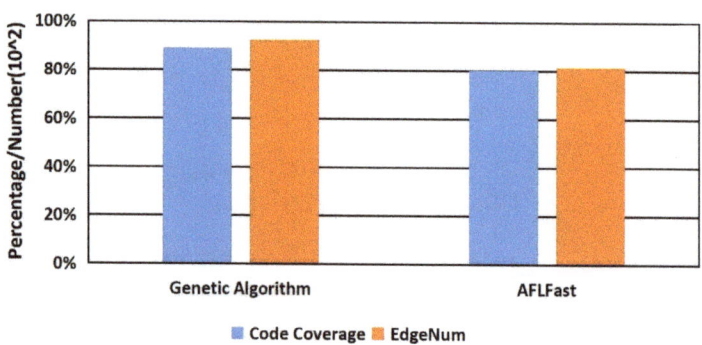

Figure 9. Comparative Test Results Of Genetic Algorithm And AFLFast.

In Figure 9, the ordinate dimension of code coverage is a percentage, and the dimension of the sequence number of edges is $value \times 10^2$. As can be seen from Figure 9, compared with AFLFast, the genetic algorithm has significant performance advantages in code coverage and the number of edge sequences. The genetic algorithm finds 9246 edge sequences for FFmpeg, while AFLFast only finds 8137 edge sequences. Because of the positive correlation between the number of edges and code coverage, the code coverage of the genetic algorithm is better than that of AFLFast.

So far, we have effectively solved the first problem, that is, the CVDF DYNAMIC theoretical model is effective. For the bi-LSTM neural network part of CVDF DYNAMIC, Figure 7 shows that our model achieves ideal training results. For the part of genetic algorithm generating test cases in CVDF DYNAMIC, our test cases have performance advantages over AFLFast in terms of code coverage and number of edges.

4.3. Performance Comparison between CVDF DYNAMIC and Existing Fuzzy Testing Tools

For Q2, we use NeuFuzz, which is also based on a neural network to guide the generation of fuzzy testing samples, and AFLFast tools for comparative testing. In order to facilitate testing and comparison, we use widely used evaluation metrics in vulnerability mining and neural networks, including false positive rate (FPR), true positive rate (TPR) and accuracy rate (ACC).

Firstly, the common definitions of vulnerability evaluation index are given.

TP (true positive): True positive samples are samples with their own vulnerabilities and are correctly identified.

FP (false positive): False positive samples are samples that do not contain vulnerabilities and are not correctly identified.

FN (false negative): False negative samples are samples that contain vulnerabilities and are not correctly identified.

TN (true negative): True negative samples are samples that do not contain vulnerabilities and are correctly identified.

The specific forms of FPR, TPR and ACC are as follows:

$$TPR = \frac{TP}{TP + FN}$$

$$FPR = \frac{FP}{FP + TN}$$

$$ACC = \frac{TP + TN}{TP + FP + TN + FN}$$

On the other hand, in order to intuitively show the performance advantages of the bi-LSTM neural network and genetic algorithm integration, we also add two evaluation indices, which are code coverage and path depth detection ability, and use the dataset constructed in this paper to test it. The experimental results are shown in Table 2.

Table 2. Comparison Test Results Of CVDF DYNAMIC With Other Tools.

Evaluation Indicator Tool	FPR	TPR	ACC	Code Coverage	WDC
CVDF DYNAMIC	5.6%	92.3%	88.9%	89.6%	2.76
NeuFuzz	10.2%	79.8%	83.4%	24.7%	2.78
VDiscover	8.5%	86.7%	85.8%	86.5%	2.33
AFLFast	11.2%	88.7%	82.9%	80.1%	1.94

It can be seen from Table 2 that CVDF DYNAMIC has performance advantages over other fuzzy testing tools. This is because CVDF DYNAMIC combines the advantages of neural network and genetic algorithm and is superior to other tools in comprehensive performance. However, other tools are also very advanced fuzzy testing tools, so they also have good performance in contrast testing. CVDF DYNAMIC and NeuFuzz are very close to each other in terms of other evaluation indices, except code coverage. However, CVDF DYNAMIC has obvious advantages over NeuFuzz in code coverage because it combines the advantages of the bi-LSTM neural network and genetic algorithm. It should also be pointed out here that the author of NeuFuzz explains that NeuFuzz focuses on seed mutation and test case generation under critical execution path rather than code coverage. However, CVDF DYNAMIC is still in the leading position in comprehensive performance.

4.4. Performance Overhead of CVDF DYNAMIC and Effectiveness of Sample Set Reduction

For Q3, we consider the performance cost of CVDF DYNAMIC and the effectiveness of sample set reduction from the number of sample sets before and after reduction, the time of fuzzy testing before and after reduction, the compression ratio and other evaluation indicators.

From Table 3, it can be seen that the compression algorithm greatly reduces the number of samples, and the compression rate reaches 54.6%. However, because the compressed sample set basically retains the key path, the execution time has decreased to some extent, but it is not as obvious as the compression rate. The code coverage and WDC evaluation index of the compressed sample set are identical with the original sample set. It shows that the compression of the test case sample set has no loss of performance, and then proves the significance and necessity of the sample set compression.

Table 3. Index Comparison Of Sample Set Before And After Compression.

	Number of Samples	Compression Ratio	Execution Times/s	Code Coverage	WDC
Initial sample set	6308	54.6%	46,184	89.6%	2.76
Compressed sample set	2864		32,428	89.6%	2.76

We use a random sampling method to form 6 initial sample sets with the scales of 1000, 2000, 3000, 4000, 5000 and 6000. The execution efficiency and time of the initial sample set and of the compressed sample set are compared, and the results are shown in Figure 10.

As can be seen from Figure 10, with the increase in the initial sample set size, the execution time efficiency after compression is gradually improved compared with that before compression.

Finally, this paper compares the compression ratio and test time of the sample set between the CVDF DYNAMIC heuristic genetic algorithm and the greedy-based approximation algorithm. The experimental results are shown in Figures 11 and 12.

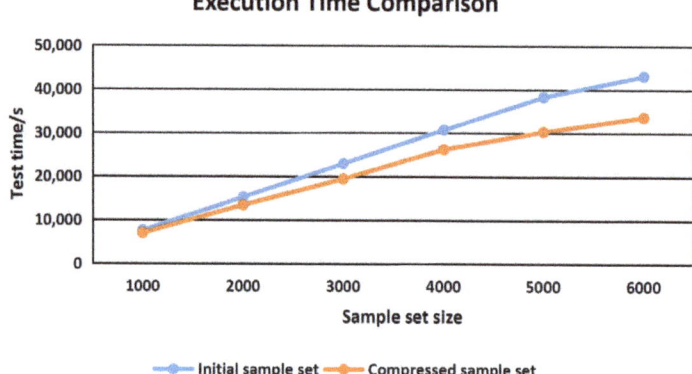

Figure 10. Execution Time Comparison.

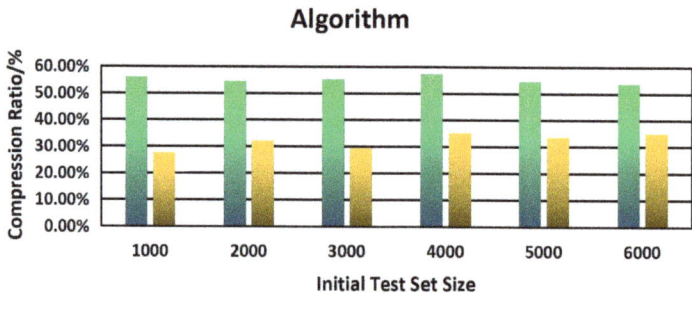

Figure 11. Comparison Of Compression Ratio Between Heuristic Genetic Algorithm And Approximation Algorithm.

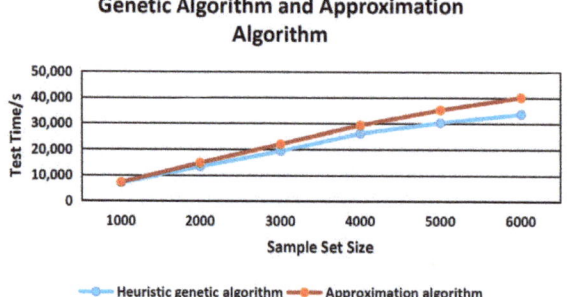

Figure 12. Comparison of Test Time Between Heuristic Genetic Algorithm And Approximation Algorithm.

It can be seen that the compression ratio based on the heuristic genetic algorithm has obvious advantages in different size sample sets compared with an approximation algorithm. With the increase in sample size, the test time of the heuristic genetic algorithm is more and more advanced.

5. Discussion on Security and Privacy of CVDF DYNAMIC Model

Because CVDF DYNAMIC combines the bi-LSTM neural network and the genetic algorithm to generate fuzzy testing samples, the final sample set is a mixed sample set, and the sample set has no label for classification. Therefore, it is very difficult to deduce the sensitive training data of CVDF DYNAMIC through the final sample set generated by CVDF DYNAMIC. On the other hand, in the description of experiment part 4.1, the training data of CVDF DYNAMIC comes from the vulnerability databases of many different countries or companies. Some of these databases are open access and some are private, but CVDF DYNAMIC adopts mixed training for datasets from different sources in the training process and randomly selects 80% as the training set and 20% as the testing set in the mixed datasets. Therefore, even if the attacker obtains the CVDF DYNAMIC datasets through reverse derivation, it is also very difficult to further distinguish private data from the middle. However, the bi-LSTM neural network adopted by CVDF DYNAMIC is a mature neural network structure, and there are corresponding scientific studies to attack this neural network structure. The security of the bi-LSTM neural network structure still needs to be strengthened in the future.

6. Conclusions

Existing fuzzy testing tools and methods only focus on the code coverage or the test case generation on the critical path. It is difficult to take both the code coverage and path depth detection ability into account. Therefore, this paper proposes CVDF DYNAMIC, a fuzzy testing sample generation framework based on the bi-LSTM and the genetic algorithm.

By combining the genetic algorithm and the bi-LSTM neural network, the framework has the ability of code coverage and path depth detection and has excellent comprehensive performance. This paper also proposes path depth detection ability, which is an evaluation metrics of vulnerability detection ability under critical execution path. Meanwhile, a heuristic genetic algorithm is used for simplifying the sample set. Finally, the experimental results show that CVDF DYNAMIC is feasible and effective, and its performance is improved compared with existing fuzzy testing tools, such as AFLFast and NeuFuzz in several evaluation indices. (FPR, TPR, ACC, Code Coverage and WDC). The reduction in the sample set further improves the efficiency of the CVDF DYNAMIC test case generation. In the future, we will further optimize the performance of CVDF DYNAMIC by optimizing the neural network structure in CVDF DYNAMIC and perfecting the iterative rules of genetic algorithm and integrate more fuzzy testing sample generation methods to further improve the code coverage and path depth detection ability.

Author Contributions: Conceptualization, M.M.; methodology, M.M.; software, M.M.; validation, M.M.; formal analysis, M.M.; investigation, M.M.; resources, L.H.; data curation, M.M.; writing—original draft preparation, M.M.; writing—review and editing, L.H.; visualization, M.M.; supervision, Y.Q.; project administration, L.H.; funding acquisition, L.H. All authors have read and agreed to the published version of the manuscript.

Funding: This paper is supported by the National Natural Science Foundation of China: 6217071437, 62072200, 62127808.

Institutional Review Board Statement: Not applicable.

Informed Consent Statement: Not applicable.

Data Availability Statement: Data available on request due to restrictions eg privacy or ethical. The data presented in this study are available on request from the corresponding author. The data are not publicly available due to [Data privacy issues].

Acknowledgments: The authors thank the anonymous reviewers for their insightful suggestions on this work.

Conflicts of Interest: The authors declare no conflict of interest.

References

1. Zou, Q.; Zhang, T.; Wu, R.; Ma, J.; Li, M.; Chen, C.; Hou, C. From automation to intelligence: Survey of research on vulnerability discovery technique. *J. Tsinghua Univ. (Sci. Technol.)* **2018**, *58*, 1079–1094. [CrossRef]
2. Borzacchiello, L.; Coppa, E.; Demetrescu, C. FUZZOLIC: Mixing fuzzing and concolic execution. *Comput. Secur.* **2021**, *108*, 102368. [CrossRef]
3. Liang, J.; Jiang, Y.; Wang, M.; Jiao, X.; Chen, Y.; Song, H.; Choo, K.R. DeepFuzzer: Accelerated Deep Greybox Fuzzing. *IEEE Trans. Dependable Secur. Comput.* **2019**, *18*, 2675–2688. [CrossRef]
4. Huang, H.; Yao, P.; Wu, R.; Shi, Q.; Zhang, C. PANGOLIN: Incremental Hybrid Fuzzing with Polyhedral Path Abstraction. In Proceedings of the 2020 IEEE Symposium on Security and Privacy, San Francisco, CA, USA, 18–21 May 2020. [CrossRef]
5. Gan, S.; Zhang, C.; Chen, P.; Zhao, B.; Qin, X.; Wu, D.; Chen, Z. GreyOne: Data Flow Sensitive Fuzzing. In Proceedings of the 29th USENIX Security Symposium, Boston, MA, USA, 12–14 August 2020. Available online: https://www.usenix.org/conference/usenixsecurity20/presentation/gan (accessed on 25 January 2022).
6. Bohme, M.; Pham, V.-T.; Roychoudhury, A. Coverage-Based Greybox Fuzzing as Markov Chain. *IEEE Trans. Softw. Eng.* **2019**, *5*, 489–506. [CrossRef]
7. Chen, Y.; Jiang, Y.; Ma, F.; Liang, J.; Wang, M.; Zhou, C.; Jiao, X.; Su, Z. EnFuzz: Ensemble Fuzzing with Seed Synchronization among Diverse Fuzzers. In Proceedings of the 28th USENIX Security Symposium, Santa Clara, CA, USA, 14–16 August 2019. Available online: https://www.usenix.org/conference/usenixsecurity19/presentation/chen-yuanliang (accessed on 24 January 2022).
8. Wang, Y.; Wu, Z.; Wei, Q.; Wang, A.Q. NeuFuzz: Efficient Fuzzing with Deep Neural Network. *IEEE Access* **2019**, *7*, 36340–36352. [CrossRef]
9. Lin, P.; Hong, Z.; Li, Y.; Wu, L. A priority based path searching method for improving hybrid fuzzing. *Comput. Secur.* **2021**, *105*, 102242. [CrossRef]
10. Chen, P.; Chen, H. Angora: Efficient Fuzzing by Principled Search. In Proceedings of the 2018 IEEE Symposium on Security and Privacy, San Francisco, CA, USA, 20–24 May 2018. [CrossRef]
11. Zhang, X.; Li, Z. Survey of Fuzz Testing Technology. *Comput. Sci.* **2016**, *43*, 5. [CrossRef]
12. Zhang, Y.; Zhao, L.; Jin, Y. Sensitive Region Prediction based on neuralnetwork in Fuzzy Test Algorithm Research. *J. Cyber Secur.* **2020**, *5*, 1. [CrossRef]
13. Xie, X.F.; Li, X.H.; Chen, X.; Meng, G.Z.; Liu, Y. Hybrid testing based on symbolic execution and fuzzing. *Ruan Jian Xue Bao/J. Softw.* **2019**, *30*, 3071–3089. (In Chinese)
14. Xu, P.; Liu, J.; Lin, B.; Sun, H.; Lei, B. Generation of fuzzing test case based on recurrent neural networks. *Appl. Res. Comput.* **2019**, *36*, 2679–2685. [CrossRef]
15. Nagy, S.; Hicks, M. Full-speed Fuzzing: Reducing Fuzzing Overhead through Coverage-guided Tracing. In Proceedings of the 2019 IEEE Symposium on Security and Privacy (SP), San Francisco, CA, USA, 19–23 May 2019. [CrossRef]
16. Yang, M.F.; Huo, W.; Zou, Y.Y.; Yin, J.W.; Liu, B.X.; Gong, X.R.; Jia, X.Q.; Zou, W. Programmable fuzzing technology. *Ruan Jian Xue Bao/J. Softw.* **2018**, *29*, 1258–1274. (In Chinese)
17. Godefroid, P.; Peleg, H.; Singh, R. Learn&Fuzz: Maching Learning for Input Fuzzing. In Proceedings of the 2017 32nd IEEE/ACM International Conference on Automated Software Engineering (ASE), Urbana, IL, USA, 30 October–3 November 2017.
18. Li, M.-L.; Huang, H.; Lu, Y.-L.; Zhu, K.-L. SymFuzz: Vulnerability Detection Technology under Complex Path Conditions. *Comput. Sci.* **2021**, *48*, 7. [CrossRef]
19. Cheng, L.; Zhang, Y.; Zhang, Y.; Wu, C.; Li, Z.; Fu, Y.; Li, H. Optimizing seed inputs in fuzzing with machine learning. In Proceedings of the 2019 IEEE/ACM 41st International Conference on Software Engineering: Companion Proceedings (ICSE-Companion), Montreal, QC, Canada, 25–31 May 2019.
20. Ma, J.; Zhang, T.; Li, Z.; Zhang, J. Improved fuzzy analysis methods. *J. Tsinghua Univ. (Sci. Technol.)* **2016**, *56*, 478–483. [CrossRef]
21. Aschermann, C.; Schumilo, S.; Abbasi, A.; Holz, T. IJON: Exploring Deep State Spaces via Fuzzing. In Proceedings of the 2020 IEEE Symposium on Security and Privacy, San Francisco, CA, USA, 18–21 May 2020. [CrossRef]
22. Perl, H.; Arp, D.; Fahl, S. VCCFinder: Finding Potential Vulnerabilities in Open-Source Projects to Assist Code Audits. In Proceedings of the CCS '15: Proceedings of the 22nd ACM SIGSAC Conference on Computer and Communications Security, Denver, CO, USA, 12–16 October 2015. [CrossRef]
23. Stephens, N.; Grosen, J.; Salls, C. Driller: Augmenting Fuzzing Through Selective Symbolic Execution. *NDSS* **2016**, *16*, 21–24. [CrossRef]

24. She, D.; Pei, K.; Epstein, D.; Yang, J.; Ray, B.; Jana, S. NEUZZ: Efficient Fuzzing with Neural Program Smoothing. In Proceedings of the 2019 IEEE Symposium on Security and Privacy, San Francisco, CA, USA, 19–23 May 2019 [CrossRef]
25. Chen, Y.; Zhong, R.; Hu, H.; Zhang, H.; Yang, Y.; Wu, D.; Lee, W. One Engine to Fuzz'em All: Generic Language Processor Testing with Semantic Validation. In Proceedings of the 2021 IEEE Symposium on Security and Privacy, San Francisco, CA, USA, 24–27 May 2021. [CrossRef]
26. Zhang, Z.; You, W.; Tao, G.; Aafer, Y.; Liu, X.; Zhang, X. STOCHFUZZ: Sound and Cost-effective Fuzzing of Stripped Binaries by Incremental and Stochastic Rewriting. In Proceedings of the 2021 IEEE Symposium on Security and Privacy, San Francisco, CA, USA, 24–27 May 2021. [CrossRef]
27. Yue, T.; Wang, P.; Tang, Y.; Wang, E.; Yu, B.; Lu, K.; Zhou, X. EcoFuzz: Adaptive Energy-Saving Greybox Fuzzing as a Variant of the Adversarial Multi-Armed Bandit. In Proceedings of the 29th USENIX Security Symposium, Boston, MA, USA, 12–14 August 2020. Available online: https://www.usenix.org/conference/usenixsecurity20/presentation/yue (accessed on 17 January 2022).
28. Zong, P.; Lv, T.; Wang, D.; Deng, Z.; Liang, R.; Chen, K. FuzzGuard: Filtering out Unreachable Inputs in Directed Grey-Box Fuzzing through Deep Learning. In Proceedings of the 29th USENIX Security Symposium, Boston, MA, USA, 12–14 August 2020. Available online: https://www.usenix.org/conference/usenixsecurity20/presentation/zong (accessed on 17 January 2022).
29. Österlund, S.; Razavi, K.; Bos, H.; Giuffrida, C. ParmeSan: Sanitizer-guided Greybox Fuzzing. In Proceedings of the 29th USENIX Security Symposium, Boston, MA, USA, 12–14 August 2020. Available online: https://www.usenix.org/conference/usenixsecurity20/presentation/osterlund (accessed on 17 January 2022).
30. Oleksenko, O.; Trach, B. Mark Silberstein, Christof Fetzer, SpecFuzz: Bringing Spectre-type vulnerabilities to the surface. In Proceedings of the 29th USENIX Security Symposium, Boston, MA, USA, 12–14 August 2020. Available online: https://www.usenix.org/conference/usenixsecurity20/presentation/oleksenko (accessed on 5 January 2022).
31. Lee, G.; Shim, W.; Lee, B. Constraint-Guided Directed Greybox Fuzzing. In Proceedings of the 30th USENIX Security Symposium, Vancouver, BC, Canada, 11–13 August 2021. Available online: https://www.usenix.org/conference/usenixsecurity21/presentation/lee-gwangmu (accessed on 13 January 2022).
32. Salls, C.; Jindal, C.; Corina, J.; Kruegel, C.; Vigna, G. Token-Level Fuzzing. In Proceedings of the 30th USENIX Security Symposium, Vancouver, BC, Canada, 11–13 August 2021. Available online: https://www.usenix.org/conference/usenixsecurity21/presentation/salls (accessed on 10 January 2022).
33. Liu, X.; Xie, L.; Wang, Y.; Zou, J.; Xiong, J.; Ying, Z.; Vasilakos, A.V. Privacy and Security Issues in Deep Learning: A Survey. *IEEE Access* **2021**, *9*, 4566–4593. [CrossRef]
34. Mollah, M.B.; Azad, M.A.K.; Vasilakos, A. Secure data sharing and searching at the edge of cloud-assisted internet of things. *IEEE Cloud Comput.* **2017**, *4*, 34–42. [CrossRef]
35. Yi, G.; Yang, X.; Huang, P.; Wang, Y. A Coverage-Guided Fuzzing Framework based on Genetic Algorithm for Neural Networks. In Proceedings of the 2021 8th International Conference on Dependable Systems and Their Applications (DSA), Yinchuan, China, 5–6 August 2021. [CrossRef]
36. Lin, G.; Guan, J. An Adaptive Memetic Algorithm for Solving the Set Covering Problem. *J. Zhejiang Univ. (Sci. Ed.)* **2016**, *43*, 168–174. [CrossRef]
37. Alyahya, K.; Rowe, J.E. Landscape Analysis of a Class of NP-Hard Binary Packing Problems. *Evol. Comput.* **2019**, *27*, 47–73. [CrossRef] [PubMed]
38. Hoesen, D.; Purwarianti, A. Investigating Bi-LSTM and CRF with POS Tag Embedding for Indonesian Named Entity Tagger. In Proceedings of the 2018 International Conference on Asian Language Processing (ALP), Bandung, Indonesia, 15–17 November 2018.
39. Zhao, Y.W.; Wang, J.; Guo, M.Z.; Zhang, Z.L.; Yu, G.X. Prediction of protein function based on 0–1 matrix decomposition. *Sci. Sin. Inf. Sci.* **2019**, *49*, 1159–1174. (In Chinese)
40. Github. Available online: https://github.com/ (accessed on 31 December 2021).
41. NIST Test Suites. Available online: https://samate.nist.gov/SRD/testsuite.php (accessed on 31 December 2021).
42. FFmpeg. FFmpeg [EB/OL]. (2016-10-01) [2017-04-13]. Available online: https://www.ffmpeg.org/ (accessed on 31 December 2021).

Article

Towards a Hybrid Machine Learning Model for Intelligent Cyber Threat Identification in Smart City Environments

Najla Al-Taleb [1] and Nazar Abbas Saqib [2,*]

1. Department of Computer Science, College of Computer Science and Information Technology, Imam Abdulrahman Bin Faisal University, P.O. Box 1982, Dammam 31441, Saudi Arabia; 2190500053@iau.edu.sa
2. SAUDI ARAMCO Cybersecurity Chair, Department of Networks and Communications, College of Computer Science and Information Technology, Imam Abdulrahman Bin Faisal University, P.O. Box 1982, Dammam 31441, Saudi Arabia
* Correspondence: nasaqib@iau.edu.sa

Abstract: The concept of a smart city requires the integration of information and communication technologies and devices over a network for the better provision of services to citizens. As a result, the quality of living is improved by continuous analyses of data to improve service delivery by governments and other organizations. Due to the presence of extensive devices and data flow over networks, the probability of cyber attacks and intrusion detection has increased. The monitoring of this huge amount of data traffic is very difficult, though machine learning algorithms have huge potential to support this task. In this study, we compared different machine learning models used for cyber threat classification. Our comparison was focused on the analyzed cyber threats, algorithms, and performance of these models. We have identified that real-time classification, accuracy, and false-positive rates are still the major issues in the performance of existing models. Accordingly, we have proposed a hybrid deep learning (DL) model for cyber threat intelligence (CTI) to improve threat classification performance. Our model was based on a convolutional neural network (CNN) and quasi-recurrent neural network (QRNN). The use of QRNN not only resulted in improved accuracy but also enabled real-time classification. The model was tested on BoT-IoT and TON_IoT datasets, and the results showed that the proposed model outperformed the other models. Due to this improved performance, we emphasize that the application of this model in the real-time environment of a smart system network will help in reducing threats in a reasonable time.

Keywords: cyber threat intelligence; privacy; smart city; machine learning; deep learning; CNN; QRNN

Citation: Al-Taleb, N.; Saqib, N.A. Towards a Hybrid Machine Learning Model for Intelligent Cyber Threat Identification in Smart City Environments. *Appl. Sci.* **2022**, *12*, 1863. https://doi.org/10.3390/app12041863

Academic Editors: Leandros Maglaras, Helge Janicke and Mohamed Amine Ferrag

Received: 21 December 2021
Accepted: 28 January 2022
Published: 11 February 2022

Publisher's Note: MDPI stays neutral with regard to jurisdictional claims in published maps and institutional affiliations.

Copyright: © 2022 by the authors. Licensee MDPI, Basel, Switzerland. This article is an open access article distributed under the terms and conditions of the Creative Commons Attribution (CC BY) license (https://creativecommons.org/licenses/by/4.0/).

1. Introduction

The transformation of cities into smart cities is on the rise, where technologies such as the Internet of Things (IoT) and cyber–physical systems (CPS) are connected through networks for the better provision of quality services to citizens [1]. The smart city concept refers to urban systems that are integrated with information and communication technologies (ICTs) to improve city services in terms of monitoring, management, and control to be more efficient and effective [2]. A smart city contains a huge number of sensors that continuously generate a tremendous amount of sensitive data such as location coordinates, credit card numbers, and medical records [3]. These data are transmitted through a network to data centers for processing and analysis so that appropriate decisions, such as managing traffic and energy, can be made in a smart city [4]. The resource limitations of technological infrastructure expose smart cities to cyber attacks [5]. For instance, sensors that generate data and devices that handle the data in a smart city have vulnerabilities that can be exploited by cybercriminals. Consequently, citizens' privacy and lives can be at risk when collected data for analysis and decision making are manipulated, which makes people intimidated by smart cities [1].

A smart city environment collects a tremendous amount of private and sensitive data and depends on ICT, which makes smart cities target for different cyber attacks, such as distributed denial of service (DDoS), using IoT devices by infecting them with bots and launch an attack against a target [6–9]. Cyber threat intelligence (CTI) can provide secure environments for smart cities, where it can rely on cloud services to monitor possible threats in real time and take appropriate prevention measures without human intervention [10–15]. Moreover, CTI can provide a light security mechanism, as it is not implemented on smart city devices; rather, it monitors attacks through the cloud to obtain information about recent threat behavior and indicator of compromise (IoC), and it reports this information to connected smart city systems. Different techniques and machine learning (ML) models have been proposed to analyze cyber threats for CTI such as deep learning (DL) models [16,17], random forest (RF) [18], and K-NN [19]. Nevertheless, artificial intelligence (AI)-based models can have a high false-positive rates (FPRs) and low true-positive rates (TPRs) if the attack traffic is not profiled and modeled well enough [20]. This limits real-time classification efficiency and degrades smart city network security. To address this issue, improve threat analysis, and lower FPRs, we propose a hybrid DL model that is based on a convolutional neural network (CNN) and quasi-recurrent neural network (QRNN). The proposed model can automatically learn spatial features using CNN and temporal features using QRNN without human intervention. The CNN model can automatically select the relevant features from the dataset and reduce the irrelevant features to improve classification performance [21]. For cyber threat analysis, several works have shown the efficiency of CNN for feature selection, such as [20,22]. The QRNN model performs computation in parallel, which improves computation time while maintaining sequence modeling [23]. Thus, this hybrid model (CNN–QRNN) can help improve real-time analysis in CTI while providing a high accuracy and low FPR. Therefore, the proposed model can improve CTI performance for smart cities. We evaluated our proposed model with two IoT network traffic datasets. The evaluation results demonstrate the effectiveness of our proposed model. The main contributions of this study are summarized as follows:

- We propose a hybrid DL model that consists of QRNN and CNN to improve cyber threat analysis accuracy, lower FPR, and provide real-time analysis.
- We evaluated our proposed model on two datasets that were simulated to represent a realistic IoT environment.

The rest of this paper is structured as follows. In Section 2, we discuss related work by comparing and analyzing different threat classification schemes that have been proposed in the literature. The proposed model is presented in Section 3. The implementation of the proposed model is discussed in Section 4, the experiment results and analysis are presented in Section 5, and conclusions are presented in Section 6.

2. Related Work

In recent years, different studies have proposed mechanisms to predict and analyze cyber attacks in smart city environments. The authors of [24] proposed an ML-based detection mechanism that focused on classifying DDoS patterns to protect a smart city from them. In [25], the authors studied how IoT devices can affect smart city cyber security; the authors proposed a detection mechanism that depends on the selected features to improve the threat detection for IoT. The results of the proposed system showed high accuracy, but the dataset, KDD CUP 99, did not represent the behavior of IoT network attacks. Soe et al. [21] proposed an algorithm to improve prediction accuracy by selecting the optimal features for each type of attack in an IoT environment. The authors used ML models to evaluate the proposed feature selection algorithm, which was able to accurately predict the threats. However, the proposed algorithm selected a static set of features for each type of attack, which could be easily bypassed if exposed to the threat environment. In [26], the authors used a DL model to select the best features for threat prediction to improve the detection time in an IoT environment. The proposed model selects a set of features that are fed into feed-forward neural networks (FFNNs) to detect cyber threats and classify threat types. However, the proposed model showed limited accuracy in predicting information theft data.

In [19], the authors discussed how to use the ML model to rapidly and efficiently detect and classify IoT network attacks. The authors performed an experimental study by implementing various ML models and evaluating their performance. In [27], the authors proposed a hybrid ML model to detect IoT network attacks including that of the zero-day. The proposed model mainly consists of two stages: the first stage classifies the traffic into two categories (normal or attack), and the second stage classifies the type of attacks using SVM. Similarly, in [28], the authors proposed a hybrid ML model to detect and classify IoT network attacks in real time. The first layer of the proposed model uses a decision tree classifier to detect malicious behavior and the second layer classifies the type of attack using random forest (RF). In [29], the authors investigated the remote-control threat of connected cars and used an ML model to predict threats. The authors proposed a proactive anomaly detection mechanism that profiled the behavior of the autonomous connected cars using a recursive Bayesian estimator. To evaluate the effectiveness of the proposed method, the authors designed a dataset for connected cars using hypothetical events routes and global positioning system coordinates, and they then modeled the data to predict the anomalies' behavior. Lee et al. [30] proposed a technique, based on DL models, that transforms the multitude of security events into individual event profiles. The authors discussed how anomaly-based detection can be costly since it can trigger many false alerts. Therefore, they focused on improving security information and event management system by using DL to reduce the cost to differentiate between true and false alerts. In [31], the authors proposed a hybrid ML method to detect cyber threats. The authors focused on how to improve detection accuracy to handle an attacker's methods to evade detection tools. To evaluate the proposed method, the authors used different datasets including KDD Cup and UNSW-NB15. In [32], the authors discussed how to improve the threat analysis and classification, including novel attacks. The authors proposed a model based on a stacked autoencoder to enhance and automate feature selection to classify the threats.

Various scientific studies have proposed a hybrid DL model to improve threat analysis and classification. In [33], the authors proposed an improved version of grey wolf optimization (GWO) and a CNN. In the proposed hybrid model, the first GWO model is used to select the features and the second CNN model is used for threat classification. Other studies have used a hybrid DL model that is based on CNNs and RNNs for spatial and temporal feature extraction to improve attack classification. In [34], the authors used a CNN for feature selection since it could provide fast feature selection to support real-time analysis. For threat classification, the authors used one of the variants of the LSTM model: weight-dropped LSTM (WDLSTM). The proposed hybrid model showed good performance in terms of execution time. Vinayakumar et al. [35] studied the effect of CNN in threat classification and intrusion detection system (IDS). The authors investigated different hybrid DL models with CNNs including CNN-LSTM, CNN-GRU, and CNN-RNN, and the model implementing CNN-LSTM outperformed the other models. Moreover, the authors highlighted that selecting a minimum set of features for threat classification degraded the performance of the classification. Therefore, DL models can perform well in terms of feature selection. In [36], the authors proposed a hierarchical model based on CNN-LSTM. The authors used stacked CNN layers for spatial features learning using image classification and then stacked LSTM for temporal features learning. Similarly, in [20], the authors proposed an LuNet model based on CNN-LSTM. The authors discussed how stacking LSTM layers after CNN layers could drop some of the temporal features. Thus, the authors proposed the LuNet block, which consists of LSTM layer stacked after the CNN layer, and they then stacked the LuNet block in multiple layers to improve classification performance and lower the FPR.

As shown in Table 1, different network traffic benchmark datasets have been used to analyze the low-level IoC such as UNSW-NB15, NSL-KDD, and KDD CUP 99. For IoT attack classification, the BoT-IoT dataset has been used in multiple studies to evaluate the performance of proposed models. Different ML and DL models, such as the SVM, CNN, and LSTM, have been used to analyze threats and provide accurate results, and the CNN-LSTM hybrid model has been used in multiple studies to improve threat classification performance.

Table 1. Comparison between proposed attack classification methods.

Ref	Cyber Threats	Algorithm	Data Sources	Accuracy	FPR
[24]	DDoS	Restricted Boltzmann machine and FFNN	Simulated smart water system dataset	97.5%	-
[21]	Information theft, reconnaissance, and DDoS	J48	BoT-IoT UNSW	-	0.41
[26]	Information theft, reconnaissance, and DDoS	FFNN	BoT-IoT UNSW	-	-
[19]	DDoS, DoS, data exfiltration, keylogging, OS fingerprinting, and service scan	K-nearest neighbors (K-NN)	BoT-IoT UNSW	99.00%	-
[27]	DDoS, DoS, keylogging, and reconnaissance	C5-SVM	BoT-IoT UNSW	99.97%	0.001
[28]	DDoS, DoS, data exfiltration, keylogging, OS fingerprinting, and service scan	Decision tree-RF	BoT-IoT UNSW	99.80%	-
[29]	Remote car control	Recursive Bayesian estimation	Route data for connected cars	-	-
[30]	DoS, probe, R2L, and U2R	FCNN, CNN, and LSTM	Network events	94.7%	0.049
[31]	Tor traffic (anonymous IP)	C4.5, Multilayer perceptron (MLP), SVM, and linear discriminant analysis (LDA)	UNB-CIC TOR Network Traffic dataset	100	0
[31]	Worms, DoS, backdoors, reconnaissance, exploits, analysis, generic, fuzzers, and shellcode	C4.5, Multilayer perceptron (MLP), SVM, and linear discriminant analysis (LDA)	UNSW-NB15	97.84%	0.23
[32]	Injection, Flooding, Impersonation	Stacked auto-encoder (SAE)	AWID-CLS-R	98.66%	-
[33]	DoS, probe, R2L, and U2R	GWO-CNN	DARPA1998	97.92%	3.60
[33]	DoS, probe, R2L, and U2R	GWO-CNN	KDD CUP 99	98.42%	2.22
[34]	Worms, DoS, backdoors, reconnaissance, exploits, analysis, generic, fuzzers, and shellcode	CNN-LSTM	UNSW-NB15	98.43%	-
[35]	DoS, probe, R2L, and U2R	CNN-LSTM	KDD CUP 99	98.7%	0.005
[36]	DoS, probe, R2L, U2R, BruteForce SSH, DDoS, and infiltrating	CNN-LSTM	ISCX2012	99.69%	0.22
[36]	DoS, probe, R2L, U2R, BruteForce SSH, DDoS, and infiltrating	CNN-LSTM	DARPA1998	99.68%	0.07
[20]	Worms, DoS, backdoors, reconnaissance, exploits, analysis, generic, fuzzes, and shellcode	CNN-LSTM	UNSW-NB15	84.98%	1.89
[20]	DoS, probe, R2L, and U2R	CNN-LSTM	NSL-KDD	99.05%	0.65

In terms of the CTI for smart cities, multiple papers, including [24,25], have analyzed the threats pattern based on network traffic. Additionally, in [37], the authors proposed a trustworthy privacy-preserving secured framework (TP2SF) for smart cities; the authors used the optimized gradient tree boosting system (XGBoost) and blockchain, and they evaluated the proposed framework on two datasets: BoT-IoT and TON_IoT. DDoS is one of the challenging threats in a smart city that has been studied by different researchers, who have proposed methods to analyze IP addresses and track the sources to prevent this attack or to identify the behavior of the network when there is overload traffic. Data theft, which can be described as privacy and identity theft, is another threat that has been studied by various researchers. Data theft threats include reconnaissance, information theft, probe, R2L, and U2R, which may lead to the exposure of various vulnerabilities that can help in launching data theft attacks such as sniffing passwords and unauthorized access. Some of the proposed models for smart cites set a fixed threshold to detect attacks, which is not effective and can raise a lot of false alarms that affect the power consumption of the connected systems. In smart cities, the normal behavior of a system can change due to the increasing number of connected devices, so some researchers have achieved high accuracy but bad performance in terms of FPR.

Even though different researchers have proposed models to enhance threat classification for IoT environments, many aspects still require improvement. One of the limitations that is common between different methods is performance time. Low-level IoCs that are collected from network traffic have been used to analyze the threats in various papers to provide timely information to the CTI knowledge base and update the detection and prevention information for all systems connected to the CTI. However, to enhance classification performance, various models have multiple stacked ML model layers. Therefore, it may take time to train a model and classify threats while not taking advantage of these IoCs. Secondly, when some models are not provided with enough data for each type of threat, threat traffic cannot be profiled and modeled well enough. Consequently, ML models can have high FPRs. Furthermore, some models only provide accurate results when their system has precise details of threats. Consequently, the system is not able to recognize threats that do not have enough data for model training, which affects classification accuracy.

Moreover, we observed that few papers have addressed diverse patterns for threat analysis while considering time, accuracy, and FPR. Several works have proposed hybrid models based on the CNN and LSTM to learn spatial and temporal data. However, LSTM is computationally complex and requires a long time for analysis [38]. The QRNN model is a type of RNN that allows for sequence modeling by implementing computation in parallel while maintaining the data's long- and short-term sequence dependencies [23]. We could not find a work that used the QRNN model to improve cyber threat classification time while demonstrating high accuracy. Thus, in this work, we propose a hybrid DL model for CTI for smart cities that addresses the abovementioned challenges and uses the QRNN model. The proposed hybrid model can improve threat classification accuracy and lower the FPR in a reasonable time. Therefore, it can predict different attacks to protect citizens' data and enhance the security of smart cities.

3. Proposed Model

In this section, we discuss the proposed hybrid DL model in terms of its structure, the selected DL algorithms, and relevant theoretical concepts. The selected DL models (CNN and QRNN) can be used to classify a threat type in real time while providing a low FPR. The architecture of the proposed model is presented in Figure 1.

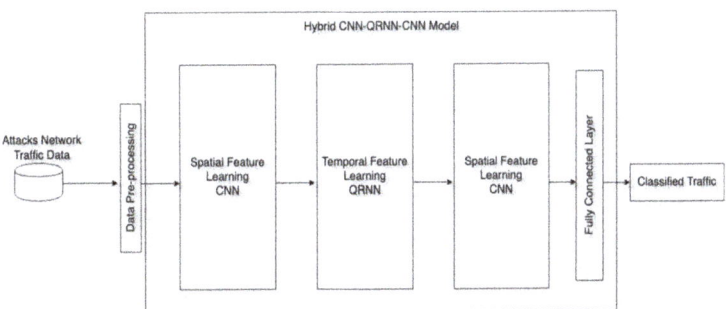

Figure 1. The architecture of the proposed hybrid model.

A CNN is an extension of a neural network [39] and it is effective at extracting features at a low level from the source data, especially spatial features [40].

CNNs are used widely in image processing due to their ability to automate feature extraction [41]. Additionally, CNNs have demonstrated their effectiveness in many fields such as biomedical text analysis and malware classification [30]. Based on the shape of the input data, a CNN can be classified into different types including a two-dimensional (2D) CNN, which uses data such as images, and a one-dimensional (1D) CNN, which uses data such as text. A CNN consists of a convolution layer, pooling layer, fully connected (FC) layer, and activation function [42]. The convolution layer is fundamental building block in CNNs that takes two sets of information as inputs and performs a mathematical operation with these inputs. The two sets of information are the data and a filter, which can be referred to as kernel. The filter is applied to an entire dataset to produce a feature map [41]. Each CNN filter extracts a set of features that are aggregated to a new feature map as output [30]. The pooling layer is implemented to reduce feature map dimensions and to remove irrelevant data to improve learning [20]. The output of the pooling layer is fed into the FC layer to classify the data [43].

The LSTM-RNN is one of the most powerful neural network models that is used in cyber security due to its ability to accurately model temporal sequences and their long-term dependencies [44]. However, LSTM usually takes a longer time for model training and high computation cost [45]. The QRNN model [23] was designed to overcome the RNN limitations in terms of each timestep's computation dependency on the previous timestep, which limits the power of parallelism. The QRNN combines the benefits of the CNN and RNN by using convolutional filters on the input data and allowing the long-term sequence dependency to store the data of previous timestamps [23]. The computation structure of the QRNN is presented in Figure 2. The QRNN consists of convolutional layers and recurrent pooling function, which allow the QRNN to work faster than LSTM due to its a 16-times-increase in speed while achieving the same accuracy as LSTM [46]. The convolutional and pooling layers allow for the parallel computation of the batch and feature dimensions [23]. The QRNN has been used in different applications such as video classification [45], speech synthesis [46], and natural language processing [47].

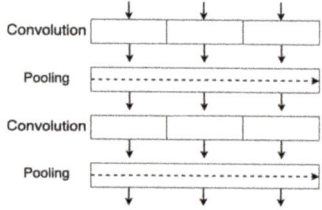

Figure 2. The computation structure of the QRNN.

Our hybrid DL model consists of a 1D convolutional layer, 1D max-pooling layer, a QRNN, and FC layers. The first 1D convolutional layer selects the spatial features and produces a feature map that will be processed by the activation function. The Rectified Linear Unit (ReLU) activation function is used in the convolutional layers because of its rapid convergence of gradient descent, which made it a good choice for our proposed model [41]. Then, the feature map is processed by the second layer that uses the max-pooling operation. The max-pooling operation selects the maximum value in the pooling operation [41]. The pooling layer reduces dimensionality and removes irrelevant features. The output of the CNN model retains the temporal feature that is extracted by the QRNN model. Figure 3 provides details of our proposed model and shows that we used two QRNN layers to extract the temporal features. In the two layers of the QRNN, the hidden size represents the number of the hidden units and the output dimension. The hidden units can be selected based on the value of the number of features [45]. One of the problems of a neural network is overfitting, which means that a model learns the data too well. Consequently, the model is not able to identify variants in new data [22]. We added a dropout layer to prevent overfitting.

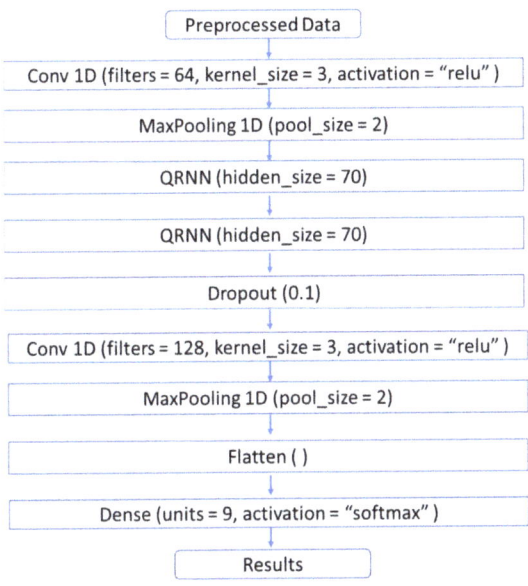

Figure 3. Illustration of the details of the proposed model.

Then, a 1D convolutional layer and max-pooling layer are used to extract more spatial-temporal features. The output of the CNN model is passed to the Flatten layer, which is a fully connected input layer that transforms the output of the pooling layer into one vector to be an input for the next layer [48]. Finally, the dense layer, which is also a fully connected layer, with the SoftMax activation function is used to classify the threats by calculating the probabilities for each class [34].

4. Implementation

In this section, we describe the datasets that we selected to evaluate the proposed model. Additionally, we discuss the data preprocessing steps, model parameter selection process, and selected evaluation metrics.

4.1. Datasets

In this work, we selected the BoT-IoT and TON-IoT datasets because they have been simulated to represent realistic IoT environments such as smart homes and cities. The datasets had a heterogeneity of simulated IoT devices including weather-monitoring systems, smart lights, smart thermostats, and a variety of cyber threats.

4.1.1. BoT-IoT Dataset

In previous studies, different datasets, such as KDD99, ISCX, and CICIDS2017, have been used to evaluate ML models; however, few datasets have been produced to reflect realistic IoT network traffic. These datasets were either not diverse enough in terms of attacks or not realistic in terms of the testbed [19]. Therefore, Koroniotis et al. [49] designed the BoT-IoT dataset to address these limitations. The BoT-IoT dataset is used in forensic analysis and to evaluate IDS. The dataset contains normal IoT traffic and different types of attack traffic with subcategories for each type, which are listed in Table 2. Reconnaissance is one of the privacy threats, and it allows a threat actor to collect data about a victim via port scanning and OS fingerprinting, among other ways. Information theft includes data theft by unauthorized access and keylogging. On the other hand, a DoS threat affects the availability of services and can damage systems, which make it one of the biggest threats to smart cities. In this dataset, UDP, TCP, and HTTP protocols were used to perform both DoS and DDoS attacks.

Table 2. Attack categories in BoT-IoT dataset.

Attack	Attack Subcategory	Number of Instances
Reconnaissance	Service scan	73,168
	OS fingerprinting	17,914
DoS	TCP	615,800
	UDP	1,032,975
	HTTP	1485
DDoS	TCP	977,380
	UDP	948,255
	HTTP	989
Information theft	Keylogging	73
	Data theft	6

4.1.2. TON_IoT Dataset

The ToN_IoT dataset [50] is one of the newest cyber security datasets; it as collected from a testbed network for industry 4.0 IoT and Industrial IoT (IIoT), which makes it suitable to evaluate CTI for a smart city. We used the TON_IoT train–test dataset, which is in the CSV format. The dataset contains a total of 461,043 instances and 9 types of attacks, which are presented in Table 3 along with the number of instances for each type.

Table 3. Attack categories in TON_IoT dataset.

Attack	Number of Instances
DoS	20,000
DDoS	20,000
Scanning	20,000
Ransomware	20,000
Backdoor	20,000
Injection	20,000
Cross-Site Scripting (XSS)	20,000
Password	20,000
Man-In-The-Middle (MITM)	1043

4.2. Data Preprocessing

Since we were interested in evaluating CTI for threat classification, we deleted the normal traffic from the datasets. Additionally, in the BoT-IoT dataset, we omitted the pkSeqID feature since it represented an identifier for the traffic records. The datasets contains some categorical features that could not be processed by the neural network. Thus, we converted the nominal values into numeric using sklearn LabelEncoder. LabelEncoder converts categorical values into numerical values [22]. We implemented sklearn StandardScaler to scale the data. For training and evaluation, several papers have split the dataset into training and testing, with a ratio of 20% for testing s in [19] and 30% for testing in [21]. However, due to the size of the BoT-IoT dataset and the resource constraints of our device, we divided the data into training and testing sets, with a ratio of 35% for testing, while having the same ratio of classes in both parts by using the stratify parameter.

4.3. Model Implementation

The parameters of the hybrid model were obtained during the training phase by trial and error including the number of CNN filters, the number of QRNN hidden units, and the dropout rate. As mentioned in different studies [35], kernel size values of 3 and 5 are the most common, so we used kernel size 3 with both datasets in our experiment. A filter can help in extracting more details from a dataset by increasing the number of filters [51]. Thus, for the first CNN layer, we used 64 filters, and for the other CNN, we used 128 filters. Additionally, we set the value of the batch size for the training at 128 and the value of the number of epochs at 10. The details and the selected parameters of the hybrid DL model are presented in Figure 3.

4.4. Evaluation Tools and Metrics

Different evaluation metrics were used in this work to evaluate the performance of the proposed model including accuracy, FPR, TPR, precision, recall, and F-Score. Accuracy represents the ratio of correctly classified threats to the total number of classified threats, so it demonstrates how accurate an model in classifying threats [52]. The FPR represents the ratio of misclassified data as a different type of threat, and the TPR represents a model's ability to correctly classify threats. A low FPR and a high TPR demonstrate the ability of a model to correctly classify cyber threats [53]. Precision, recall, and F-Score were used to evaluate the overall performance of the proposed model; a high value of precision indicates a low FPR, and recall represents a model's ability to correctly classify threats. Equations (1)–(6) represent the evaluation metrics, where FP is false positive, TP is true positive, TN is true negative, and FN is false negative.

$$\text{Accuracy} = \frac{TP + TN}{TP + TN + FP + FN} \quad (1)$$

$$\text{FPR} = \frac{FP}{FP + TN} \quad (2)$$

$$\text{TPR} = \frac{TP}{TP + FN} \quad (3)$$

$$\text{Precision} = \frac{TP}{TP + FP} \quad (4)$$

$$\text{Recall} = \frac{TP}{TP + FN} \quad (5)$$

$$\text{F-Score} = \frac{2(\text{Precision} \times \text{Recall})}{\text{Precision} + \text{Recall}} \quad (6)$$

5. Results and Discussion

5.1. Results and Analysis

This section presents the results and analysis for model implementation. We used Jupyter Notebook software with the Python programming language. We used the Keras and scikitlearn packages for data pre-processing and implementing the proposed model. We trained the proposed model on a MacBook Air with an Intel Core i5 CPU 1.6 GHz processor and 8 GB RAM. Additionally, we implemented different state-of-the-art ML models on the datasets to compare their performance with that of our proposed model. Figure 4 presents the confusion matrix of our proposed model on the BoT-IoT dataset. The results show that the model correctly classified most of the cyber threat categories. Furthermore, to illustrate the quality of the proposed model, the receiver operating characteristic (ROC) curve is plotted in Figure 5 for the BoT-IoT dataset.

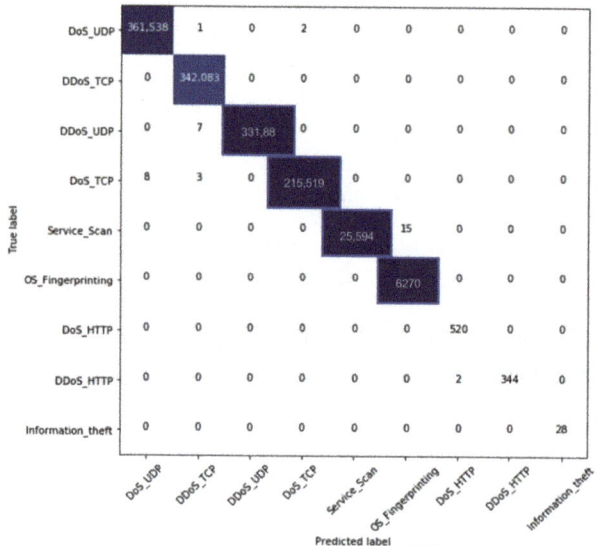

Figure 4. Confusion matrix based on the BoT-IoT dataset.

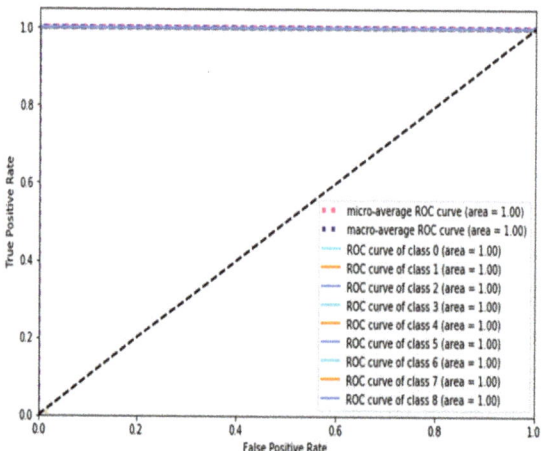

Figure 5. ROC curve of using our proposed model on the BoT-IoT dataset.

Figure 6 presents the confusion matrix of our proposed model on the TON_IoT dataset, and the ROC curve is presented in Figure 7. Both ROC curves show that our proposed model achieved the highest value of 1. Thus, our proposed model performed very well with all the classes.

The results of our proposed model on the testing datasets are presented in Table 4.

Table 4. Results of cyber threat classification on both datasets.

Dataset	Accuracy%	TPR%	FPR
BoT-IoT	99.99	99.92	0.0003
TON_IoT	99.99	99.99	0.001

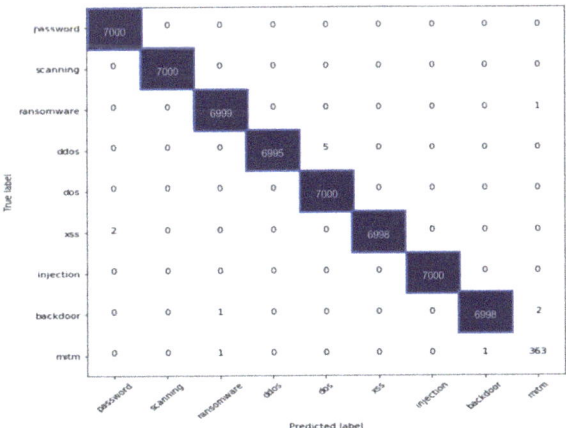

Figure 6. Confusion matrix based on the TON_IoT dataset.

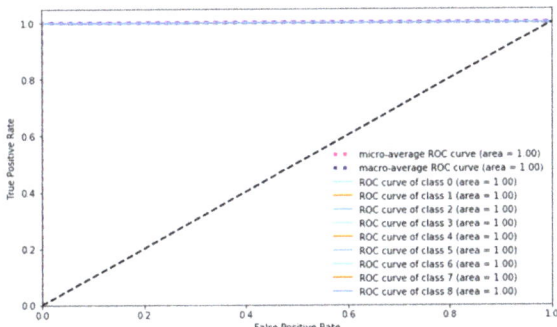

Figure 7. ROC curve of using our proposed model on the TON_IoT dataset.

As shown in Table 4, the proposed model achieved high accuracy, with an average of 99.99% on both datasets. The TPR reached averages of 99.92% with the BoT-IoT dataset and 99.99% with the TON_IoT dataset. The proposed model achieved a low FPR of 0.0003 with the BoT-IoT dataset and 0.001 with the TON_IoT dataset. Thus, the proposed model showed good performance in classifying the threats with both datasets. Moreover, to demonstrate the effectiveness of the QRNN, we implemented our proposed model with LSTM instead of the QRNN to compare performance. Cybersecurity threats are very critical [54–56], and the results shown in Tables 5 and 6 highlight that our proposed approach could be very effective in dealing with them.

Table 5. Comparison of our proposed model while using LSTM and QRNN based on BoT-IoT dataset.

Model	Accuracy	Precision	Recall	F-Score	Avg. Training Time per Epoch	Classification Time
With LSTM	99.99%	100%	100%	100%	1717.4 s	326 s
With QRNN	99.99%	100%	100%	100%	1299.1 s	251 s

Table 6. Comparison of our proposed model while using LSTM and the QRNN based on TON_IoT dataset.

Model	Accuracy	Precision	Recall	F-Score	Avg. Training Time per Epoch	Classification Time
With LSTM	99.99%	100%	100%	100%	86.3 s	16 s
With QRNN	99.99%	100%	100%	100%	66.5 s	13 s

According to the results in Tables 5 and 6, our proposed model with the QRNN showed the same performance as our proposed model with LSTM in terms of accuracy, precision, recall, and F-Score. In terms of time, the proposed model with the QRNN showed better performance for training the model and testing. The average training time per epoch demonstrated that the QRNN performed faster than LSTM in terms of training the model on both datasets, with a 418.3 s difference on the BoT-IoT dataset and a 19.8 s difference on the TON_IoT dataset. Additionally, for the classification time on the test dataset, the QRNN model performed faster than LSTM, with a 75 s difference on the BoT-IoT dataset and a 3 s difference on the TON_IoT dataset. The QRNN showed its effectiveness in increasing the speed of the model while providing a high accuracy and low FPR. Therefore, the model can be used for real-time CTI. We further compared the performance of our proposed model on the BoT-IoT and TON_IoT datasets against the state-of-the-art models for the multi-class classification of threats. The results of these comparisons are shown in Tables 7 and 8.

Table 7. Comparison of our proposed model with state-of-the-art models based on the BoT-IoT dataset.

Model	Accuracy%	Precision%	Recall%	F-Score%
K-NN [19]	99.00	99.00	99.00	99.00
Hybrid IDS [27]	99.97	-	-	95.7
RF [28]	99.80	99.00	99.00	98.80
RF [37]	99.99	79.76	62.98	65.08
TP2SF [37]	99.99	99.97	94.92	97.08
Our model	99.99	100	100	100

Table 8. Comparison of our proposed model with state-of-the-art models based on the TON_IoT dataset.

Model	Accuracy%	Precision%	Recall%	F-Score%
RF [37]	97.81	87.55	85.43	86.41
TP2SF [37]	98.84	97.23	94.03	95.28
Our model	99.99	100	100	100

As shown in Tables 7 and 8, though K-NN [19] and RF [28] showed good performance for recall and F-score on the BoT-IoT dataset, our proposed model outperformed the state-of-the-art models on both datasets. Additionally, we implemented different ML models to compare their performance with that of our model. The accuracy, TPR, and FPR values of each model are given are Tables 9 and 10. Our model performed better than the other four models, with accuracy measured as 99.99% on both datasets and low FPR values of 0.0003 on the BoT-IoT dataset and 0.001 on the TON_IoT dataset. The LSTM model showed good performance in terms of accuracy and FPR, while the GRU showed a high TPR compared to the LSTM on the BoT-IoT dataset. On the TON_IoT dataset, the GRU performed poorly compared to the other models.

Table 9. Comparison of our proposed model with other ML models based on BoT-IoT dataset.

Model	Accuracy%	TPR%	FPR
MLP	99.98	86.42	0.002
CNN	99.98	88.13	0.001
GRU	99.98	96.06	0.001
LSTM	99.99	94.69	0.0004
Our model	99.99	99.92	0.0003

Table 10. Comparison of our proposed model with other ML models based on TON_IoT dataset.

Model	Accuracy%	TPR%	FPR
MLP	99.67	99.51	0.03
CNN	99.88	99.75	0.01
GRU	97.85	96.95	0.27
LSTM	99.83	99.79	0.02
Our model	99.99	99.99	0.001

5.2. Theoretical and Practical Implications

This work describes a model that can correctly classify cyber threats with a low FPR while considering time performance. Thus, the proposed model can improve decision making for risk mitigation so that appropriate protection measures against cyber attacks in smart cities can be taken [57,58]. Additionally, this model will benefit organizations and services providers in smart cities because of the high costs of implementing and maintaining cyber security solutions [59]. The organizations and service providers in smart cities can take accurate proactive measures against detected cyber attacks such as data breaches, which will help in saving costs [60]. Furthermore, our proposed model can be implemented in the cloud to monitor cyber security and collect and update cyber threat data from the connected systems in smart cities.

6. Conclusions

A smart city facilitates the life of its citizens by providing better services than non-smart cities. Due to the extensive presence of digital data, smart cities are also vulnerable to various types of attacks. Machine-learning-based cyber threat intelligence can secure smart city environments by monitoring attacks and analyzing data threats in order to take prevention measures. In this paper, we have proposed a hybrid deep learning model to classify threats. The proposed model uses a CNN and a QRNN to improve feature extraction, increases classification accuracy, and lower the FPR. We evaluated our model on the BoT-IoT and TON_IoT datasets, and our results showed the effectiveness of our model in improving classification accuracy and lowering the FPR. In addition, the results showed that the QRNN model could improve classification time performance while providing high accuracy and lower FPR than LSTM. Thus, the proposed model for CTI for smart cities can accurately analyze and classify data in real time.

One of the limitations of this work is the authors' use of datasets. Due to the security and privacy of smart city citizens, it was difficult to evaluate the proposed model on real-time data. Additionally, for implementation, we evaluated the model as a centralized system. In future work, we can implement the proposed model in a distributed environment with parallel training to improve classification performance.

Author Contributions: Data curation, N.A.-T.; Formal analysis, N.A.-T.; Funding acquisition, N.A.S.; Investigation, N.A.S. and N.A.-T.; Project administration, N.A.S.; Resources, N.A.S.; Supervision, N.A.S.; Validation, N.A.-T.; Writing—original draft, N.A.-T. All authors have read and agreed to the published version of the manuscript.

Funding: The authors would like to thank SAUDI ARAMCO Cybersecurity Chair, Imam Abdulrahman Bin Faisal University for funding this project.

Acknowledgments: The authors would like to thank Attaur-Rahman and Sujata Dash for their feedback on an earlier non-peer-reviewed version of the manuscript which was shared on the Arxiv repository.

Conflicts of Interest: The authors declare no conflict of interest.

References

1. AlZaabi, K.A.J.A. The Value of Intelligent Cybersecurity Strategies for Dubai Smart City. In *Smart Technologies and Innovation for a Sustainable Future*; Springer International Publishing: Cham, Switzerland, 2019; pp. 421–445, ISBN 9783030016593.
2. Behzadan, V.; Munir, A. Adversarial Exploitation of Emergent Dynamics in Smart Cities. In Proceedings of the 2018 IEEE International Smart Cities Conference (ISC2), Kansas City, MO, USA, 16–19 September 2018; pp. 1–8.
3. Butt, T.A.; Afzaal, M. Security and Privacy in Smart Cities: Issues and Current Solutions. In *Smart Technologies and Innovation for a Sustainable Future*; Springer International Publishing: Cham, Switzerland, 2019; pp. 317–323, ISBN 9783030016593.
4. Lee, J.; Kim, J.; Seo, J. Cyber attack scenarios on smart city and their ripple effects. In Proceedings of the 2019 International Conference on Platform Technology and Service (PlatCon), Jeju, Korea, 28–30 January 2019; pp. 1–5.
5. Ahmad, F.; Adnane, A.; Franqueira, V.N.L.; Kurugollu, F.; Liu, L. Man-In-The-Middle Attacks in Vehicular Ad-Hoc Networks: Evaluating the Impact of Attackers' Strategies. *Sensors* **2018**, *18*, 4040. [CrossRef] [PubMed]
6. Alibasic, A.; Junaibi, R.A.; Aung, Z.; Woon, W.L.; Omar, M.A. Cybersecurity for Smart Cities: A Brief Review. In *International Workshop on Data Analytics for Renewable Energy Integration*; Springer: Cham, Switzerland, 2017; pp. 22–30.
7. Braun, T.; Fung, B.C.M.; Iqbal, F.; Shah, B. Security and privacy challenges in smart cities. *Sustain. Cities Soc.* **2018**, *39*, 499–507. [CrossRef]
8. Cui, L.; Xie, G.; Qu, Y.; Gao, L.; Yang, Y. Security and Privacy in Smart Cities: Challenges and Opportunities. *IEEE Access* **2018**, *6*, 46134–46145. [CrossRef]
9. Kettani, H.; Cannistra, R.M. On Cyber Threats to Smart Digital Environments. In Proceedings of the 2nd International Conference on Smart Digital Environment, Rabat, Morocco, 18–20 October 2018; ACM: New York, NY, USA, 2018; pp. 183–188.
10. Sookhak, M.; Tang, H.; Yu, F.R. Security and Privacy of Smart Cities: Issues and Challenges. In Proceedings of the 2018 IEEE 20th International Conference on High Performance Computing and Communications; IEEE 16th International Conference on Smart City; IEEE 4th International Conference on Data Science and Systems (HPCC/SmartCity/DSS), Exeter, UK, 28–30 June 2018; pp. 1350–1357.
11. Liu, M.; Xue, Z.; He, X.; Chen, J. Cyberthreat-Intelligence Information Sharing: Enhancing Collaborative Security. *IEEE Consum. Electron. Mag.* **2019**, *8*, 17–22. [CrossRef]
12. Zhang, K.; Ni, J.; Yang, K.; Liang, X.; Ren, J.; Shen, X.S. Security and Privacy in Smart City Applications: Challenges and Solutions. *IEEE Commun. Mag.* **2017**, *55*, 122–129. [CrossRef]
13. Abu, S.; Selamat, S.R.; Ariffin, A.; Yusof, R. Cyber Threat Intelligence—Issue and Challenges. *Indones. J. Electr. Eng. Comput. Sci.* **2018**, *10*, 371–379.
14. Conti, M.; Dehghantanha, A.; Dargahi, T. Cyberthreat intelligence: Challenges and opportunities. In *Cyber Threat Intelligence*; Springer: Cham, Switzerland, 2018; pp. 1–6.
15. Myat, K.; Win, N.; Myo, Y.; Khine, K. Information Sharing of Cyber Threat Intelligence with their Issue and Challenges. *Int. J. Trend Sci. Res. Dev.* **2019**, *3*, 878–880.
16. Diro, A.A.; Chilamkurti, N. Distributed attack detection scheme using deep learning approach for Internet of Things. *Futur. Gener. Comput. Syst.* **2018**, *82*, 761–768. [CrossRef]
17. Berman, D.S.; Buczak, A.L.; Chavis, J.S.; Corbett, C.L. A Survey of Deep Learning Methods for Cyber Security. *Information* **2019**, *10*, 122. [CrossRef]
18. Abawajy, J.; Huda, S.; Sharmeen, S.; Hassan, M.M.; Almogren, A. Identifying cyber threats to mobile-IoT applications in edge computing paradigm. *Futur. Gener. Comput. Syst.* **2018**, *89*, 525–538. [CrossRef]
19. Alsamiri, J.; Alsubhi, K. Internet of Things Cyber Attacks Detection using Machine Learning. *Int. J. Adv. Comput. Sci. Appl.* **2019**, *10*, 627–634. [CrossRef]
20. Wu, P.; Guo, H. LuNet: A Deep Neural Network for Network Intrusion Detection. In Proceedings of the 2019 IEEE Symposium Series on Computational Intelligence (SSCI), Xiamen, China, 6–9 December 2019; pp. 617–624.
21. Soe, Y.N.; Feng, Y.; Santosa, P.I.; Hartanto, R.; Kouichi, S. Towards a Lightweight Detection System for Cyber Attacks in the IoT Environment Using Corresponding Features. *Electronics* **2020**, *9*, 144. [CrossRef]
22. Wu, P.; Guo, H.; Moustafa, N. Pelican: A Deep Residual Network for Network Intrusion Detection. In Proceedings of the 2020 50th Annual IEEE/IFIP International Conference on Dependable Systems and Networks Workshops (DSN-W), Valencia, Spain, 29 June–2 July 2020.
23. Bradbury, J.; Merity, S.; Xiong, C.; Socher, R. Quasi-Recurrent Neural Networks. *arXiv* **2017**, arXiv:1611.01576.
24. Elsaeidy, A.; Munasinghe, K.S.; Sharma, D.; Jamalipour, A. A Machine Learning Approach for Intrusion Detection in Smart Cities. In Proceedings of the 2019 IEEE 90th Vehicular Technology Conference (VTC2019-Fall), Honolulu, HI, USA, 22–25 September 2019; pp. 1–5.

25. Li, D.; Deng, L.; Lee, M.; Wang, H. IoT data feature extraction and intrusion detection system for smart cities based on deep migration learning. *Int. J. Inf. Manag.* **2019**, *49*, 533–545. [CrossRef]
26. Ge, M.; Fu, X.; Syed, N.; Baig, Z.; Teo, G.; Robles-kelly, A. Deep Learning-based Intrusion Detection for IoT Networks. In Proceedings of the 2019 IEEE 24th Pacific Rim International Symposium on Dependable Computing (PRDC), Kyoto, Japan, 1–3 December 2019.
27. Khraisat, A.; Gondal, I.; Vamplew, P.; Kamruzzaman, J.; Alazab, A. A Novel Ensemble of Hybrid Intrusion Detection System for Detecting Internet of Things Attacks. *Electronics* **2019**, *8*, 1210. [CrossRef]
28. Ullah, I.; Mahmoud, Q.H. A Two-Level Flow-Based Anomalous Activity Detection System for IoT Networks. *Electronics* **2020**, *9*, 530. [CrossRef]
29. Al-Khateeb, H.; Epiphaniou, G.; Reviczky, A.; Karadimas, P.; Heidari, H. Proactive Threat Detection for Connected Cars Using Recursive Bayesian Estimation. *IEEE Sens. J.* **2018**, *18*, 4822–4831. [CrossRef]
30. Lee, J.; Kim, J.; Kim, I.; Han, K. Cyber Threat Detection Based on Artificial Neural Networks Using Event Profiles. *IEEE Access* **2019**, *7*, 165607–165626. [CrossRef]
31. Sornsuwit, P.; Jaiyen, S. A New Hybrid Machine Learning for Cybersecurity Threat Detection Based on Adaptive Boosting. *Appl. Artif. Intell.* **2019**, *33*, 462–482. [CrossRef]
32. Thing, V.L.L. IEEE 802.11 Network Anomaly Detection and Attack Classification: A Deep Learning Approach. In Proceedings of the 2017 IEEE Wireless Communications and Networking Conference (WCNC), San Francisco, CA, USA, 19–22 March 2017; pp. 1–6.
33. Garg, S.; Kaur, K.; Kumar, N.; Kaddoum, G.; Zomaya, A.Y.; Ranjan, R. A Hybrid Deep Learning-Based Model for Anomaly Detection in Cloud Datacenter Networks. *IEEE Trans. Netw. Serv. Manag.* **2019**, *16*, 924–935. [CrossRef]
34. Hassan, M.M.; Gumaei, A.; Alsanad, A.; Alrubaian, M.; Fortino, G. A hybrid deep learning model for efficient intrusion detection in big data environment. *Inf. Sci.* **2020**, *513*, 386–396. [CrossRef]
35. Vinayakumar, R.; Kp, S.; Poornachandran, P. Applying Convolutional Neural Network for Network Intrusion Detection. In Proceedings of the 2017 International Conference on Advances in Computing, Communications and Informatics (ICACCI), Udupi, India, 13–16 September 2017; pp. 1222–1228.
36. Wang, W.; Sheng, Y.; Wang, J.; Zeng, X.; Ye, X.; Huang, Y.; Zhu, M. HAST-IDS: Learning Hierarchical Spatial-Temporal Features Using Deep Neural Networks to Improve Intrusion Detection. *IEEE Access* **2018**, *6*, 1792–1806. [CrossRef]
37. Kumar, P.; Gupta, G.P.; Tripathi, R. TP2SF: A Trustworthy Privacy-Preserving Secured Framework for sustainable smart cities by leveraging blockchain and machine learning. *J. Syst. Archit.* **2021**, *115*, 101954. [CrossRef]
38. Niu, X.; Ma, J.; Wang, Y.; Zhang, J.; Chen, H.; Tang, H. A Novel Decomposition-Ensemble Learning Model Based on Ensemble Empirical Mode Decomposition and Recurrent Neural Network for Landslide Displacement Prediction. *Appl. Sci.* **2021**, *11*, 4684. [CrossRef]
39. LeCun, Y.; Bengio, Y.; Hinton, G. Deep learning. *Nature* **2015**, *521*, 436–444. [CrossRef]
40. Ferrag, M.A.; Maglaras, L.; Moschoyiannis, S.; Janicke, H. Deep learning for cyber security intrusion detection: Approaches, datasets, and comparative study. *J. Inf. Secur. Appl.* **2020**, *50*, 102419. [CrossRef]
41. Hasan, M.N.; Toma, R.N.; Nahid, A.; Islam, M.M.M.; Kim, J. Electricity Theft Detection in Smart Grid Systems: A CNN-LSTM Based Approach. *Energies* **2019**, *12*, 3310. [CrossRef]
42. Kwon, D.; Natarajan, K.; Suh, S.C.; Kim, H.; Kim, J. An Empirical Study on Network Anomaly Detection using Convolutional Neural Networks. In Proceedings of the In 2018 IEEE 38th International Conference on Distributed Computing Systems (ICDCS), Vienna, Austria, 2–6 July 2018; pp. 1595–1598.
43. Liu, H.; Lang, B.; Liu, M.; Yan, H. Knowledge-Based Systems CNN and RNN based payload classification methods for attack detection. *Knowl.-Based Syst.* **2019**, *163*, 332–341. [CrossRef]
44. Khan, A.; Sarfaraz, A. RNN-LSTM-GRU based language transformation. *Soft Comput.* **2019**, *23*, 13007–13024. [CrossRef]
45. Bolelli, F.; Baraldi, L.; Pollastri, F.; Grana, C. A Hierarchical Quasi-Recurrent approach to Video Captioning. In Proceedings of the 2018 IEEE International Conference on Image Processing, Applications and Systems (IPAS), Sophia Antipolis, France, 12–14 December 2018; pp. 162–167.
46. Wang, M.; Wu, X.; Wu, Z.; Kang, S.; Tuo, D.; Li, G.; Su, D.; Yu, D.; Meng, H. Quasi-fully Convolutional Neural Network with Variational Inference for Speech Synthesis. In Proceedings of the ICASSP 2019–2019 IEEE International Conference on Acoustics, Speech and Signal Processing (ICASSP), Brighton, UK, 12–17 May 2019; pp. 7060–7064.
47. Huang, J.; Feng, Y. Optimization of Recurrent Neural Networks on Natural Language Processing. In Proceedings of the Proceedings of the 2019 8th International Conference on Computing and Pattern Recognition, New York, NY, USA, 23–25 October 2019; pp. 39–45.
48. Yao, D.; Wen, M.; Liang, X.; Fu, Z.; Zhang, K.; Yang, B. Energy Theft Detection with Energy Privacy Preservation in the Smart Grid. *IEEE Internet Things J.* **2019**, *6*, 7659–7669. [CrossRef]
49. Koroniotis, N.; Moustafa, N.; Sitnikova, E.; Turnbull, B. Towards the Development of Realistic Botnet Dataset in the Internet of Things for Network Forensic Analytics: Bot-IoT Dataset. *Futur. Gener. Comput. Syst.* **2018**, *100*, 779–796. [CrossRef]
50. Moustafa, N. TON_IoT Datasets. In Proceedings of the IEEE Dataport, Brisbane, Australia, 16 October 2019. [CrossRef]

51. Safa, H.; Nassar, M.; Al Orabi, W.A.R. Benchmarking Convolutional and Recurrent Neural Networks for Malware Classification. In Proceedings of the 2019 15th International Wireless Communications & Mobile Computing Conference (IWCMC), Tangier, Morocco, 24–28 June 2019; pp. 561–566.
52. Shafiq, M.; Tian, Z.; Sun, Y.; Du, X.; Guizani, M. Selection of effective machine learning algorithm and Bot-IoT attacks traffic identification for internet of things in smart city. *Futur. Gener. Comput. Syst.* **2020**, *107*, 433–442. [CrossRef]
53. Vinayakumar, R.; Alazab, M.; Member, S.; Soman, K.P. Deep Learning Approach for Intelligent Intrusion Detection System. *IEEE Access* **2019**, *7*, 41525–41550. [CrossRef]
54. Obaidan, F.A.; Saeed, S. Digital Transformation and Cybersecurity Challenges: A Study of Malware Detection Using Machine Learning Techniques. In *Handbook of Research on Advancing Cybersecurity for Digital Transformation*; IGI Global: Pennsylvania, PA, USA, 2021; pp. 203–226.
55. Naeem, H.; Ullah, F.; Naeem, M.R.; Khalid, S.; Vasan, D.; Jabbar, S.; Saeed, S. Malware detection in industrial internet of things based on hybrid image visualization and deep learning model. *Ad. Hoc. Netw.* **2020**, *105*, 102154. [CrossRef]
56. Khadam, U.; Iqbal, M.M.; Saeed, S.; Dar, S.H.; Ahmad, A.; Ahmad, M. Advanced security and privacy technique for digital text in smart grid communications. *Comput. Electr. Eng.* **2021**, *93*, 107205. [CrossRef]
57. Yamin, M.M.; Katt, B.; Nowostawski, M. Serious games as a tool to model attack and defense scenarios for cyber-security exercises. *Comput. Secur.* **2021**, *110*, 102450. [CrossRef]
58. Poleto, T.; Carvalho VD, H.D.; Silva AL, B.D.; Clemente TR, N.; Silva, M.M.; Gusmão AP, H.D.; Costa, A.P.C.S.; Nepomuceno, T.C.C. Fuzzy cognitive scenario mapping for causes of cybersecurity in telehealth services. *Healthcare* **2021**, *9*, 1504. [CrossRef]
59. Shayan, S.; Kim, K.P.; Ma, T.; Nguyen, T.H.D. The first two decades of smart city research from a risk perspective. *Sustainability* **2020**, *12*, 9280. [CrossRef]
60. Kumar, S.; Biswas, B.; Bhatia, M.S.; Dora, M. Antecedents for enhanced level of cyber-security in organisations. *J. Enterp. Inf. Manag.* **2021**, *34*, 1597–1629. [CrossRef]

Article

Few-Shot Network Intrusion Detection Using Discriminative Representation Learning with Supervised Autoencoder

Auwal Sani Iliyasu [1,*], Usman Alhaji Abdurrahman [2] and Lirong Zheng [2]

[1] School of Computer Science and Engineering, South China University of Technology, Guangzhou 510006, China
[2] School of Information Science and Technology, Fudan University, Shanghai 200433, China; aausman18@fudan.edu.cn (U.A.A.); lrzheng@fudan.edu.cn (L.Z.)
* Correspondence: engrausan@gmail.com

Abstract: Recently, intrusion detection methods based on supervised deep learning techniques (DL) have seen widespread adoption by the research community, as a result of advantages, such as the ability to learn useful feature representations from input data without excessive manual intervention. However, these techniques require large amounts of data to generalize well. Collecting a large-scale malicious sample is non-trivial, especially in the modern day with its constantly evolving landscape of cyber-threats. On the other hand, collecting a few-shot of malicious samples is more realistic in practical settings, as in cases such as zero-day attacks, where security agents are only able to intercept a limited number of such samples. Hence, intrusion detection methods based on few-shot learning is emerging as an alternative to conventional supervised learning approaches to simulate more realistic settings. Therefore, in this paper, we propose a novel method that leverages discriminative representation learning with a supervised autoencoder to achieve few-shot intrusion detection. Our approach is implemented in two stages: we first train a feature extractor model with known classes of malicious samples using a discriminative autoencoder, and then in the few-shot detection stage, we use the trained feature extractor model to fit a classifier with a few-shot examples of the novel attack class. We are able to achieve detection rates of 99.5% and 99.8% for both the CIC-IDS2017 and NSL-KDD datasets, respectively, using only 10 examples of an unseen attack.

Keywords: network intrusion detection; few-shot learning; deep learning; discriminative autoencoder

1. Introduction

Cyber defense is a continuous process that entails tasks, such as prevention, detection, and recovery, which are applied at various system levels. Network intrusion detection is a branch of cyber security that deals with the detection of attacks at the network layer level.

Network intrusion detection techniques can be broadly divided into two types: signature-based and anomaly-based methods [1]. Signature-based methods operate by matching incoming network traffic against a predefined set of known attack signatures. Thus, they perform well in detecting previously known attack signatures; however, signature-based methods fail to detect novel attacks [2]. On the other hand, anomaly-based methods, which entail machine learning methods, operate by modeling normal network traffic data and then flag any network traffic that deviates from the model pattern as an anomaly. However, these approaches sometimes lead to too many false alarm rates (FARs).

Network intrusion detection using machine learning methods has been studied for a long time, with many commercial intrusion detection systems (IDSs) using machine learning algorithms as part of their detection engines [3].

Recently, technologies such as cloud computing, IoT, and 5G have led to an explosion in the volume and diversity of network traffic, which provide fertile ground for applying deep learning (DL) techniques. Deep learning techniques are end-to-end learning models, capable of learning highly complex non-linear functions, which enable them to learn powerful representations directly from input data [4]. Thus, recent research on intrusion detection system (IDS) methods are mostly focused on this area [5].

However, network IDSs based on supervised deep learning techniques require huge amounts of labeled data in order to generalize well. Collecting a large-scale malicious sample to train DL classifiers is prohibitively expensive, and subject to obsolescence as the landscape is constantly evolving. Regardless, unsupervised anomaly-based methods provide an alternative towards generalization of an unseen malicious sample, and these approaches are highly susceptible to false alarm rates [6]. Hence, there has been an increase in interest from the research community towards approaches that require a handful of samples to achieve detection. Since collecting a few samples of malicious traffic is more realistic in a practical settings, which, for instance, can be realized from a few successfully detected intrusions from a deployed detection system, few-shot learning is emerging as an alternative to conventional supervised learning methods to simulate more realistic settings.

Few-shot learning measures the challenging issue of a model's ability to generalize new tasks using limited data [7]. This was addressed recently, based on the idea of meta-learning or "learning to learn" [8–12]. The meta-learning paradigm consist of two disjointed stages: meta-training and meta-testing. Each of the meta stages consists of a number of classification tasks with limited training data that require fast adaptability by the learner. The goal is to leverage the meta-training stage to learn transferable knowledge from a set of tasks that will enable fast adaptability to novel tasks in the testing stage.

However, recently, it has been established that good learned representations are very powerful for few-shot classification tasks, and perform on par with, or slightly worse than, the current set of complicated meta-learning algorithms [13–15]. Therefore, in this paper we propose a simple framework that relies on learning good representations to achieve few-shot intrusion detection. Our approach consists of a linear model trained on top of a pre-trained feature extractor model. The feature extractor model is trained to learn good representations using a discriminative autoencoder.

In contrast to conventional autoencoders, which are purely unsupervised representation learning methods, discriminative autoencoders are a form of supervised autoencoders that leverages the class information of their inputs. Thus, they combine both reconstruction and classification errors in their objective functions. This makes the representations learned by discriminative autoencoders more discriminative and more suitable for classification tasks [16,17].

The remainder of the paper is organized as follows: Section 2 presents the related work, Section 3 presents our problem formulation of few-shot intrusion detection using discriminative autoencoders, Section 4 presents the results and discussion of our experiments, and Section 5 concludes the paper.

2. Related Works

Traditionally, networks are defended against intrusion using signature-based techniques, whereby incoming network traffic is compared against commonly known attack patterns. These approaches perform well against previously known attacks, but fail to detect novel attacks.

Classical machine learning (ML) methods provide an upgrade over traditional signature-based techniques. These methods exploit various features of network traffic, which enable them to detect attack signatures without explicit rule specifications [18]. Thus, popular classical ML approaches, such as K-nearest neighbor (KNN) [19], support vector machines (SVM) [20], decision tree (DT) [21], and random forest (RF) [22], have all been employed as network-based IDSs.

For example, Kutrannont et al. [23] proposed a KNN-based IDS. KNN operates based on the assumption that a sample belongs to the class where most of its top K-neighbors reside. Therefore, parameter K affects the performance of the model. In their work, Kutrannont et al. proposed the integration of a simplified neighborhood classification using a percentage instead of group rankings. Taking into account the unevenness of data distribution, the improve rule selects a fixed percentage (50%) of neighboring samples as neighbors and its efficiency is enhanced via parallel processing using a graphical processing unit (GPU). The algorithm performs well on sparse data, achieving an accuracy of 99.30%.

Goeschel et al. [24] employed a combination of SVM, decision tree (DT), and naïve Bayes classifiers. The SVM was first trained to perform a binary classification to separate data instances into benign and malicious classes. The malicious classes are then categorized into specific classes of attacks using a DT classifier. However, since DT can only separate known classes of attacks, they further employed a naïve Bayes classifier to identify unknown attacks types. This hybrid method achieved an accuracy of 99.62% and a false alarm rate of 1.57%.

Malik et al. [25] proposed an IDS using random forest (RF) and particle swarm optimization (PSO). They trained the IDS in two stages: feature selection and classification. The PSO serves as feature selection algorithm, which is used to select appropriate features for classifying attacks, while the RF is used as a classifier. They evaluated their approach using the KDD cup99 dataset, and achieved detection rates of 99.92%, 99.49%, and 88.46% on DoS, Probe, and U2R attack classes.

Recently, there has been widespread adoption of DL techniques for network-based IDSs for sizeable numbers of datasets. These techniques can operate directly on raw data, learn features, and perform classifications. Hence, they achieve better performances when compared to classical machine learning methods [26]. Deep learning models, such as multi-layer perceptron (MLP), convolutional neural network (CNN), autoencoders (AE), recurrent neural network (RNN), as well as deep generative networks, such as the deep belief network (DBN) and generative adversarial networks (GANs) have all been applied in the context of network-intrusion detection [27,28].

Min et al. [29] proposed an IDS named TR-IDS, which leverages both statistical features as well as payload features. They employed a CNN to extract important features from the payload. To accomplish this, they first encoded each byte in the payload in to a word vector using skip-gram word embedding, and then applied the CNN to extract the features. The extracted features were then combined with the statistical features generated from each network flow, which included fields from the packet header and statistical attributes of the entire flow. The features were then used to train a random forest classifier, which achieves an accuracy of 99.13%.

In the work by Yin et al. [30], a recurrent neural network (RNN) was directly applied for intrusion detection tasks. The RNN model achieved better performance on a NSL-KDD dataset when compared with classical ML techniques consisting of support vector machines and random forest.

Wang et al. employed a combination of CNN and long short-term memory (LSTM). Intuitively, the CNN learns the low-level spatial features of network traffic, while the LSTM learns the high-level temporal features of the data. The learned features enable the model to improve the false alarm rate of an IDS [31].

In another work, Al Qatf et al. employed a sparse autoencoder (AE) for dimensionality reduction and the reduced features are then retrained using the SVM classifier. This enables the model to outperform classical machine learning methods [32].

Similarly, in a recent work by Narayana et al., a hybrid methodology involving a sparse autoencoder, DNN, and LSTM was employed. In the first stage, the autoencoder is trained in an unsupervised fashion with smoothed l1 regularization to enforce sparsity. This enables the autoencoder to learn sparse representations, which are then used to train the MLP and LSTM classifiers in the second stage. The model performs better than conventional deep learning classifiers in terms of detection rates and low false positive rates [33].

Another hybrid intrusion detection method that employs both classical machine learning and deep learning techniques was proposed by Le, et al. They first built a feature selection model termed a sequence forward selection (SFS) algorithm (SFSD) and a decision tree. The SFSD algorithm selects the best subset of features, which are then used in the second part to train various forms of RNN (traditional RNN, LSTM and gated recurrent neural network (GRU)). The model achieves significant improvements in detection rates when compared with classical methods [34].

However, these techniques require huge amounts of labeled data during training in order to generalize well. The dynamic nature of the modern-day cyber-threat landscape makes it unfeasible or prohibitively expensive to acquire sufficient enough malicious samples to train deep learning classifiers. Therefore, a trend is developing towards techniques that require only a few shot of malicious examples to achieve detection.

For example, Hindy et al. proposed an intrusion detection model using one-shot learning. The main idea of one-shot learning is to learn patterns and similarities from previously seen classes that enable classifying unseen classes using only one instance. Thus, one-shot learning is an instance of few-shot learning, whereby the number of examples is restricted to only a single example [35]. To model an IDS using one-shot learning, Hindy et al. employed a Siamese neural network, a form of neural network consisting of twin networks. The Siamese network is trained using two pairs of instances to learn patterns and similarities instead of fitting the model to fixed classes. Therefore, during the training stage, the Siamese network learns patterns and discriminate between benign traffic and different classes of a known cyber-attacks. At the evaluation stage, a new traffic instance is compared against all known classes (used during training) without any form of additional training. Although the approach provides a simple framework for one-shot learning, in general, they achieve lower detection rates relative to other works [36].

In another work, Xu et al. proposed an intrusion detection method using few-shot learning. They employed a deep neural network architecture (DNN) named FC-Net, which is composed of two parts: a feature extraction network and a comparison network. FC-Net is trained using a meta-learning approach consisting of two disjointed stages of meta-training and meta-testing. In the meta-training phase, the feature extraction network of FC-Net is trained using several meta-tasks, where a meta-task is comprised of a binary classification between an attack category and benign traffic. This enables FC-Net to learn a pair of feature maps, which are then used by the comparison network in the meta-testing stage to determine whether a new traffic instance belongs to the different classes of attacks learned during training [37]. However, one drawback with their approach is it requires a complex DNN architecture and computationally intensive optimization procedures.

3. Our Proposed Few-Shot Intrusion Detection Method

Supervised learning approaches for network intrusion detection require all categories of attacks to be known in advance, with a sufficient number of training examples available for each category. The basic task is to use a classifier, f, to infer labels for network traffic samples, N. The number of samples, N, is often very large and is simply composed of two groups: a training set and a test set. Contrary to this, in real-world settings, new attacks frequently emerge, and only subset of categories are known beforehand, with few examples per category. Therefore, in such scenarios, where the number of samples, N, is small, the problem is considered as a few-shot classification. Applying the conditions of a supervised learning method to this problem will encounter overfitting.

Few-shot learning is popularly addressed based on the meta-learning paradigm, which is composed of meta-training and meta-testing. Each one of the meta stages consists of a number of classification tasks, where each task describes a pair: training (support) and testing (query). The meta-training set is described as $T = \left\{ \left(D_i^{train}, D_i^{test} \right) \right\}_{i=1}^{I}$ and the meta-testing set is $S = \left\{ \left(D_q^{train}, D_q^{test} \right) \right\}_{q=1}^{Q}$, with each dataset containing pairs of data points and their ground-truth labels, i.e., $D^{train} = \left\{ (x_t, y_t) \right\}_{t=1}^{T}$ and $D^{test} = \left\{ (x_q, y_q) \right\}_{q=1}^{Q}$, which are sampled from the same distribution. The objective is to leverage the meta-training stage to learn good representations, which will enable it to adapt quickly to unseen tasks in the meta-testing stage, using powerful optimization techniques.

In a network intrusion detection context, a task, T, can be simply defined as a binary classification between a normal network traffic sample and a category of malicious samples. Supposing that there are five different network traffic samples, O, A, B, C and E such that, sample O is a benign network traffic, samples A, B, C indicate known categories of attacks with sufficient examples, while the remaining sample, E, refers to a newly found category of attack with a few examples. The goal is to identify the new attack sample, S, with as few examples as possible. Then, three different tasks can be constructed, T_1, T_2 and T_3 where T_1, T_2 and T_3 define a binary classification task between a normal sample O and attack categories A, B and C. T_1, T_2 and T_3 constitute the meta-training set, while the meta-test set consists of a normal sample, O, and the novel class, E, which has few examples. The idea is to leverage the meta-training stage to learn transferable knowledge from T_1, T_2 and T_3 that will enable a classifier to accomplish task T_4 (a binary classification between normal sample O and attack category E) with as few examples as possible during the meta-testing phase. Thus, in our case, a discriminative autoencoder was employed to acquire such transferable knowledge.

Feature Extraction with Discriminative Autoencoder

Autoencoders have been proved to be powerful models for learning representations in an unsupervised fashion. However, discriminative autoencoders are a form of autoencoders that, in addition to residual errors, considers class information of the input in its objective function. This ensures that more powerful and discriminative representations are learned than those learned by conventional autoencoders.

We adopted the discriminative autoencoder proposed in [38], which, in its setup, uses data from two distributions, termed positive (X^+) and negative (X^-), with their labeled information. The discriminative autoencoder then learns a manifold that is good at reconstructing the data from the positive distribution, while ensuring that those of the negative distributions are pushed away from the manifold. This enables it to learn robust patterns and similarities that separate the two distributions.

In our case, the two distributions, X^+ and X^-, can be generated from benign network traffic classes and malicious traffic classes.

Let $l(x)$ denote the label of an example, x, with $l(x) \in \{-1, 1\}$ and $d(x)$ is the distance of that example to the manifold, with $d(x) = \|x - \bar{x}\|$. Then, the loss function is described as:

$$L(X^+ \cup X^-) = \sum_{x \in X^+ \cup X^-} \max(0, l(x) \cdot (d(x) - 1)) \quad (1)$$

Thus, to train the discriminative autoencoder, we merged all the meta-training tasks D_t^{train} from T into a single training set, D^{new}, of seen classes:

$$D^{new} = \cup \left\{ D_1^{train}, ..., D_t^{train}, ..., D_T^{train} \right\} \quad (2)$$

We trained the discriminative autoencoder during the meta-training stage (Algorithm 1). After training, the decoder part of the model was discarded, while the encoder module, which then served as our feature extractor was retained. The encoder was then employed in a fixed state (no fine-tuning) in the meta testing stage. The meta testing stage consists of the task of identifying a novel class of attack, which has few examples.

For a given task $\left(D_q^{train}, D_q^{test} \right)$ sampled from the meta-testing set, S, we trained a classifier, f, on top of the extracted features to recognize the unseen classes using the training dataset, D_q^{train} (Algorithm 2).

Algorithm 1 Discriminative Autoencoder Training

Input: meta-training dataset D containing, n normal data samples and m malicious samples, $l(x)$ the label of the dataset with $l(x) \in \{1, 0\}$,
Output: encoder f_e and a decoder f_d
$\theta_e \leftarrow$ initialize encoder parameters
$\theta_d \leftarrow$ initialize decoder parameters
Repeat
for $i = 1$ to k **do**
Draw a batch of k samples $x^{(1)}, \ldots, x^{(k)}$ from the dataset D
$z^i = f_e(x^i)$
$\hat{x}^i = f_d(z^i)$
$L_{DAE} = \frac{1}{k}\sum_{i=1}^{k} \max\left(0, l(x) \cdot \left(\| x^i - \hat{x}^i \| - 1\right)\right)$
end for
// update parameters with gradients
$\theta_e \leftarrow \theta_e - \nabla_{\theta_e} L_{DAE}$
$\theta_d \leftarrow \theta_d - \nabla_{\theta_d} L_{DAE}$
until convergence of parameters θ_e, θ_d

Algorithm 2 Few-Shot Detection

Input: meta-testing dataset D containing, n normal data samples and m few malicious samples with $n \gg m$, $l(x)$ the label of the dataset with $l(x) \in \{1, 0\}$, trained encoder f_e
Output: classifier c_l, prediction l_{pred}
$\theta_c \leftarrow$ initialize classifier parameters
Repeat
for $i =$ to k **do**
Draw a batch of k samples $x^{(1)}, \ldots, x^{(k)}$ from the dataset D
$f^i_{extract} = f_E(x^i)$
$l_{pred} = C_l(f^i_{extract})$
$L_C = binarycrossentropy(l, l_{pred})$
end for
// update parameters with gradients
$\theta_C \leftarrow \theta_C - \nabla_{\theta_C} L_C$
until convergence of parameters θ_C

4. Evaluation

4.1. Evaluation Metrics

Accuracy, recall (detection rate) and precision are common metrics normally used when evaluating a classifier. Accuracy, which indicates the percentage of correctly classified data items, is a poor metric for a task where the dataset is largely skewed in favor of normal samples. Similarly, a high detection rate implies fewer chances of a model missing alarming anomalies, while a high precision highlights the ability of a model to classify normal data. We consider detection rate (DR) to be more significant in our case than precision; for example, if a model predicts a normal sample as an attack, it is easier to correct the prediction of the model through domain knowledge, since there are few attack samples. However, if the model fails to detect an attack, it is difficult to find the attack in such a huge dataset. Nonetheless, we consider all three metrics; when comparing our approach with baseline models, we only considered DR since it is more significant in our situation.

4.2. Datasets

We evaluated the performance of our approach against baseline models using the following network intrusion detection datasets: CIC-IDS2017 and NSL-KDD. The CIC-IDS2017 dataset reflects recent attacks, and, to some extent, it satisfies the criteria for reliable intrusion detection datasets proposed by [39], which are anonymity, attack diversity, complete capture, complete interaction, complete network configuration, available protocol, complete traffic feature set, meta data, heterogeneity, and labeling.

The dataset was developed at the Canadian Institute of Cybersecurity of the University of New Brunswick (UNB) in 2017. The dataset comprises raw PCAP files, as well as 80 statistical features generated from the PCAP files, which were captured on different days of a week (from Monday to Friday). The dataset considers several attacks and sub-attacks, as depicted in Table 1.

Table 1. Attack composition of the CIC-IDS2017 datasets.

Attack Class	Subclasses	Number of Records
DoS	Dos Hulk	231,073
	Dos Slow loris	5796
	Dos Golden Eye	10,293
	DoS Http Test	5499
Web Attack	Brute force	1507
	XSS	652
	SQL injection	21
Ports Scan	-	158,930
DDoS	-	41,835
Botnet	-	1966
FTP-Patator	-	7938
SSH-Patator	-	5897
Infiltration	-	36
Benign	-	2,358,036

The NSL-KDD dataset was also generated at the Canadian Institute of Cybersecurity. This dataset was purposely created to solve the problem of the original KDDcup'99 dataset, which has about 78% and 75% of the training and testing set duplicated, respectively [40]. The NSL-KDD rectifies this problem and still retains the original 41 features.

Table 2 depicts a breakdown of some of the attacks that exist in the dataset with more than five samples.

Table 2. Attack composition of the NSL-KDD datasets (training set).

Attack Type	Subclasses	Number of Records
DoS	Neptune	41,214
	Smurf	2646
	Tear drop	892
	Back	956
Probe	Ip sweep	3599
	Nmap	1493
	Port sweep	2931
	Satan	3633
User to root (U2R)	Buffer overflow	30
	Load module	9
	rootkit	10
Remote login (R2L)	Guess password	53
	Warezmaster	890
	Imap	11
	Multihop	7

4.3. Implementation

We employed tensorflow 2.6 as the deep learning framework to implement our approach. We implemented the feature extractor module (discriminative autoencoder), and the classifier was a standard multi-layer perceptron (MLP) neural network. Table 3 present the architectural details of the modules.

Table 3. Architectural details of our models.

Model	Architectural Details	
Encoder	Layer 1	Dense, output: 128, activation: Relu
	Layer 2	Dense, output: 64, activation: Relu
	Layer 3	Dense, output: 32, activation: Relu
	Layer 4	Dense, output: 8, activation: Relu
Decoder	Layer 1	Dense, output: 32, activation: Relu
	Layer 2	Dense, output: 32, activation: Relu
	Layer 3	Dense, output: No, of features, activation: sigmoid
Classifier	Layer 1	Dense, output:1, activation: sigmoid

4.4. Experiments and Results

We conducted the following experiments to evaluate our few-shot detection approach.

4.4.1. Experiment 1: Detecting Mutants of Existing Attacks

The objective of this experiment was to evaluate our approach in detecting mutants of existing attacks. Malicious users develop mutants of existing attacks to evade detection, as conventional supervised learning models detect such mutants poorly, especially when there is significant variation with the known existing attack.

To accomplish this, we extracted categories of attacks, which we believed to comprise variants of one another in both the NSL-KDD and CIC-2017IDS datasets. Tables 1 and 2 depict the compositions of such attack mutants.

To evaluate our approach in detecting a particular mutant, we utilized all the other mutants of the attack as a data source for the meta-training stage. For instance, if the attack mutant we wanted to detect was DoS hulk, then our meta-training data source will comprise all the other DoS subclasses (golden eye, slow loris, slow http test). The DoS hulk would then serve as the data source for the meta-testing stage. We then trained our feature extractor using the meta-training set. Thereafter, in the few-shot detection stage, we selected 10 samples of the meta-testing set and trained our classifier on top of our feature extractor. We applied the same logic to all attacks in both the NSL-KDD and CIC-IDS2017 datasets. However, we conducted this experiment on classes with sufficient numbers of samples. Therefore, on the NSL-KDD set we were able to perform the experiment on DoS and probe classes. While on the CIC-IDS2017 dataset, only the DoS class has a sufficient number of samples.

Figures 1–3 display the results of our experiment on both the CIC-IDS2017 and NSL-KDD datasets. As can be seen from the results, our model achieves the highest detection rate of 99.9% on DoS attack mutant Neptune (Figure 1), followed by 90.0% and 80.0% on tear drop and smurf attack, respectively. These are good results, considering that we employed few-shots examples to train our classifier. However, our model achieves a DR of 0.0% on DoS attack mutant back. In fact, our model scores 0.0% on all the other performance metrics (accuracy and precision). This indicates that the DoS attack mutant back has significant variations with other DoS attack subclasses.

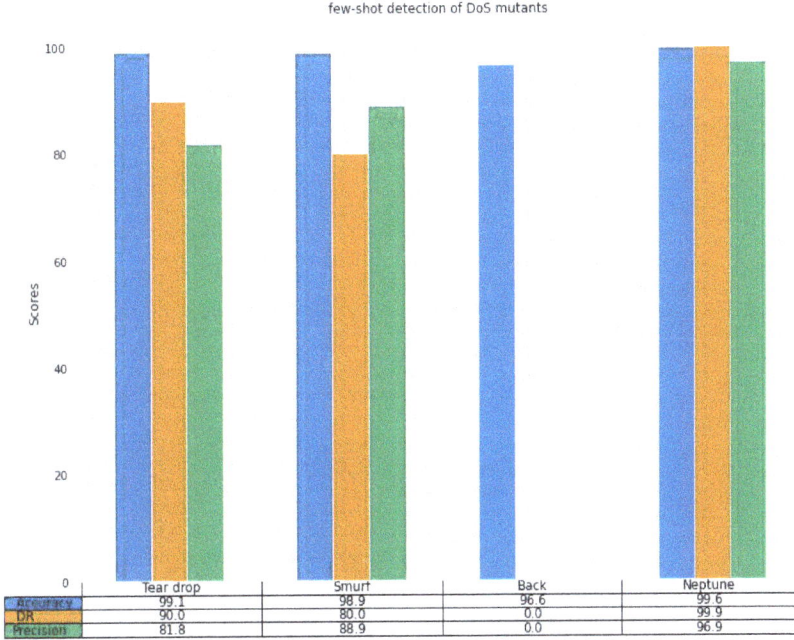

Figure 1. Results of few-shot detection of DoS attack mutant for NSL-KDD dataset.

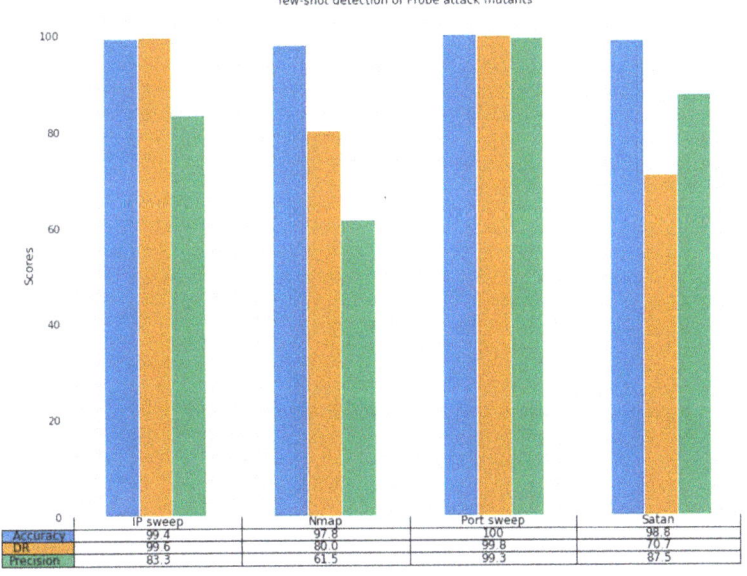

Figure 2. Result of few-shot detection of probe attack mutants for NSL-KDD dataset.

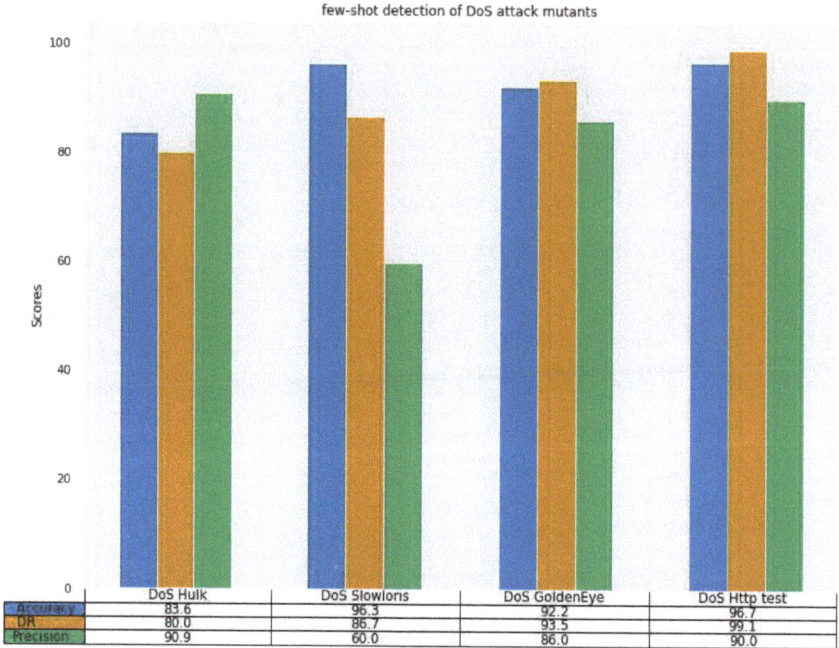

Figure 3. Result of few-shot detection on DoS attack mutant for CIC-IDS2017 dataset.

For the probe attack mutants (Figure 2), our model also performs well, achieving the highest DR score of 99.8% on the port sweep attack, and a DR score of 99.6%, 80.0%, and 70.7% for the port sweep, Nmap and Satan, respectively. All the attack mutants achieved a good DR, which indicated that the attack mutants had many attributes in common, which our feature extractor model was able to discover.

Similarly, with the CIC-IDS2017 dataset, our model's performance was good. We achieved the highest DR score of 99.1% on DoS http test attack (Figure 3), and DR scores of 93.5%, 86.7%, and 80.0% on DoS golden eye, DoS slow loris and DoS hulk attacks, respectively. The results follow a similar trend as in the NSL-KDD dataset.

We also compared our model's DR with baseline models, and we employed a combination of classical and deep learning algorithms as our baselines. For deep learning, we employed a multi-layer perceptron (MLP), while for the classical ML models we employed K-nearest neighbor (KNN), decision tree, support vector machine (SVM) and random forest. All the baseline models were trained using the full dataset.

Figures 4–6 depict the results of such comparisons. As can be seen from results, there was not much of a difference in terms of performance between the DL models, ML models, and our model. It is well known that classical ML models perform well on small and medium datasets. The highest DR score was 100% on the Neptune DoS mutant (Figure 4), which was achieved by all baselines, while our model scored 99.6% (despite having been trained with only 10 examples), which is quite on par with the baselines results. The lowest DR score was for the back attack mutant, where our model's DR score as 0.0%. However, the lowest DR score from the baselines was achieved by SVM, which scored 40.4% on the tear drop attack.

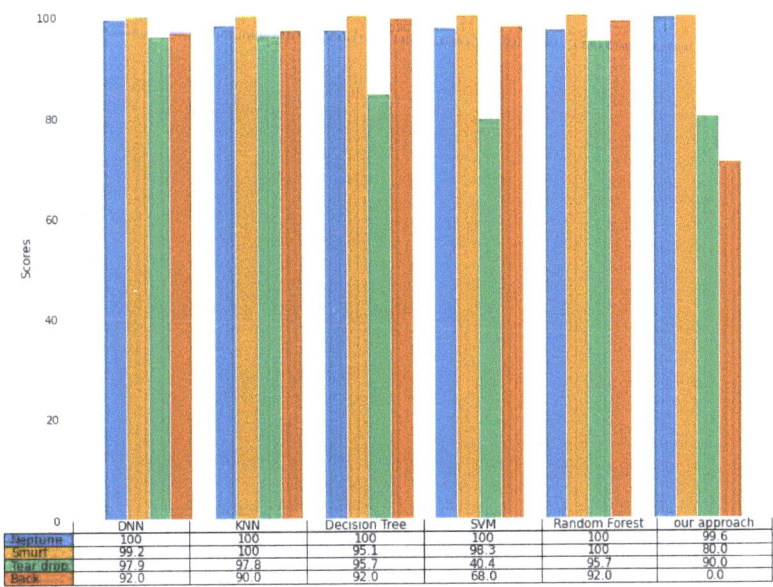

Figure 4. Detection rate (DR) comparison of various classification methods on DoS subclasses for NSL-KDD datasets.

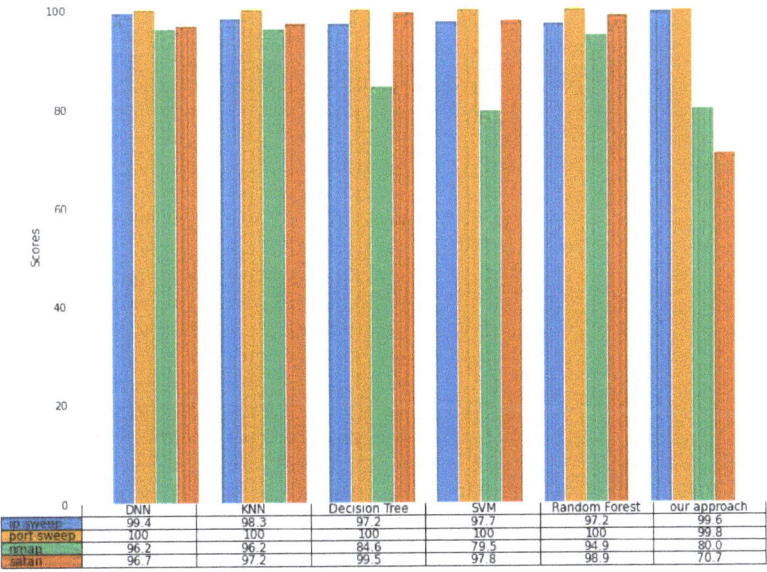

Figure 5. Detection rate (DR) comparison of various classification methods on probe attack subclasses for NSL-KDD datasets.

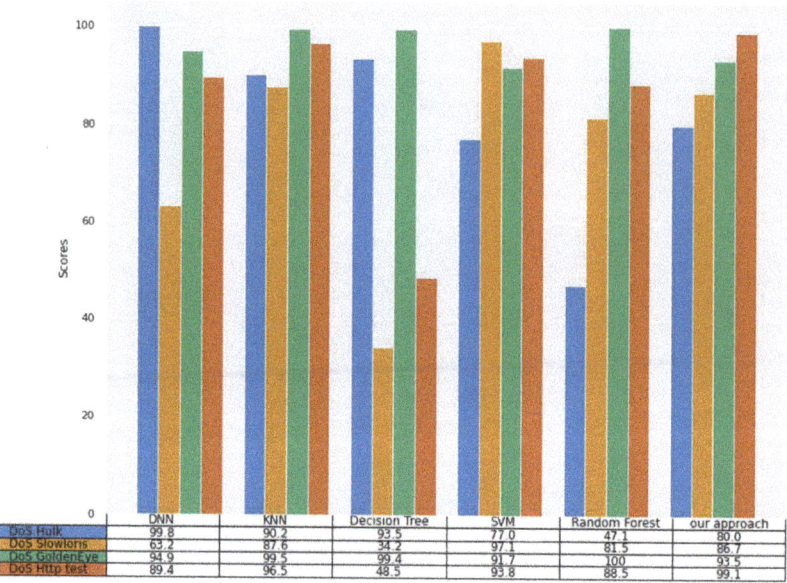

Figure 6. Detection rate (DR) comparison of various classification methods on DoS subclasses for CIC-IDS2017 datasets.

In addition, for the probe attack class, the result followed similar trend, the highest DR score achieved was 100%, which was achieved by the baseline models on the port sweep attack mutant (Figure 5), while our model was slightly worse with a DR score of 99.8% on the same attack class.

Our model also recorded the lowest DR score, 70.7%, on the Satan attack mutant, while the lowest DR score achieved by the baselines was 79.5%, which was scored by SVM on the Nmap attack mutant. Overall, there was no significant difference in performance with our model; our model was either slightly worse or on par with the baselines.

Similarly, Figure 6 presents the results of the comparison with baselines models for the CIC-IDS2017 dataset. The highest performing model was random forest, which achieved a DR of 100% for the DoS golden eye attack mutant. The lowest DR score was 48.5% which was achieved by the decision tree model for the DoS http test attack mutant. Similar to the NSL-KDD dataset, our model performed competitively with the baseline models, where it achieved DRs of 99.6%, 99.8%, 80.0%, and 70.7% on IP sweep, port sweep, Nmap and Satan, respectively.

4.4.2. Experiment: General Anomaly Detection

The objective here was to evaluate our approach in detecting broader classes of attacks that were not seen before. For example, the NSL-KDD dataset can be broadly categorized into the following classes: DoS, Probe, R2L and U2R. Similarly, the CIC-IDS2017 dataset, when classified broadly, is made up of the following classes: DoS, port scan, heart bleed, Botnet, SSH-Patator, FTP-Patator, and web-based attacks

To perform the experiment, we applied a similar logic as in experiment 1. For instance, to detect the probe class of attack in the NSL-KDD dataset, our meta-training set consisted of all other attack classes (DoS, R2L, and U2R). While the meta-testing set comprised the attack we wanted to classify, which was the probe attack in this case. We applied the same procedure as for the CIC-IDS2017 dataset.

Figures 7 and 8 present the results of our experiments when the number of samples selected to train our classifier was limited 10 for both NSL-KDD and CIC-IDS2017, respectively.

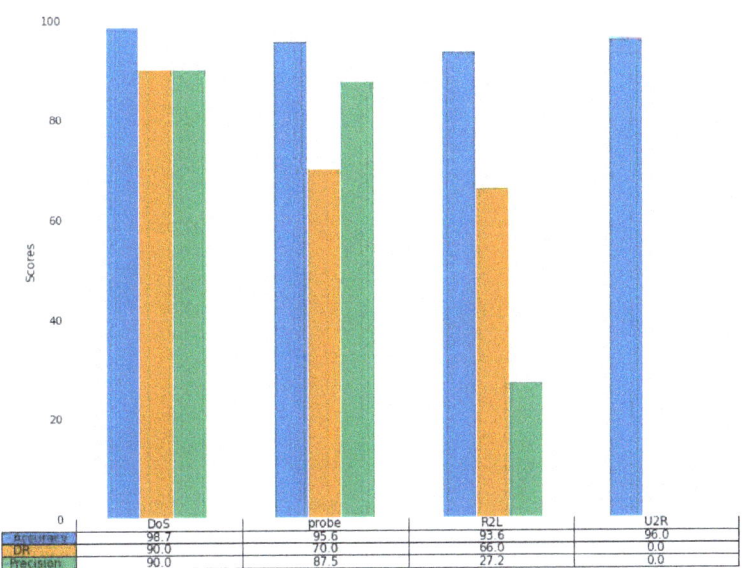

Figure 7. Result of experiment 2: general anomaly detection for NSL-KDD dataset.

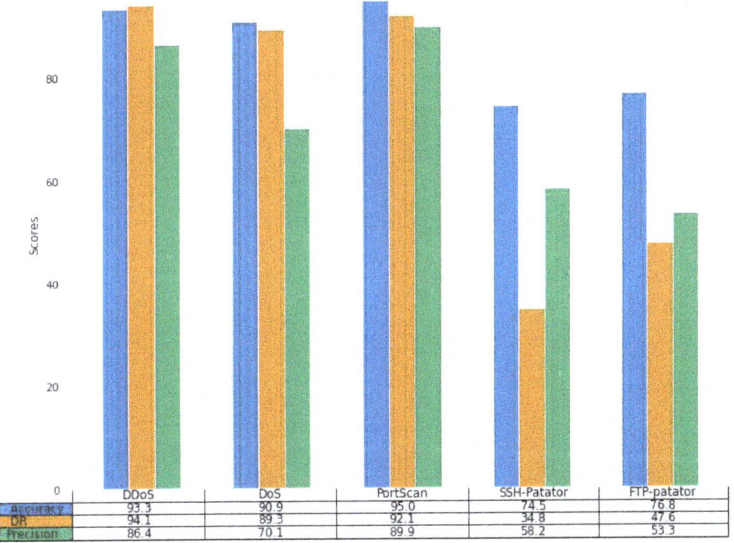

Figure 8. Result of experiment 2: general anomaly detection for CIC-IDS2017 dataset.

For the NSL-KDD dataset, (Figure 7), our approach performed reasonably well, especially for DoS and probe classes. Our model achieved the highest DR score of 90% with the DoS class and 70% on probe class. However, overall, our model's performance was low compared to the results of experiment 1. Some attacks were poorly detected. For example, our model's DR scores was 66% for the R2L class, and it completely failed to detect the U2R attack class, scoring a DR of 0.0%.

Similarly, the result of our experiment on CIC-IDS2017 (Figure 8) follows similar trend. There is drop in performance, our model performance was not as good as that of experiment 1. The highest DR score was 94.1%, which was achieved for the DDoS attack class. Our model poorly detects the Patator attack classes, scoring 34.8% and 47.6%, respectively.

4.5. Discussion

Section 4.4 presents a performance evaluation of our approach. We designed two different experiments to evaluate our approach. The first experiment was designed to evaluate detecting mutants of existing attacks (subclasses of attacks). This was designed to addresses situations where attackers introduce some variations of a known existing attack to evade detection. We hypothesized that finding good representations using existing known similar attacks will enable the model to detect a novel mutant of that attack by using a few examples since attack mutants share similar attributes. Thus, our feature extractor would be able to discover the manifold of a broader class of such attacks.

Figures 1–3 in Section 4.4 present the results of our experiment, which prove our hypothesis. As we were able to detect several mutants with an excellent detection rate using only a few shot examples. We achieved up to 99.9% DR on the Neptune DoS mutant (Figure 1), and, overall, we achieved an excellent detection rate. This shows that our feature extractor was able to discover the manifold of optimal representations of such a class.

To further validate the efficacy of our feature extractor model, we increased the number of samples required to train our classifier in the few-shot detection stage from 10 to 20 samples. However, as can be seen in Tables 4–6, respectively, the DR varies slightly, despite doubling the number of samples. This is contrary to conventional supervised learning approaches, where the DR increases significantly with an increase in the number of training examples.

Table 4. Effect of increasing the number of samples on DoS attacks in the NSL-KDD dataset.

Attack Type	No. of Samples = 10 DR (%)	No. Samples = 20 DR (%)
Neptune	99.9	100
Smurf	80.0	81.9
Tear drop	90.0	93.5
Back	0.0	0.0

Table 5. Effect of increasing the number of samples on probe attacks in the NSL-KDD dataset.

Attack Type	No. of Samples = 10 DR (%)	No. Samples = 20 DR (%)
IP sweep	99.6	99.9
Port sweep	99.8	100
Nmap	80.0	82.0
Satan	70.7	71.3

Table 6. Effect of increasing the number of samples on DoS attacks in the CIC-IDS2017 dataset.

Attack Type	No. of Samples = 10 DR (%)	No. Samples = 20 DR (%)
DoS hulk	80.0	81.1
DoS golden eye	93.5	95.0
DoS http test	99.1	99.8
DoS slow loris	86.7	88.0

The more powerful the representations, the smaller the number of training samples required. This concludes that our feature extractor model is able to discover the manifold of optimal representations, which causes the DR to slightly depends on the number of examples.

We also compared our approach with the baseline models, which comprised the DL model and the classical ML, as shown in Figures 4–6. The baseline models were trained on the full dataset.

As can be observed from the results, our approach performed competitively with the baseline models. On all the datasets, our model was on par with, or slightly worse than, the baseline models, despite being trained on few-shot examples.

The second experiment was designed to discover whether it was possible to learn representations that would be useful in detecting any class of attack, when re-trained with few-shot examples of that attack. The results of the experiment are presented in Figures 6 and 7. Our model performed reasonably well in detecting certain classes of attacks. For instance, our model achieved detection rates of 90% and 89% for DoS classes in both the NSL-KDD and CIC-IDS2017 datasets. This shows that our feature extractor still learns some useful representations to detect these classes of attacks. However, it performs poorly in detecting other classes, such as R2L and U2R, from NSL-KDD (Figure 7), where the detection rate was 66% and 0.0%, respectively. Similarly, it failed to achieve good detection rates for both FTP-Patator and SSH-Patator attack classes for the CIC-IDS2017 dataset (Figure 8), scoring 34.8% and 47.6%, respectively. Therefore, as expected, there was a drop in overall performance for our model compared to the results of experiment 1. This was due to the fact that our feature extractor tries to discover a singular representation for identification, based on the different classes of attacks observed in the meta-training stage. However, it is difficult to discover such representations due to the diverse nature of the attacks in the meta-training set. Since attacks differ in purpose and implementation, for instance, the DoS attack class tries to shut down traffic flow to and from a target system, a U2R attack tries to gain access to the system or network, an R2L attack tries to gain access to the remote machine, while attacks such as probe try to get information from a network, we assumed that these attack types are too diverse to allow our feature extractor to learn singular representations for identification that can enable excellent detection when trained with a few examples.

5. Conclusions

Network intrusion detection using machine learning methods has been studied for a long time, with many commercial intrusion detection systems (IDSs) using machine learning algorithms as part of their detection engines. However, machine learning-based IDSs are susceptible to false alarm rates, which makes the field an active area of research.

Recently, DL methods have been widely applied in network-based IDSs due to their success in fields such as natural language processing (NLP) and computer vision. However, to achieve a better detection rate, DL methods require sizeable volumes of datasets. Collecting large-scale datasets is non-trivial, especially in the cybersecurity domain where the landscape is constantly changing. Hence, few-shot network intrusion detection is emerging as an alternative to conventional supervised DL methods. The concept is popularly addressed based on a meta-learning paradigm, whereby transferable knowledge is learned in some related tasks using complex optimization techniques, which enables generalization at test time with limited examples.

However, in this paper, we propose a simple framework for few-shot network intrusion detection. Our approach relies on learning powerful representations, and is implemented in two stages. We first train a feature extractor model using discriminative representation learning with a supervised autoencoder, and we then train a classifier on top of the feature extractor, which is able to generalize with a few examples.

To validate our approach, we evaluated our model using two publicly available intrusion detection datasets. Our proposed method achieved excellent detection rates in detecting mutants of existing attacks. However, though our approach achieves good detection rates for certain classes of attacks in the general anomaly detection scenario it performs poorly for others. This is due to the diverse nature of attacks, and it is difficult to learn singular representations that can enable generalization with only a few examples.

Therefore, based on the results of the experiments conducted, our approach is more suited for detecting specific classes of attack or mutants of an existing attack. In addition, it is safe to say that our model can be used in situations like zero-day attacks, since, even in such a scenario, a few samples of attacks can be obtained, which will be sufficient enough to train our model to detect similar occurrences of the same attack or its variants in the future.

Author Contributions: Conceptualization, A.S.I. and U.A.A.; methodology, A.S.I. and U.A.A.; software, A.S.I.; validation, U.A.A. and L.Z.; formal analysis, L.Z.; investigation, L.Z.; resources, L.Z.; data curation, U.A.A.; writing-original draft preparation, A.S.I.; writing-review and editing, U.A.A.; visualization, A.S.I.; supervision, L.Z.; project administration, L.Z., funding acquisition, L.Z. All authors have read and agreed to the published version of the manuscript.

Funding: This research was supported by the China Scholarship Council (CSC) 2018GXZ021733.

Institutional Review Board Statement: Not applicable.

Informed Consent Statement: Not applicable.

Data Availability Statement: The datasets used in this work were collected by the Canadian Institute of Cybersecurity (CIC) and are publicly available at https://www.unb.ca/cic/datasets, accessed on 24 October 2021.

Conflicts of Interest: The authors declare no conflict of interest.

References

1. Scarfone, K.A.; Mell, P.M. *Guide to Intrusion Detection and Prevention Systems (IDPS)*; NIST Special Publication 800–94; National Institute of Standards and Technology: Gaithersburg, MD, USA, 2007. [CrossRef]
2. Kruegel, C.; Toth, T. Using Decision Trees to Improve Signature-Based Intrusion Detection. In *Recent Advances in Intrusion Detection*; Vigna, G., Kruegel, C., Jonsson, E., Eds.; Springer: Berlin/Heidelberg, Germany, 2003; Volume 2820, pp. 173–191. [CrossRef]
3. Mell, P.M.; Hu, V.; Lippmann, R.; Haines, J.; Zissman, M. *An Overview of Issues in Testing Intrusion Detection Systems*; NIST Interagency/Internal Report (NISTIR)—7007; National Institute of Standards and Technology: Gaithersburg, MD, USA, 2003. [CrossRef]
4. LeCun, Y.; Bengio, Y.; Hinton, G. Deep learning. *Nature* **2015**, *521*, 436–444. [CrossRef] [PubMed]
5. Gamage, S.; Samarabandu, J. Deep learning methods in network intrusion detection: A survey and an objective comparison. *J. Netw. Comput. Appl.* **2020**, *169*, 102767. [CrossRef]
6. Ziai, A. Active Learning for Network Intrusion Detection. In *Data Science. Theory, Algorithms, and Applications*; Verma, G.K., Badal, S., Bourennane, S., Ramos, A.C.B., Eds.; Springer: Cham, Switzerland, 2021; p. 11.
7. Wang, Y.; Yao, Q.; Kwok, J.; Ni, L.M. Generalizing from a Few Examples: A Survey on Few-Shot Learning. *arXiv* **2020**, arXiv:1904.05046. Available online: http://arxiv.org/abs/1904.05046 (accessed on 4 December 2021). [CrossRef]
8. Finn, C.; Abbeel, P.; Levine, S. Model-Agnostic Meta-Learning for Fast Adaptation of Deep Networks. *arXiv* **2017**, arXiv:170303400. Available online: http://arxiv.org/abs/1703.03400 (accessed on 4 December 2021).
9. Bertinetto, L.; Henriques, J.F.; Torr, P.H.S.; Vedaldi, A. Meta-Learning with Differentiable Closed-Form Solvers. *arXiv* **2019**, arXiv:1805.08136. Available online: http://arxiv.org/abs/1805.08136 (accessed on 4 December 2021).
10. Wang, Y.-X.; Hebert, M. Learning to Learn: Model Regression Networks for Easy Small Sample Learning. In *Computer Vision—ECCV 2016*; Leibe, B., Matas, J., Sebe, N., Welling, M., Eds.; Springer International Publishing: Cham, Switzerland, 2016; Volume 9910, pp. 616–634. [CrossRef]
11. Wang, Y.-X.; Hebert, M. Learning from Small Sample Sets by Combining Unsupervised Meta-Training with CNNs. In Proceedings of the 30th Conference on Neural Information Processing Systems, Barcelona, Spain, 5–10 December 2016; 9p.
12. Li, Z.; Zhou, F.; Chen, F.; Li, H. Meta-SGD: Learning to Learn Quickly for Few-Shot Learning. *arXiv* **2017**, arXiv:170709835. Available online: http://arxiv.org/abs/1707.09835 (accessed on 6 December 2021).
13. Dhillon, G.S.; Chaudhari, P.; Ravichandran, A.; Soatto, S. A Baseline for Few-Shot Image Classification. *arXiv* **2020**, arXiv:190902729. Available online: http://arxiv.org/abs/1909.02729 (accessed on 6 December 2021).
14. Tian, Y.; Wang, Y.; Krishnan, D.; Tenenbaum, J.B.; Isola, P. Rethinking Few-Shot Image Classification: A Good Embedding Is All You Need? *arXiv* **2020**, arXiv:200311539. Available online: http://arxiv.org/abs/2003.11539 (accessed on 6 December 2021).

15. Ouali, Y.; Hudelot, C.; Tami, M. Spatial Contrastive Learning for Few-Shot Classification. *arXiv* **2021**, arXiv:201213831. Available online: http://arxiv.org/abs/2012.13831 (accessed on 4 December 2021).
16. Gogna, A.; Majumdar, A. Discriminative Autoencoder for Feature Extraction: Application to Character Recognition. *Neural Process. Lett.* **2019**, *49*, 1723–1735. [CrossRef]
17. Du, F.; Zhang, J.; Ji, N.; Hu, J.; Zhang, C. Discriminative Representation Learning with Supervised Auto-encoder. *Neural Process. Lett.* **2019**, *49*, 507–520. [CrossRef]
18. Tsai, C.-F.; Hsu, Y.-F.; Lin, C.-Y.; Lin, W.-Y. Intrusion detection by machine learning: A review. *Expert Syst. Appl.* **2009**, *36*, 11994–12000. [CrossRef]
19. Liao, Y.; Vemuri, V.R. Use of K-Nearest Neighbor classifier for intrusion detection. *Comput. Secur.* **2002**, *21*, 439–448. [CrossRef]
20. Li, W.; Liu, Z. A method of SVM with Normalization in Intrusion Detection. *Procedia Environ. Sci.* **2011**, *11*, 256–262. [CrossRef]
21. Kumar, M.; Hanumanthappa, M.; Kumar, T.V.S. Intrusion Detection System using decision tree algorithm. In Proceedings of the IEEE 14th International Conference on Communication Technology, Chengdu, China, 9–11 November 2012; pp. 629–634. [CrossRef]
22. Farnaaz, N.; Jabbar, M.A. Random Forest Modeling for Network Intrusion Detection System. *Procedia Comput. Sci.* **2016**, *89*, 213–217. [CrossRef]
23. Kuttranont, P.; Boonprakob, K.; Phaudphut, C.; Permpol, S.; Aimtongkhamand, P.; KoKaew, U.; Waikham, B.; So-In, C. Parallel KNN and Neighborhood Classification Implementations on GPU for Network Intrusion Detection. *J. Telecommun. Electron. Comput. Eng.* **2017**, *9*, 29–33.
24. Goeschel, K. Reducing false positives in intrusion detection systems using data-mining techniques utilizing support vector machines, decision trees, and naive Bayes for off-line analysis. In Proceedings of the IEEE SoutheastCon 2016, Norfolk, VA, USA, 30 March–3 April 2016; pp. 1–6. [CrossRef]
25. Malik, A.J.; Shahzad, W.; Khan, F.A. Network intrusion detection using hybrid binary PSO and random forests algorithm: Network intrusion detection using hybrid binary PSO. *Secur. Commun. Netw.* **2015**, *8*, 2646–2660. [CrossRef]
26. Liu, H.; Lang, B. Machine Learning and Deep Learning Methods for Intrusion Detection Systems: A Survey. *Appl. Sci.* **2019**, *9*, 4396. [CrossRef]
27. Alom, M.Z.; Bontupalli, V.; Taha, T.M. Intrusion detection using deep belief networks. In Proceedings of the 2015 National Aerospace and Electronics Conference (NAECON), Dayton, OH, USA, 15–19 June 2015; pp. 339–344. [CrossRef]
28. Chen, H.; Jiang, L. Efficient GAN-based method for cyber-intrusion detection. *ArXiv* **2019**, ArXiv:190402426. Available online: http://arxiv.org/abs/1904.02426 (accessed on 31 May 2021).
29. Min, E.; Long, J.; Liu, Q.; Cui, J.; Chen, W. TR-IDS: Anomaly-Based Intrusion Detection through Text-Convolutional Neural Network and Random Forest. *Secur. Commun. Netw.* **2018**, *2018*, 4943509. [CrossRef]
30. Yin, C.; Zhu, Y.; Fei, J.; He, X. A Deep Learning Approach for Intrusion Detection Using Recurrent Neural Networks. *IEEE Access* **2017**, *5*, 21954–21961. [CrossRef]
31. Wang, W.; Sheng, Y.; Wang, J.; Zeng, X.; Ye, X.; Huang, Y.; Zhu, M. HAST-IDS: Learning Hierarchical Spatial-Temporal Features Using Deep Neural Networks to Improve Intrusion Detection. *IEEE Access* **2018**, *6*, 1792–1806. [CrossRef]
32. Al-Qatf, M.; Lasheng, Y.; Al-Habib, M.; Al-Sabahi, K. Deep Learning Approach Combining Sparse Autoencoder With SVM for Network Intrusion Detection. *IEEE Access* **2018**, *6*, 52843–52856. [CrossRef]
33. Rao, K.N.; Rao, K.V.; Prasad Reddyand, P.V.G.D. A hybrid Intrusion Detection System based on Sparse autoencoder and Deep Neural Network. *Comput. Commun.* **2021**, *180*, 77–88. [CrossRef]
34. Le, T.-T.-H.; Kim, Y.; Kim, H. Network Intrusion Detection Based on Novel Feature Selection Model and Various Recurrent Neural Networks. *Appl. Sci.* **2019**, *9*, 1392. [CrossRef]
35. Bertinetto, L.; Henriques, J.F.; Valmadre, J.; Torr, P.; Vedaldi, A. Learning feed-forward one-shot learners. In Proceedings of the 30th Conference on Neural Information Processing Systems (NIPS 2016), Barcelona, Spain, 5–10 December 2016; 9p.
36. Hindy, H.; Tachtatzis, C.; Atkinson, R.; Brosset, D.; Bures, M.; Andonovic, I.; Michie, C.; Bellekens, X. Leveraging Siamese Networks for One-Shot Intrusion Detection Model. *arXiv* **2021**, arXiv:200615343. Available online: http://arxiv.org/abs/2006.15343 (accessed on 4 December 2021).
37. Xu, C.; Shen, J.; Du, X. A Method of Few-Shot Network Intrusion Detection Based on Meta-Learning Framework. *IEEE Trans. Inf. Forensics Secur.* **2020**, *15*, 3540–3552. [CrossRef]
38. Razakarivony, S.; Jurie, F. Discriminative Autoencoders for Small Targets Detection. In Proceedings of the 22nd International Conference on Pattern Recognition (ICPR), Stockholm, Sweden, 24–28 August 2014.
39. Sharafaldin, I.; Lashkari, A.H.; Ghorbani, A.A. Toward Generating a New Intrusion Detection Dataset and Intrusion Traffic Characterization. In Proceedings of the 4th International Conference on Information Systems Security and Privacy (ICISSP 2018), Funchal, Portugal, 22–24 January 2018; pp. 108–116.
40. Tavallaee, M.; Bagheri, E.; Lu, W.; Ghorbani, A.A. A detailed analysis of the KDD CUP 99 data set. In Proceedings of the IEEE Symposium on Computational Intelligence for Security and Defense Applications, Ottawa, ON, Canada, 8–10 July 2009; pp. 1–6. [CrossRef]

Article

A Study on Reversible Data Hiding Technique Based on Three-Dimensional Prediction-Error Histogram Modification and a Multilayer Perceptron

Chih-Chieh Hung [1], Chuang-Chieh Lin [2], Hsien-Chu Wu [3,*] and Chia-Wei Lin [1]

[1] Department of Management Information Systems, National Chung Hsing University, Taichung City 402202, Taiwan; smalloshin@nchu.edu.tw (C.-C.H.); cwlin@nchu.edu.tw (C.-W.L.)
[2] Department of Computer Science and Information Engineering, Tamkang University, New Taipei City 251301, Taiwan; josephcclin@gms.tku.edu.tw
[3] Department of Computer Science and Information Engineering, National Chin-Yi University of Science and Technology, Taichung City 411030, Taiwan
* Correspondence: wuhc@ncut.edu.tw

Citation: Hung, C.-C.; Lin, C.-C.; Wu, H.-C.; Lin, C.-W. A Study on Reversible Data Hiding Technique Based on Three-Dimensional Prediction-Error Histogram Modification and a Multilayer Perceptron. *Appl. Sci.* **2022**, *12*, 2502. https://doi.org/10.3390/app12052502

Academic Editors: Leandros Maglaras, Helge Janicke and Mohamed Amine Ferrag

Received: 13 January 2022
Accepted: 23 February 2022
Published: 28 February 2022

Publisher's Note: MDPI stays neutral with regard to jurisdictional claims in published maps and institutional affiliations.

Copyright: © 2022 by the authors. Licensee MDPI, Basel, Switzerland. This article is an open access article distributed under the terms and conditions of the Creative Commons Attribution (CC BY) license (https://creativecommons.org/licenses/by/4.0/).

Abstract: In the past few years, with the development of information technology and the focus on information security, many studies have gradually been aimed at data hiding technology. The embedding and extraction algorithms are mainly used by the technology to hide the data that requires secret transmission into a multimedia carrier so that the data transmission cannot be realized to achieve secure communication. Among them, *reversible data hiding* (RDH) is a technology for the applications that demand the secret data extraction as well as the original carrier recovery without distortion, such as remote medical diagnosis or military secret transmission. In this work, we hypothesize that the RDH performance can be enhanced by a more accurate pixel value predictor. We propose a new RDH scheme of prediction-error expansion (PEE) based on a multilayer perceptron, which is an extensively used artificial neural network in plenty of applications. The scheme utilizes the correlation between image pixel values and their adjacent pixels to obtain a well-trained multilayer perceptron so that we are capable of achieving more accurate pixel prediction results. Our data mapping method based on the three-dimensional prediction-error histogram modification uses all eight octants in the three-dimensional space for secret data embedding. The experimental results of our RDH scheme show that the embedding capacity greatly increases and the image quality is still well maintained.

Keywords: reversible data hiding; three-dimensional prediction-error histogram modification; multilayer perceptron

1. Introduction

1.1. Background

With the rapid development of information technology, the internet has been ubiquitous in the world. Thanks to the development of optical communication systems (see [1] for more discussions), people can easily communicate with each other and share multimedia messages, including texts, sound, images, videos, etc. Obviously, the internet provides much more impact on human society than any other medium, while at the same time, issues regarding information security have received considerable and critical attention.

Data hiding is an available technique to deal with secure communication so that the secure data is imperceptibly embedded without drawing attention [2]. The multimedia is used as a cover carrier to hide secret data which will be transmitted in the internet. Reversible data hiding (RDH) not only guarantees the safe transmission of data content but also recovers the hidden data as well as the cover images [3,4]. However, most of these RDH algorithms bring permanent distortions to the original carrier during the embedding process, and these distortions are unacceptable in certain applications [5]. In order to achieve information hiding and distortion-free recovery of the original carrier, distortion-free reversible data hiding is considered [4]. This technique enables the receiver to both extract the embedded data correctly and acquire the original carrier without distortion. Generally, RDH is a fragile hiding technology, which is different from digital watermarking. When implementing the RDH mothod, the distortion that occurs during the transmission of the carrier should be avoided. According to the embedding method, the existing image RDH algorithm can be divided into spatial domain, transform domain, and encryption domain RDH scheme [6]. In this paper, we focus on RDH of spatial domain. Its embedding and extraction frameworks are shown in Figure 1. In the embedding side, the sender embeds secret data into the cover image by a reversible embedding algorithm. In the extraction side, the receiver extracts the secret data embedded in the stego-image by a reversible extraction algorithm and achieves distortion-free recovery image exactly the same as the original image. The performance of RDH algorithm depends on two conflicting factors as trade-offs: the embedding distortion between the cover image and the stego-image and the embedding capacity (EC). For the former factor, PSNR (peak-to-noise-ratio) is widely used (refer to [7] for more discussion). A higher PSNR value means that the stego-image is more similar to the original one. For the latter factor, EC stands for the number of bits which can be embedded into the cover image. Therefore, we favor an RDH algorithm which brings higher EC and lower PSNR, while a trade-off of them is usually considered to fit specific applications [6].

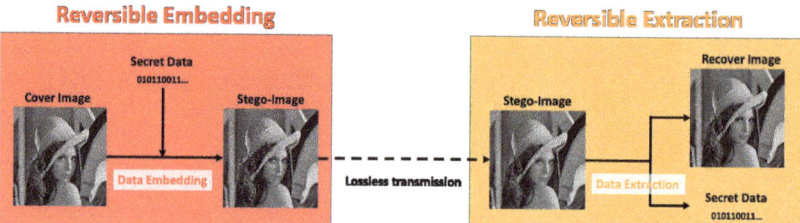

Figure 1. The embedding and extraction frameworks of the spatial domain RDH.

1.2. Prediction-Error Expansion

In this subsection, we introduce the RDH paradigm we mainly follow: the *prediction-error expansion* (PEE) approach, which was first proposed by Thodi and Rodriguez [8]. PEE is a kind of histogram-shifting technique for which histograms of the feature elements (e.g., pixel values, errors between cover pixel values and their predicted values) are shifted to prepare vacant positions for embedding the secret bits. Since the most frequent feature elements determines the EC, and moreover, peaks of the prediction-error histograms usually center at zero, PEE has the advantage over the other histogram-shifting techniques in the spatial domain, especially for the cover image with flat pixel value histogram [9].

PEE can exploit spatial redundancy in the image. The correlation of local neighborhood of each pixel is taken into consideration. Following a certain order of scanning the original image, a *predictor* is used to make prediction of each pixel. Denote by $\hat{x}$ the predicted value of a pixel x. The prediction-error of x is defined as $e_x = x - \hat{x}$. One can *expand* the prediction-error e_x to be $e_x^* = f(e_x, m)$ for some shifting operation f and a to-be-embedded bit $m \in \{0, 1\}$. When the context is clear, we omit the parameter m in f to make formula concise. In the stego-image this pixel will be $\tilde{x} = \hat{x} + e_x^*$. As illustrated in [5],

$$e_x^* = \begin{cases} e_x + m, & \text{if } e_x = 0 \\ e_x - m, & \text{if } e_x = -1 \\ e_x + 1, & \text{if } e_x > 0 \\ e_x - 1, & \text{if } e_x < -1 \end{cases},$$

where $m \in \{0,1\}$ is a to-be-embedded bit. At pixel x at which $e_x \in \{0,-1\}$, a secret bit is embedded, while for pixel x at which $e_x \notin \{0,-1\}$, the pixel value is shifted by 1 or -1. With prediction-errors at hand, the *prediction-error histogram* (PEH) can be created as $h(a) = |\{i : e_i = a\}|$ for each prediction-error a. Specifically, PEE can be implemented as *histogram modification* of the PEH, that is, expanding the bins of -1 and 0 and shifting the other bins to create space to ensure the reversibility. Such a paradigm has been extended to 2D-PEH (e.g., [10]), where the PEH is defined by $h_2(a,b) = |i : (e_{2i-1}, e_{2i}) = (a,b)|$, and also 3D-PEH (e.g., [5]), where the PEH is defined as $h_3(a,b,c) = |i : (e_{3i-2}, e_{3i-1}, e_{3i}) = (a,b,c)|$. Here we have $f : \mathbf{Z}^3 \mapsto \mathcal{P}(\mathbf{Z}^3)$ to be a mapping which realizes RDH, where $\mathcal{P}(A)$ denotes the power set of a set A, such that $f(p)$ represents the set of marked prediction-errors for a prediction-error p (e.g., $p = (e_{3i-2}, e_{3i-1}, e_{3i})$ for some i). As long as $f(p) \neq \emptyset$ for any prediction-error p and $f(p) \cap f(q) = \emptyset$ for every two prediction-errors p and q, the reversibility of the mapping can be guaranteed. PEE has attracted considerable attention [5,8–27] since it can maintain low embedding distortion while at the same time provide sufficiently large payload in terms of high EC.

1.3. Our Contribution

As the illustrating example of PEE shows, EC depends on the prediction accuracy of the pixels. When PEE is applied, the data bits are embedded only when the prediction-error is -1 or 0. Hence, we have the following hypothesis.

Hypothesis 1. *As the prediction accuracy is improved, the performance of the PEE techniques for RDH is enhanced.*

In this paper, we devote our efforts in validating this hypothesis. Specifically, *we aim at improving the prediction accuracy in PEE using deep an artificial neural network (ANN)*, which has been developed rapidly and extensively studied in the past decade. We propose a novel method based on a *multilayer perceptron* (MLP), which is a well-known ANN consisting of multiple sequential fully connected layers and providing nonlinear mapping between input data and output data with nonlinear activation functions. Moreover, we consider *eight octants in the three-dimensional space for embedding*, which makes better use of space (c.f. [5] which considers only the first octant for the embedding). We conduct experiments by applying our proposed method on six test images, including Lena, Baboon, Boat, Peppers, Airplane (F-16), and House. The experimental results well support our hypothesis. The EC greatly increases and is 1.9–9.8 times of previous methods. On the other hand, the image quality is still well maintained in terms of low PSNR, which is competitive compared with previous work.

Remark 1. *Our MLP consists of layers of nodes. The nodes between consecutive layers are fully connected by weighted edges. Each node receives input from nodes on the previous layer and sends output by passing the aggregated input to a nonlinear activation function. It has been shown that the well-trained MLP can be used to approximate any smooth and measurable function [28]. The MLP has been proven to be an effective alternative to more traditional statistical techniques [29]. Recently, the MLP has been widely used in many different fields of research (e.g., see [30–34] for more details). Our proposed method applies MLP to the* pixel prediction *phase of prediction-error histogram modification. We train the MLP network and use it to derive more accurate pixel prediction. Unlike other statistical techniques, the MLP makes no prior assumptions on the data distribution and can be accurately applied even when new or unseen data appear. These features of the MLP make it an attractive alternative when developing numerical models and choosing between statistical methods.*

2. Related Work and Comparisons between the Methods

Shi et al. [6] reviewed the recent advances on RDH in the past two decades, including various RDH schemes in image spatial domain, RDH for compressed images, robust RDH which aims at recovering hidden message from the lossily compressed image, RDH for encrypted images and RDH for video and audio. The RDH in image spatial domain is the most investigated subject and strongly related to this paper. We summarize progresses on this subject as below.

1. Lossless compression-based methods.
   Most early RDH was implemented based on lossless compression [35–42]. Partial space is released by lossless compressing a feature set of the original image, and the data is embedded using the released space to achieve RDH. The performance of this method depends on the lossless compression algorithm used and the selection of compressed feature sets. The experimental results suggest that the algorithm based on lossless compression will result in greater distortion and poorer embedding effect than the subsequent RDH method.

2. Integer-transform-based methods.
   Integer-transform-based methods can be seen in [36,39,41]. In this type of method, the original image is initially divided, so that multiple adjacent pixels can form an embedding unit. Subsequently, the secret information is embedded into each unit using integer transform. However, this type of method usually uses the average value of a pixel block to predict each pixel in the block, so that the image redundancy cannot be well utilized. Moreover, its algorithm cannot control the maximum modification range of each pixel so that the embedded distortion cannot be controlled effectively. Due to two defects mentioned above, the embedding performance of the integer transform-based methods is limited. The performance of this type of method has been significantly improved compared to the lossless compression-based methods; however, it still cannot achieve good embedding performance.

3. Two-phase embedding with location maps.
   There are RDH schemes proceeds with two-phases (e.g., [43–45]) using location maps which map each pixel to a certain value and also ensure the reversibility of the cover image. In [44], Malik et al. considered even-valued and odd-valued pixels separately and embed the secret data bit for each pixel of the cover image by changing its value by at most 1. Their work improves previous complementary embedding strategy by Chang and Kieu [43] which uses vertical embedding and horizontal embedding separately in two phases. Kumar et al. considered even-valued and odd-valued pixels with location maps as well while the cover image is divided into non-overlapping 2-by-2 blocks of pixels and the secret bits are converted into 2-bit segments and embedded into the blocks by increasing or decreasing the pixel value of the corresponding block by at most 1. Since the second phase embedding has the affect as complement of the first phase embedding, this kind of approach persist the stego-image's quality while doubling the EC.

4. Histogram modification-based methods.
   In this type of method, the original image is mapped to space with a lower dimension at the beginning by using the redundancy of the image. Then generate a histogram by counting the distribution of the low-dimensional space. Finally, the reversible embedding is realized by modifying the histogram. The earliest method having a great impact is proposed by Ni et al. in 2006 [46]. In this method, the secret data is embedded into the pixels with the highest frequency in the image histogram by expanding the histogram. The stego-image with this method maintains high image quality, but the embedding rate is low. Therefore, Lee et al. [47] improved the method of [46], which uses the image difference histogram that the shape rule is similar to Laplace distribution. The histogram of the method experiences a very high peak and rapidly dropping; therefore, it can have a better embedding capacity while maintaining image quality.

2.1. Further Discussion on Histograph Modification-Based Approaches

The method of Ni et al. [46] constitutes a rough framework and foundation for RDH based on histogram modification, and hence has been further developed in the follow-up research [5,8–27,30,48–51], In these studies, a histogram is first generated from the prediction error of pixels, and then it is modified by expansion or shifting to achieve reversible embedding. Currently, such methods, modifying the *prediction error histogram* (PEH), are collectively perceived as prediction error expansion (PEE). RDH based on histogram modification has the following two advantages:

- Using histograms, especially PEHs, can effectively utilize image redundancy.
- Modifying the histogram by expansion or shifting can control the maximum modification range of each pixel and the embedding distortion effectively.

From the above points of view, the methods based on histogram modification, especially PEE based on PEH modification, have better embedding performance than other methods. Therefore, we focus on histogram generation and three-dimensional histogram modification. Note that the current RDH methods based on histogram modification mainly include the following aspects:

- Generation method of histogram.
  Combined with PEE, the methods of this research direction mainly aim to generate a sharp and rapidly dropping PEH by using better image prediction methods, e.g., the methods of [12,13,19,20,23,24].
- Modification method of histogram.
  Different from the early expansion methods [8,9,16,24] using a peak in histogram, several authors [15,25–27] proposed methods to expand the histogram by adaptively selecting with the frequency of pixels in the image histogram. These methods can significantly reduce the embedding distortion of PEE.
- Selection of embedding location.
  This type of method firstly selects the image area that is more suitable for reversible embedding (usually smooth areas), and then uses the selected area as a new carrier for RDH. The effect of these methods are remarkable. Combining with PEE can effectively reduce the embedding distortion of PEE. Its idea was first proposed by Kamstra et al. [18], and many subsequent works have also applied this method as an auxiliary means to further optimize the embedding performance.
- High-dimensional histogram modification.
  Several authors [10,21] proposed the methods based on high-dimensional histogram modification. They map high-dimensional redundant features of images to two-dimensional space, and then modify the two-dimensional histogram to achieve reversible embedding. In recent works [5,14,17], the methods based on three-dimensional or high-dimensional histogram modification are proposed. By mapping the redundant features of the image to a higher-dimensional space, the embedding capacity is increased and the image quality is maintained. This type of method can greatly improve the embedding performance of existing PEE algorithms.
- Multi-histogram modification.
  In [11,22], the reversible embedding methods based on using multi-histograms are proposed. Compared with the method of using a single histogram, the use of multiple histograms has greater flexibility and can further improve the performance of PEE algorithms.
- PEH for color images.
  In [51], Zhan et al. applied 3D-PEH to color images. Their approach is to predict the pixel values of each RGB channel of a color image and establish the 3D prediction-error histogram. Their results yield low distortion for color images.

Below we summarize two recent progress on the other perspectives on the histogram modification-based methods.

- Histogram-shifting-imitated technique based on human visual system (HVS).
  Kumar et al. take human visual system into consideration [52] and improves previous work using histogram-shifting-imitated reversible data hiding method in [53]. Since human eyes are more sensitive to the changes in lower intensity pixels than higher ones, this approach divide the intensity levels into four groups of equal size and embed less bits in the low intensity pixels for less conceived distortion of the stego-image so that the visual imperceptibility is improved.
- Pixel Value Ordering (PVO).
  Li et al. [20] proposed the pixel value ordering (PVO) technique which is an advancement of PEE. When the cover images are divided into blocks, PVO first sorts pixel values in each block and then computes minimum, maximum, second-minimum and second maximum pixels which are used for data embedding depending on the minimum and maximum prediction errors in the blocks. PVO changes the pixel values only by at most 1; hence, it generates high quality stego-images. Kaur et al. [54] propose RDH technique using PVO and pairwise PEE to improve EC while retain the quality of the stego image. The embedding strategy is performed in two-phases on three-pixel blocks. Pixels are traversed in a zig-zag way and then sorted based on their rhombus means. The key of PVO for increasing EC is that smaller prediction errors are derived after pixels are sorted. Kaur et al. [55] also considered RDH based on PVO for roughly texture images. For more thorough survey on RDH approaches based on PVO can refer to the survey in [54].

2.2. Comparisons and Highlight of Our Approach

According to above discussions, we list the general comparisons of RDH methods in Table 1. As for the histogram modification-based framework which has attracted much attention and is strongly related to our work and covers the PEE paradigm we mainly follow, we highlight in Table 2 our proposed approach by comparisons with other approaches of this type, such as Ni et al. [46], Lee et al. [47], Li et al. [21], and Cai et al. [5], using average experimental results for six gray-scale images. As Table 2 shows, the embedding capacity of our method is much more than the four other methods, while the image quality is a bit sacrificed due to slightly larger image distortion, though it is tolerable since the PSNR is still close to 50dB. Our results reveal that, due to much better prediction accuracy of pixel values, our method is capable of achieving high embedding capacity while suffering only slight image distortion.

Table 1. General comparison of RDH methods. ×: poor; Δ: unable to control effectively/limited; ○: good; ◉: even better.

RDH Method Types	Image Quality	Embedding Capacity
Lossless-compression	×	×
Integer-transform	Δ	Δ
Two-phased embedding+location maps	○	○
Histogram modification	○	◉

Table 2. Comparisons of histogram modification-based RDH methods. The image quality is measured by average PSNR (dB) when maximum embedding capacity is attained. Average embedding capacity are measured in bits.

Methods	Characteristics	Image Quality (PSNR (dB))	Embedding Capacity (bits)
Ni et al. [46]	first histogram modification/baseline	53.04	4923.67
Lee et al. [47]	image-difference histogram	51.75	9729.00
Li et al. [21]	2D-PEH modification	51.07	24,612.50
Cai et al. [5]	3D-PEH modification (1st octant)	63.72	9444.17
Our method	**3D-PEH + MLP Prediction (8 octants)**	**48.55**	**48,344.17**

3. The Proposed Approach

In this section, we introduce our reversible data hiding scheme based on 3D-PEH modification and a MLP as the pixel value predictor. As the hypothesis in Section 1 states, we expect the performance of such a RDH scheme can be greatly enhanced by an accurate MLP predictor. The characteristics of the correlation between the image pixel value and the neighboring pixels is used, so that the accuracy of pixel prediction can be hopefully improved due to a better trained MLP model. This then leads to increased embedding capacity. Overall, our proposed method includes four parts: the pre-processing phase, the training and prediction phase, the embedding and shifting phase, and the extraction and recovery phase. The flowchart of the proposed method is shown in Figure 2. We specify the four phases in the following subsections.

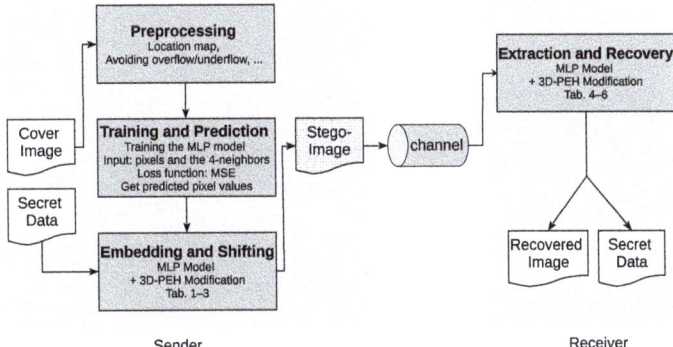

Figure 2. The flowchart of our method.

3.1. The Pre-Processing Phase

The pixel values of the cover image will be modified by +1 or −1 when the secret data embed based on 3D-PEH. Therefore, in order to avoid the overflow and underflow, the cover image will be pre-processed. Amend the pixel with value 0 to 1, and the pixel with value 255 to 254. Meanwhile, a location map is created to record these modified pixel positions. The location map is a binary sequence, which can be losslessly compressed to reduce its size. Then the secret data and the compressed location map are combined (hereinafter referred to as secret data); thereby, the pre-processing phase has been completed. After that, they will be embedded in the pre-processed cover image together.

3.2. The Training and Prediction Phase

The PEE method aims at the correlations between the pixels to derive accurate predictions where the prediction-errors are modified separately. However, the traditional PEE method uses the same algorithm to predict pixels for all images. This results in poor prediction accuracy and the prediction error increases as the image is relatively complex. Therefore, our proposed method, which leverages the power of trained MLP model, can predict the pixels of the cover image and significantly reduces the prediction-error so that the embedding capacity can be hopefully increased.

In the MLP training stage, except for pixels located in borders, the pixels are scanned from left to right and top to bottom to derive the cover sequence $(y_1, \ldots, y_n)$. Consider the four-neighbor tuple $(x_{top}, x_{bottom}, x_{left}, x_{right})$ of a given pixel y_i, shown in the left part of the Figure 3. The four-neighbor tuple is used as input data of the neural network, and the desired output value is y_i.

The structure of an MLP neural network has one input layer, two hidden layers, and one output layer, as shown in Figure 3. The input of four-neighbor tuples $(x_{top}, x_{bottom}, x_{left}, x_{right})$ from the cover image is fed into the input layer of the MLP. Between the input and output layers, there have 100 and 200 neurons in two hidden layers, respectively. After the information income is processed by the network, the output layer of the neural network provides one output $\tilde{y}_i$ as the predicted value by the MLP, and the corresponding y_i in the cover image is used as reference data. We use the mean squared error (MSE) as the loss function which is calculated by taking the average squared difference between the predicted pixel value and the reference pixel value. The MSE function is defined as the Equation (1). Apparently, there is no prediction errors if and only if the MSE value is 0.

$$\text{MSE} = \frac{1}{N} \cdot \sum_{i=1}^{N}(\tilde{y}_i - y_i)^2. \tag{1}$$

Here, N is the number of data points, $\tilde{y}_i$ is the value returned by the model, and y_i is the actual value for data point. Based on those input and reference data, the MLP network is then trained with the loss function such that the edge weights of the MLP are optimized to best associate given neighborhoods with the reference pixel values.

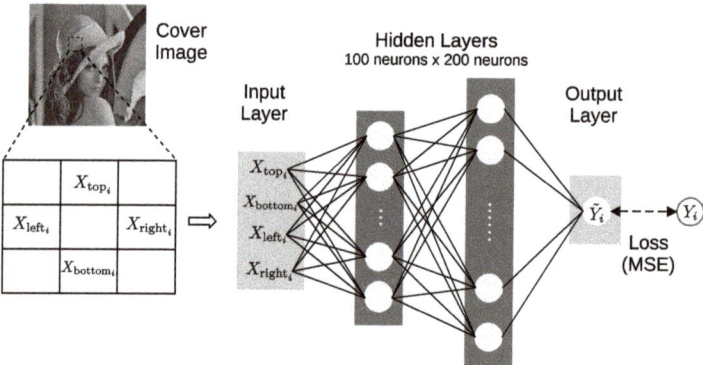

Figure 3. The structure of our MLP neural network.

3.3. The Embedding and Shifting Phase

After the training and prediction phase is completed, the scheme enters the embedding and shifting phase. In order to embed the binary secret data in the cover image, the three-dimensional PEH (3D-PEH) modification is used for embedding and shifting. However, in the previous work on 3D-PEH modification, only the points located in the first octant of the three-dimensional coordinate system are modified. This way of hiding secrets did not make use of most of the space in the three-dimensional coordinate system for embedding; hence, the embeddable pixels are relatively less and a less embedding capacity of images is made. Instead, our proposed method embed secret data in eight octants of the three-dimensional space, so that we possibly exploit much more space than previous approaches.

We adopt rhombus prediction and double-layered embedding, the same as the way used in [5,24], for the implementation of the proposed method to generate non-overlapping prediction-error triple $(e_x, e_y, e_z) = (e_{3i-2}, e_{3i-1}, e_{3i})$ for feasible i (i.e., each pixel in the triple has four neighboring pixels). A 3D-PEH is generated by counting each non-overlapping prediction error triple, and the data embedding is realized by the obtained 3D-PEH modification using the designed reversible mapping. The data embedding procedure is briefly described as follows.

First, adopt double-layered embedding to divide the cover image into two sets denoted as "star" and "dot" (as shown in Figure 4a). The star and dot sets are embedded with half of the secret data, separately. Except for the pixels located in borders, the pixels of the star or dot set are scanned from left to right and top to bottom to derive the cover sequence $(p_1, \ldots, p_n)$. The scan orders for star and dot pixels are shown in Figure 4b,c.

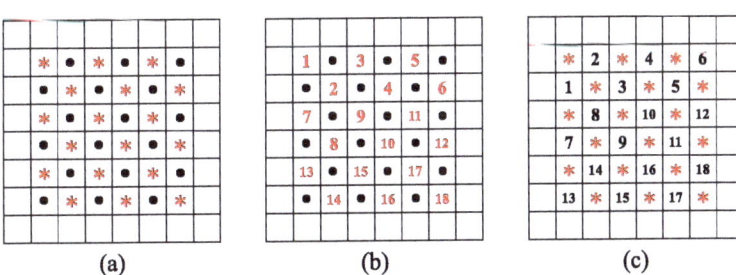

Figure 4. (**a**) Star/dot pixels partition. (**b**) Scan order for star pixels. (**c**) Scan order for dot pixels.

Then, the 4-neighbor pixels of each p_i are introduced to the trained MLP to obtain its predicted value $\hat{p}_i$. The predicted value is used to determine the prediction-error sequence $(e_1, \ldots, e_n)$, and the sequence is divided into the prediction-error triples e_x, e_y, e_z. The prediction-error e_i can be obtained as

$$e_i = p_i - \hat{p}_i. \tag{2}$$

Lastly, modify each prediction-error triple (e_x, e_y, e_z) to be (e_x^*, e_y^*, e_z^*) and get $(\tilde{p}_x, \tilde{p}_y, \tilde{p}_z) = (\hat{p}_x + e_x^*, \hat{p}_y + e_y^*, \hat{p}_z + e_z^*)$ to embed data based on the 3D-PEH in the method shown in Tables 3–5. The 3D-PEH mapping method is divided into seven types: Type *A* to Type *G*.

Table 3. Type A–C of the marked values of prediction-error triple (e_x, e_y, e_z) and cover pixel triple p_x, p_y, p_z in different types of the proposed method with *embedding* as the data embedding operations on (e_x, e_y, e_z).

Type	(e_x, e_y, e_z)	Secret Bits	EC (bits)	(e_x^*, e_y^*, e_z^*)	$(\tilde{p}_x, \tilde{p}_y, \tilde{p}_z)$
A	$(e_x, e_y, e_z) = (0,0,0)$	[0], [0], [0] [0], [0], [1] [0], [1], [0] [0], [1], [1] [1], [0], [0] [1], [0], [1]	3	$(0,0,0)$ $(0,0,1)$ $(0,1,0)$ $(0,-1,0)$ $(1,0,0)$ $(0,0,-1)$	(p_x, p_y, p_z) $(p_x, p_y, p_z + 1)$ $(p_x, p_y + 1, p_z)$ $(p_x, p_y - 1, p_z)$ $(p_x + 1, p_y, p_z)$ $(p_x, p_y, p_z - 1)$
A	$(e_x, e_y, e_z) = (0,0,0)$	[1], [1]	2	$(-1,0,0)$	$(p_x - 1, p_y, p_z)$
B	$(e_x, e_y, e_z) = (\pm 1, \pm 1, \pm 1)$	[0] [1]	1	$(\pm 1, \pm 1, \pm 1)$ $(\pm 2, \pm 2, \pm 2)$	(p_x, p_y, p_z) $(p_x \pm 1, p_y \pm 1, p_z \pm 1)$
C	$e_x \neq 0, (e_y, e_z) = (0,0)$	[1], [1], [1]	3	$(e_x \pm 1, -1, 0)$	$(p_x \pm 1, p_y - 1, p_z)$
C	$e_x \neq 0, (e_y, e_z) = (0,0)$	[0], [0] [0], [1] [1], [0] [1], [1]	2	$(e_x \pm 1, 0, 0)$ $(e_x \pm 1, 0, 1)$ $(e_x \pm 1, 1, 0)$ $(e_x \pm 1, 0, -1)$	$(p_x \pm 1, p_y, p_z)$ $(p_x \pm 1, p_y, p_z + 1)$ $(p_x \pm 1, p_y + 1, p_z)$ $(p_x \pm 1, p_y, p_z - 1)$
C	$e_y \neq 0, (e_x, e_z) = (0,0)$	[1], [1], [1]	3	$(-1, e_y \pm 1, 0)$	$(p_x - 1, p_y \pm 1, p_z)$
C	$e_y \neq 0, (e_x, e_z) = (0,0)$	[0], [0] [0], [1] [1], [0] [1], [1]	2	$(0, e_y \pm 1, 0)$ $(0, e_y \pm 1, 1)$ $(1, e_y \pm 1, 0)$ $(0, e_y \pm 1, -1)$	$(p_x, p_y \pm 1, p_z)$ $(p_x, p_y \pm 1, p_z + 1)$ $(p_x \pm 1, p_y \pm 1, p_z)$ $(p_x, p_y \pm 1, p_z - 1)$
C	$e_z \neq 0, (e_x, e_y) = (0,0)$	[1], [1], [1]	3	$(-1, 0, e_z \pm 1)$	$(p_x - 1, p_y, p_z \pm 1)$
C	$e_z \neq 0, (e_x, e_y) = (0,0)$	[0], [0] [0], [1] [1], [0] [1], [1]	2	$(0, 0, e_z \pm 1)$ $(0, 1, e_z \pm 1)$ $(1, 0, e_z \pm 1)$ $(0, -1, e_z \pm 1)$	$(p_x, p_y, p_z \pm 1)$ $(p_x, p_y + 1, p_z \pm 1)$ $(p_x + 1, p_y, p_z \pm 1)$ $(p_x, p_y - 1, p_z \pm 1)$

Table 4. Type D–F of the marked values of prediction-error triple (e_x, e_y, e_z) and cover pixel triple p_x, p_y, p_z in different types of the proposed method with *embedding* as the data embedding operations on (e_x, e_y, e_z).

Type	(e_x, e_y, e_z)	Secret Bits	EC (bits)	(e_x^*, e_y^*, e_z^*)	$(\tilde{p}_x, \tilde{p}_y, \tilde{p}_z)$
D	$(e_x, e_y, e_z) = (0, \pm 1, \pm 1)$	[0], [0] [0], [1] [1], [0] [1], [1]	2	$(0, \pm 1, \pm 1)$ $(0, \pm 2, \pm 2)$ $(1, \pm 2, \pm 2)$ $(-1, \pm 2, \pm 2)$	(p_x, p_y, p_z) $(p_x, p_y \pm 1, p_z \pm 1)$ $(p_x + 1, p_y \pm 1, p_z \pm 1)$ $(p_x - 1, p_y \pm 1, p_z \pm 1)$
D	$(e_x, e_y, e_z) = (\pm 1, 0, \pm 1)$	[0], [0] [0], [1] [1], [0] [1], [1]	2	$(\pm 1, 0, \pm 1)$ $(\pm 2, 0, \pm 2)$ $(\pm 2, 1, \pm 2)$ $(\pm 2, -1, \pm 2)$	(p_x, p_y, p_z) $(p_x \pm 1, p_y \pm 1, p_z)$ $(p_x \pm 1, p_y \pm 1, p_z + 1)$ $(p_x \pm 1, p_y - 1, p_z \pm 1)$
D	$(e_x, e_y, e_z) = (\pm 1, \pm 1, 0)$	[0], [0] [0], [1] [1], [0] [1], [1]	2	$(\pm 1, \pm 1, 0)$ $(\pm 2, \pm 2, 0)$ $(\pm 2, \pm 2, 1)$ $(\pm 2, \pm 2, -1)$	(p_x, p_y, p_z) $(p_x \pm 1, p_y \pm 1, p_z)$ $(p_x \pm 1, p_y \pm 1, p_z + 1)$ $(p_x \pm 1, p_y \pm 1, p_z - 1)$
E	$e_x = 0, e_y, e_z \notin \{0, \pm 1\}$	[0], [0] [0], [1]	2	$(0, e_y \pm 1, e_z \pm 1)$ $(1, e_y \pm 1, e_z \pm 1)$	$(p_x, p_y \pm 1, p_z \pm 1)$ $(p_x + 1, p_y \pm 1, p_z \pm 1)$
E	$e_x = 0, e_y, e_z \notin \{0, \pm 1\}$	[1]	1	$(-1, e_y \pm 1, e_z \pm 1)$	$(p_x - 1, p_y \pm 1, p_z \pm 1)$
E	$e_y = 0, e_x, e_z \notin \{0, \pm 1\}$	[0], [0] [0], [1]	2	$(e_x \pm 1, 0, e_z \pm 1)$ $(e_x \pm 1, 1, e_z \pm 1)$	$(p_x \pm 1, p_y, p_z \pm 1)$ $(p_x \pm 1, p_y + 1, p_z \pm 1)$
E	$e_y = 0, e_x, e_z \notin \{0, \pm 1\}$	[1]	1	$(e_x \pm 1, -1, e_z \pm 1)$	$(p_x \pm 1, p_y - 1, p_z \pm 1)$
E	$e_z = 0, e_x, e_y \notin \{0, \pm 1\}$	[0], [0] [0], [1]	2	$(e_x \pm 1, e_y \pm 1, 0)$ $(e_x \pm 1, e_y \pm 1, 1)$	$(p_x \pm 1, p_y \pm 1, p_z)$ $(p_x \pm 1, p_y \pm 1, p_z + 1)$
E	$e_z = 0, e_x, e_y \notin \{0, \pm 1\}$	[1]	1	$(e_x \pm 1, e_y \pm 1, -1)$	$(p_x \pm 1, p_y \pm 1, p_z - 1)$
F	$|e_x| > 1, (e_y, e_z) = (\pm 1, \pm 1)$	[0] [1]	1	$(e_x \pm 1, \pm 1, \pm 1)$ $(e_x \pm 1, \pm 2, \pm 2)$	$(p_x \pm 1, p_y, p_z)$ $(p_x \pm 1, p_y \pm 1, p_z \pm 1)$
F	$|e_y| > 1, (e_x, e_z) = (\pm 1, \pm 1)$	[0] [1]	1	$(\pm 1, e_y \pm 1, \pm 1)$ $(\pm 2, e_y \pm 1, \pm 2)$	$(p_x, p_y \pm 1, p_z)$ $(p_x \pm 1, p_y \pm 1, p_z \pm 1)$
F	$|e_z| > 1, (e_x, e_y) = (\pm 1, \pm 1)$	[0] [1]	1	$(\pm 1, \pm 1, e_z \pm 1)$ $(\pm 2, \pm 2, e_z \pm 1)$	$(p_x, p_y, p_z \pm 1)$ $(p_x \pm 1, p_y \pm 1, p_z \pm 1)$

Table 5. Type G of the marked values of prediction-error triple (e_x, e_y, e_z) and cover pixel triple p_x, p_y, p_z in different types of the proposed method with *shifting* as the data embedding operations on (e_x, e_y, e_z).

Type	(e_x, e_y, e_z)	Secret Bits	EC (bits)	(e_x^*, e_y^*, e_z^*)	$(\tilde{p}_x, \tilde{p}_y, \tilde{p}_z)$
G	$e_x, e_y, e_z \neq 0$, and $(e_x, e_y, e_z) \notin \text{Type } B, F$	–	–	$(e_x \pm 1, e_y \pm 1, e_z \pm 1)$	$(p_x \pm 1, p_y \pm 1, p_z \pm 1)$

Figure 5 visualizes the mapping how the secret data are embedded. The goal of such visualization is to provide an intuitive way to verify the reversibility of the our proposed method. First of all, there are seven types of embedding in the proposed method, the mapping relationship of Type A, B, ..., and G can be visualized as shown in Figure 5. An arrow with the starting point x to the end point y represents the data x transforms to the data y in this mapping. That is, the prediction-error groups e_x, e_y, e_z and the cover pixel groups p_x, p_y, p_z are modified by type A to type F according to the condition of the secret which will be embedded. For example, Type A could hide data by transforming $(0,0,0)$ into $(0,0,0), (0,0,1), (0,1,0), (0,-1,0), (1,0,0), (0,0,-1)$, and $(-1,0,0)$. Therefore, Figure 5a shows the six arrows which starts from $(0,0,0)$ to the destinations $(0,0,0), (0,0,1), (0,1,0), (0,-1,0), (1,0,0), (0,0,-1)$, and $(-1,0,0)$, respectively. Therefore, one can check if the mapping for data hiding is revertible by checking if one point in the mapping diagram can be reached by multiple points.

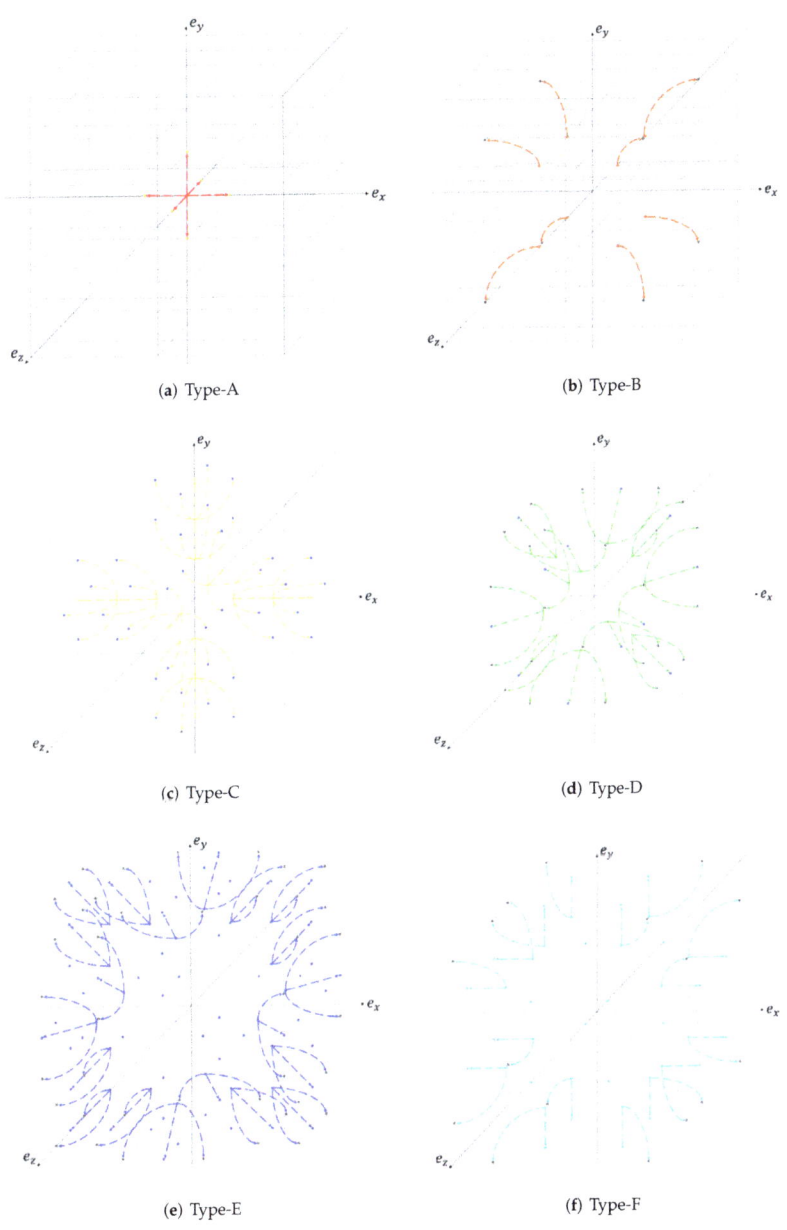

(a) Type-A   (b) Type-B

(c) Type-C   (d) Type-D

(e) Type-E   (f) Type-F

Figure 5. *Cont.*

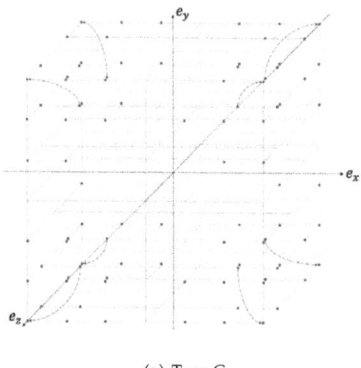

(g) Type-G

Figure 5. The 3D-PEH mappings for the proposed scheme.

After the embedding and shifting phase, the stego-image embedded with secret data will be obtained. Then, the stego-image and the trained MLP model are sent to the receiver side through the communication channel.

Example 1. *Consider the cover image $P = \{210, 99, 131, 65, 72, 162, 17, 19, 25, 161, 25, 71, 86, 95, 47\}$, the secret bits $S = \{0,0,0,1,1,0,1,1,1,1,0\}$, and the prediction error $E = \{0,0,0,0,0,0,1,1,-1,1,0,0,0,1,0\}$.*

- Step 1:
    1. Get the three bits from $E = \{\mathbf{0,0,0}, 0,0,0,1,1,-1,1,0,0,0,1,0\}$: $(e_x, e_y, e_z) = (0,0,0)$. This is a Type-A case.
    2. Get three bits from $S = \{\mathbf{0,0,0},1,1,0,1,1,1,1,0\}$ if the first two bits are $[0], [0]$. Since the secret bits are $[0], [0], [0]$, we have $(e_x^*, e_y^*, e_z^*) = (0,0,0)$.
    3. Get three units from $P = \{\mathbf{210, 99, 131}, 65, 72, 162, 17, 19, 25, 161, 25, 71, 86, 95, 47\}$ and derive $(\tilde{p}_x, \tilde{p}_y, \tilde{p}_z) = (210, 99, 131) + (0,0,0) = (210, 99, 131)$.

    The results of this step are $E^* = \{0,0,0,\ldots\}$ and $\tilde{p}_x = \{210, 99, 131, \ldots\}$.

- Step 2:
    1. Get three bits from $E = \{0,0,0,\mathbf{0,0,0},1,1,-1,1,0,0,0,1,0\}$: $(e_x, e_y, e_z) = (0,0,0)$. This is a Type-A case.
    2. Get two bits from $S = \{0,0,0,\mathbf{1,1},0,1,1,1,1,0\}$. Since the secret bits are $[1], [1]$, we have $(e_x^*, e_y^*, e_z^*) = (-1, 0, 0)$.
    3. Get three units from $P = \{210, 99, 131, \mathbf{65, 72, 162}, 17, 19, 25, 161, 25, 71, 86, 95, 47\}$ and derive $(\tilde{p}_x, \tilde{p}_y, \tilde{p}_z) = (65, 72, 162) + (-1, 0, 0) = (64, 72, 162)$.

    The results of this step are $E^* = \{0,0,0,-1,0,0,\ldots\}$ and $\tilde{p}_x = \{210, 99, 131, 64, 72, 162, \ldots\}$.

- Step 3:
    1. Get three bits from $E = \{0,0,0,0,0,0,\mathbf{1,1,-1},1,0,0,0,1,0\}$: $(e_x, e_y, e_z) = (1, 1, -1)$. This is a Type-B case.
    2. Get one bit from $S = \{0,0,0,1,1,\mathbf{0},1,1,1,1,0\}$. Since the secret bit is $[0]$, we have $(e_x^*, e_y^*, e_z^*) = (1, 1, -1)$.
    3. Get three units from $P = \{210, 99, 131, 65, 72, 162, \mathbf{17, 19, 25}, 161, 25, 71, 86, 95, 47\}$ and derive $(\tilde{p}_x, \tilde{p}_y, \tilde{p}_z) = (17, 19, 25) + (0, 0, 0) = (17, 19, 25)$.

    The results of this step are $E^* = \{0,0,0,-1,0,0,1,1,-1,\ldots\}$ and $\tilde{p}_x = \{210, 99, 131, 64, 72, 162, 17, 19, 25, \ldots\}$.

- Step 4:
    1. Get three bits from $E = \{0,0,0,0,0,0,1,1,-1,\mathbf{1,0,0},0,1,0\}$: $(e_x, e_y, e_z) = (1,0,0)$. This is a Type-C case.
    2. Get three bits from $S = \{0,0,0,1,1,0,\mathbf{1,1,1},1,0\}$ if the secret bits are $[1], [1], [1]$. Since the secret bit is $[1], [1], [1]$, we have $(e_x^*, e_y^*, e_z^*) = (2,-1,0)$.
    3. Get three units from $P = \{210, 99, 131, 65, 72, 162, 17, 19, 25, \mathbf{161, 25, 71}, 86, 95, 47\}$. and derive $(\tilde{p}_x, \tilde{p}_y, \tilde{p}_z) = (161, 25, 71) + (1, -1, 0) = (162, 24, 71)$

    The results of this step are $E^* = \{0,0,0,-1,0,0,1,1,-1,2,-1,0,\ldots\}$ and $\tilde{p}_x = \{210, 99, 131, 64, 72, 162, 17, 19, 25, 162, 24, 71, \ldots\}$.

- Step 5:
    1. Get three bits from $E = \{0,0,0,0,0,0,1,1,-1,1,0,0,\mathbf{0,1,0}\}$: $(e_x, e_y, e_z) = (0,1,0)$. This is a Type-C case.
    2. Get two bits from $S = \{0,0,0,1,1,0,1,1,1,\mathbf{1,0}\}$ if the secret bits are not $[1], [1], [1]$. Since the secret bit is $[1], [0]$, we have $(e_x^*, e_y^*, e_z^*) = (1,2,0)$.
    3. Get three units from $P = \{210, 99, 131, 65, 72, 162, 17, 19, 25, 161, 25, 71, \mathbf{86, 95, 47}\}$. and derive $(\tilde{p}_x, \tilde{p}_y, \tilde{p}_z) = (86, 95, 47) + (1, 1, 0) = (87, 96, 47)$.

    The results of this step are $E^* = \{0,0,0,-1,0,0,1,1,-1,2,-1,0,1,2,0\}$ and $\tilde{p}_x = \{210, 99, 131, 64, 72, 162, 17, 19, 25, 87, 96, 47\}$.

3.4. The Extraction and Recovery Phase

Through the communication channel, the stego-image and the trained MLP model are received. Next, we consider the secret data extraction from the stego-image and the stego-image recovery. The scheme then enters the extraction and recovery phase.

In the extraction and recovery stage, the procedure of the secret data extraction and the stego-image recovery is similar to the procedure of embedding and shifting. The secret data extraction process is briefly described as follows.

First, rhombus prediction and double-layered embedding is adopted to divide the stego-image into two sets denoted as "star" and "dot" (as shown in Figure 4a), and half of the secret data will be extracted from the star and dot sets, respectively. Except for the pixels located in borders, the pixels of the star or dot set are scanned from top-left to bottom-right to derive the stego sequence $(p'_1, \ldots, p'_n)$.

Then, the 4-neighbor dots of each p'_i are introduced to the trained MLP to obtain its predicted value $\hat{p}'_i$. The predicted value is used to determine the prediction-error sequence $(e'_1, \ldots, e'_n)$, and the sequence is divided into the prediction-error triples (e'_x, c'_y, c'_z). The prediction-error e'_i can be obtained as

$$e'_i = p'_i - \hat{p}'_i.$$

Finally, each recovered triple (p'_x, p'_y, p'_z) is extracted based on the 3D-PEH as the method shown in Table 6–8. The 3D-PEH recovery method is divided into seven types: Type A' to Type G'. Besides, (e'_x, e'_y, e'_z) should be the prediction-errors between the "marked pixels" (in the stego-image) and the prediction of the "marked pixels". When the prediction-error e'_i is 1, the recovered value is $p_i = p'_i - 1$, and when the prediction-error e'_i is -1, the recovered value is $p_i = p'_i + 1$.

The secret data bits are extracted by type A' to type G' according to the condition of the prediction-error group and the stego pixel group (p'_x, p'_y, p'_z) is recovered to the recover pixel groups that have the same pixel values as the cover pixel groups (p_x, p_y, p_z). In addition, the type G' has no embedded data bits, so only recover the stego pixel groups to the recover pixel groups without secret data extraction.

Table 6. Type A'–C' of the The extracted secret bits and the recovered values of prediction-error triple (e'_x, e'_y, e'_z) and stego pixel triple (p'_x, p'_y, p'_z) in different types of the proposed method with *embedding* as the data embedding operations on (e'_x, e'_y, e'_z).

Type	(e'_x, e'_y, e'_z)	Extracted Secret Bits	(e_x, e_y, e_z)	(p_x, p_y, p_z)
A'	$(e'_x, e'_y, e'_z) = (0,0,0)$	$[0], [0], [0]$	$(0,0,0)$	(p'_x, p'_y, p'_z)
	$(e'_x, e'_y, e'_z) = (0,0,1)$	$[0], [0], [1]$		$(p'_x, p'_y, p'_z - 1)$
	$(e'_x, e'_y, e'_z) = (0,1,0)$	$[0], [1], [0]$		$(p'_x, p'_y - 1, p'_z)$
	$(e'_x, e'_y, e'_z) = (0,-1,0)$	$[0], [1], [1]$		$(p'_x, p'_y + 1, p'_z)$
	$(e'_x, e'_y, e'_z) = (1,0,0)$	$[1], [0], [0]$		$(p'_x - 1, p'_y, p'_z)$
	$(e'_x, e'_y, e'_z) = (0,0,-1)$	$[1], [0], [1]$		$(p'_x, p'_y, p'_z + 1)$
	$(e'_x, e'_y, e'_z) = (0,-1,0)$	$[1], [1]$		$(p'_x, p'_y + 1, p'_z + 1)$
B'	$(e'_x, e'_y, e'_z) = (\pm 1, \pm 1, \pm 1)$	$[0]$	$(\pm 1, \pm 1, \pm 1)$	(p'_x, p'_y, p'_z)
	$(e'_x, e'_y, e'_z) = (\pm 2, \pm 2, \pm 2)$	$[1]$		$(p'_x \pm 1, p'_y \pm 1, p'_z \pm 1)$
C'	$\|e'_x\| > 1, (e'_y, e'_z) = (-1,0)$	$[1], [1], [1]$	$(e'_x \pm 1, 0, 0)$	$(p'_x \pm 1, p'_y + 1, p'_z)$
	$\|e'_x\| > 1, (e'_y, e'_z) = (0,0)$	$[0], [0]$		$(p'_x \pm 1, p'_y, p'_z)$
	$\|e'_x\| > 1, (e'_y, e'_z) = (0,1)$	$[0], [1]$		$(p'_x \pm 1, p'_y, p'_z - 1)$
	$\|e'_x\| > 1, (e'_y, e'_z) = (1,0)$	$[1], [0]$		$(p'_x \pm 1, p'_y - 1, p'_z)$
	$\|e'_x\| > 1, (e'_y, e'_z) = (0,-1)$	$[1], [1]$		$(p'_x \pm 1, p'_y, p'_z + 1)$
C'	$\|e'_y\| > 1, (e'_x, e'_z) = (-1,0)$	$[1], [1], [1]$	$(0, e'_y \pm 1, 0)$	$(p'_x + 1, p'_y, p'_z)$
	$\|e'_y\| > 1, (e'_x, e'_z) = (0,0)$	$[0], [0]$		$(p'_x, p'_y \pm 1, p'_z)$
	$\|e'_y\| > 1, (e'_x, e'_z) = (0,1)$	$[0], [1]$		$(p'_x, p'_y \pm 1, p'_z - 1)$
	$\|e'_y\| > 1, (e'_x, e'_z) = (1,0)$	$[1], [0]$		$(p'_x - 1, p'_y \pm 1, p'_z)$
	$\|e'_y\| > 1, (e'_x, e'_z) = (0,-1)$	$[1], [1]$		$(p'_x, p'_y \pm 1, p'_z + 1)$
C'	$\|e'_z\| > 1, (e'_x, e'_y) = (-1,0)$	$[1], [1], [1]$	$(0, 0, e'_z \pm 1)$	$(p'_x + 1, p'_y, p'_z \pm 1)$
	$\|e'_z\| > 1, (e'_x, e'_y) = (0,0)$	$[0], [0]$		$(p'_x, p'_y, p'_z \pm 1)$
	$\|e'_z\| > 1, (e'_x, e'_y) = (0,1)$	$[0], [1]$		$(p'_x, p'_y - 1, p'_z \pm 1)$
	$\|e'_z\| > 1, (e'_x, e'_y) = (1,0)$	$[1], [0]$		$(p'_x - 1, p'_y, p'_z \pm 1)$
	$\|e'_z\| > 1, (e'_x, e'_y) = (0,-1)$	$[1], [1]$		$(p'_x, p'_y + 1, p'_z \pm 1)$

Table 7. Type D'–F' of the the extracted secret bits and the recovered values of prediction-error triple (e'_x, e'_y, e'_z) and stego pixel triple (p'_x, p'_y, p'_z) in different types of the proposed method with *embedding* as the data embedding operations on (e'_x, e'_y, e'_z).

Type	(e'_x, e'_y, e'_z)	Extracted Secret Bits	(e_x, e_y, e_z)	(p_x, p_y, p_z)
D'	$(e'_x, e'_y, e'_z) = (0, \pm 1, \pm 1)$	$[0], [0]$	$(0, \pm 1, \pm 1)$	(p'_x, p'_y, p'_z)
	$(e'_x, e'_y, e'_z) = (0, \pm 2, \pm 2)$	$[0], [1]$		$(p'_x, p'_y \pm 1, p'_z \pm 1)$
	$(e'_x, e'_y, e'_z) = (1, \pm 2, \pm 2)$	$[1], [0]$		$(p'_x - 1, p'_y \pm 1, p'_z \pm 1)$
	$(e'_x, e'_y, e'_z) = (-1, \pm 2, \pm 2)$	$[1], [1]$		$(p'_x + 1, p'_y \pm 1, p'_z \pm 1)$
D'	$(e'_x, e'_y, e'_z) = (\pm 1, 0, \pm 1)$	$[0], [0]$	$(\pm 1, 0, \pm 1)$	(p'_x, p'_y, p'_z)
	$(e'_x, e'_y, e'_z) = (\pm 2, 0, \pm 2)$	$[0], [1]$		$(p'_x \pm 1, p'_y, p'_z \pm 1)$
	$(e'_x, e'_y, e'_z) = (\pm 2, 1, \pm 2)$	$[1], [0]$		$(p'_x \pm 1, p'_y - 1, p'_z \pm 1)$
	$(e'_x, e'_y, e'_z) = (\pm 2, -1, \pm 2)$	$[1], [1]$		$(p'_x \pm 1, p'_y + 1, p'_z \pm 1)$
D'	$(e'_x, e'_y, e'_z) = (\pm 1, \pm 1, 0)$	$[0], [0]$	$(\pm 1, \pm 1, 0)$	(p'_x, p'_y, p'_z)
	$(e'_x, e'_y, e'_z) = (\pm 2, \pm 2, 0)$	$[0], [1]$		$(p'_x \pm 1, p'_y \pm 1, p'_z)$
	$(e'_x, e'_y, e'_z) = (\pm 2, \pm 2, 1)$	$[1], [0]$		$(p'_x \pm 1, p'_y \pm 1, p'_z - 1)$
	$(e'_x, e'_y, e'_z) = (\pm 2, \pm 2, -1)$	$[1], [1]$		$(p'_x \pm 1, p'_y \pm 1, p'_z + 1)$
E'	$e'_x = 0, \|e'_y\| > 1$	$[0], [0]$	$(0, e'_y \pm 1, e'_z \pm 1)$	$(p'_x, p'_y \pm 1, p'_z \pm 1)$
	$e'_x = 1, \|e'_z\| > 1$	$[0], [1]$		$(p'_x - 1, p'_y \pm 1, p'_z \pm 1)$
	$e'_x = -1, (e'_y, e'_z) \neq (\pm 2, \pm 2)$	$[1]$		$(p'_x, p'_y \pm 1, p'_z \pm 1)$

Table 7. Cont.

Type	(e'_x, e'_y, e'_z)	Extracted Secret Bits	(e_x, e_y, e_z)	(p_x, p_y, p_z)
E'	$e'_y = 0, \|e'_x\| > 1$ $e'_y = 1, \|e'_y\| > 1$ $e'_y = -1, (e'_x, e'_z) \neq (\pm 2, \pm 2)$	[0], [0] [0], [1] [1]	$(e'_x \pm 1, 0, e'_z \pm 1)$	$(p'_x \pm 1, p'_y, p'_z \pm 1)$ $(p'_x \pm 1, p'_y - 1, p'_z \pm 1)$ $(p'_x \pm 1, p'_y + 1, p'_z \pm 1)$
E'	$e'_z = 0, \|e'_x\| > 1$ $e'_z = 1, \|e'_y\| > 1$ $e'_z = -1, (e'_x, e'_y) \neq (\pm 2, \pm 2)$	[0], [0] [0], [1] [1]	$(e'_x \pm 1, e'_y \pm 1, 0)$	$(p'_x \pm 1, p'_y \pm 1, p'_z)$ $(p'_x \pm 1, p'_y \pm 1, p'_z - 1)$ $(p'_x \pm 1, p'_y \pm 1, p'_z + 1)$
F'	$\|e'_x\| > 2, (e'_y, e'_z) = (\pm 1, \pm 1)$ $\|e'_x\| > 2, (e'_y, e'_z) = (\pm 2, \pm 2)$	[0] [1]	$(e'_x \pm 1, \pm 1, \pm 1)$	$(p'_x \pm 1, p'_y, p'_z)$ $(p'_x \pm 1, p'_y \pm 1, p'_z \pm 1)$
F'	$\|e'_y\| > 2, (e'_x, e'_z) = (\pm 1, \pm 1)$ $\|e'_y\| > 2, (e'_x, e'_z) = (\pm 2, \pm 2)$	[0] [1]	$(\pm 1, e'_y \pm 1, \pm 1)$	$(p'_x, p'_y \pm 1, p'_z)$ $(p'_x \pm 1, p'_y \pm 1, p'_z \pm 1)$
F'	$\|e'_z\| > 2, (e'_x, e'_y) = (\pm 1, \pm 1)$ $\|e'_z\| > 2, (e'_x, e'_y) = (\pm 2, \pm 2)$	[0] [1]	$(\pm 1, \pm 1, e'_z \pm 1)$	$(p'_x, p'_y, p'_z \pm 1)$ $(p'_x \pm 1, p'_y \pm 1, p'_z \pm 1)$

Table 8. Type G' of the The extracted secret bits and the recovered values of prediction-error triple (e'_x, e'_y, e'_z) and stego pixel triple p'_x, p'_y, p'_z in different types of the proposed method with *no embedded data bit* on (e'_x, e'_y, e'_z).

Type	(e'_x, e'_y, e'_z)	Extracted Secret Bits	(e_x, e_y, e_z)	(p_x, p_y, p_z)
G'	$\|e'_x\| > 1, \|e'_y\| > 1, \|e'_z\| > 1,$ $(e'_x, e'_y, e'_z) \notin \text{Type } B, F$	no embedded data bit	$(e'_x \pm 1, e'_y \pm 1, e'_z \pm 1)$	$(p'_x \pm 1, p'_y \pm 1, p'_z \pm 1)$

Through the extraction and recovery phase, the secret data and the recovered image are obtained.

Example 2. Let $P' = \{210, 99, 131, 64, 72, 162, 17, 19, 25, 162, 24, 71, 87, 96, 47\}$, and $E' = \{0, 0, 0, -1, 0, 0, 1, 1, -1, 2, -1, 0, 1, 2, 0\}$.

- Step 1:
  1. Get the three bits from $E' = \{\mathbf{0, 0, 0}, -1, 0, 0, 1, 1, -1, 2, -1, 0, 1, 2, 0\}$: $(e'_x, e'_y, e'_z) = (0, 0, 0)$. This is a Type-$A'$ case. The extracted secret bits are $(0, 0, 0)$.
  2. Get three bits from $P' = \{\mathbf{210, 99, 131}, 64, 72, 162, 17, 19, 25, 162, 24, 71, 87, 96, 47\}$. Then, we can derive $(p_x, p_y, p_z) = (210, 99, 131)$.

  The results of this step are $S = \{0, 0, 0, \ldots\}$ and $P = \{210, 99, 131, \ldots\}$.

- Step 2:
  1. Get the three bits from $E' = \{0, 0, 0, \mathbf{-1, 0, 0}, 1, 1, -1, 2, -1, 0, 1, 2, 0\}$: $(e'_x, e'_y, e'_z) = (-1, 0, 0)$. This is a Type-$A'$ case. The extracted secret bits are $[1], [1]$.
  2. Get three bits from $P' = \{210, 99, 131, \mathbf{64, 72, 162}, 17, 19, 25, 162, 24, 71, 87, 96, 47\}$. Then, we can derive $(p_x, p_y, p_z) = (64 + 1, 72, 162) = (65, 72, 162)$.

  The results of this step are $S = \{0, 0, 0, 1, 1, \ldots\}$ and $P = \{210, 99, 131, 65, 72, 162 \ldots\}$.

- Step 3:
  1. Get the three bits from $E' = \{0, 0, 0, -1, 0, 0, \mathbf{1, 1, -1}, 2, -1, 0, 1, 2, 0\}$: $(e'_x, e'_y, e'_z) = (1, 1, -1)$. This is a Type-$B'$ case. The extracted secret bits are $[0]$.
  2. Get three bits from $P' = \{210, 99, 131, 64, 72, 162, \mathbf{17, 19, 25}, 162, 24, 71, 87, 96, 47\}$. Then, we can derive $(p_x, p_y, p_z) = (17, 19, 25)$.

  The results of this step are $S = \{0, 0, 0, 1, 1, 0, \ldots\}$ and $P = \{210, 99, 131, 65, 72, 162, 17, 19, 25 \ldots\}$.

- Step 4:
    1. Get the three bits from $E' = \{0,0,0,-1,0,0,1,1,-1,\mathbf{2},-\mathbf{1},\mathbf{0},1,2,0\}$: $(e'_x, e'_y, e'_z) = (2,-1,0)$. This is a Type-$C'$ case. The extracted secret bits are $[1],[1],[1]$.
    2. Get three bits from $P' = \{210,99,131,64,72,162,17,19,25,\mathbf{162},\mathbf{24},\mathbf{71},87,96,47\}$. Then, we can derive $(p_x, p_y, p_z) = (162-1, 24+1, 71) = (161, 25, 71)$.

    The results of this step are $S = \{0,0,0,1,1,0,1,1,1,\ldots\}$ and $P = \{210,99,131,65,72,162,17,19,25,161,25,71\ldots\}$.

- Step 5:
    1. Get the three bits from $E' = \{0,0,0,-1,0,0,1,1,-1,2,-1,0,\mathbf{1},\mathbf{2},\mathbf{0}\}$: $(e'_x, e'_y, e'_z) = (1,2,0)$. This is a Type-$C'$ case. The extracted secret bits are $[1],[0]$.
    2. Get three bits from $P' = \{210,99,131,64,72,162,17,19,25,162,24,71,\mathbf{87},\mathbf{96},\mathbf{47}\}$. Then, we can derive $(p_x, p_y, p_z) = (87-1, 96-1, 47) = (86, 95, 47)$.

    The results of this step are $S = \{0,0,0,1,1,0,1,1,1,0\}$ and $P = \{210,99,131,65,72,162,17,19,25,161,25,71,86,95,47\}$.

4. Computational Complexity

Assume that the image has height M and width N respectively. In the pre-processing phase (where a lossless compression is used for the location map; however, we can assume that it can be done in time linearly in the number of pixels if we do not require the space usage as small as possible), training-and-prediction phase, embedding-and-shifting phase and extraction and recovery phase, the computational complexity is basically $O(MN)$ because there are $O(MN)$ pixels to be scanned for a constant number of times. We remark here that though the structure of the MLP neural network is fixed so that this part contributes a constant factor in the complexity, such a constant factor hidden in the asymptotic notation can actually be huge. More specifically, for each input data point (i.e., a set of four pixels) fed to the input layer of the MLP neural network in one iteration, there are 100×200 multiplications required to compute the activation of all the neurons.

5. Experimental Results

The experimental results are shown in this section. Six grayscale images of size 512-by-512, including Lena, Baboon, Boat, Peppers, Airplane (F-16), and House, are used in our experiments. The cover images and the stego-images which are embedded 10,000 bits of secret data are shown in Figure 6. In addition, the variations in image quality under different embedding capacities are compared (as shown in Figure 7). The most common strategy to measure the image quality is the calculation of Peak Signal to Noise Ratio (PSNR) function which is defined as

$$\text{PSNR} = 10 \cdot \log_{10}\left(\frac{255 \cdot 255}{\text{MSE}}\right),$$

$$\text{MSE} = \frac{1}{MN} \cdot \sum_{i=1}^{M}\sum_{j=1}^{N}(x_{i,j} - x'_{i,j})^2.$$

The results of the testing image (Lena) is presented in Figure 7. In addition, from the line chart can be observed that when the embedding capacity is less than 60,000 bits, the PSNR will decrease steadily. However, when the embedding capacity is more than 60,000 bits, PSNR will begin to decline relatively quickly.

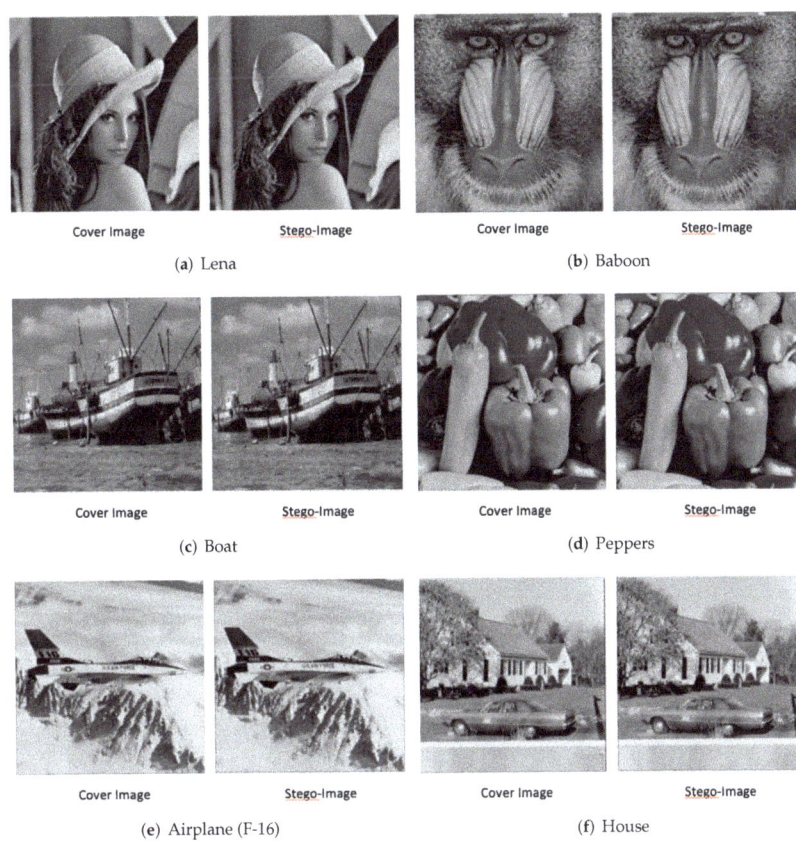

Figure 6. The cover images and the stego-images (embed 10,000 bits).

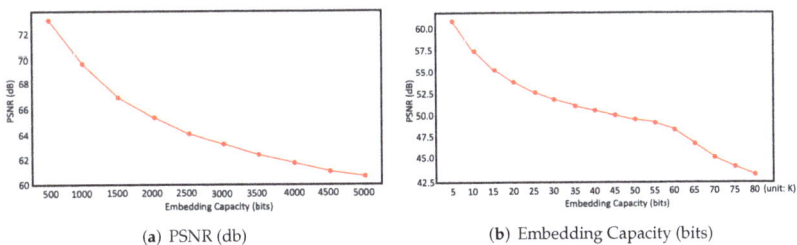

Figure 7. PSNR (dB) and embedding capacity (bits) of the proposed scheme, for image Lena.

5.1. Performance Comparison between the Proposed Method and Baseline Approaches

In this subsection, the proposed method is compared with the previously mentioned schemes. The compared results divide into two parts: maximum embedding capacity and embedding capability in different embedding capacities. The comparison results show that the proposed method has better embedding capacity, and the image qualities are still maintained well.

5.1.1. Maximum Embedding Capacity

We compared the embedding capacity and the image quality when the cover image was embedded once from beginning to end. The comparison is between the proposed method and the methods of Ni et al. [46], Lee et al. [47], Li et al. [21], and Cai et al. [5]. Shown in Table 9 is the comparison of maximum embedding capacity for six test images between the proposed method and the other schemes. In addition, the Table 10 is the comparison of PSNR for maximum embedding capacity between the proposed method and the other schemes.

Table 9. Comparison of maximum embedding capacity (bits) for six test images between the proposed method and the methods of Ni et al. [46], Lee et al. [47], Li et al. [21], and Cai et al. [5].

Image	Ni et al.	Lee et al.	Li et al.	Cai et al.	Our Method
Lena	2785	10,139	24,255	7964	**59,751**
Baboon	2717	4069	9885	990	**19,136**
Boat	5796	7193	17,295	2923	**37,938**
Peppers	2753	8591	19,687	3040	**35,402**
Aiprplain (F-16)	8155	15,797	39,843	21,300	**66,465**
House	7336	12,585	36,710	20,448	**71,373**
Average	4923.67	9729.00	24,612.50	9444.17	**48,344.17**

From the results in Table 9, whether in a smooth image (like image Lena) or in a complex image (like image Baboon), the proposed method has a better embedding capacity.

According to Table 10, the average PSNR of the stego-image among the previous schemes [5,21,46,47] and the proposed method are 53.04 dB, 51.75 dB, 51.61 dB, 63.72 dB, and 48.55 dB, respectively. Clearly, the larger the embedding capacity is, the lower the quality of the image we get. Although the PSNR of the proposed method is lower than other methods, the embedding capacity of it is much more than other methods. According to the above results, when the cover image is only embedded once, our proposed method can have the maximum embedding capacity and maintain good image quality.

Table 10. Comparison of PSNR (dB) between the proposed method and the methods of Ni et al. [46], Lee et al. [47], Li et al. [21], and Cai et al. [5] for maximum embedding capacity.

Image	Ni et al.	Lee et al.	Li et al.	Cai et al.	Our Method
Lena	53.70	51.72	51.60	62.16	48.64
Baboon	51.96	51.35	51.34	70.84	48.26
Boat	51.97	51.52	51.47	66.26	48.41
Peppers	52.49	51.60	51.51	65.92	48.39
Aiprplain (F-16)	54.21	52.16	51.90	58.27	48.75
House	53.91	52.14	51.83	58.86	48.84
Average	53.04	51.75	51.07	63.72	48.55

5.1.2. Embedding Capability in Different Embedding Capacities

In this section, the variations in image quality under different embedding capacities between the proposed method and the methods of Ni et al. [46], Lee et al. [47], Li et al. [21], and Cai et al. [5] are compared. The image quality comparison for six test images in different embedding capacities between the proposed method and the other schemes are shown in Tables 11–13. In addition, the performance comparisons between the proposed method and other related researches are shown in Figure 8 as line graphs.

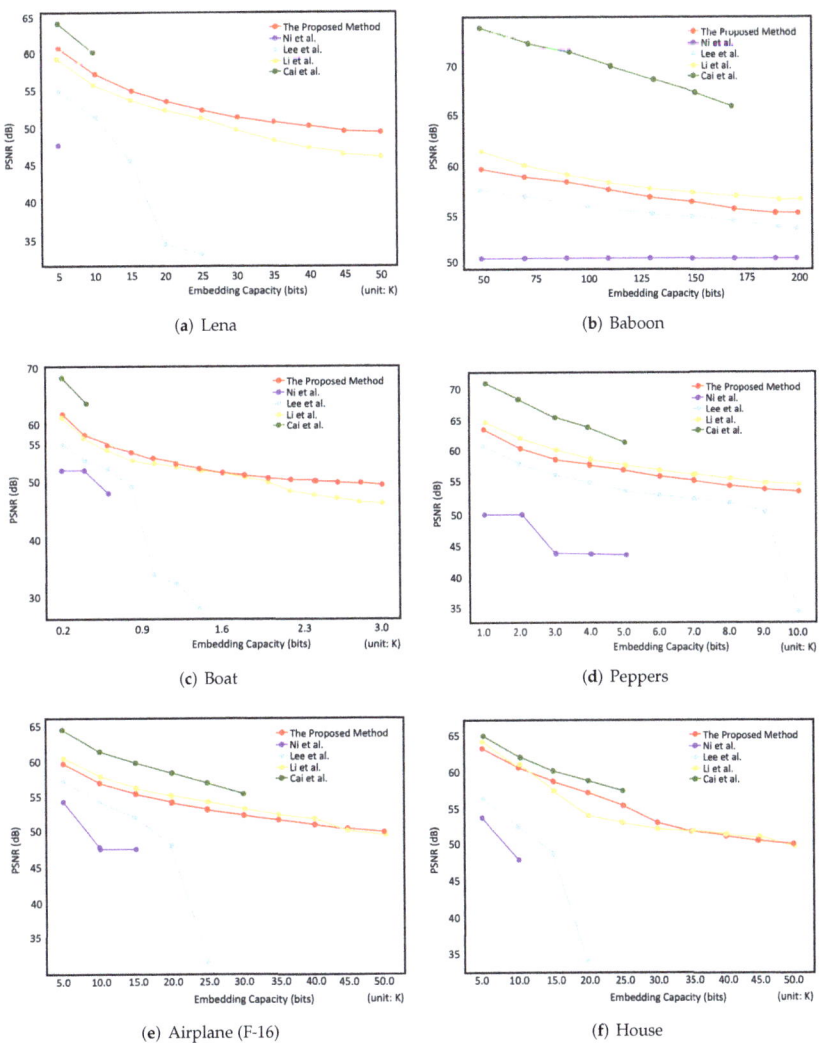

Figure 8. Performance comparisons among the proposed method and other approaches on different images.

Table 11. Comparison of PSNR (dB) between the proposed method and the methods of Ni et al. [46], Lee et al. [47], Li et al. [21], and Cai et al. [5] for a capacity of 1000 bits.

Image	Ni et al.	Lee et al.	Li et al.	Cai et al.	Our Method
Lena	53.75	63.17	66.84	71.30	**69.62**
Baboon	50.47	55.81	58.59	70.78	**57.90**
Boat	52.07	60.28	64.63	70.96	**64.97**
Peppers	50.14	61.14	64.93	70.94	**63.61**
Aiprplain (F-16)	54.46	63.29	66.68	71.28	**66.21**
House	54.11	67.52	72.41	72.32	**68.51**
Average	52.50	61.87	65.68	71.26	**65.14**

Table 12. Comparison of PSNR (dB) between the proposed method and the methods of Ni et al. [46], Lee et al. [47], Li et al. [21], and Cai et al. [5] for a capacity of 5,000 bits.

Image	Ni et al.	Lee et al.	Li et al.	Cai et al.	Our Method
Lena	47.81	55.14	59.50	54.14	60.59
Baboon	44.41	47.23	53.89	–	52.52
Boat	51.99	52.98	56.55	–	57.27
Peppers	44.08	53.96	57.82	61.52	57.00
Airplane (F-16)	54.31	57.34	60.50	64.29	59.86
House	53.98	56.60	64.53	65.28	63.48
Average	49.43	53.88	58.80	63.81	58.45

Table 13. Comparison of PSNR (dB) between the proposed method and the methods of Ni et al. [46], Lee et al. [47], Li et al. [21], and Cai et al. [5] for a capacity of 10,000 bits.

Image	Ni et al.	Lee et al.	Li et al.	Cai et al.	Our Method
Lena	–	51.79	55.90	60.32	57.15
Baboon	–	–	50.93	–	50.62
Boat	48.24	46.25	53.35	–	54.00
Peppers	–	48.66	54.53	–	54.11
Airplane (F-16)	48.10	54.47	57.87	61.43	57.18
House	48.35	52.80	61.09	62.25	60.68
Average	48.23	50.79	55.61	61.33	55.62

5.2. Comparison between the Proposed Method and the Different Embedding Methods with Different Octant Embed Number

In this subsection, the variations in image quality under different embedding capacities between the proposed method and the different embedding methods are compared. The different embedding methods are generated by reducing the octant embed number of the 3D-PEH in the proposed method. The comparison results are shown in Figure 9.

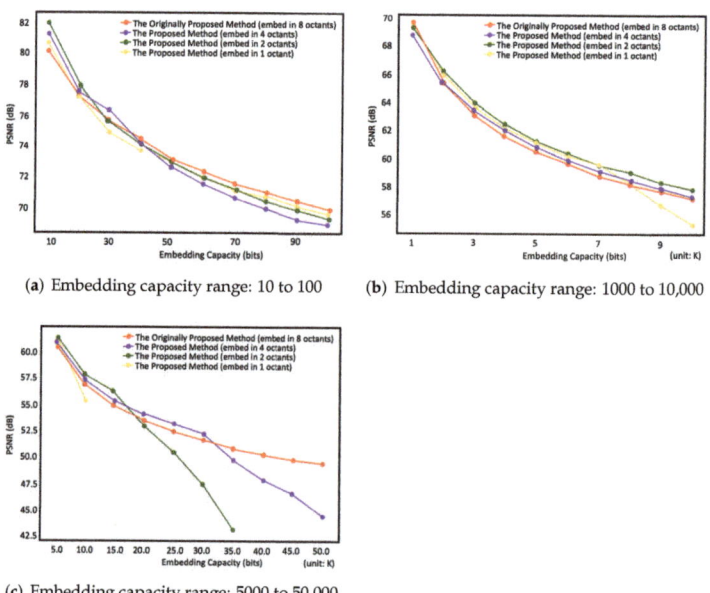

(a) Embedding capacity range: 10 to 100

(b) Embedding capacity range: 1000 to 10,000

(c) Embedding capacity range: 5000 to 50,000

Figure 9. PSNR with different embedding methods and capacity ranges for the image Lena.

According to the above results, when the bits of embedded secret data are few, the distortion of the image can be slightly reduced by embedding the secret in fewer octants. Thus, the reducing effect is limited. Conversely, when the bits of embedded secret data is larger, the better quality of the image can be kept by embedding secret in more octants of the 3D-PEH. It can be expected that the more bits of embedded secret data, the larger gap between different embedding methods occurs. Therefore, we consider embedding secret data in eight octants in the proposed method.

6. Conclusions

Machine learning, especially deep learning, has made significant progress in many research areas and applications such as visual recognition, image classification and image processing, etc. However, to the best of our knowledge, no deep learning approaches have been successfully applied to RDH schemes which require images to be completely restored and secret information to be extracted. This motivates us to apply such approaches to RDH. In this paper, we propose a reversible data hiding scheme based on three-dimensional prediction-error histogram modification and MLP networks. We utilize a trained MLP neural network to predict pixel values and combining with PEE to achieve RDH. In addition, the proposed method of modifying the three-dimensional prediction-error histogram can better utilize the space in the three-dimensional coordinates for data embedding. Evaluation of the quality and embedding capacity of the stego-images shows that the proposed method still maintains a good PSNR and increases the maximum embedding capacity which is 1.9–9.8 times of previous methods. Nevertheless, the proposed method still has its disadvantages. Specifically, training the neural network and predicting pixels bit-by-bit are both time-consuming. Developing methods to enhance the efficiency of the proposed method, such as reducing the training time and predicting multiple bits at once, deserves to be further investigated in future works. Moreover, this work focused on proposing a novel reversible data hiding scheme which trains multilayer perceptrons by utilizing the correlation between image pixel values and their adjacent pixels so that the accurate pixel predictions can be achieved. There should be a trade-off between the performance and the fragility. For a future research direction, it is worthy to discuss the impact of fragility caused by transmission errors.

Author Contributions: Conceptualization, H.-C.W.; methodology, H.-C.W., C.-C.H. and C.-W.L.; software, C.-W.L.; validation, C.-C.H., C.-C.L., H.-C.W. and C.-W.L.; formal analysis, C.-C.L. and C.-W.L.; investigation, C.-C.L. and C.-W.L.; resources, H.-C.W.; data curation, C.-W.L.; writing—original draft preparation, C.-C.H. and C.-W.L.; writing—review and editing, C.-C.H. and C.-C.L.; supervision, H.-C.W.; project administration, H.-C.W.; funding acquisition, C.-C.H. and H.-C.W. All authors have read and agreed to the published version of the manuscript.

Funding: This research is supported by the Taiwan Ministry of Science and Technology under grant no. MOST 110-2221-E-005 -045, no. MOST 110-2222-E-032-002-MY2, and no. MOST 110-2221-E-167-002.

Institutional Review Board Statement: Not applicable.

Informed Consent Statement: Not applicable.

Data Availability Statement: Data is contained within the article.

Conflicts of Interest: The authors declare no conflicting interest regarding the publication of this work.

References

1. Yu, J.; Zhang, J. Recent progress on high-speed optical transmission. *Digit. Commun. Netw.* **2016**, *2*, 65–76. [CrossRef]
2. Wang, C.; Wu, N.; Tsai, C.; Hwang, M. A high quality steganographic method with pixel-value differencing and modulus function. *J. Syst. Softw.* **2008**, *81*, 150–158. [CrossRef]
3. Alattar, A.M. Reversible watermark using the difference expansion of a generalized integer transformation. *IEEE Trans. Image Process.* **2004**, *13*, 1147–1156. [CrossRef] [PubMed]
4. Shi, Y.Q. Reversible data hiding. In Proceedings of the International Workshop on Digital Watermarking (IWDW'04), Seoul, Korea, 30 October–1 November 2004; pp. 1–12.

5. Cai, S.; Li, X.; Liu, J.; Guo, Z. A new reversible data hiding scheme exploiting high-dimensional prediction-error histogram. In Proceedings of the IEEE International Conference on Image Processing (ICIP'16), Phoenix, AZ, USA, 25–28 September 2016; pp. 2732–2736.
6. Shi, Y.Q.; Li, X.; Zhang, X.; Wu, H.T.; Ma, B. Reversible data hiding: Advances in the past two decades. *IEEE Access* **2016**, *4*, 3210–3237. [CrossRef]
7. Almohammad, A.; Ghinea, G. Stego image quality and the reliability of PSNR. In Proceedings of the 2nd International Conference on Image Processing Theory, Tools and Applications (IPTA'10), Paris, France, 7–10 July 2010; pp. 215–220.
8. Thodi, D.M.; Rodriguez, J.J. Expansion embedding techniques for reversible watermarking. *IEEE Trans. Image Process.* **2007**, *16*, 721–730. [CrossRef] [PubMed]
9. Honga, W.; Chen, T.S.; Shiua, C.W. Reversible data hiding for high quality images using modification of prediction errors. *J. Syst. Softw.* **2009**, *82*, 1833–1842. [CrossRef]
10. Ou, B.; Li, X.; Zhao, Y.; Ni, R.; Shi, Y.Q. Pairwise prediction-error expansion for efficient reversible data hiding. *IEEE Trans. Image Process.* **2013**, *22*, 5010–5021. [CrossRef] [PubMed]
11. Caciula, I.; Coltuc, D. Improved control for low bit-rate reversible watermarking. In Proceedings of the IEEE International Conference on Acoustics, Speech and Signal Processing (ICASSP'14), Florence, Italy, 4–9 May 2014; pp. 7425–7429.
12. Dragoi, I.C.; Coltuc, D. Local-prediction-based difference expansion reversible watermarking. *IEEE Trans. Image Process.* **2014**, *23*, 1779–1790. [CrossRef]
13. Dragoi, I.; Coltuc, D. On local prediction based reversible watermarking. *IEEE Trans. Image Process.* **2015**, *24*, 1244–1246. [CrossRef]
14. Gui, X.; Li, X.; Yang, B. High-dimensional histogram utilization for reversible data hiding. In Proceedings of the International Workshop on Digital Watermarking (IWDW'14), Taipei, Taiwan, 1–4 October 2014; pp. 243–253.
15. Hwang, H.J.; Kim, H.J.; Sachnev, V.; Joo, S.H. Reversible watermarking method using optimal histogram pair shifting based on prediction and sorting. *KSII Trans. Internet Inf. Syst.* **2010**, *4*, 655–670. [CrossRef]
16. Hu, Y.; Lee, H.K.; Li, J. DE-based reversible data hiding with improved overflow location map. *IEEE Trans. Circuits Syst. Video Technol.* **2008**, *19*, 250–260.
17. Jiang, R.; Zhang, W.; Hou, D.; Wang, H.; Yu, N. Reversible data hiding for 3D mesh models with three-dimensional prediction-error histogram modification. *Multimed. Tools Appl.* **2018**, *77*, 5263–5280. [CrossRef]
18. Kamstra, L.; Heijmans, H.J.A.M. Reversible data embedding into images using wavelet techniques and sorting. *IEEE Trans. Image Process.* **2005**, *14*, 2082–2090. [CrossRef]
19. Luo, L.; Chen, Z.; Chen, M.; Zeng, X.; Xiong, Z. Reversible image watermarking using interpolation technique. *IEEE Trans. Inf. Forensics Secur.* **2009**, *5*, 187–193.
20. Li, X.; Li, J.; Li, B.; Yang, B. High-fidelity reversible data hiding scheme based on pixel-value-ordering and prediction-error expansion. *Signal Process.* **2013**, *93*, 198–205. [CrossRef]
21. Li, X.; Zhang, W.; Gui, X.; Yang, B. A novel reversible data hiding scheme based on two-dimensional difference-histogram modification. *IEEE Trans. Inf. Forensics Secur.* **2013**, *8*, 1091–1110.
22. X. Li, W.Z.; Gui, X.; Yang, B. Efficient reversible data hiding based on multiple histograms modification. *IEEE Trans. Inf. Forensics Secur.* **2015**, *10*, 2016–2027. [CrossRef]
23. Qin, C.; Chang, C.C.; Huang, Y.H.; Liao, L.T. An inpainting-assisted reversible steganographic scheme using a histogram shifting mechanism. *IEEE Trans. Circuits Syst. Video Technol.* **2012**, *23*, 1109–1118. [CrossRef]
24. Sachnev, V.; Kim, H.J.; Nam, J.; Suresh, S.; Shi, Y.Q. Reversible watermarking algorithm using sorting and prediction. *IEEE Trans. Circuits Syst. Video Technol.* **2009**, *19*, 989–999. [CrossRef]
25. Xuan, G.; Shi, Y.Q.; Chai, P.; Cui, X.; Ni, Z.C.; Tong, X.F. Optimum histogram pair based image lossless data embedding. In Proceedings of the International Workshop on Digital Watermarking (IWDW'07), Guangzhou, China, 3–5 December 2007; pp. 264–278.
26. Xuan, G.; Tong, X.; Teng, J.; Zhang, X.; Shi, Y.Q. Optimal histogram-pair and prediction-error based image reversible data hiding. In Proceedings of the International Workshop on Digital Watermarking (IWDW'12), Shanghai, China, 31 October–3 November 2012; pp. 368–383.
27. Wang, J.; Ni, J.; Zhang, X.; Shi, Y.Q. Rate and distortion optimization for reversible data hiding using multiple histogram shifting. *IEEE Trans. Cybern.* **2016**, *47*, 315–326. [CrossRef]
28. Hornik, K.; Stinchcombe, M.; White, H. Multilayer feedforward networks are universal approximators. *Neural Netw.* **1989**, *2*, 359–366. [CrossRef]
29. Schalkoff, R.J. *Pattern Recognition: Statistical, Structural and Neural Approaches*; Wiley: New York, NY, USA, 2007.
30. Salgado, C.M.; Dam, R.S.F.; Salgado, W.L.; Werneck, R.R.A.; Pereira, C.M.N.A.; Schirru, R. The comparison of different multilayer perceptron and general regression neural networks for volume fraction prediction using MCNPX code. *Appl. Radiat. Isot.* **2020**, *162*, 109170. [CrossRef] [PubMed]
31. Sharifzadeh, F.; Akbarizadeh, G.; Kavian, Y.S. Ship classification in SAR images using a new hybrid CNN–MLP classifier. *J. Indian Soc. Remote Sens.* **2019**, *47*, 551–562. [CrossRef]
32. Heidari, A.A.; Faris, H.; Aljarah, I.; Mirjalili, S. An efficient hybrid multilayer perceptron neural network with grasshopper optimization. *Soft Comput.* **2019**, *23*, 7941–7958. [CrossRef]

33. Fath, A.H.; Madanifar, F.; Abbasi, M. Implementation of multilayer perceptron (MLP) and radial basis function (RBF) neural networks to predict solution gas-oil ratio of crude oil systems. *Petroleum* **2020**, *6*, 80–91. [CrossRef]
34. Janani, V.; Maadhuryaa, N.; Pavithra, D.; Sree, S.R. Dengue prediction using (MLP) multilayer perceptron—A machine learning approach. *Int. J. Res. Eng. Sci. Manag.* **2020**, *3*, 226–231.
35. Fridrich, J.; Goljan, M.; Du, R. Invertible authentication. In *Security and Watermarking of Multimedia Contents III*; SPIE: Bellingham, WA, USA, 2001; Volume 4314, pp. 197–208.
36. Goljan, M.; Fridrich, J.; Du, R. Distortion-free data embedding for images. In Proceedings of the International Workshop on Information Hiding (IHW'01), Pittsburgh, PA, USA, 25–27 April 2001; pp. 27–41.
37. Goljan, M.; Fridrich, J.; Du, R. Lossless data embedding: New paradigm in digital watermarking. *EURASIP J. Adv. Signal Process.* **2002**, *2002*, 185–196.
38. Xuan, G.; Chen, J.; Zhu, J.; Shi, Y.Q.; Ni, Z.; Su, W. Lossless data hiding based on integer wavelet transform. In Proceedings of the IEEE Workshop on Multimedia Signal Processing (MMSP'02), St. Thomas, VI, USA, 9–11 December 2002; pp. 312–315.
39. Xuan, G.; Zhu, J.; Chen, J.; Shi, Y.Q.; Ni, Z.; Su, W. Distortionless data hiding based on integer wavelet transform. *Electron. Lett.* **2002**, *38*, 1646–1648. [CrossRef]
40. Celik, M.U.; Sharma, G.; Tekalp, A.M.; Saber, E. Reversible data hiding. In Proceedings of the IEEE International Conference on Image Processing (ICIP'02), Rochester, NY, USA, 22–25 September 2002; pp. 157–160.
41. Celik, M.; Sharma, G.; Tekalp, A.; Saber, E. Lossless generalized-LSB data embedding. *IEEE Trans. Image Process.* **2005**, *14*, 253–266. [CrossRef] [PubMed]
42. Celik, M.U.; Sharma, G.; Tekalp, A.M. Lossless watermarking for image authentication: A new framework and an implementation. *IEEE Trans. Image Process.* **2006**, *15*, 1042–1049. [CrossRef] [PubMed]
43. Chang, C.C.; Kieu, T.D. A reversible data hiding scheme using complementary embedding strategy. *Inf. Sci.* **2010**, *180*, 3045–3058. [CrossRef]
44. Malik, A.; Singh, S.; Kumar, R. Recovery based high capacity reversible data hiding scheme using even-odd embedding. *Multimed. Tools Appl.* **2018**, *77*, 15803–15827. [CrossRef]
45. Kumar, R.; Chand, S.; Singh, S. A reversible data hiding scheme using pixel location. *Int. Arab J. Inf. Technol.* **2018**, *15*, 763–768.
46. Ni, Z.; Shi, Y.Q.; Ansari, N.; Su, W. Reversible data hiding. *IEEE Trans. Circuits Syst. Video Technol. (TCSVT'06)* **2006**, *16*, 354–362.
47. Lee, S.K.; Suh, Y.H.; Ho, Y.S. Reversible image authentication based on watermarking. In Proceedings of the IEEE International Conference on Multimedia and Expo (ICME'06), Toronto, ON, Canada, 9–12 July 2006; pp. 1321–1324.
48. Lin, S.J.; Chung, W.H. The scalar scheme for reversible information-embedding in gray-scale signals: Capacity evaluation and code constructions. *IEEE Trans. Inf. Forensics Secur.* **2012**, *7*, 1155–1167. [CrossRef]
49. Zhang, X. Reversible data hiding with optimal value transfer. *IEEE Trans. Multimed.* **2012**, *15*, 316–325. [CrossRef]
50. Zhang, W.; Hu, X.; Li, X.; Yu, N. Recursive histogram modification: Establishing equivalency between reversible data hiding and lossless data compression. *IEEE Trans. Image Process.* **2013**, *22*, 2775–2785. [CrossRef]
51. Zhan, Y.; Su, Y.; Wang, X.; Pei, Q. Three-dimensional prediction-error histograms based reversible data hiding algorithm for color images. *Multimed. Tools Appl.* **2019**, *78*, 35289–35311. [CrossRef]
52. Kumar, R.; Chand, S.; Singh, S. An improved histogram-shifting-imitated reversible data hiding based on HVS characteristics. *Multimed. Tools Appl.* **2018**, *77*, 13445–13457. [CrossRef]
53. Wang, Z.H.; Lee, C.F.; Chang, C.Y. Histogram-shifting-imitated reversible data hiding. *J. Syst. Softw.* **2013**, *86*, 315–323. [CrossRef]
54. Kaur, G.; Singh, S.; Rani, R.; Kumar, R. A comprehensive study of reversible data hiding (RDH) schemes based on pixel value ordering (PVO). *Arch. Comput. Methods Eng.* **2020**, *28*, 3517–3568. [CrossRef]
55. Kaur, G.; Singh, S.; Rani, R. PVO based reversible data hiding technique for roughly textured images. *Multidimens. Syst. Signal Process.* **2021**, *32*, 533–558. [CrossRef]

Article

A Novel Model for Vulnerability Analysis through Enhanced Directed Graphs and Quantitative Metrics

Ángel Longueira-Romero [1,2,*], Rosa Iglesias [1], Jose Luis Flores [1] and Iñaki Garitano [2]

[1] Ikerlan Technology Research Centre, Basque Research and Technology Alliance (BRTA), 20500 Arrasate, Spain; riglesias@ikerlan.es (R.I.); jlflores@ikerlan.es (J.L.F.)
[2] Department of Electronics and Computing, Mondragon Unibertsitatea, 20500 Mondragón, Spain; igaritano@mondragon.edu
* Correspondence: alongueira@ikerlan.es; Tel.: +34-9-4371-2400

Abstract: The rapid evolution of industrial components, the paradigm of Industry 4.0, and the new connectivity features introduced by 5G technology all increase the likelihood of cybersecurity incidents. Such incidents are caused by the vulnerabilities present in these components. Designing a secure system is critical, but it is also complex, costly, and an extra factor to manage during the lifespan of the component. This paper presents a model to analyze the known vulnerabilities of industrial components over time. The proposed Extended Dependency Graph (EDG) model is based on two main elements: a directed graph representation of the internal structure of the component, and a set of quantitative metrics based on the Common Vulnerability Scoring System (CVSS). The EDG model can be applied throughout the entire lifespan of a device to track vulnerabilities, identify new requirements, root causes, and test cases. It also helps prioritize patching activities. The model was validated by application to the OpenPLC project. The results reveal that most of the vulnerabilities associated with OpenPLC were related to memory buffer operations and were concentrated in the *libssl* library. The model was able to determine new requirements and generate test cases from the analysis.

Keywords: CPE; CVE; CVSS; CWE; CAPEC; directed graph; IACS; cybersecurity; vulnerability assessment; security metrics; IEC 62443; OpenPLC

Citation: Longueira-Romero, Á.; Iglesias, R.; Flores, J.L.; Garitano, I. A Novel Model for Vulnerability Analysis through Enhanced Directed Graphs and Quantitative Metrics. *Sensors* **2022**, *22*, 2126. https://doi.org/10.3390/s22062126

Academic Editors: Leandros Maglaras, Helge Janicke and Mohamed Amine Ferrag

Received: 4 February 2022
Accepted: 7 March 2022
Published: 9 March 2022

Publisher's Note: MDPI stays neutral with regard to jurisdictional claims in published maps and institutional affiliations.

Copyright: © 2022 by the authors. Licensee MDPI, Basel, Switzerland. This article is an open access article distributed under the terms and conditions of the Creative Commons Attribution (CC BY) license (https://creativecommons.org/licenses/by/4.0/).

1. Introduction

Industrial components are the driving force of almost every industrial field, such as automotive, energy production, and transportation [1–6]. These types of components are rapidly evolving [7,8] and increasing in number [9]. This increase is related to several factors: (1) the reuse of open-source hardware and software, (2) new connectivity features, and (3) more complex systems.

Open-source hardware and software, and Commercial Off-The-Shelf (COTS) components are being integrated to speed up their development [10–12]. COTS are easy to use, but they can introduce vulnerabilities, creating potential entry points for attackers [13,14].

Industrial components are providing more advanced connectivity features, enabling new automation applications, services, and data exchange. This new connectivity, boosted by the fifth generation (5G) of wireless technology for cellular networks, will further open the window of exposure to any threat [6,9,15,16].

The complexity of industrial systems is also increasing with the integration of new trends, such as the Internet of Things (IoT) [16–19], cloud computing, Artificial Intelligence (AI) [19,20], and big data. The extensive use of these technologies further opens the windows for attackers [21–26]. Complexity is a critical aspect of industrial components design because it is closely related to the number of vulnerabilities [27,28].

This scenario points to security as a key aspect of industrial components. Moreover, numerous attacks have been reported targeting industrial enterprises across the globe since 2010 [29]. An exponential rise in such attacks is predicted for future years [30,31].

Although great efforts are being made to develop new and better ways to analyze vulnerabilities [32,33], to measure them (e.g., Common Vulnerabilities and Exposures (CVE) [34], Common Vulnerability Scoring System (CVSS) [35–37], or Common Weakness Enumeration (CWE) [38,39]), or to aggregate them [40], to the best of our knowledge, existing models do not cover the entire life cycle of industrial components. Performing a vulnerability analysis at a single point in time (e.g., during development or when a product has been released) is not enough for industrial components, and their long lifespan has to be considered [41,42]. Furthermore, both software and hardware should be considered, given the strong bonding between hardware and software in industrial components [43–46].

In the present paper, we propose an Extended Dependency Graph (EDG) model that performs continuous vulnerability assessment to determine the source and nature of vulnerabilities and enhance security throughout the entire life cycle of industrial components. The proposed model is built on a directed graph-based structure, and a set of metrics based on globally accepted security standards.

This paper is structured as follows: First, the related work is reviewed in Section 2. Then, the main pieces of the proposed model are defined in Section 3. Second, to demonstrate the potential of this proposal, the proposed model is applied to a real use case in Section 4. Finally, conclusions and future work of this research are described in Section 5.

2. Related Work

This section will review the current status of vulnerability assessment. This review aims to find similar approaches from the literature, including the current standard and metrics.

2.1. Vulnerability Analysis in Security Standards

Industry is currently making a significant effort to incorporate security aspects into the development of industrial components, which has led to a set of standards, such as the Common Criteria and ISA/IEC 62443. This review is focused on how these standards conduct vulnerability analysis, the use of metrics, their management of the life cycle of the device, the techniques that they propose, and the security evaluation of both software and hardware.

2.1.1. ISA/IEC 62443

ISA/IEC 62443 constitutes a series of standards, technical reports, and related information that define the procedures and requirements for implementing electronically secure Industrial Automation and Control Systems (IACSs) [47]. As expressed by this standard, security risk management shall jointly and collaboratively be addressed by all the entities involved in the design, development, integration, and maintenance of the industrial and/or automation solution (including subsystems and components) to achieve the required security level [48].

This joint effort is reflected in the organization of the documents of the standard, which is divided into four parts:

1. Part 1—General: Provides background information such as security concepts, terminology, and metrics;
2. Part 2—Policies and procedures: Addresses the security and patch management policies and procedures;
3. Part 3—System: Provides system development requirements and guidance;
4. Part 4—Component: Provides product development and technical requirements, which are intended for product vendors.

The ISA/IEC 62443-4-1 technical document is divided into eight practices, which specify the secure product development life cycle requirements for both the development and the maintenance phases [49]. The "Practice 5—Security verification and validation testing" (SVV) section of this document specifies that a process shall be employed to identify and characterize potential security vulnerabilities in the product, including known and unknown vulnerabilities [50,51]. Two requirements in Practice 5 are in charge of the task of analyzing vulnerabilities, as follows:

- Requirement SVV-3. Vulnerability Testing [49]. This requirement states that a process shall be employed to perform tests that focus on identifying and characterizing potential and known security vulnerabilities in the product (i.e., fuzz testing, attack surface analysis, black box known vulnerability scanning, software composition analysis, and dynamic runtime resource management testing).
- Requirement SVV-4. Penetration Testing [49]. This requirement states that a process shall be employed to identify and characterize security-related issues via tests that focus on discovering and exploiting security vulnerabilities in the product (i.e., penetration testing).

Although the ISA/IEC 62443-4-1 document considers the possibility of analyzing and characterizing the vulnerabilities of an industrial component, it does not propose a technique to perform this task but instead refers to other standards for vulnerability handling processes [52]. In addition, it does not indicate how the data obtained from the analysis should be interpreted, and it does not define metrics or reference values for the current state of compliance with the requirement. Finally, it does not take into account neither the dependencies among the assets of the industrial component (dependency trees) or their evolution of the number of vulnerabilities over time.

2.1.2. Common Criteria

The Common Criteria (CC) for Information Technology Security Evaluation (ISO/IEC 15408) is an international standard that has a long tradition in computer security certification [53]. CC is a framework that provides assurance that the processes of specification, implementation, and evaluation of a computer security product have been conducted in a rigorous, standard, and repeatable manner at a level that is commensurate with the target environment for use.

To describe the rigor and depth of an evaluation, the CC defines seven Evaluation Assurance Levels (EALs) on an increasing scale [53], from EAL1 (the most basic) to EAL7 (the most stringent security level). It is important to notice that the EAL levels do not measure security itself. Instead, emphasis is given to functional testing, confirming the overall security architecture and design, and performing some testing techniques (depending on the EAL to be achieved).

The CC defines five tasks in the vulnerability assessment class, which manage the deepness of the vulnerability assessment. The higher the EAL to be achieved, the greater the number of tasks in the list to be performed [54]:

1. Vulnerability survey,
2. Vulnerability analysis,
3. Focused vulnerability analysis,
4. Methodical vulnerability analysis, and
5. Advanced methodical vulnerability analysis.

Every task checks for the presence of publicly known vulnerabilities. Penetration testing is also performed. The main difference among the five levels of vulnerability analysis described here is the deepness of the analysis of known vulnerabilities and the penetration testing.

The CC scheme defines the general activities, but it does not specify how to perform them, therefore no technique for analyzing vulnerabilities is proposed. The evaluator decides the most appropriate techniques for each test in each scenario and for each device, which adds a large degree of subjectivity to the evaluation. Furthermore, dependencies among vulnerabilities and assets are not considered in the analysis. Moreover, the CC does not define a procedure to manage the life cycle of the device. In other words, when updated, the whole device has to be reevaluated [55–58]. Finally, although the usage of metrics is encouraged by the CC, it does not propose any explicitly defined metric to be used during the evaluation.

2.2. Vulnerability Analysis Methodologies

Vulnerability analysis is a key step towards the security evaluation of a device. Consequently, many research efforts have been focused on solving this issue. In this subsection, the most relevant works related to vulnerability analysis are reviewed.

Homer et al. [59] present a quantitative model for computer networks that objectively measures the likelihood of a vulnerability. Attack graphs and individual vulnerability metrics, such as CVSS and probabilistic reasoning are applied to produce a sound risk measurement. However, the main drawback is that their work is only applicable to computer networks. Although they propose new metrics based on the CVSS for probabilistic calculations, they do not integrate standards such as CAPEC to enhance their approach centered on possible attacks and privilege escalation. They also fail to establish a relationship among existing vulnerabilities, and they fail to obtain the source problem causing each vulnerability.

Zhang et al. [60,61] developed a quantitative model that can be used to aggregate vulnerability metrics in an enterprise network based on attack graphs. Their model measures the likelihood that breaches can occur within a given network configuration, taking into consideration the effects of all possible interplays between vulnerabilities. This research is centered on computer networks, using attack graphs. Although the proposed model is capable of managing shared dependencies and cycles, only CVSS-related metrics are used. Moreover, this model assumes that the attacker knows all of the information in the generated attack graphs. Finally, the method that they proposed for the aggregation of metrics is not valid for vulnerability analysis, because the dependency between vulnerabilities reflected in attacks graphs are is not trivially obtained.

George et al. [30] propose a graph-based model to address the security issues in Industrial IoT (IIoT) networks. Their model is useful because it represents the relationships among entities and their vulnerabilities, serving as a security framework for the risk assessment of the network. Risk mitigation strategies are also proposed. Finally, the authors discuss a method to identify the strongly connected vulnerabilities. However, the main drawback of this work is that each node of the generated attack graph represents a vulnerability instead of representing a device or an asset of that device. This leads to a loss of information in the analysis because there is no way to know which vulnerability belongs to which device. Moreover, these methods need to know the relationships among present vulnerabilities in the devices. This information is not trivially obtained, and a human in the loop is needed. The proposals of [62,63] follow a similar graph-based approach to study the effects of cascade failures in the power grid and a subway network.

Poolsappasit et al. [64] propose a risk management framework using Bayesian networks that enables a system administrator to quantify the chances of network compromise at various levels. The authors are able to model attacks on the network, and also to integrate standardized information of the vulnerabilities involved, such as their CVSS score. Although their proposed model lends itself to dynamic analysis during the deployed phase of the network, these results can only be applied to computer networks where the relationship among the existing vulnerabilities is known. Meanwhile, the prior probabilities that are used in the model are assigned by network administrators, and hence are subjective. The proposed model also has some issues related to scalability.

Muñoz-González et al. [65] propose the use of efficient algorithms to make an exact inference in Bayesian attack graphs, which enables static and dynamic network risk assessments. This model is able to compute the likelihood of a vulnerability and can be extended to include zero-day vulnerabilities, attacker's capabilities, or dependencies between vulnerability types. Although this model is centered on studying possible attacks, it fails to integrate standards (such as CAPEC) that are related to attack patterns. Moreover, the generated graphs are focused on privilege escalation, trust, and users, rather than including information about vulnerabilities and the analyzed device.

Liu et al. [66] carry out a detailed assessment of vulnerabilities in IoT-based critical infrastructures from the perspectives of applications, networking, operating systems, software, firmware, and hardware. They highlight the three key critical infrastructure IoT-based cyber-physical systems (i.e., smart transportation, smart manufacturing, and smart grid). They also provide a broad collection of attack examples upon each of the key applications. Finally, the authors provide a set of best practices and address the necessary steps to enact countermeasures for any generic IoT-based critical infrastructure system. Nevertheless, their proposal is focused on attacks and countermeasures, and it leaves aside the inner analysis of the targets. Continuous evaluation over time is not considered in this proposal, and no enhancements of the development process are generated. On the other hand, Pascale et al. [67] proposed the analysis in both spatial and temporal dimensions for intrusion detection.

Hu et al. [68] propose a network security risk assessment method that is based on the Improved Hidden Markov Model (I-HMM). The proposed model reflects the security risk status in a timely and intuitive manner, and it detects the degree of risk that different hosts pose to the network. Although this is a promising approach, it is centered on computer networks and is at a higher abstraction level. No countermeasure or enhancement in the development process is proposed or generated.

Zografopoulos et al. [13] provide a comprehensive overview of the Cyber-Physical System (CPS) security landscape, with an emphasis on Cyber-Physical Energy Systems (CPES). Specifically, they demonstrate a threat modeling methodology to accurately represent the CPS elements, their interdependencies, as well as the possible attack entry points and system vulnerabilities. They present a CPS framework that is designed to delineate the hardware, software, and modeling resources that are required to simulate the CPS. They also construct high-fidelity models that can be used to evaluate the system's performance under adverse scenarios. The performance of the system is assessed using scenario-specific metrics. Meanwhile, risk assessment enables system vulnerability prioritization, while factoring in the impact on the system's operation. Although this research work is comprehensive, it is focused on enhancing the existing adversary and attack modeling techniques of CPSs of the energy industry. Moreover, their model does not integrate the internal structure of the target of evaluation, and it does not take both software and hardware into account for the evaluation. Continuous evaluation over time is not considered. Finally, they do not propose countermeasures or any kind of mechanism to enhance the security or the development of the CPSs.

Most of the works reviewed here are more focused on modeling threats and attacks, instead of using their results to propose enhancements during other steps in the life cycle of CPS (e.g., development, and maintenance). It is worth noting that they are still more focused on software evaluation, while hardware is usually neglected in their proposals.

As shown in this review, most of the research has adopted dependency trees, attack graphs, or directed graphs as the main tool to manage and assess vulnerabilities in computer networks. Graphs are an efficient technique to represent the relationships between entities, and they can also effectively encode the vulnerability relations in the network. Furthermore, the analysis of the graph can reveal the security-relevant properties of the network. For fixed infrastructure networks, graphical representations, such as attack graphs, are developed to represent the possible attack paths by exploiting the vulnerability relationships. For these reasons, vulnerability analysis techniques based on directed graphs are frequently found in the literature [69]. However, despite their potential, these analysis techniques have been relegated to vulnerability analysis in computer networks. Graph-based analysis has rarely been applied to industrial components.

2.3. Security Metrics

Standards of measurement and metrics are a powerful tool to manage security and for making decisions [70–72]. If carefully designed and chosen, metrics can provide a quantitative, repeatable, and reproducible value. This value is selected to be related to the property of interest of the systems under test (e.g., number and distribution of vulnerabilities). The use of metrics enables results to be compared over time, and among different devices. In addition, they can also be used to systematically improve the security level of a system or to predict this security level at a future point in time.

Although the capabilities of metrics have been demonstrated, they are not free of drawbacks. In our previous research work [72], we performed a systematic review of the literature and standards. To detect possible gaps, our objective was to find which types of metrics have been proposed and in which fields have been applied. This research work concludes that, in general, standards encourage the use of metrics, but they do not usually propose any specific set of metrics. If metrics are proposed, then they are conceived to be applied at a higher level (i.e., organization level), and then cannot be applied to industrial components. This type of metric is usually related to measuring the return on security investment, security budget allocation, and reviewing security-related documentation.

Our previous results also highlight that scientific papers have focused their efforts on software-related metrics: 77.5% of the analyzed metrics were exclusively applicable to software (e.g., lines of code, number of functions, and so on), whereas only 0.6% were related exclusively to hardware (e.g., side-channel vulnerability factor metric). In addition, 14.8% of them could be applied to both software and hardware (e.g., the historically exploited vulnerability metric that measures the number of vulnerabilities exploited in the past), and the remaining 7.1% are focused on other aspects, such as user usability. This shows that there is a clear lack of hardware security metrics in the literature, and the main contributions are centered on software security.

Other research works also reveal common problems across security metrics [73,74]:

- Hardly any security metric has a solid theoretical foundation or empirical evidence in support of the claimed correlation.
- Many security metrics lack an adequate description of the scale, unit, and reference values to compare and interpret the results.
- Only a few implementations or programs were available to test these security metrics and only one of the analyzed papers performed some kind of benchmarking or comparison with similar metrics.
- The information provided in the analyzed papers is insufficient to understand whether the proposed metrics are applicable in a given context, or how to use them.

Under this scenario, it seems reasonable that future research should be focused on the development of a convincing theoretical foundation, empirical evaluation, and systematic improvement of existing approaches, in an attempt to solve the lack of widely accepted solutions. In this research work, metrics constitute a key element. They are developed to analyze the distribution of vulnerabilities and to track their evolution over time.

3. Proposed Approach

In this research work, we propose an EDG model for the continuous assessment of vulnerabilities over time in industrial components. The proposed model is intended to:

- Identify the root causes and nature of vulnerabilities, which will enable the extraction of new requirements and test cases.
- Support the prioritization of patching.
- Track vulnerabilities during the whole lifespan of industrial components.
- Support the development and maintenance of industrial components.

To accomplish this task, the proposed model comprises two basic elements: (1) the model itself, which is capable of representing the internal structure of the system under test; (2) a set of metrics, which allow conclusions to be drawn about the origin, distribution, and severity of vulnerabilities. Both the model and metrics are very flexible and exhibit some properties that make them suitable for industrial components, and can also be applied to enhance the ISA/IEC 62443 standard.

The content in this section is distributed into four sections, namely:

1. Model: The proposed model is explained, together with the systems in which it can be applied and the algorithms that are used to build it.
2. Metrics: Metrics are a great tool to measure the state of the system and to track its evolution. The proposed metrics and their usage are described in this section.
3. Properties: The main features of the proposed model and metrics (e.g., granularity of the analysis, analysis over time, and patching policy prioritization support) are described in detail.
4. Applicability: Even though the reviewed standards exhibit some gaps, the proposed model aims to serve as the first step towards generating a set of tools to perform a vulnerability analysis in a reliable and continuous way. This last section will discuss the requirements of the ISA/IEC 62443-4-1 that can be enhanced using our model.

3.1. Description of the Model

The proposed model is based on directed graphs. It requires knowledge of the internal structure of the device to be evaluated (i.e., the assets, both hardware and software, that comprise it and the relationships between them). This section defines the most basic elements that make up the model, the algorithms to build it for any given system, and its graphical representation.

Definition 1. *A System Under Test (SUT) (following the denomination in the ISA/IEC 62443 standard [47], the SUT may be an industrial component, a part of an industrial component, a set of industrial components, a unique technology that may never be made into a product, or a combination of these) is now represented by an Extended Dependency Graph (EDG) model $G = (\langle A, V \rangle, E)$ that is based on directed graphs, where A and V represent the nodes of the graphs, and E represents its edges or dependencies:*

- $A = \{a_1, \ldots, a_n\}$ *represents the set of assets in which the SUT can be decomposed, where n is the total number of obtained assets. An asset a is any component of the SUT that supports information-related activities and includes both hardware and software [75–77]. Each asset is characterized by its corresponding Common Platform Enumeration (CPE) [78–80] identifier, while its weaknesses are characterized by the corresponding CWE identifier. In the EDG model, the assets are represented by three types of nodes in the directed graphs (i.e., root nodes, asset nodes, and cluster).*
- $V = \{v_1, \ldots, v_q\}$ *represents the set of known vulnerabilities that are present in each asset of A, where q is the total number of vulnerabilities. They are characterized by the corresponding CVE and CVSS values. In the EDG model, vulnerabilities are represented using two types of nodes in the directed graphs (i.e., known vulnerability nodes and clusters).*
- $E = \{e_{ij} | \forall i, j \in \{1, \ldots, n+q\}$ *such that $i \neq j\}$ represents the set of edges or dependencies among the assets, and between assets and vulnerabilities. e_{ij} indicates that a dependency relation is established from asset a_i to asset a_j. Dependencies are represented using two different types of edges in the EDG (i.e., normal dependency and deprecated asset/updated vulnerability edges).*

In other words, the EDG model can represent a system, from its assets to its vulnerabilities, and its dependencies as a directed graph. Assets and vulnerabilities are represented as nodes, whose dependencies are represented as arcs in the graph. The information in the EDG is further enhanced by introducing metrics.

The EDG model of a given SUT will include four types of node and two types of dependency. The graphical representation for each element is shown in Table 1. Figure 1 shows an example of a simple EDG and its basic elements. All of the elements that make up an EDG will be explained in more detail below:

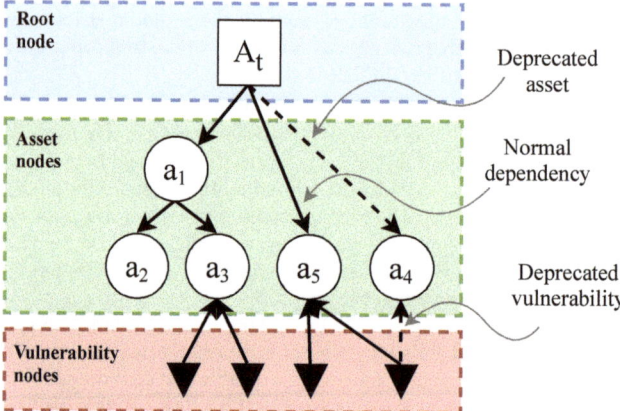

Figure 1. Basic elements of an EDG. Note that clusters are not displayed in this figure. For clusters, see Figure 2. For metric definitions, see Section 3.2.

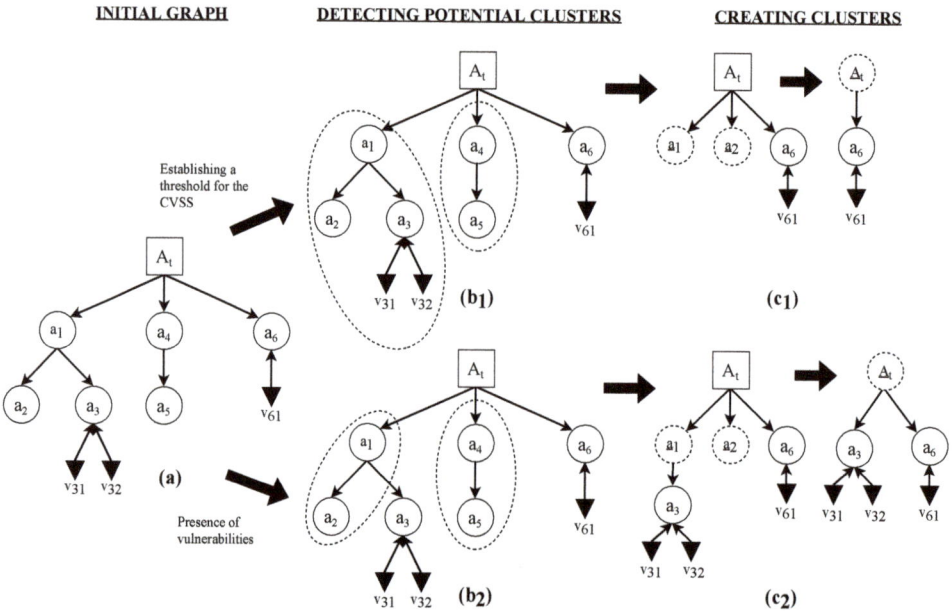

Figure 2. Creating clusters. Application of the two proposed criteria to the creation of clusters to simplify the graph, where (**a**) represents the initial EDG: (1) Establishing a threshold to select which vulnerability stays outside the cluster (upper side). In step (**b$_1$**), potential clusters are detected according to the established threshold, while in (**c$_1$**) the final EDG with the generated clusters is shown. The severity value (CVSS) for v_{31} and v_{32} is supposed to be lower than the establish threshold. (2) Choosing the absence of vulnerabilities as the criterion to create clusters (lower side). In step (**b$_2$**), nodes with no vulnerability are detected. In (**c$_2$**), the final EDG with the generated clusters is shown.

Table 1. Overview of the information that is necessary to define each of the EDG elements.

Symbol	Notation	Meaning	Values
☐	$A(t)$	Root Node / Device Node	$CPE_{current}$
◯	$a(t)$	Asset Node	$CPE_{previous}, CPE_{current}, CWE_{a_i}(t)$
⬭	$\underline{a}(t)$	Cluster	$\{CPE_{previous}, CPE_{current}, CWE_{a_i}(t)\}, \{CVE_{a_i}(t), CVSS_{v_i}(t), CAPEC_{w_i}(t)\}, \{Dependencies\}$
▼	$v(t)$	Known Vulnerability Node	$CVE_{a_i}(t), CVSS_{v_i}(t), CAPEC_{w_i}(t)$
⟶	$e(t)$	Dependency Relation	—
--→	$e(t)$	Updated Asset / Patched Vulnerability	—

3.1.1. Types of Node

The EDG model uses four types of nodes:

- Root nodes represent the SUT,
- Asset nodes represent each one of the assets of the SUT,
- Known vulnerability nodes represent the vulnerabilities in the SUT, and
- Clusters summarize the information in a subgraph.

Root nodes (collectively, set G_R) are a special type of node that represents the whole SUT. Any EDG starts in a root node and each EDG will only have one single root node, with an associated timestamp (t) that indicates when the last check for changes was done. This timestamp is formatted following the structure defined in the ISO 8601 standard for date and time [81].

Asset nodes (collectively, set G_A) represent the assets that comprise the SUT. The EDG model does not impose any restrictions on the minimum number of assets that the graph must have. However, the SUT can be better monitored over time when there is a higher number of assets. Moreover, the results and conclusions obtained will be much more accurate. Nevertheless, each EDG will have as many asset nodes as necessary, and the decomposition of assets can go as far and to as low-level as needed.

Each asset node node will be characterized by the following set of values:

- $CPE_{current}$: Current value for the CPE. This points to the current version of the asset it refers to.
- $CPE_{previous}$: Value of the CPE that identifies the previous version of this asset. This will be used by the model to trace back all the versions of the same asset over time, from the current version to the very first version.
- $CWE_{a_i}(t)$: Set of all the weaknesses that are related to the vulnerabilities present in the asset. The content of this list can vary depending on the version of the asset.

Figure 3 illustrates how the tracking of the versions of an asset using CPE works. On the one hand, version a_i is the current version of asset a. It contains its current CPE value and the CPE of its previous version. On the other hand, a_2 and a_1 are previous versions of asset a. The last value of a_1 points to a null value. This indicates that it is the last value in the chain, and therefore the very first version of the asset a.

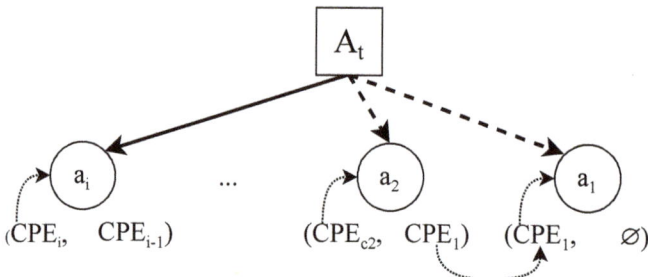

Figure 3. Tracking dependencies between the previous and current CPE values for asset a.

Known vulnerability nodes (collectively, set G_V) represent a known vulnerability present in the asset that it relates to. Each asset will have a known vulnerability node for each known vulnerability belonging to that asset. Assets alone cannot tell how severe or dangerous the vulnerabilities might be, so unique characterization of vulnerabilities is crucial [30].

To identify each known vulnerability node, each will be characterized by the following set of features (formally defined in Section 3.2:

- $CVE_{a_i}(t)$: This serves as the identifier of a vulnerability of asset a_i.
- $CVSS_{v_i}(t)$: This metric assigns a numeric value to the severity of vulnerability v_i. Each CVE has a corresponding CVSS value.
- $CAPEC_{w_i}(t)$: Each vulnerability (CVE) is a materialization of a weakness (CWE) w_i that can be exploited using a concrete attack pattern. In many cases, each CWE has more than one Common Attack Pattern Enumeration and Classification (CAPEC) [82,83] associated. Consequently, this field is a set that contains all the possible attack patterns that can exploit the vulnerability that is being analyzed.

Clusters (collectively, set G_S) are a special type of node that summarizes and simplifies the information contained in a subgraph in an EDG. Figure 2 shows how the clusters work.

To identify each cluster, and to be able to recover the information that they summarize, each is characterized by the data that define each of the elements that they contain: $\{CPE_{previous}, CPE_{current}, CWE_{a_i}(t)\}$, $(CVE_{a_i}(t), CVSS_{v_i}(t), \{CAPEC_{w_i}(t)\})$, and their dependencies.

Two types of criteria can be used to create clusters and to simplify the obtained graph Figure 2:

1. Absence of vulnerabilities: Using this criterion, clusters will group all nodes that contain no associated vulnerabilities.
2. CVSS score below a certain threshold: With this criterion, a threshold for the CVSS scores will be chosen. Nodes whose CVSS score is less than the defined threshold will be grouped into a cluster.

3.1.2. Types of Edge

In the EDG model, edges play a key role in representing dependencies. Two types of edge can be identified:

- Normal dependencies relate two assets, or an asset and a vulnerability. They represent that the destination element depends on the source element. Collectively, they are known as set G_D.
- Deprecated asset or patched vulnerability dependencies indicate when an asset or a vulnerability is updated or patched. They represent that the destination element used to depend on the source element. Collectively, they are known as set G_U.

The possibility of representing old dependencies brings the opportunity to reflect the evolution of the SUT over time. When a new version of an asset is released, or a vulnerability is patched, the model will be updated. Their dependencies will change from a normal dependency to a deprecated asset or vulnerability dependency to reflect that change.

3.1.3. Conditions of Application of EDGs

The EDG model is applicable to SUTs that meet the following set of criteria:

- Software and hardware composition: In our approach, the model is created by means of a white-box analysis. The absence of or impossibility to perform a white-box analysis limits the ability to create an accurate model. Some knowledge about the internal structure and code is expected. This information is usually only known by the manufacturer of the component unless the component is publicly available or open-source. It should be also possible to decompose the SUT into simpler assets to generate a relevant EDG.
- Existence of publicly known vulnerabilities: The EDG model focuses on known vulnerabilities. This is not critical because many industrial components use commercial or open-source elements. The SUT must be composed of assets for which public information is available. If the majority of SUT assets are proprietary, or the SUT is an ad hoc development that is never exposed, then the generated EDG will not evolve. Therefore, the analysis will not be relevant.

3.1.4. Steps to Build the Model

This section explains the process and algorithms that were used to build the corresponding EDG of a given SUT. The main scenarios that can be found are also described.

Before extracting useful information about the SUT, the directed graph associated with the SUT has to be built. This comprises several steps, which are described in the following paragraphs (see the flowchart in Figures 4 and 5):

Step 1—Decompose the SUT into assets. For the model to work properly, it relies on the SUT being able to be decomposed into assets. With this in mind, the first step involves obtaining the assets of the SUT, either software or hardware. In the CC, this process is called modular decomposition of the SUT [53]. Ideally, every asset should be represented in the decomposition process, but this is not compulsory for the model to work properly. Each one of the assets obtained in this step will be represented as an asset node. In this step, the dependencies among the obtained assets are also added as normal dependencies.

Step 2—Assign a CPE to each asset. Once the assets and their dependencies have been identified, the next task is to assign the corresponding CPE identifier to each asset. If there is no publicly available information of a certain asset, and therefore, it does not have a CPE identifier, then it is always possible to generate one using the fields described in the CPE naming specification documents [79] for internal use in the model.

Step 3—Add known vulnerabilities to the assets. In this step, the vulnerabilities ($CVE_{a_i}(t)$) of each asset are set. This is done by consulting public databases of known vulnerabilities [34,84] looking for existing vulnerabilities for each asset. When a vulnerability is found, it is added to the model of the SUT, including its dependencies. If there were no known vulnerabilities in an asset, then the asset would become the last leaf of its branch. In this step, the corresponding value of the CVSS of each vulnerability is also added to the model.

Step 4—Assign to each asset its weaknesses and possible CAPECs. After the vulnerabilities, the corresponding weaknesses to each vulnerability ($CWE_{a_i}(t)$) are added, along with the corresponding attack patterns ($CAPEC_{w_i}(t)$) for each weakness. If there is no known vulnerability in an asset, then there will be no weaknesses. Meanwhile, it would be possible to have a known vulnerability in an asset, but no known weakness or attack pattern for that vulnerability. Finally, more than one CAPEC can be assigned to the same weakness. Consequently, it would be common to have a set of possible CAPECs that can be used to exploit the same weakness. It is worth noting that not all of them could be applied in every scenario.

Step 5—Computing Metrics and tracking the SUT. At this point, the EDG of the SUT is completed with all the public information that can be gathered. This last step is to calculate the metrics defined (for further information, see Section 3.2), generate the corresponding reports and track the state of the SUT for possible updates in the information of the model. This step is always triggered when the SUT is updated. This can imply that a new asset can appear, an old asset can disappear, an old vulnerability can be patched, or a new one can appear in the SUT. All of these scenarios will be reflected in the model as they arise during its life cycle.

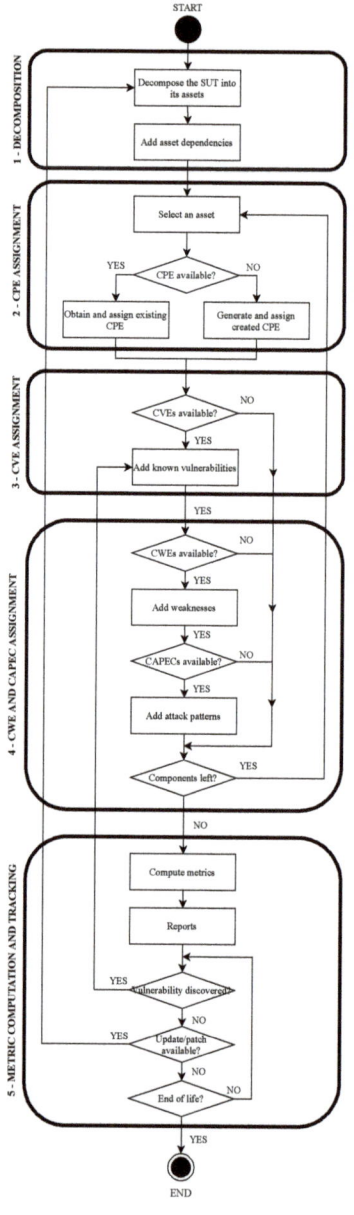

Figure 4. Algorithm to generate the initial EDG of a given SUT.

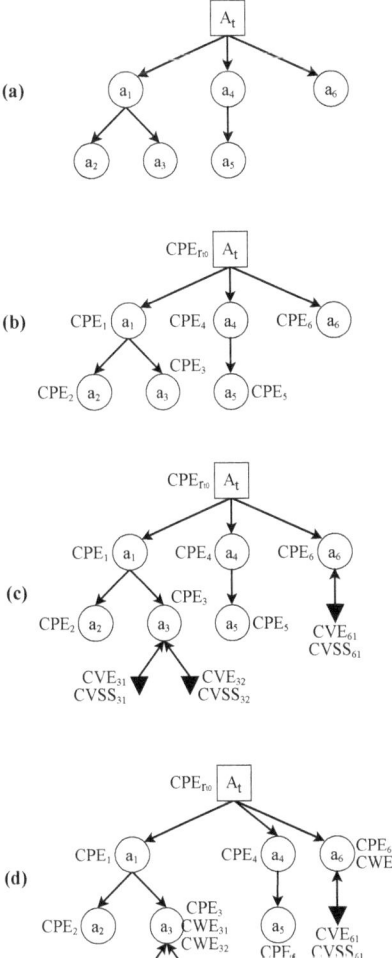

Figure 5. Example of the process of building the EDG model of a given SUT A. (**a**) Decompose of the SUT into assets. (**b**) Assign a CPE to each asset. (**c**) Add known vulnerabilities. (**d**) Add weaknesses and attack patterns.

3.2. Security Metrics

The EDG model that was proposed in the previous sections is by itself capable of representing the internal structure of the SUT, and it can display it graphically for the user. This representation not only includes the internal assets of the SUT, but also captures their relationships, existing vulnerabilities, and weaknesses. Moreover, assets, vulnerabilities, and weaknesses are easily identified using their corresponding CPE, CVE, and CWE values, respectively. Altogether, this constitutes a plethora of information that the model can use to improve the development and maintenance steps of the SUT, enhance its security, and track its status during its whole life cycle. Metrics are a great tool to integrate these features into the model.

Metrics can serve as a tool to manage security, make decisions, and compare results over time. They can also be used to systematically improve the security level of an industrial component or to predict its security level at a future point in time.

In this section, the basic definitions that serve as the foundation of the metrics are described. Then, the proposed metrics are introduced to complement the functionality of the EDG model. The main feature of these metrics is that they all depend on time as a variable, so it is possible to capture the actual state of the SUT, track its evolution over time, and compare the results.

3.2.1. Basic Definitions

In this section, the basic concepts on which the definitions of the metrics will be based are formalized.

Definition 2. *The set of all possible weaknesses at a time t is represented as $CWE(t)$, where*

$$CWE(t) = \{cwe_1, \ldots, cwe_m\} \quad (1)$$

and m is the total number of weaknesses at time t. This set contains the whole CWE database defined by MITRE [38].

Definition 3. *The set of all of the possible vulnerabilities at a time t is represented as $CVE(t)$ where*

$$CVE(t) = \{cve_1, \ldots, cve_p\} \quad (2)$$

and p is the total number of vulnerabilities. This set contains the whole CVE database defined by MITRE [34].

Definition 4. *The set of all possible attack patterns at a time t is represented as $CAPEC(t)$, where*

$$CAPEC(t) = \{capec_1, \ldots, capec_q\} \quad (3)$$

and q is the total number of attack patterns at time t. This set contains the whole CAPEC database defined by MITRE [82].

Definition 5. *The set of weaknesses of an asset a_i at a time t is defined as*

$$CWE_{a_i}(t) = \{cwe_j | cwe_j \text{ is in the asset } a_i \text{ at time } t \wedge cwe_j \in CWE(t) \\ \wedge \forall k \neq j, cwe_j \neq cwe_k\} \quad (4)$$

From this expression, the set of all the weaknesses of a particular asset throughout its life cycle is defined as

$$CWE_{a_i} = \bigcup_{t=1}^{T} CWE_{a_i}(t) \quad (5)$$

where $|CWE_{a_i}|$ is the total number of non-repeated weaknesses in its entire life cycle.

Definition 6. *The set of vulnerabilities of an asset a_i at a time t is defined as*

$$CVE_{a_i}(t) = \{cve_j | cve_j \text{ is in the asset } a_i \text{ at time } t \wedge cve_j \in CVE(t)\} \quad (6)$$

From this expression, the set of vulnerabilities of an asset throughout its entire life cycle is defined as

$$CVE_{a_i} = \bigcup_{t=1}^{T} CVE_{a_i}(t) \quad (7)$$

where $|CVE_{a_i}|$ is the total number of vulnerabilities in its entire life cycle.

Definition 7. *The set of weaknesses of a SUT A with n assets at a time t is defined as:*

$$CWE_A(t) = \bigcup_{i=1}^{n} CWE_{a_i}(t) \tag{8}$$

Definition 8. *The set of vulnerabilities of a SUT A with n assets at a time t is defined as:*

$$CVE_A(t) = \bigcup_{i=1}^{n} CVE_{a_i}(t) \tag{9}$$

Definition 9. *The set of vulnerabilities associated with the weakness cwe_j and to the asset a_i at a time t is defined as:*

$$CVE_{a_i|cwe_j}(t) = \{cve_k | cve_k \text{ associated with weakness } cwe_j \text{ and to asset } a_i \text{ at time } t\} \tag{10}$$

It is worth noting that CWE is used as a classification mechanism that differentiates CVEs by the type of vulnerability that they represent. A vulnerability will usually have only one associated weakness, and weaknesses can have one or more associated vulnerabilities [85].

Definition 10. *The partition j of an asset a_i at time t conditioned by a weakness cwe_k is defined as*

$$CVE_{a_i|cwe_k}(t) = \{cwe_l | cwe_l = cwe_k \land cwe_l \in CVE_{a_i}(t)\} \tag{11}$$

Definition 11. *The partition j of the SUT A at time t conditioned by a weakness cwe_k is defined as*

$$CVE_{A|cwe_k}(t) = \{cwe_l | cwe_l = cwe_k \land cwe_l \in CVE_A(t)\} \tag{12}$$

Definition 12. *The set of attack patterns associated to a weakness w_i at a time t is defined as*

$$CAPEC_{w_i}(t) = \{capec_j | capec_j \text{ can exploit weakness } w_i \text{ at time } \\ t \land capec_j \in CAPEC(t)\} \tag{13}$$

Definition 13. *The set of metrics that are defined in this research work based on the EDG model is defined as*

$$M = \{m_1, \ldots, m_r\} \tag{14}$$

where r is the total number of metrics. This set can be extended, defining more metrics according to the nature of the SUT.

3.2.2. Metrics

This section will describe the metrics that were defined based on the EDG model and the previous definitions. Although it might seem trivial, the most interesting feature of these metrics is that they all depend on time. Using time as an input variable for the computation of the metrics opens the opportunity to track results over time, compare them, and analyze the evolution of the status of the SUT. Furthermore, some metrics take advantage of time to generate an accumulated value, giving information about the life cycle of the SUT. Table 2 shows all of the proposed metrics, their definition, and their reference values.

Table 2. Proposed metrics for the model.

	Metric	Definition	Reference Value
VULNERABILITIES	$M_0(A) = \frac{|CVE_A(t)|}{n(t)}$	Arithmetic mean of vulnerabilities in the SUT A, where $n(t)$ is the number of assets in a SUT at a time t. M_0 shows how many vulnerabilities would be present in each asset if they were evenly distributed among the assets of the SUT. The result of M_0 can serve as a preliminary analysis of the SUT, related to the criticality of its state. From Equation (8).	$M_0 < 1$: The number of vulnerabilities is lower than the number of assets. $M_0 \geq 1$: Every asset has at least one vulnerability.
	$M_1(A, t) = |CVE_A(t)|$	Number of vulnerabilities in a SUT A at time t. From Equation (8).	Ideally, the values of M_1 should be zero (no vulnerability in A), but the lower the value of M_1, the better.
	$M_2(A) = \sum_{t=1}^{T} |CVE_A(t)| = \sum_{t=1}^{T} M_1(A, t)$	Number of vulnerabilities in a SUT A throughout its entire life cycle T. This metric computes the accumulated value of the number of vulnerabilities of a SUT throughout its entire life cycle. From Equation (8).	The lower the value of M_2, the better.
	$M_3(a_i, t) = |CVE_{a_i}(t)|$	Number of vulnerabilities in an asset a_k at time t. The values of M_3 can be useful during a vulnerability analysis, or when performing a penetration test, to identify the asset with more vulnerabilities. From Equation (6).	Ideally, the value of M_3 should be zero.
	$M_4(a_k, t) = \frac{|CVE_{a_k}(t)|}{\sum_{i=1}^{n} |CVE_{a_i}(t)|}$	Relative frequency of vulnerabilities of the asset a_k at a time t. From Equation (6).	Ideally, the value of M_4 should be zero, or at least $M_4 \leq \frac{1}{n(t)}$, being $n(t)$ the number of assets in the SUT. This value can also be expressed as the percentage of vulnerabilities of asset a_i respect to the total number of vulnerabilities in the SUT, $M_4(a_k, t) = \frac{|CVE_{a_k}(t)|}{\sum_{i=1}^{n} |CVE_{a_i}(t)|} \cdot 100$
	$M_5(a_i, cwe_j, t) = |CVE_{a_i|cwe_j}(t)|$	Multiplicity of weakness cwe_j of the asset a_i at a time t. This metric represents the number of times a weakness is present among the vulnerabilities of the asset a_i. This is possible because a vulnerability can have associated the same weakness as other vulnerabilities. From Equation (9).	Ideally, the value of M_5 should be zero, or at least, $M_5 \leq \frac{|CVE_{A|cwe_j}(t)|}{n(t)}$, being $n(t)$ the number of assets in the SUT. The value of the metric could be further narrowed by assuming that cwe_j will be present in all but one asset, so $M_5 \leq \frac{|CVE_{A|cwe_j}(t)|}{n(t)-1}$ to be in acceptable values.
	$M_6(A, cwe_j, t) = |CVE_{A|cwe_j}(t)|$	Multiplicity of weakness cwe_j of the SUT A at a time t. This metric represents the number of times a weakness is present among the vulnerabilities of the SUT A. From Equation (11).	Ideally, the value of M_6 should be zero.
WEAKNESSES	$M_7(A, t) = |CWE_A(t)|$	Number of weaknesses in a SUT A at time t. From Equation (7).	Ideally, the value of M_7 should be zero (no weakness in A), but the lower the value of M_7, the better.
	$M_8(A) = \sum_{t=1}^{T} |CWE_A(t)| = \sum_{t=1}^{T} M_7(A, t)$	Number of weaknesses in a SUT A throughout its entire life cycle T. This metric computes the accumulated value of weaknesses of a SUT throughout its entire life cycle. From Equation (7).	The lower the value of M_8, the better.

In addition to the metrics in Table 2, the model allows the definition of other types of metrics according to the analysis to be performed, and the nature of the SUT (e.g., the vulnerability evolution function for SUT A up to time t for all vulnerabilities could be defined as the linear regression of the total number of vulnerabilities in each time t for SUT A, or using any other statistical model).

3.3. Properties

Together, the EDG model and the defined metrics exhibit a series of characteristics that make them suitable for vulnerability assessment. These properties represent an advantage over the techniques reviewed in the state of the art, including automatic inference of root causes, spatial and temporal distribution of vulnerabilities, and prioritization of patching, which will be described in the following subsections.

3.3.1. Automatic Inference of Root Causes

Each CWE natively contains information that is directly related to the root cause of a vulnerability. From this information, new requirements and test cases can be proposed.

3.3.2. Spatial and Temporal Distribution of Vulnerabilities

The key feature of the proposed model is the addition of the temporal dimension in the analysis of vulnerabilities. This makes it possible to analyze the location of the vulnerabilities both in space (in which asset) and time (their recurrence), which allows us to track the state of the device throughout the whole life cycle. This approach also enables further analysis of the SUT, by updating data in the model, such as new vulnerabilities that are found or new patches that are released.

Each time that a new vulnerability is found, or an asset is patched (i.e., via an update), the initial EDG is updated to reflect those changes. An example of this process can be seen in Figure 6.

At time t_0, the initial graph of the SUT A is depicted in Figure 6. Because there is no vulnerability at that time, this graph can be simplified using the cluster notation, with just a cluster containing all assets. At time t_1, a new vulnerability that affects the asset a_2 is discovered. At time t_2, the asset a_2 is updated. This action creates a new version of asset a_2, asset a_3. Because the vulnerability was not corrected in the new update, both versions contain the vulnerability that was initially presented in asset a_2. Finally, at time t_3, the asset a_3 is updated to its new version a_4, and the vulnerability is corrected.

This approach enables a further analysis of the SUT, including updated data, according to new vulnerabilities that are found or new patches that are released.

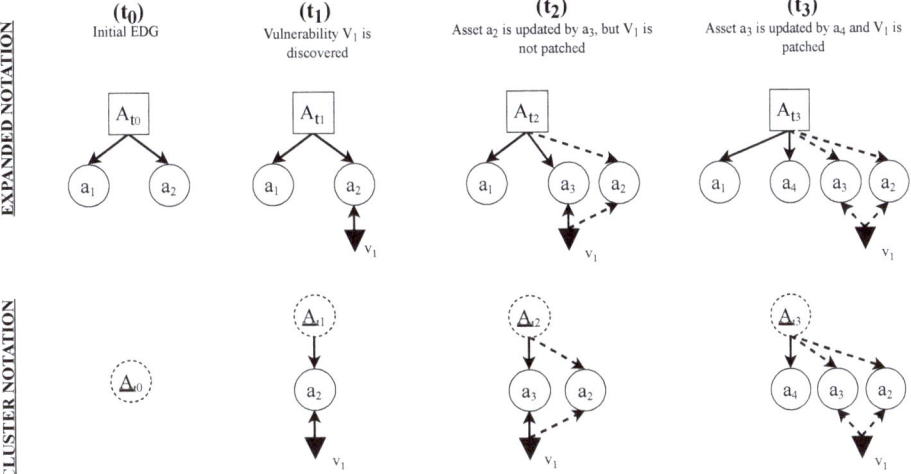

Figure 6. Representation of the temporal behavior in the graphical model using the two kinds of dependencies of the model. It is worth mentioning that these graphs could be further simplified by taking advantage of the cluster notation, as shown at the bottom of this figure.

3.3.3. Patching Policies Prioritization Support

The proposed model is not only able to include known vulnerabilities associated with an asset, but it also provides a relative importance sorting of vulnerabilities by CVSS. Relying on the resulting value, it is possible to assist in the vulnerability patching prioritization process. Furthermore, the presence of an existing exploit for a known vulnerability can be also be taken into account, when deciding which vulnerabilities need to be patched first. A high CVSS value combined with an available exploit for a given vulnerability is a priority when patching.

4. Real Use Case

In this section, we applied the EDG model to analyze the vulnerabilities of the Open-PLC project. For the sake of simplicity, the use case focuses on version one (V1) of OpenPLC. We centered the analysis on two of the assets that compose this version of the project: `libssl` and `nodejs`.

OpenPLC is the first functional standardized open-source Programmable Logic Controller (PLC), both in software and hardware [86–89]. It was mainly created for research purposes in the areas of industrial and home automation, the Internet of Things (IoT), and SCADA. Given that it is the only controller that provides its entire source code, it represents an engaging low-cost industrial solution—not only for academic research but also for real-world automation [90,91].

4.1. Structure of OpenPLC

The OpenPLC project consists of three parts:

1. Runtime: It is the software that plays the same role as the firmware in a traditional PLC. It executes the control program. The runtime can be installed in a variety of embedded platforms, such as the Raspberry Pi, and in Operating Systems (OSs) such as Windows or Linux.
2. Editor: An application that runs on a Windows or Linux OS that is used to write and compile the control programs that will be later executed by the runtime.
3. HMI Builder: This software is to create web-based animations that will reflect the state of the process, in the same manner as a traditional HMI.

When installed, the OpenPLC runtime executes a built-in webserver that allows OpenPLC to be configured and new programs for it to run to be uploaded. In this use case, we focused the analysis on the runtime of OpenPLC V1.

4.2. Setup Through the Analysis

Ubuntu Linux was selected as the platform to install the runtime of OpenPLC V1. Ubuntu Linux provides comprehensive documentation, previous versions are accessible, and software dependencies can easily be obtained.

To make the analysis fair, a contemporary operating system was selected, according to the version of Ubuntu that was available at the release time of OpenPLC V1. The Long Term Support (LTS) version was chosen because industry tends to work with the most stable version available of any software and security updates are provided for a longer time. OpenPLC V1 was released in 2016/02/05, so we found that Ubuntu 14.04 LTS was the most suitable version [92]. The setup consisted of OpenPLC installed on 14.04 LTS Ubuntu Linux in a virtual machine. All configuration options were by default.

4.3. Building the EDG

We built the entire EDG for OpenPLC V1, which can be found in Appendix B. Nevertheless, for the sake of clarity, we restricted this analysis in two ways: (1) focusing on two assets, `libssl` and `nodejs`; (2) integrating only security updates (discarding updates that introduced more functionalities). Table 3 shows the updates and their date of availability for both `libssl` [93] and `nodejs` [94] for Ubuntu 14.04 LTS. There were two security updates available for the amd64 architecture for each asset. Figure 7 illustrates step by step the partials EDG graphs, and Figure 8 shows the final EDG with all the updates merged in a single graph.

Table 3. Update information of both `libssl` and `nodejs`.

Asset	1st Update	Solved Vulnerabilities (CVSS)	2nd Update	Solved Vulnerabilities (CVSS)
libssl	2014/04/07	CVE-2014-0076 (1.9) CVE-2014-0160 (5.0)	2018/12/06	CVE-2018-5407 (1.9) CVE-2018-0734 (4.3)
nodejs	2014/03/27	—	2018/08/10	CVE-2016-5325 (4.3)

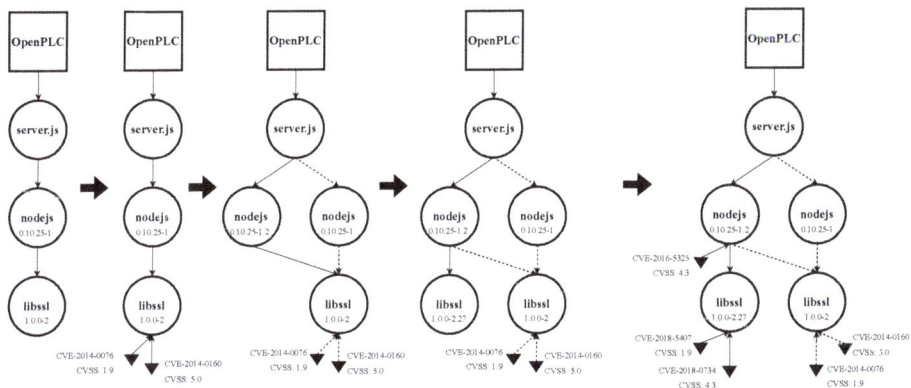

Figure 7. Temporal evolution of the EDG for OpenPLC V1 for both libss and nodejs.

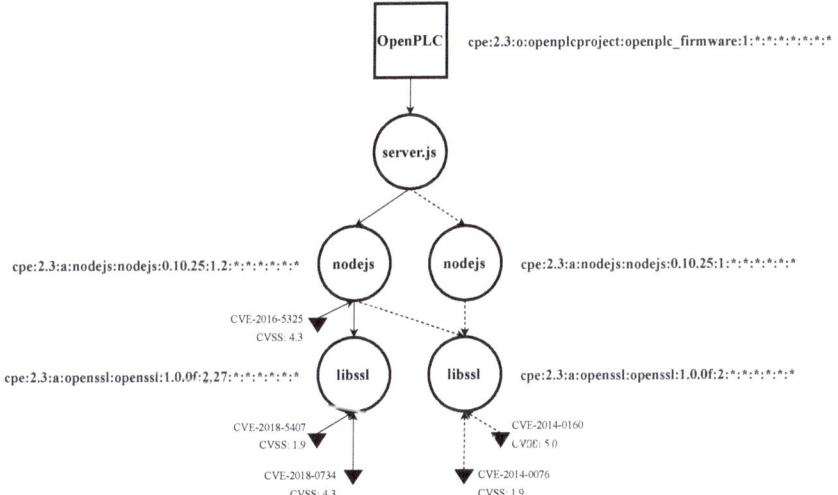

Figure 8. Final EDG for libssl and nodejs integrating all the updates for Ubuntu Linux 14.04 for amd64 architecture.

4.4. Analysis of the EDG

Using Figure 8 as reference, we can analyze the obtained EDG:

1. Analysis of the induced EDG model: The structure, assets, and dependencies are the focus of this first step.
   We can observe that libssl is used by nodejs, and they are not at the same level of the hierarchy. So vulnerabilities could propagate upwards through the EDG.
2. Vulnerability analysis: Vulnerability number, distribution, and severity are analyzed in this step. A proposal for vulnerability prioritization is also generated.
   We can highlight that nodejs had one vulnerability discovered after its first update, whereas libssl had vulnerabilities in both periods of time. We could argue that, as nodejs is the most accessible asset from the exterior, its vulnerabilities should be first addressed, even though the associated CVSS is not the highest one.
3. Weaknesses analysis: Finally, the root cause of each vulnerability is found. In this step, new requirements, test cases, and training activities are proposed based on the results of the analysis.

Table 4 shows the root cause for each vulnerability. Using this data, new requirements, test cases, and training activities were proposed (see Appendix C).

Table 4. Relationship between vulnerabilities and weaknesses for both `libssl` and `nodejs`.

CVE	CVSS	CWE	Description
CVE-2014-0076	1.9	CWE-310	Cryptographic Issues
CVE-2014-0160	7.5	CWE-119	Improper Restriction of Operations within the Bounds of a Memory Buffer
CVE-2016-5325	6.1	CWE-113	Improper Neutralization of CRLF Sequences in HTTP Headers ('HTTP Response Splitting')
CVE-2018-0734	5.9	CWE-327	Use of a Broken or Risky Cryptographic Algorithm
CVE-2018-5407	4.7	CWE-203	Observable Discrepancy
		CWE-200	Exposure of Sensitive Information to an Unauthorized Actor

5. Conclusions and Future Work

Vulnerability analysis is a critical task which ensures the security of industrial components. The EDG model that we propose performs continuous vulnerability assessment throughout the entire life cycle of industrial components. The model is built on a directed graph-based structure and a set of metrics based on globally accepted security standards. Metrics can be used by the model to improve the development process of the SUT, enhance its security, and track its status. The key feature of the proposed model is the addition of the temporal dimension in the analysis of vulnerabilities. The location of vulnerabilities can be analyzed in both space (in which asset) and time (their recurrence), which allows the state of the device to be tracked throughout the whole life cycle.

The model was successfully applied to the OpenPLC use case, which demonstrated its advantages, applicability, and potential. The use case showed that the model can assist in updating management activities, applying patching policies, launching training activities, and generating new test cases, and requirements. This has significant implications for cybersecurity evaluators, as it can serve as a starting point for identifying vulnerabilities, weaknesses, and attack patterns.

Further research will enhance the EDG by adding a mathematical model to aggregate the values of the CVSS metric for each asset, and a value for the whole SUT. This will enable the comparison of different SUTs over time. More improvements will be made in the prioritization of patching, taking into account the context and the functionalities of the SUT. Finally, historical information about the developers can be integrated into the EDG model to predict future vulnerabilities.

Author Contributions: Conceptualization, Á.L.-R., J.L.F. and I.G.; data curation, Á.L.-R.; formal analysis, Á.L.-R. and J.L.F.; funding acquisition, R.I. and J.L.F.; investigation, Á.L.-R.; methodology, Á.L.-R.; project administration, Á.L.-R.; resources, R.I., J.L.F. and I.G.; software, Á.L.-R.; supervision, Á.L.-R. and J.L.F.; validation, Á.L.-R., J.L.F. and I.G.; visualization, Á.L.-R. and J.L.F.; writing—original draft, Á.L.-R.; writing—review and editing, Á.L.-R., Rosa Iglesias, J.L.F. and I.G. All authors have read and agreed to the published version of the manuscript.

Funding: This work was partially supported by both the Department of Economic Development and Infrastructures, and the Ayudas Cervera para Centros Tecnológicos grant of the Spanish Centre for the Development of Industrial Technology (CDTI). This work was partially funded by REMEDY project (KK-2021/00091), EGIDA project (CER-20191012), and H2020 (957212).

Institutional Review Board Statement: Not applicable.

Informed Consent Statement: Not applicable.

Data Availability Statement: Not applicable.

Conflicts of Interest: The authors declare no conflict of interest.

Abbreviations

The following abbreviations are used in this manuscript:

AI	Artificial Intelligence
CC	Common Criteria
CAPEC	Common Attack Pattern Enumeration and Classification
COTS	Commercial Off-The-Shelf
CPE	Common Platform Enumeration
CPS	Cyber-Physical System
CVE	Common Vulnerabilities and Exposures
CVSS	Common Vulnerability Scoring System
CWE	Common Weakness Enumeration
EAL	Evaluation Assurance Level
EDG	Extended Dependency Graph
ES	Embedded System
IACS	Industrial Automation Control System
IoT	Internet Of Things
PLC	Programmable Logic Controller
SUT	System Under Test

Appendix A. Applicability in the Context of ISA/IEC 62443

In this section, the potential application of the proposed EDG model to the existing security standards is described. The proposed EDG model can be used isolated by itself, or in combination with other techniques that complement the analysis. In this sense, the EDG model can be used to enhance some tasks in the security evolution processes defined by security standards.

The ISA/IEC 62443-4-1 standard specifies 47 process requirements for the secure development of products used in industrial automation and control systems [49]. Thus, the EDG model was developed to enhance the execution of one of those requirements defined by the standard: the "SVV-3: Vulnerability testing" requirement, serving as a support for the execution of Practice 5—Security Verification and Validation testing. According to the SVV-3 requirement, both known and unknown vulnerability analysis has to be performed. The EDG model proposed in this research work is intended to support the identification of known vulnerabilities, their dependencies, and the possible consequences of their propagation, yielding the opportunity to analyze them systematically. Nevertheless, more requirements of the ISA/IEC 62443 can be mapped to one or more of the metrics defined in this research work. Using this relationship, it is possible to apply the EDG model to enhance the analysis and review of the following requirements:

Appendix A.1. Security Requirements—2: Threat Model (SR-2)

"A process shall be employed to ensure that all products have a threat model specific to the current development scope of the product. The threat model shall be reviewed and verified periodically" [49]. The proposed EDG model can serve as an abstraction of the threat model that has to be obtained. Moreover, the standard states that this threat model has to be reviewed periodically for updates. Given that the EDG of a given SUT evolves with every update, the threat model would be always up-to-date. Potential threats and their severity using the CVSS can also be analyzed with this proposal. Finally, these results can be used to enhance the risk assessment of the SUT.

Appendix A.2. Security Management—13: Continuous Improvement (SM-13)

"A process shall be employed for continuously improving the secure development life cycle" [49]. The EDG model can be used to identify recurrent issues in the development of an industrial component, due to its ability to track the state of a SUT over time. Consider the scenario where a piece of code contains an unknown vulnerability. For example, this code can implement a communication protocol or the generation of a cryptographic key.

If this piece of code is recurrently integrated into many types of devices, then when they are released to the market, the end-users can identify that vulnerability and report it to the product supplier. The EDG can reflect the presence of that vulnerability. If an EDG is done for each type of device, then this problem can be detected beforehand. Using the CWE, the root problem can be detected. With this information, new training and corrective actions can be proposed to avoid this issue.

Appendix A.3. Specification of Security Requirements—5: Security Requirements Review (SR-5)

"A process shall be employed to ensure that security requirements are reviewed, updated, and approved" [49]. As before, taking advantage of the previous scenario, the information extracted from the generated EDG model can be used to propose new requirements or to update the existing requirements.

Appendix A.4. Security Verification and Validation Testing—4: Penetration Testing (SVV-4)

"A process shall be employed to identify and characterize security-related issues via tests that focus on discovering and exploiting security vulnerabilities in the product" [49]. The EDG model facilitates the identification of possible entry points to the SUT when carrying out a penetration test. In addition, existing attack patterns (CAPEC) and weaknesses (CWE) can serve as a starting point to discover unknown vulnerabilities and exploits.

Appendix A.5. Management of Security-Related Issues—3: Assessing Security-Related Issues (DM-3)

"A process shall be employed for analyzing security-related issues in the product" [49]. When a new vulnerability is detected, end-users will report it to the product suppliers. Then, the corresponding EDG model of that SUT will be updated to reflect that change. This information, in addition to that previously contained in the EDG, can be used to obtain the severity value of the discovered vulnerability using the CVSS. This also facilitates the identification of root causes, related security issues, or the impact.

Table A1. Mapping between the developed metrics and the requirements they refer in the ISA/IEC 62443. SR (Security Requirements), SM (Security Management), SVV (Security Validation and Verification), DM (Management of Security-Related Issues).

Metric	SR-2	SR-5	SM-13	SVV-4	DM-3
$M_0(A) = \frac{|CVE_A(t)|}{n(t)}$	■	■	■	■	■
$M_1(A,t) = |CVE_A(t)|$	■	■	■	■	■
$M_2(A) = \sum_{t=1}^{T} |CVE_A(t)| = \sum_{t=1}^{T} M_1(A,t)$	□	■	■	□	□
$M_3(A,t) = |CVE_{a_i}(t)|$	■	■	■	■	□
$M_4(a_k,t) = \frac{|CVE_{a_k}(t)|}{\sum_{i=1}^{n} |CVE_{a_i}(t)|}$	□	■	■	□	□
$M_5(a_i, cwe_j, t) = |CVE_{a_i|cwe_j}(t)|$	■	■	■	■	□
$M_6(A, cwe_j, t) = |CVE_{A|cwe_j}(t)|$	■	■	□	■	■
$M_7(A,t) = |CWE_A(t)|$	■	□	□	■	■
$M_8(A) = \bigcup_{t=1}^{T} |CWE_A(t)| = \bigcup_{t=1}^{T} M_7(A,t)$	□	■	■	□	□

Finally, the ISA/IEC 62443-4-2 document defines four types of components of an IACS (i.e., software applications, embedded devices, host devices, network devices) [95]. The proposed model is capable of representing the inherent complexity of each of them.

Appendix B. EDG for OpenPLC V1

This appendix contains the generated EDG for OpenPLC V1.

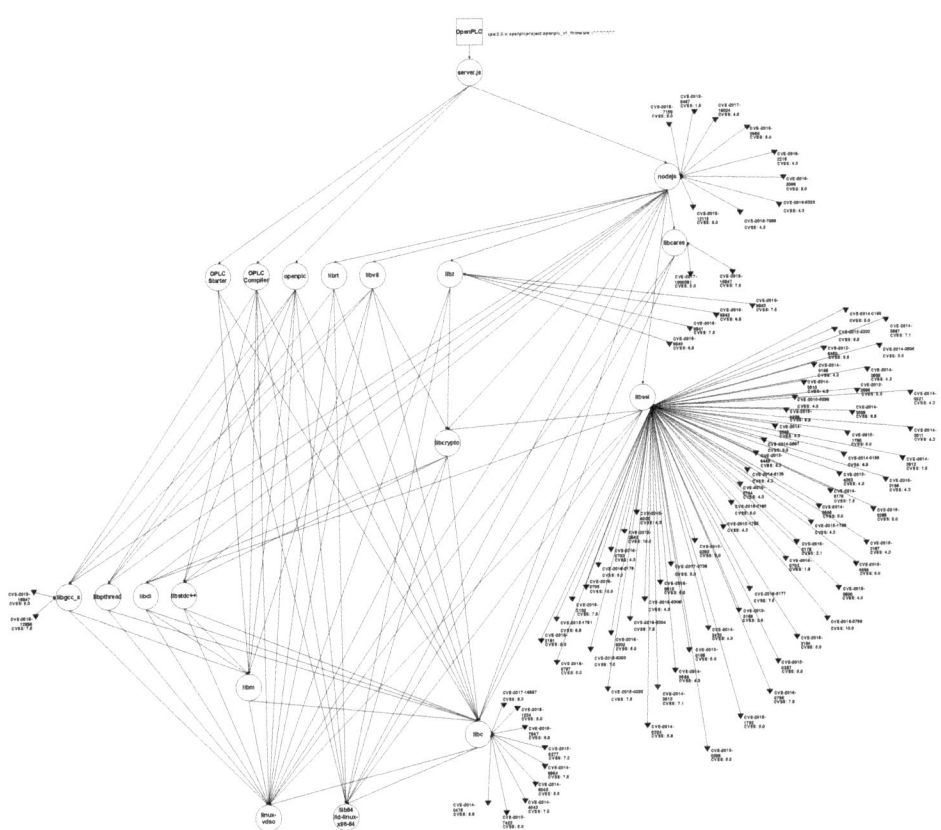

Figure A1. EDG for OpenPLC V1. Notice that, for simplicity, CWE and CAPEC values are omitted, and only the CPE identifier of the SUT is shown.

Appendix C. Proposed Requirements, Training, and Test Cases

In this appendix, we show the generated requirements, training, and test cases from the EDG model of OpenPLC V1.

Table A2. An example of generated requirements for OpenPLC V1.

CWE ID	Requirements
CWE-119	Use languages that perform their own memory management.
CWE-119	Use libraries or frameworks that make it easier to handle numbers without unexpected consequences. Examples include safe integer handling packages such as SafeInt (C++) or IntegerLib (C or C++).
CWE-119, CWE-200	Use a CPU and operating system that offers Data Execution Protection (NX) or its equivalent.
CWE-190, CWE-200	Ensure that all protocols are strictly defined, such that all out-of-bounds behaviors can be identified simply, and require strict conformance to the protocol.

Table A2. *Cont.*

CWE ID	Requirements
CWE-310	Clearly specify which data or resources are valuable enough that they should be protected by encryption. Require that any transmission or storage of this data/resource should use well-vetted encryption algorithms. Up-to-date algorithms must be used, and the entropy of the keys must be sufficient for the application.
CWE-113	Use an input validation framework such as Struts or the OWASP ESAPI Validation API.
CWE-113	Assume all input is malicious. Use an "accept known good" input validation strategy, i.e., use a list of acceptable inputs that strictly conform to specifications. Reject any input that does not strictly conform to specifications, or transform it into something that does.
CWE-113	Hard-code the search path to a set of known-safe values (such as system directories), or only allow them to be specified by the administrator in a configuration file. Do not allow these settings to be modified by an external party.
CWE-119	Run or compile the software using features or extensions that automatically provide a protection mechanism that mitigates or eliminates buffer overflows.
CWE-119	Replace unbounded copy functions with analogous functions that support length arguments, such as strcpy with strncpy. Create these if they are not available.

Table A3. Example of proposed training for OpenPLC V1.

CWE ID	Training
CWE-113, CWE-119	Identification of all potentially relevant properties of an input (length, type of input, the full range of acceptable values, missing or extra inputs, syntax, consistency across related fields).
CWE-113, CWE-119	Input validation strategies.
CWE-113, CWE-119, CWE-200	Allowlists and Denylists.
CWE-113, CWE-119	Character encoding compatibility.
CWE-113, CWE-119	Buffer overflow detection during compilation (e.g., Microsoft Visual Studio /GS flag, Fedora/Red Hat FORTIFY_SOURCE GCC flag, StackGuard, and ProPolice).
CWE-113, CWE-119CWE-200	Secure functions, such as *strcpy* with *strncpy*. Create these if they are not available.
CWE-113, CWE-119CWE-190	Secure programming: memory management.
CWE-113, CWE-119	Understand the programming language's underlying representation and how it interacts with numeric calculation.
CWE-113, CWE-119	System compartmentalization.
CWE-200, CWE-310	Certificate management.
CWE-200, CWE-310	Certificate pinning.
CWE-310	Encryption integration (do not develop custom or private cryptographic algorithms).
CWE-310	Secure up-to-date cryptographic algorithms.
CWE-200	Shared resource management.
CWE-200	Thread-safe functions.

Table A4. Example of generated test cases for OpenPLC V1.

Capec ID	Test Cases
CAPEC-119	Check for buffer overflows through manipulation of environment variables. This test leverages implicit trust often placed in environment variables.
CAPEC-119	Static analysis of the code: secure functions and buffer overflow.
CAPEC-119	Feed overly long input strings to the program in an attempt to overwhelm the filter (by causing a buffer overflow) and hoping that the filter does not fail securely (i.e. the user input is let into the system unfiltered)
CAPEC-119	This test uses symbolic links to cause buffer overflows. The evaluator can try to create or manipulate a symbolic link file such that its contents result in out-of-bounds data. When the target software processes the symbolic link file, it could potentially overflow internal buffers with insufficient bounds checking.
CAPEC-119	Static analysis of the code: secure functions and buffer overflow.

References

1. Qingyu, O.; Fang, L.; Kai, H. High-Security System Primitive for Embedded Systems. In Proceedings of the 2009 International Conference on Multimedia Information Networking and Security, Wuhan, China, 18–20 November 2009; Volume 2, pp. 319–321. [CrossRef]
2. Chen, T.M.; Abu-Nimeh, S. Lessons from Stuxnet. *Computer* **2011**, *44*, 91–93. [CrossRef]
3. Vai, M.; Nahill, B.; Kramer, J.; Geis, M.; Utin, D.; Whelihan, D.; Khazan, R. Secure architecture for embedded systems. In Proceedings of the 2015 IEEE High Performance Extreme Computing Conference (HPEC), Waltham, MA, USA, 15–17 September 2015; pp. 1–5. [CrossRef]
4. Ten, C.W.; Manimaran, G.; Liu, C.C. Cybersecurity for Critical Infrastructures: Attack and Defense Modeling. *IEEE Trans. Syst. Man Cybern.-Part A Syst. Hum.* **2010**, *40*, 853–865. [CrossRef]
5. Gressl, L.; Steger, C.; Neffe, U. Design Space Exploration for Secure IoT Devices and Cyber-Physical Systems. *ACM Trans. Embed. Comput. Syst.* **2021**, *20*, 1–24. [CrossRef]
6. Gupta, M.; Abdelsalam, M.; Khorsandroo, S.; Mittal, S. Security and Privacy in Smart Farming: Challenges and Opportunities. *IEEE Access* **2020**, *8*, 34564–34584. [CrossRef]
7. Mumtaz, S.; Alsohaily, A.; Pang, Z.; Rayes, A.; Tsang, K.F.; Rodriguez, J. Massive Internet of Things for Industrial Applications: Addressing Wireless IIoT Connectivity Challenges and Ecosystem Fragmentation. *IEEE Ind. Electron. Mag.* **2017**, *11*, 28–33. [CrossRef]
8. Ojo, M.O.; Giordano, S.; Procissi, G.; Seitanidis, I.N. A Review of Low-End, Middle-End, and High-End Iot Devices. *IEEE Access* **2018**, *6*, 70528–70554. [CrossRef]
9. Shafique, K.; Khawaja, B.A.; Sabir, F.; Qazi, S.; Mustaqim, M. Internet of Things (IoT) for Next-Generation Smart Systems: A Review of Current Challenges, Future Trends and Prospects for Emerging 5G-IoT Scenarios. *IEEE Access* **2020**, *8*, 23022–23040. [CrossRef]
10. Ponta, S.E.; Plate, H.; Sabetta, A. Detection, assessment and mitigation of vulnerabilities in open source dependencies. *Empir. Softw. Eng.* **2020**, *25*, 3175–3215. [CrossRef]
11. Hejderup, J.I.; Van Deursen, A.; Mesbah, A. In Dependencies We Trust: How Vulnerable are Dependencies in Software Modules? Ph.D. Thesis, Department of Software Technology, TU Delft, Delft, The Netherlands, 2015. http://resolver.tudelft.nl/uuid:3a15293b-16f6-4e9d-b6a2-f02cd52f1a9e (accessed on 27 January 2022).
12. Pashchenko, I.; Plate, H.; Ponta, S.E.; Sabetta, A.; Massacci, F. Vulnerable Open Source Dependencies: Counting Those That Matter. In Proceedings of the 12th International Symposium on Empirical Software Engineering and Measurement (ESEM), Oulu, Finland, 11–12 October 2018. https://dl.acm.org/doi/10.1145/3239235.3268920
13. Zografopoulos, I.; Ospina, J.; Liu, X.; Konstantinou, C. Cyber-Physical Energy Systems Security: Threat Modeling, Risk Assessment, Resources, Metrics, and Case Studies. *IEEE Access* **2021**, *9*, 29775–29818. [CrossRef]
14. McLaughlin, S.; Konstantinou, C.; Wang, X.; Davi, L.; Sadeghi, A.R.; Maniatakos, M.; Karri, R. The Cybersecurity Landscape in Industrial Control Systems. *Proc. IEEE* **2016**, *104*, 1039–1057. [CrossRef]
15. Mathew, A. Network Slicing in 5G and the Security Concerns. In Proceedings of the 2020 Fourth International Conference on Computing Methodologies and Communication (ICCMC), Erode, India, 11–13 March 2020; pp. 75–78. [CrossRef]

16. Christidis, K.; Devetsikiotis, M. Blockchains and Smart Contracts for the Internet of Things. *IEEE Access* **2016**, *4*, 2292–2303. [CrossRef]
17. Hassija, V.; Chamola, V.; Saxena, V.; Jain, D.; Goyal, P.; Sikdar, B. A Survey on IoT Security: Application Areas, Security Threats, and Solution Architectures. *IEEE Access* **2019**, *7*, 82721–82743. [CrossRef]
18. Ayaz, M.; Ammad-Uddin, M.; Sharif, Z.; Mansour, A.; Aggoune, E.H.M. Internet-of-Things (IoT)-Based Smart Agriculture: Toward Making the Fields Talk. *IEEE Access* **2019**, *7*, 129551–129583. [CrossRef]
19. Fuller, A.; Fan, Z.; Day, C.; Barlow, C. Digital Twin: Enabling Technologies, Challenges and Open Research. *IEEE Access* **2020**, *8*, 108952–108971. [CrossRef]
20. Xin, Y.; Kong, L.; Liu, Z.; Chen, Y.; Li, Y.; Zhu, H.; Gao, M.; Hou, H.; Wang, C. Machine Learning and Deep Learning Methods for Cybersecurity. *IEEE Access* **2018**, *6*, 35365–35381. [CrossRef]
21. Benias, N.; Markopoulos, A.P. A review on the readiness level and cyber-security challenges in Industry 4.0. In Proceedings of the 2017 South Eastern European Design Automation, Computer Engineering, Computer Networks and Social Media Conference (SEEDA-CECNSM), Kastoria, Greece, 23–25 September 2017; pp. 1–5. [CrossRef]
22. Matsuda, W.; Fujimoto, M.; Aoyama, T.; Mitsunaga, T. Cyber Security Risk Assessment on Industry 4.0 using ICS testbed with AI and Cloud. In Proceedings of the 2019 IEEE Conference on Application, Information and Network Security (AINS), Pulau Pinang, Malaysia, 19–21 November 2019; pp. 54–59. [CrossRef]
23. Culot, G.; Fattori, F.; Podrecca, M.; Sartor, M. Addressing Industry 4.0 Cybersecurity Challenges. *IEEE Eng. Manag. Rev.* **2019**, *47*, 79–86. [CrossRef]
24. Lezzi, M.; Lazoi, M.; Corallo, A. Cybersecurity for Industry 4.0 in the current literature: A reference framework. *Comput. Ind.* **2018**, *103*, 97–110. [CrossRef]
25. Ustundag, A.; Cevikcan, E. *Industry 4.0: Managing The Digital Transformation*; Springer International Publishing: Berlin/Heidelberg, Germany, 2018. https://link.springer.com/book/10.1007/978-3-319-57870-5. [CrossRef]
26. Thames, L.; Schaefer, D. (Eds.) *Cybersecurity for Industry 4.0*; Springer International Publishing: Berlin/Heidelberg, Germany, 2017. [CrossRef]
27. Medeiros, N.; Ivaki, N.; Costa, P.; Vieira, M. Software Metrics as Indicators of Security Vulnerabilities. In Proceedings of the 2017 IEEE 28th International Symposium on Software Reliability Engineering (ISSRE), Toulouse, France, 23–26 October 2017; pp. 216–227. [CrossRef]
28. Alenezi, M.; Zarour, M. On the Relationship between Software Complexity and Security. *Int. J. Softw. Eng. Appl.* **2020**, *11*, 51–60. Available online: https://aircconline.com/abstract/ijsea/v11n1/11120ijsea04.html (accessed on 27 January 2022). [CrossRef]
29. Langner, R. Stuxnet: Dissecting a Cyberwarfare Weapon. *IEEE Secur. Priv.* **2011**, *9*, 49–51. [CrossRef]
30. George, G.; Thampi, S.M. A Graph-Based Security Framework for Securing Industrial IoT Networks From Vulnerability Exploitations. *IEEE Access* **2018**, *6*, 43586–43601. [CrossRef]
31. Papp, D.; Ma, Z.; Buttyan, L. Embedded systems security: Threats, vulnerabilities, and attack taxonomy. In Proceedings of the 2015 13th Annual Conference on Privacy, Security and Trust (PST), Izmir, Turkey, 21–23 July 2015; pp. 145–152. [CrossRef]
32. Nielsen, B.B.; Torp, M.T.; Møller, A. Modular Call Graph Construction for Security Scanning of Node.Js Applications. In *Proceedings of the 30th ACM SIGSOFT International Symposium on Software Testing and Analysis*; Association for Computing Machinery: New York, NY, USA, 2021; pp. 29–41. [CrossRef]
33. Sawilla, R.E.; Ou, X. Identifying Critical Attack Assets in Dependency Attack Graphs. In *Computer Security—ESORICS 2008*; Jajodia, S., Lopez, J., Eds.; Springer: Berlin/Heidelberg, Germany, 2008; pp. 18–34. Available online: https://link.springer.com/chapter/10.1007/978-3-540-88313-5_2#citeas (accessed on 27 January 2022).
34. MITRE Corporation. CVE—Common Vulnerability and Exposures. Available online: https://cve.mitre.org/index.html (accessed on 27 January 2022).
35. MITRE Corporation. CVE—Common Vulnerabilities and Exposures: Definitions. Available online: https://cve.mitre.org/about/terminology.html (accessed on 21 January 2022).
36. National Institute for Standards and Technology (NIST). National Vulnerability Database NVD—Vulnerability List. Available online: https://nvd.nist.gov/vuln/full-listing (accessed on 27 January 2022).
37. FIRST—global Forum of Incident Response and Security Teams. Common Vulnerability Scoring System (CVSS). Available online: https://www.first.org/cvss/ (accessed on 27 January 2022).
38. MITRE Corporation. CWE—Common Weakness Enumeration. Available online: https://cwe.mitre.org/index.html (accessed on 27 January 2022).
39. MITRE Corporation. CWE—Common Weakness Enumeration: Definitions. Available online: https://cwe.mitre.org/about/faq.html (accessed on 27 January 2022).
40. Jiang, Y.; Atif, Y.; Ding, J. Cyber-Physical Systems Security Based on a Cross-Linked and Correlated Vulnerability Database. In *Critical Information Infrastructures Security*; Nadjm-Tehrani, S., Ed.; Springer International Publishing: Cham, Switzerland, 2020; pp. 71–82. Available online: https://link.springer.com/chapter/10.1007/978-3-030-37670-3 (accessed on 27 January 2022).
41. Kleidermacher, D.; Kleidermacher, M. Practical Methods for Safe and Secure Software and Systems Development. In *Embedded Systems Security*; Kleidermacher, D., Kleidermacher, M., Eds.; Newnes: Oxford, UK, 2012. [CrossRef]
42. Andreeva, O.; Gordeychik, S.; Gritsai, G.; Kochetova, O.; Potseluevskaya, E.; Sidorov, S.; Timorin, A. *Industrial Control Systems Vulnerabilities Statistics*; Technical Report; Karpersky: Moscow, Russia, 2016. [CrossRef]

43. Hwang, D.; Schaumont, P.; Tiri, K.; Verbauwhede, I. Securing embedded systems. *IEEE Secur. Priv.* **2006**, *4*, 40–49. [CrossRef]
44. Viega, J.; Thompson, H. The State of Embedded-Device Security (Spoiler Alert: It's Bad). *IEEE Secur. Priv.* **2012**, *10*, 68–70. [CrossRef]
45. Marwedel, P. Embedded Systems Foundations of Cyber-Physical Systems, and the Internet of Things. In *Embedded System Design*; Springer Nature: Cham, Switzerland, 2018. [CrossRef]
46. Arpaia, P.; Bonavolontà, F.; Cioffi, A.; Moccaldi, N. Reproducibility Enhancement by Optimized Power Analysis Attacks in Vulnerability Assessment of IoT Transducers. *IEEE Trans. Instrum. Meas.* **2021**, *70*, 1–8. [CrossRef]
47. IEC 62443; Industrial Communication Networks—Network and System Security. IEC Central Office: Geneva, Switzerland, 2010.
48. Mugarza, I.; Flores, J.L.; Montero, J.L. Security Issues and Software Updates Management in the Industrial Internet of Things (IIoT) Era. *Sensors* **2020**, *20*, 7160. [CrossRef] [PubMed]
49. IEC 62443; Security for Industrial Automation and Control Systems—Part 4-1: Secure Product Development Lifecycle Requirements. International Electrotechnical Commission: Geneva, Switzerland, 2018.
50. Avizienis, A.; Laprie, J.; Randell, B.; Landwehr, C. Basic concepts and taxonomy of dependable and secure computing. *IEEE Trans. Dependable Secur. Comput.* **2004**, *1*, 11–33. [CrossRef]
51. He, W.; Li, H.; Li, J. Unknown Vulnerability Risk Assessment Based on Directed Graph Models: A Survey. *IEEE Access* **2019**, *7*, 168201–168225. [CrossRef]
52. ISO/IEC 30111:2019; Information Technology—Security Techniques—Vulnerability Handling Processes. International Organization for Standardization: Geneva, Switzerland, 2019. Available online: https://www.iso.org/standard/69725.html (accessed on 27 January 2022).
53. Common Criteria (CC). The Common Criteria for Information Technology Security Evaluation—Introduction and General Model. Available online: https://www.commoncriteriaportal.org/files/ccfiles/CCPART1V3.1R5.pdf (accessed on 27 January 2022).
54. Common Criteria (CC). Part 3: Security Assurance Components. Available online: https://commoncriteriaportal.org/files/ccfiles/CCPART3V3.1R5.pdf (accessed on 27 January 2022).
55. Herrmann, D. *Using the Common Criteria for IT Security Evaluation*; Auerbach Publications, Boca Raton, FL, USA, 2002; pp. 1–289. [CrossRef]
56. Matheu, S.N.; Hernandez-Ramos, J.L.; Skarmeta, A.F. Toward a Cybersecurity Certification Framework for the Internet of Things. *IEEE Secur. Priv.* **2019**, *17*, 66–76. [CrossRef]
57. Mellado, D.; Fernández-Medina, E.; Piattini, M. A common criteria based security requirements engineering process for the development of secure information systems. *Comput. Stand. Interfaces* **2007**, *29*, 244–253. [CrossRef]
58. Hohenegger, A.; Krummeck, G.; Baños, J.; Ortega, A.; Hager, M.; Sterba, J.; Kertis, T.; Novobilsky, P.; Prochazka, J.; Caracuel, B.; et al. Security certification experience for industrial cyberphysical systems using Common Criteria and IEC 62443 certifications in certMILS. In Proceedings of the 2021 4th IEEE International Conference on Industrial Cyber-Physical Systems (ICPS), Victoria, BC, Canada, 10–12 May 2021; pp. 25–30. [CrossRef]
59. Homer, J.; Ou, X.; Schmidt, D. A Sound and Practical Approach to Quantifying Security Risk in Enterprise Networks. Technical Report. 2009. Available online: https://www.cse.usf.edu/~xou/publications/tr_homer_0809.pdf (accessed on 27 January 2022).
60. Zhang, S.; Ou, X.; Singhal, A.; Homer, J. *An Empirical Study of a Vulnerability Metric Aggregation Method*; Technical Report; Kansas State University: Manhattan, KS, USA, 2011. Available online: https://www.cse.usf.edu/~xou/publications/stmacip11.pdf (accessed on 27 January 2022).
61. Homer, J.; Zhang, S.; Ou, X.; Schmidt, D.; Du, Y.; Rajagopalan, S.R.; Singhal, A. Aggregating vulnerability metrics in enterprise networks using attack graphs. *J. Comput. Secur.* **2013**, *21*, 561–597. [CrossRef]
62. Li, S.; Chen, Y.; Wu, X.; Cheng, X.; Tian, Z. Power Grid-Oriented Cascading Failure Vulnerability Identifying Method Based on Wireless Sensors. *J. Sens.* **2021**, *2021*, 8820413. [CrossRef]
63. Liu, B.; Zhu, G.; Li, X.; Sun, R. Vulnerability Assessment of the Urban Rail Transit Network Based on Travel Behavior Analysis. *IEEE Access* **2021**, *9*, 1407–1419. [CrossRef]
64. Poolsappasit, N.; Dewri, R.; Ray, I. Dynamic Security Risk Management Using Bayesian Attack Graphs. *IEEE Trans. Dependable Secur. Comput.* **2012**, *9*, 61–74. [CrossRef]
65. Muñoz-González, L.; Sgandurra, D.; Barrère, M.; Lupu, E.C. Exact Inference Techniques for the Analysis of Bayesian Attack Graphs. *IEEE Trans. Dependable Secur. Comput.* **2019**, *16*, 231–244. [CrossRef]
66. Liu, X.; Qian, C.; Hatcher, W.G.; Xu, H.; Liao, W.; Yu, W. Secure Internet of Things (IoT)-Based Smart-World Critical Infrastructures: Survey, Case Study and Research Opportunities. *IEEE Access* **2019**, *7*, 79523–79544. [CrossRef]
67. Pascale, F.; Adinolfi, E.A.; Coppola, S.; Santonicola, E. Cybersecurity in Automotive: An Intrusion Detection System in Connected Vehicles. *Electronics* **2021**, *10*, 1765. [CrossRef]
68. Hu, J.; Guo, S.; Kuang, X.; Meng, F.; Hu, D.; Shi, Z. I-HMM-Based Multidimensional Network Security Risk Assessment. *IEEE Access* **2020**, *8*, 1431–1442. [CrossRef]
69. Khosravi-Farmad, M.; Bafghi, A. Bayesian Decision Network-Based Security Risk Management Framework. *J. Netw. Syst. Manag.* **2020**, *28*, 1794–1819. [CrossRef]
70. Atzeni, A.; Lioy, A. Why to adopt a security metric? A brief survey. *Adv. Inf. Secur.* **2006**, *23*, 1–12. [CrossRef]
71. Zeb, T.; Yousaf, M.; Afzal, H.; Mufti, M.R. A quantitative security metric model for security controls: Secure virtual machine migration protocol as target of assessment. *China Commun.* **2018**, *15*, 126–140. [CrossRef]

72. Longueira-Romero, A.; Iglesias, R.; Gonzalez, D.; Garitano, I.N. How to Quantify the Security Level of Embedded Systems? A Taxonomy of Security Metrics. In Proceedings of the 2020 IEEE 18th International Conference on Industrial Informatics (INDIN), Warwick, UK, 20–23 July 2020; Volume 1, pp. 153–158. [CrossRef]
73. Rudolph, M.; Schwarz, R. A Critical Survey of Security Indicator Approaches. In Proceedings of the 2012 Seventh International Conference on Availability, Reliability and Security, Prague, Czech Republic, 20–24 August 2012; pp. 291–300. [CrossRef]
74. Sentilles, S.; Papatheocharous, E.; Ciccozzi, F. What Do We Know about Software Security Evaluation? A Preliminary Study. QuASoQ@APSEC. 2018. Available online: http://ceur-ws.org/Vol-2273/QuASoQ-04.pdf (accessed on 27 January 2022).
75. Amutio, M.A.; Candau, J.; Mañas, J.A. *MAGERIT V3.0. Methodology for Information Systems Risk Analysis and Management*; Book I—The Method; National Standard; Ministry of Finance and Public Administration: Madrid, Spain, 2014.
76. Dekker, M.; Karsberg, C. *Guideline on Threats and Assets: Technical Guidance on Threats and Assets in Article 13a*; Technical Report; European Union Agency for Network and Information Security. 2015. Available online: https://www.enisa.europa.eu/publications/technical-guideline-on-threats-and-assets (accessed on 27 January 2022).
77. *ISO/IEC 13335-1:2004*; Information Technology—Security Techniques—Management of Information and Communications Technology Security—Part 1: Concepts and Models for Information and Communications Technology Security Management. International Organization for Standardization: Geneva, Switzerland, 2004.
78. National Institute for Standards and Technology (NIST). CPE—Common Platform Enumeration. Available online: https://nvd.nist.gov/products/cpe (accessed on 27 January 2022).
79. Cheikes, B.A.; Waltermire, D.; Scarfone, K.. *NIST Interagency Report 7695—Common Platform Enumeration: Naming Specification Version 2.3*; Nist interagency Report; National Institute for Standards and Technology (NIST): Gaithersburg, MD, USA, 2011. Available online: https://tsapps.nist.gov/publication/get_pdf.cfm?pub_id=909010 (accessed on 27 January 2022).
80. Parmelee, M.C.; Booth, H.; Waltermire, D.; Scarfone, K. *NIST Interagency Report 7696—Common Platform Enumeration: Name Matching Specification Version 2.3*; Nist Interagency Report; National Institute for Standards and Technology (NIST): Gaithersburg, MD, USA, 2011. Available online: https://tsapps.nist.gov/publication/get_pdf.cfm?pub_id=909008 (accessed on 27 January 2022).
81. *ISO 8601:2019*; Data and time-Representation for Information Interchange—Part 1: Basic Rules. International Organization for Standardization: Geneva, Switzerland, 2019.
82. MITRE Corporation. CAPEC—Common Attack Pattern Enumeration and Classification. Available online: https://capec.mitre.org/ (accessed on 27 January 2022).
83. MITRE Corporation. CAPEC—Common Attack Pattern Enumeration and Classification: Glossary. Available online: https://capec.mitre.org/about/glossary.html (accessed on 27 January 2022).
84. NIST—National Institute of Standards and Technology. National Vulnerability database (NVD). Available online: https://nvd.nist.gov/ (accessed on 27 January 2022).
85. Dimitriadis, A.; Flores, J.L.; Kulvatunyou, B.; Ivezic, N.; Mavridis, I. ARES: Automated Risk Estimation in Smart Sensor Environments. *Sensors* **2020**, *20*, 4617. [CrossRef]
86. Alves, T. OpenPLC Project. Available online: https://www.openplcproject.com/ (accessed on 27 January 2022).
87. Alves, T. OpenPLC V1. Available online: https://github.com/thiagoralves/OpenPLC (accessed on 27 January 2022).
88. Alves, T. OpenPLC V2. Available online: https://github.com/thiagoralves/OpenPLC_v2 (accessed on 27 January 2022).
89. Alves, T. OpenPLC V3. Available online: https://github.com/thiagoralves/OpenPLC_v3 (accessed on 27 January 2022).
90. Alves, T.R.; Buratto, M.; de Souza, F.M.; Rodrigues, T.V. OpenPLC: An open source alternative to automation. In Proceedings of the IEEE Global Humanitarian Technology Conference (GHTC 2014), San Jose, CA, USA, 10–13 October 2014; pp. 585–589. [CrossRef]
91. Alves, T.; Morris, T. OpenPLC: An IEC 61,131—3 compliant open source industrial controller for cyber security research. *Comput. Secur.* **2018**, *78*, 364–379. [CrossRef]
92. Ubuntu 14.04 and 16.04 Lifecycle Extended to Ten Years. Available online: https://ubuntu.com/blog/ubuntu-14-04-and-16-04-lifecycle-extended-to-ten-years (accessed on 27 January 2022).
93. libssl1.0.0: Trusty (14.04): Ubuntu. Available online: https://launchpad.net/ubuntu/trusty/+package/libssl1.0.0/+index (accessed on 27 January 2022).
94. nodejs: Trusty (14.04): Ubuntu. Available online: https://launchpad.net/ubuntu/trusty/+package/nodejs/+index (accessed on 27 January 2022).
95. *IEC 62443*; Security for Industrial Automation and Control Systems—Part 4-2: Technical Security Requirements for IACS Components. International Electrotechnical Commission: Geneva, Switzerland, 2019. Available online: https://www.isa.org/products/ansi-isa-62443-4-1-2018-security-for-industrial-au (accessed on 27 January 2022).

Article

Artificial Intelligence Algorithms for Malware Detection in Android-Operated Mobile Devices

Hasan Alkahtani [1] and Theyazn H. H. Aldhyani [2,*]

[1] College of Computer Science and Information Technology, King Faisal University, P.O. Box 400, Al-Ahsa 31982, Saudi Arabia; hsalkhtani@kfu.edu.sa
[2] Applied College in Abqaiq, King Faisal University, P.O. Box 400, Al-Ahsa 31982, Saudi Arabia
* Correspondence: taldhyani@kfu.edu.sa

Abstract: With the rapid expansion of the use of smartphone devices, malicious attacks against Android mobile devices have increased. The Android system adopted a wide range of sensitive applications such as banking applications; therefore, it is becoming the target of malware that exploits the vulnerabilities of the security system. A few studies proposed models for the detection of mobile malware. Nevertheless, improvements are required to achieve maximum efficiency and performance. Hence, we implemented machine learning and deep learning approaches to detect Android-directed malicious attacks. The support vector machine (SVM), k-nearest neighbors (KNN), linear discriminant analysis (LDA), long short-term memory (LSTM), convolution neural network-long short-term memory (CNN-LSTM), and autoencoder algorithms were applied to identify malware in mobile environments. The cybersecurity system was tested with two Android mobile benchmark datasets. The correlation was calculated to find the high-percentage significant features of these systems in the protection against attacks. The machine learning and deep learning algorithms successfully detected the malware on Android applications. The SVM algorithm achieved the highest accuracy (100%) using the CICAndMal2017 dataset. The LSTM model also achieved a high percentage accuracy (99.40%) using the Drebin dataset. Additionally, by calculating the mean error, mean square error, root mean square error, and Pearson correlation, we found a strong relationship between the predicted values and the target values in the validation phase. The correlation coefficient for the SVM method was $R^2 = 100\%$ using the CICAndMal2017 dataset, and LSTM achieved $R^2 = 97.39\%$ in the Drebin dataset. Our results were compared with existing security systems, showing that the SVM, LSTM, and CNN-LSTM algorithms are of high efficiency in the detection of malware in the Android environment.

Keywords: android applications; malware; machine learning; deep learning; cybersecurity

1. Introduction

In recent years, the popularity of the Android operation system has attracted the attention of malware developers, whose work has grown rapidly [1,2]. Many malware developers focus on hacking mobile devices and changing them into bots. This allows hackers to access the infected device and other connected devices and form botnets. Botnets are used to execute different malicious attacks, such as distributed denial-of-service (DDoS) attacks, sending spam, data theft, etc. The malicious botnet attacks are developed with advanced techniques (e.g., multi-staged payload or self-protection), making it difficult to identify the malware. This, in turn, poses major threats that require the design of effective approaches to detect these attacks [3].

Android botnets are used to perform attacks on the targeted devices. DDos attacks are achieved by flooding the target machine with superfluous requests and blocking legitimate requests, thus, causing a failure of the targeted system and disruption of the services [4]. Consequently, to protect against such attacks, machine learning methods are proven to be effective in detecting and tracking these threats in the internet of things [5,6]. Haystack [7] reported that a third-part of software-development companies manage 70% of the mobile application and control the personal data of users. According to the AV-TEST Security Institute [8], malicious programming increased, with 5.7 million malware Android packages detected by Kaspersky in 2020, three times more than in 2019 (2.1 million). Figure 1 summarize the increase of malware installation packages for smartphone devices in the last five years. Therefore, signature-based malicious installation packages for the extraction of malware patterns relying on their characteristics can be an effective strategy to secure mobile applications.

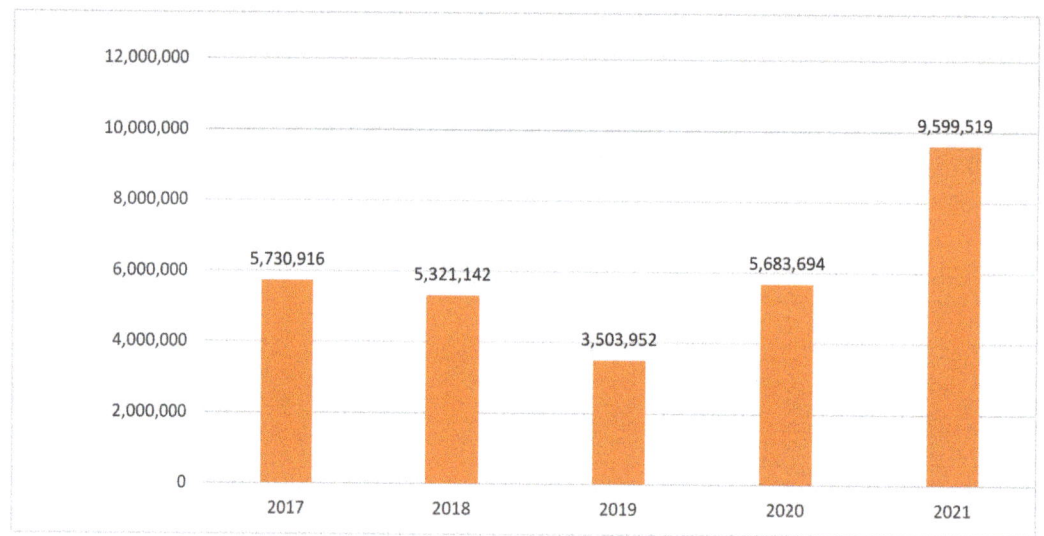

Figure 1. Malware installation packages for smartphone devices.

Malicious attacks occur in different enrolments with a variety of methods such as fuzzing, denial of service, DDoS, port scanning, and probing [9]. These attacks can be threatening to transport, application layers, or different protocols such as internet control message protocol, file transfer protocol, user datagram protocol, simple mail transfer protocol, transmission control protocol, hypertext transfer protocol, etc. Network-based intrusion detection systems can be used to deal with such attacks by scanning the network and detecting them [10].

Usually, in the Android system, security is in-built, where the sandboxing method and permission system are designed to reduce the risk of Android applications [11]. The former was developed using the Linux environment for running Android applications, which allows users to enable permission for the installation of any Android application [12]. However, when updating or upgrading mobile applications, security and privacy features such as time permission, background location, storage, etc., are changed, giving a timeframe for malware attacks. It is possible to exploit Android vulnerabilities during the application developed by users since the Google Play Store cannot detect malicious attacks after the publication of the applications [13]. The percentage of Android malware is presented in Figure 2.

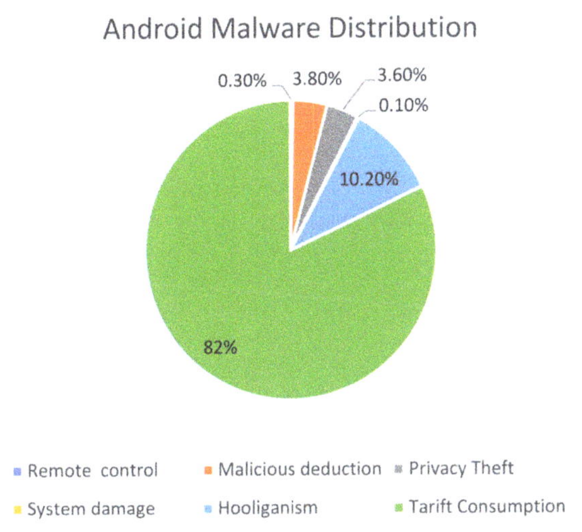

Android Malware Distribution

- Remote control: 10.20%
- Malicious deduction: 3.80%
- Privacy Theft: 3.60%
- System damage: 0.30%
- Hooliganism: 0.10%
- Tariff Consumption: 82%

Figure 2. Percentage of Android malware [14].

Intrusion detection systems are developed using machine learning and deep learning methods. However, the machine learning technique cannot cope with the huge traffic of data flooding the system. Similarly, deep learning methods fail to provide low generalization errors due to the absence of optimization. Fixed Android botnet datasets make it feasible to design detectors with high detection rates [15], but having complex traffic data hinders the obtention of an accurate prediction rate. This has motivated the development of techniques that are based on Android-malware neuro-evolution classification, thus, providing the number of layers and neurons along with the detection process [16].

The present study aimed to extract static and dynamic features from unknown applications; these features show if a particular application is "normal" or "attack". These features are used to examine the performance of several machine learning and deep learning models, including the k-nearest neighbors (KNN) [17], support vector machine (SVM) [18], convolutional neural networks (CNN) [19], dense neural networks [20], gated recurrent units (GRU), long short-term memory (LSTM) [21], and the hybrid deep learning convolutional neural networks long/short-term memory (CNN-LSTM) and convolutional neural networks/gated recurrent units CNN-GRU [22] methods.

In this study, we investigated and estimated the performance of various machine learning and deep learning algorithms in the detection of mobile malware attacks. This study offers the optimal algorithms for the monitoring of Android applications against malicious attacks. Thus, our research aims to contribute to this field with the following:

1. The development of intrusion detection in the Android system using various machine learning and deep learning algorithms.
2. The proposed system was tested and evaluated using two standard Android datasets.
3. A comparison between the tested algorithms and different state-of-the-arts models is presented.
4. The sensitivity analysis was used to find significant relationships between dataset features and the proposed classes of the datasets.

2. Background of Study

This section offers an overview of previous research related to intrusion detection systems, Android malware detection, and standard datasets of Android malicious attacks. Furthermore, it provides an overview of the machine learning and deep learning approaches applied to the design of cybersecurity systems.

The regular improvement of sophisticated Android malware families, e.g., Chamois malware, has made the task of detecting malicious attacks daunting. To tackle this, researchers developed machine learning techniques that improved the available systems. Recently, many studies have applied machine learning models for Android botnet detection, such as linear regression, KNN [23], SVM, and decision trees (DT) algorithms [24]. Some of these recent studies [25,26] used deep learning algorithms, although they do not provide a thorough understanding of their effectiveness. Therefore, the current study compares with deep learning models to examine their effectiveness in Android botnet detection with the use of the available installation support center of expertise (ISCX) botnet dataset [27–29].

Kadir et al. [30] used deep learning models to analyze Android botnet attacks in an attempt to understand the latter's hidden features. The system was evaluated using the ISCX Android botnet dataset, which contained 1929 samples. Anwar et al. [31] proposed an Android botnet detection approach based on static functions. The features of permissions, MD5 signatures, and broadcast receivers were combined and processed with machine learning algorithms. The input data collected from the ISCX dataset were 1400 from different botnet applications, with the system achieving an accuracy of 95.1% in distinguishing Android botnet attacks [32].

Several machine learning algorithms were proposed to classify normal and abnormal botnet attacks. In one study, the results indicated that the random forest approach had 0.972% precision and 0.96% recall. In [33], machine learning approaches were proposed for detecting Android botnets. The ISCX dataset consisted of 1635 benign and 1635 attacks. The random forest tree model achieved 97%. In another study [34], the DT, Naive Bayes, and random forest machine learning algorithms were used to detect Android attacks. The information gain method was used to select the significant features. The random forest algorithm achieved a 94.6% accuracy. Karim et al. [35] proposed the static analysis approach to explore the pattern of the features of Android botnet attacks. The features were compared with the intrusion application using the Drebin dataset [36]. Artificial intelligence (AI) approaches using a knowledge-based system were used to secure Android mobiles against malicious attacks [37,38]. Inspired by a meta-heuristic rule and based on fuzzy logic, intrusion detection and data mining systems were developed [39], while machine learning approaches were applied in the development of IDS applications [40–42]. The design of IDS systems employed the artificial bee colony [43], particle swarm optimization [44], grey wolf optimization [45], and artificial fish swarm [46] algorithms.

Many systems were developed based on signature-based Android malware detection approaches and behavior-based Android malware intrusion detection approaches [47]. The former is a simple detection method that manages intrusions' low degree of false positives. The latter is based on anomaly detection and is a very common method using AI algorithms to detect malicious attacks. Numerous research articles aimed to detect and classify Android malware and attacks using machine learning and deep learning approaches, such as the DT and deep learning approaches [48]. By using the generative adversarial networks algorithm [49], it was shown that traditional machine learning was successful in detecting malware in an Android environment [50].

Most of the published studies used datasets from Google Play [51], AndroZoo, Android Permission [52], Andrototal [53], Wandoujia [54], Kaggle [55], and CICMaldroid [56]. The present study aimed at developing a system to detect malware attacks in Android environments that have an in-built security system. However, there are still many Android applications with design weaknesses and security flaws that can be threatening to end-users. Therefore, it is crucial to use machine learning and deep learning algorithms to detect Android malware and vulnerability analysis to prevent the development of malware and attacks by hackers [57,58].

3. Materials and Methods

In 2008, Android was developed. With the increasing number of Android applications, companies immediately discussed and built security tools [2]. Nevertheless, the Android system is suffering from security weaknesses. In the last five years, AI approaches focused on protecting the Android system, with many researchers studying the appropriate AI approaches to obtain high accuracy. The framework of the present research is presented in Figure 3. The machine learning algorithms support vector machine (SVM), k-nearest neighbors (KNN), linear discriminant analysis (LDA) and the deep learning algorithms long short-term memory (LSTM), convolution neural network-long short-term memory (CNN-LSTM), and autoencoder algorithms were used to detect malware and attacks against Android applications. These algorithms were tested using two standard datasets. The research questions of this study were:

(1) What are the appropriate machine learning and deep learning algorithms to detect malware in Android?
(2) What are the validation accuracy, robustness, and efficiency of the proposed machine learning and deep learning models related to the detection of Android malware?

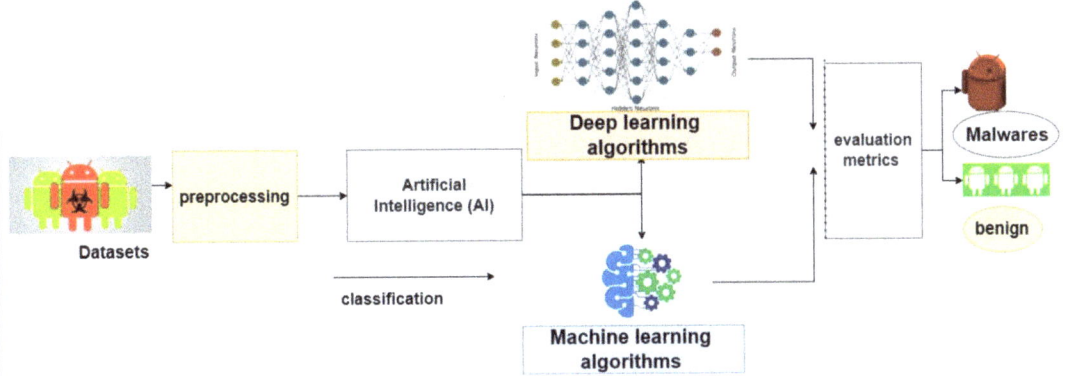

Figure 3. A generic representation of the models applied for the detection of Android malware.

3.1. Android Dataset

The experiments were conducted with two standard datasets: the Canadian Institute for Cybersecurity (CICAndMal2017) and Drebin datasets. The percentage of the classes for the entire CICAndMal201 and Drebin datasets is presented in Figure 4.

3.1.1. CICAndMal2017

The CICAndMal2017 was developed by Canadian Institute; the Cybersecurity dataset is a standard mobile malware dataset containing static and dynamic features of log files. The dataset was generated from 80 network flows using CICFlowMeter-V1 and CICFlowMeter-V3. To examine the proposed system, 667 Android malware packets consisting of 413 features were considered for the injection of malicious and normal packets. The dataset is available from this link: https://www.kaggle.com/saurabhshahane/android-permission-dataset, (accessed on 25 November 2021).

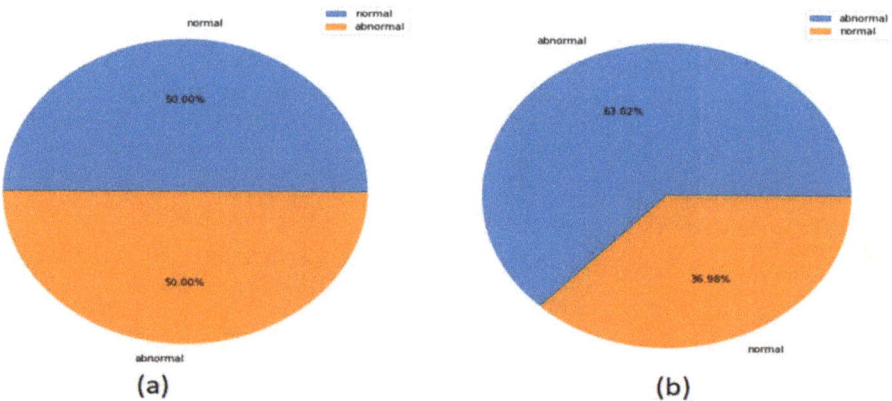

Figure 4. Percentage of classes of the datasets (**a**) CICAndMal2017 and (**b**) Drebin.

3.1.2. The Drebin Dataset

The Drebin dataset was extracted from 15,037 applications of the Drebin project, which contains 215 features and the injection of 5560 malware and 9476 normal applications. The dataset was developed by the Drebin project and published as the DroidFusion paper in the *IEEE Transactions on Cybernetics* journal [59]. The dataset was generated with different Android applications and is available through the following link: https://www.kaggle.com/shashwatwork/android-malware-dataset-for-machine-learning (accessed on 25 Novmber 2021).

3.2. Preprocessing

The Android datasets have different formats and characteristics; therefore, preprocessing is very important for managing the dataset.

Min–Max Normalization Method

Normalization is a scaling approach to shift and rescale the values of datasets. The min–max normalization method was applied to scale the data in the range between 0 and 1. The normalization method was applied for the overlap of the entire dataset using the following equation:

$$\acute{V} = \frac{V - x_{min}}{\max(A) - \min(A)} (\text{new\_max}(A) - \text{new\_min}(A)) + \text{new\_min}(A) \quad (1)$$

where, min(A) and max(A) are the minimum and maximum data, respectively, new_min(A) and new_max(A) are the new values of the minimum and maximum used for the scaling of the data, and $\acute{V}$ is the normalized data.

3.3. Classification Algorithms

In this section, the theoretical description of the machine learning and deep learning algorithms used in this research is presented.

3.3.1. K-Nearest Neighbors (KNN)

The KNN algorithm is a simple and common machine learning algorithm used to classify numbers of real-life applications by discovering neighbors. The mechanism of the KNN algorithm is finding the distance between the classes of normal values and attacks by selecting object values close to the class k-values. The algorithm starts by loading network data with the length of input data [60]. KNN is utilized to determine the k-values that are near a set of specific values in the training dataset. The majority of these k-values fall into a confirmed class. Furthermore, the input sample is classified. In this research, the Euclidean distance function (E_i) was used to find the distance between the object values. The expression of the Euclidean distance function is as follows:

$$E_i = \sqrt{(a_1 - a_2) + (b_1 - b_2)^2} \qquad (2)$$

where $a_1, a_2, b_1,$ and b_2 are variables of the input data.

3.3.2. Support Vector Machine (SVM)

SVM is a supervised machine learning algorithm developed to solve complex problems in linear and nonlinear applications. It is used to draw the hyperplane between the data points that are near the hyperplane and calculate the effect of the location and orientation of the hyperplane, called the support vector (SV) [61]. The good performance of SV is attained when the distance of the data points is close to the hyperplane. The support vector machine has a number of functions, linear and non-liner; the RBF is appropriate for separable patterns because the network data has a complex format. In this research, a Gaussian radial basis function was proposed to detect Android malware:

$$K(y, y') = \exp\left(-\frac{||y - y'||^2}{2\sigma^2}\right) \qquad (3)$$

where, y, and y' are vector features of the training data, $||y - y||^2$ is the squared Euclidean distance between the features of the training data, and σ is the parameter.

3.3.3. Linear Discriminant Analysis (LDA)

LDA is a linear machine learning algorithm used to solve applications with high dimensionality [62]. It is used to model and transform data from a high-space dimension into a low-space dimension by separating the classes of the data into two groups: normal and malicious packets. Figure 5 represent the LDA method for analyzing normal and abnormal packets, where the red line linearly separates the two classes of the data.

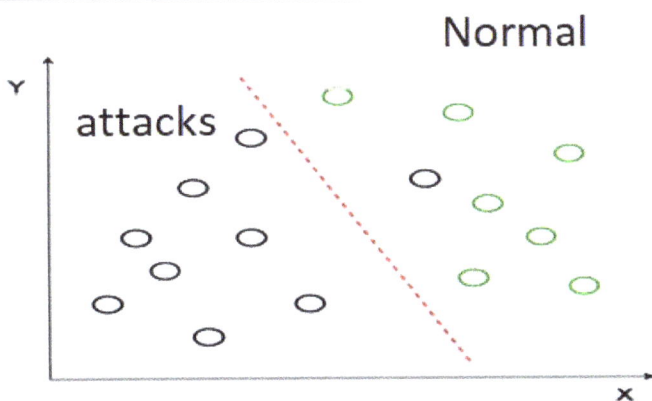

Figure 5. The linear discriminant analysis (LDA) method for analyzing datasets.

3.3.4. Deep Learning Models

CNN-LSTM is a fusion model created with the combination of CNN and LSTM; both are deep learning AI algorithms. In CNN, there are hidden neurons with trainable weights and bias parameters. It is broadly applied to analyze the data in a grid layout, making it different from other structures [63]. It is also called a feed-forward network because the input data stream in one way, from the input to the production layer [64]. Three are the main components in the CNN structure: the convolutional, pooling, and fully connected layers. For feature extraction and the reduction of dimensionality, the convolutional and pooling layers are employed. The fully connected layer is completely folded and attached to the output of the previous layer. The main architecture of the CNN model for detecting Android malware applications is displayed in Figure 6.

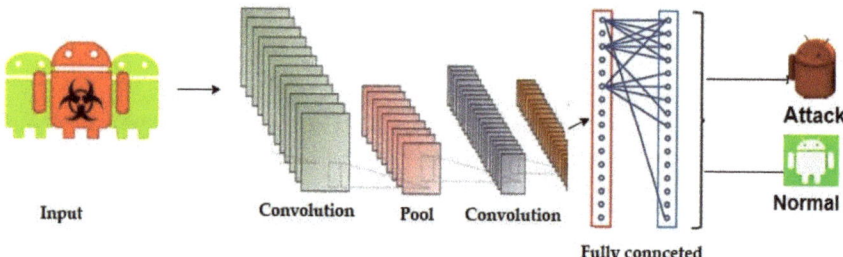

Figure 6. Structure of the CNN model.

Hochreiter et al. [65] introduced the LSTM algorithm for learning long-term data dependency. The LSTM is one type of recurrent neural network (RNN). The distinction between the LSTM and RNN techniques is the memory cells present in the LSTM structure. Every memory cell comprises four gates: the input, candidate, forget, and output gates. The forget gate categorizes the input features as to whether they must be discarded or kept. The input gate revives the memory cells in the LSTM structure, and the hidden state is always controlled by the output gate. Furthermore, LSTM uses an embedded memory block and gate mechanism that enables it to address complications related to the disappearing gradient and the explosion gradient present in the RNN learning [66]. The structure of the LSTM model is presented in Figure 7. Table 1 show the parameters of the LSTM model. It is investigated that these parameter values were significant for obtaining high performance to detect the android malware. The kernel size of convolution was 4, the max pool size id 4 for selecting significant features from the filter layer. The drop out value was 0.50 for preventing the model from overfitting; in order to optimize the model, the RSMprop optimizer function is presented. The error gradient is used batch size 150. The equations for the LSTM-related gates are defined as follows:

$$f_t = \sigma\left(W_f \cdot X_t + W_f \cdot h_{t-1} + b_f\right) \tag{4}$$

$$i_t = \sigma(W_i \cdot X_t + W_i \cdot h_{t-1} + b_i) \tag{5}$$

$$S_t = \tan h(W_c \cdot X_t + W_c \cdot h_{t-1} + b_c) \tag{6}$$

$$C_t = (i_t * S_t + f_t * S_{t-1}) \tag{7}$$

$$o_t = \sigma(W_o + X_t + W_o \cdot h_{t-1} + V_o \cdot C_t + b_o) \tag{8}$$

$$h_t = o_t + \tan h(C_t) \tag{9}$$

where X_t is the vector of the input features sent to the memory cell at a time t. W_i, W_f, W_c, W_o, and V_O represent the weight matrices, b_i, b_f, b_c, and b_o indicate the bias vectors, h_t is the point of the stated value of the memory cell at a time t, S_t and C_t are the defined values of the candidate state of the memory cell and the state of the memory cell at time t, respectively. σ and $tanh$ are activation functions, and i_t, f_t, and o_t are obtained values for the input gate, the forget gate, and the output gate at time t, respectively. h_{t-1} represents the short memory vector.

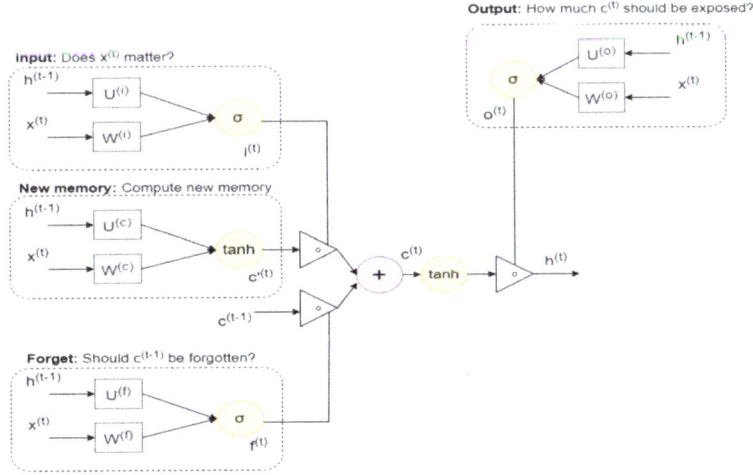

Figure 7. The structure of the LSTM technique.

Table 1. Parameters of the LSTM model.

Parameters	Values
Kernel size	4
Max pooling size	4
Drop out	0.50
Fully connected layer	32
Activation function	Relu
Optimizer	RSMprop
Epochs	10, 20
Batch size	20

The CNN-LSTM model was built, as shown in Figure 8. It was trained using the training dataset, and its hyperparameters were adjusted using the Adam optimizer and the validation dataset. The CNN-LSTM model was next implemented on the test dataset, including features of each testing record to its real class: normal or a particular class of attack [67]. The training and optimization processes of the CNN-LSTM model consisted of two one-dimensional convolution layers that cross the input vectors with 32 filters and a kernel size of 4, two fully connected dense layers composed of 256 hidden neurons, and an output layer that applies the nonlinear SoftMax activation function used for multiclass classification tasks. To overcome the model's overfitting, the global max-pooling and dropout layers were applied. The global max-pooling layer prevents overfitting of the learned features by captivating the maximum value, while the dropout layer is used to deactivate a set of specific neurons in the CNN-LSTM network. The Adam optimizer updates the weights and improves the cross-entropy loss of function. Table 2 show the parameters of the CNN-LSTM model.

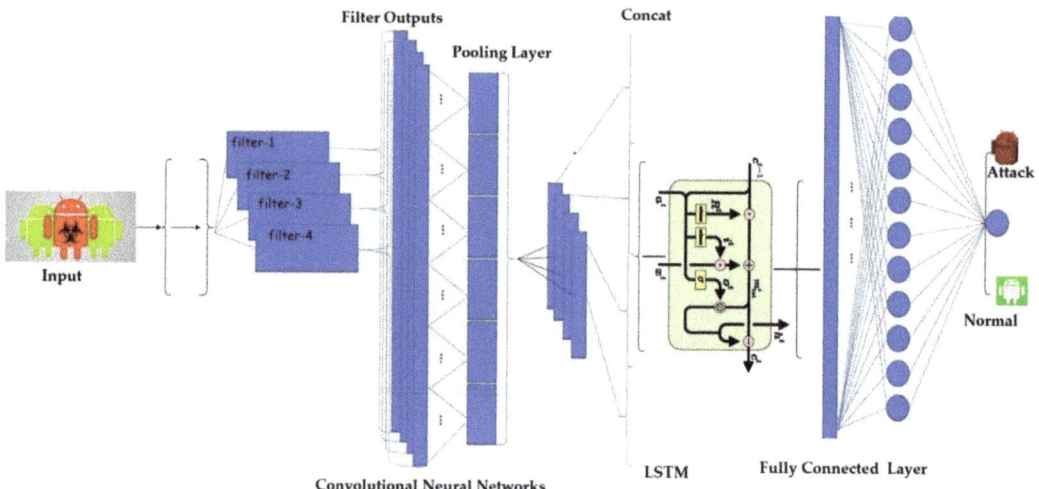

Figure 8. The structure of the CNN-LSTM model.

Table 2. Parameters of the CNN-LSTM model.

Parameters	Values
Kernel size	4
Max pooling size	4
Drop out	0.50
Fully connected layer	32
Activation function	Relu
Optimizer	RSMprop
Epochs	20
Batch size	150

3.3.5. Autoencoder (AE)

AE is a type of AI algorithm based on deep neural networks that use unsupervised learning for encoding and decoding the input data and are commonly utilized for feature extraction and denoising [68]. Two different processes are performed by AE: encoding and decoding. Hence, its structure is symmetrical. The input data are passed through three different layers: the input, latent, and output layers. These layers make up the AE architecture (Figure 9). The input and output layers have the same size, and the latent layer has a smaller size than the input layer [69]. Encoding and decoding are achieved with the following equations, respectively:

$$e = f_\theta(x) = s(Wx + b) \tag{10}$$

$$\tilde{x} = g_{\theta'}(e) = s(W'e + b') \tag{11}$$

where x is the input vector, $e \in [0, 1]^d$ represents the latent vector, and $\tilde{x} \in [0, 1]^D$ is the produced vector. From the input layer to the latent layer, the encoding process is repeated. Next, the decoding process is repeated from the latent layer to the output layer. W and W' represent the weight from the input to the latent and from the latent to the output layers, respectively. b and b' denote the bias vectors of the input layer and the latent layer. The activation functions of the latent layer neurons and the output layer neurons are represented with f_θ and $g_{\theta'}$, respectively. The weight and bias parameters are learned in the AE structure after reducing the reconstruction error. Equation (12) is used to measure the error between the reconstructed $\tilde{x}$ and the input data x for individual instances:

$$J(W, b', x, \tilde{x}) = \frac{1}{2} \|h_{w,b}(x) - \tilde{x}\|^2 \tag{12}$$

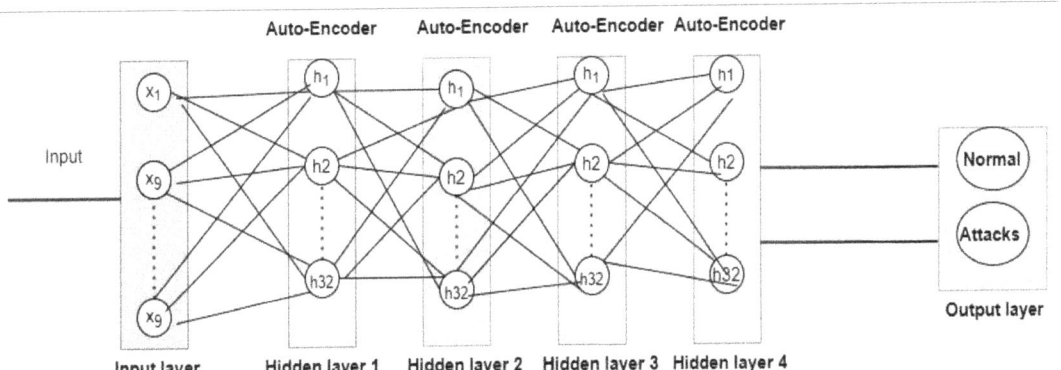

Figure 9. The structure of the auto-encoder (AE) model.

In a training dataset including D instances, the cost function is defined as follows:

$$\sum_{l=1}^{n_l-1} \sum_{i=1}^{s_l} \sum_{j=1}^{s_l+1} \left(W_{ji}^{(l)}\right)^2 = \left[\frac{1}{D} \sum_{i=1}^{D} \left(\frac{1}{2} \|h_{w,b}\left(x^{(i)} - \tilde{x}^{(i)}\right)\|^2\right)\right] + \frac{\lambda}{2} \sum_{l=1}^{n_l-1} \sum_{i=1}^{s_l} \sum_{j=1}^{s_l+1} \left(W_{ji}^{(l)}\right)^2 \tag{13}$$

where D refers to the total number of instances, s to the number of neurons in layer l, λ represents the weight attenuation parameter, and the square error is the reconstruction error of each training instance.

3.4. Performance Measurements

The statistical analysis included the calculation of the mean square error (MSE), Pearson's correlation coefficient (R), and the root-mean-square error (RMSE) to test the proposed algorithms' efficiency in detecting Android malware. The equations of these parameters are presented below:

$$MSE = \frac{1}{n} \sum_{i=1}^{n} \left(y_{i,exp} - y_{i,pred}\right)^2 \tag{14}$$

$$RMSE = \sqrt{\sum_{i=1}^{n} \frac{\left(y_{i,exp} - y_{i,pred}\right)^2}{n}} \tag{15}$$

$$R\% = \frac{n\left(\sum_{i=1}^{n} y_{i,exp} \times y_{i,pred}\right) - \left(\sum_{i=1}^{n} y_{i,exp}\right)\left(\sum_{i=1}^{n} y_{i,pred}\right)}{\sqrt{\left[n\left(\sum_{i=1}^{n} y_{i,exp}\right)^2 - \left(\sum_{i=1}^{n} y_{i,exp}\right)^2\right]\left[n\left(\sum_{i=1}^{n} y_{i,pred}\right)^2 - \left(\sum_{i=1}^{n} y_{i,pred}\right)^2\right]}} \times 100 \tag{16}$$

$$R^2 bn1 - \frac{\sum_{i=1}^{n} \left(y_{i,exp} - y_{i,pred}\right)^2}{\sum_{i=1}^{n} \left(y_{i,exp} - y_{avg,exp}\right)^2} \tag{17}$$

$$Accuracy = \frac{TP + TN}{TP + FP + FN + TN} \times 100\% \tag{18}$$

$$Specificity = \frac{TN}{TN + FP} \times 100\% \tag{19}$$

$$\text{Sensitivity} = \frac{TP}{TP+FN} \times 100\% \tag{20}$$

$$\text{Precision} = \frac{TP}{TP+FP} \times 100\% \tag{21}$$

$$\text{Fscore} = \frac{2*\text{preision}*\text{Sensitivity}}{\text{preision}+\text{Sensitivity}} \times 100\% \tag{22}$$

where $y_{i,exp}$ is the experimental value of the data point i, $y_{i,pred}$ is the predicted value of the data point i, $y_{avg,exp}$ is the average of the experimental values, R is Pearson's correlation coefficient, $y_{i,exp}$ are the Android network packets of the input data i, $y_{i,class}$ are the classes of Android malware and normal input data i, n is the total number of the input data, the true positive (TP) represents the total number of samples that are successfully classified as positive sentiment, false positive (FP) is the total number of samples that are incorrectly classified as negative sentiments, true negative (TN) denotes the total number of samples that are successfully classified as negative sentiment, and false negative (FN) represents the total number of samples that are incorrectly classified as positive sentiments.

4. Results

The investigation of the effect of the proposed models on the standard Android malware datasets was conducted using the Python programing language. The statistical analysis evaluated the results of the proposed models.

4.1. Splitting the Data

The datasets were divided into 70% training and 30% testing data. The random function for splitting the training and testing was proposed. The training phase was applied to fit the models using the Android malware datasets. The test phase was designed to validate the proposed models using new data. Table 3 show the datasets' volume.

Table 3. Volume of datasets.

Datasets	Total Volume	Training	Testing
CICAndMal2017	676	473	203
Drebin	15,031	10,521	4510

4.2. Experimental Environments

The platform used to detect intrusion in Android applications is presented in Table 4.

Table 4. Environment requirements of the proposed model.

Hardware	Software
RAM size 8 GB	Python Version 3.6
C.P.U.	Numpy Version 1.18.1
	TensorFlow library Version 2.10
	Keras library Version 2.3.1
	Matplotlib Version 3.2.0
	NumPy library Version 1.01

4.3. Model Performance

The highly efficient performance of machine learning and deep learning models guarantees the detection of Android malicious applications. The algorithms for intrusion detection were tested using two standard malware mobile datasets. The Drebin dataset contained 10,525 Android applications, and the CICAndMal2017 dataset contained 676 injections of various attack and normal packets.

4.3.1. Performance of the Machine Learning Models

In this work, the SVM, KNN, and LDA models were applied to identify Android malicious packets. The SVM algorithm achieved maximum accuracy (100%) with respect to all the performance measurements in the CICAndMal2017 dataset (Table 5). However, it achieved lower accuracy (80.71%) with the Drebin dataset.

Table 5. Results of the SVM method.

	CICAndMal2017 Dataset		
Metrics	Precision (%)	Recall (%)	F1-score (%)
Normal	100	100	100
Attacks	100	100	100
Accuracy		100	
Weighted average	100	100	100
	Drebin dataset		
Metrics	Precision (%)	Recall (%)	F1-score (%)
Normal	0.97	0.51	0.67
Attacks	0.77	0.99	0.86
Accuracy		80.71	
Weighted average	0.84	0.81	0.79

The SVM method showed the efficiency performance with the CICAndMal2017 dataset and satisfying results in the Drebin dataset. The confusion metrics of the SVM method are presented in Figure 10. In the CICAndMal2017 dataset, the percentage of the normal data classified as true negative was 45.81%, whereas the true positive represented 54.19% and were classified as malware attacks. Furthermore, the false positive and false negative data were 0, indicating that the SVM method successfully detected malicious attacks in the Drebin dataset. The confusion metrics of the SVM approach applied on the Drebin dataset were as follows: 61.56% were classified as abnormal applications, 19.15% true negatives were classified as normal applications, whereas the true positive and false negatives were 18.62% and 0.67%, respectively. We conclude that the performance of the SVM method is good since the false positive is low.

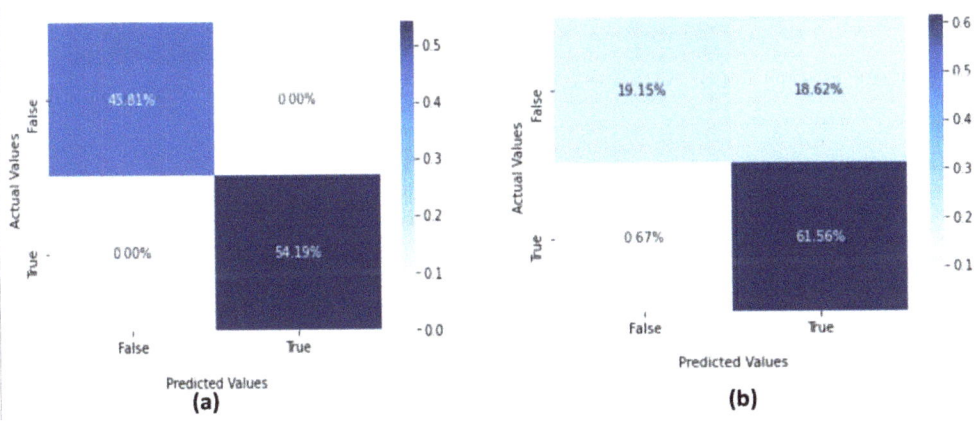

Figure 10. The confusion metrics of the SVM method using the (**a**) CICAndMal2017 and (**b**) Drebin datasets.

Table 6 summarize the performance of the KNN method in the detection of malware attacks in both datasets. We considered the scope of the KNN method with (k = 5). In the CICAndMal2017 dataset, the KNN method achieved high accuracy (90%), contrary to the Drebin dataset (81.57%).

Table 6. Results of KNN algorithm.

	CICAndMal2017		
Metrics	Precision (%)	Recall (%)	F1-score (%)
Normal	0.89	0.89	0.89
Attacks	0.91	0.91	0.91
Accuracy		0.90	
Weighted average	0.90	0.90	0.90
	Drebin dataset		
Metrics	Precision (%)	Recall (%)	F1-score (%)
Normal	0.96	0.53	0.68
Attacks	0.78	0.99	0.87
Accuracy	81.57		
Weighted average	0.85	0.82	0.80

Figure 11 show the confusion metrics for the KNN method. In the CICAndMal2017 dataset, 40.89% of the dataset was classified as true negative (normal applications), 49.26% as malware, and 4.93% as false positives (normal data classified as attacks). In the Drebin dataset, the KNN method classified 61.87% of the dataset as true positives (attacks), 19.71% as true negatives (normal), and the false positives were <0.80%. Overall, the KNN method achieved higher accuracy in the CICAndMal2017 dataset than in the Drebin dataset.

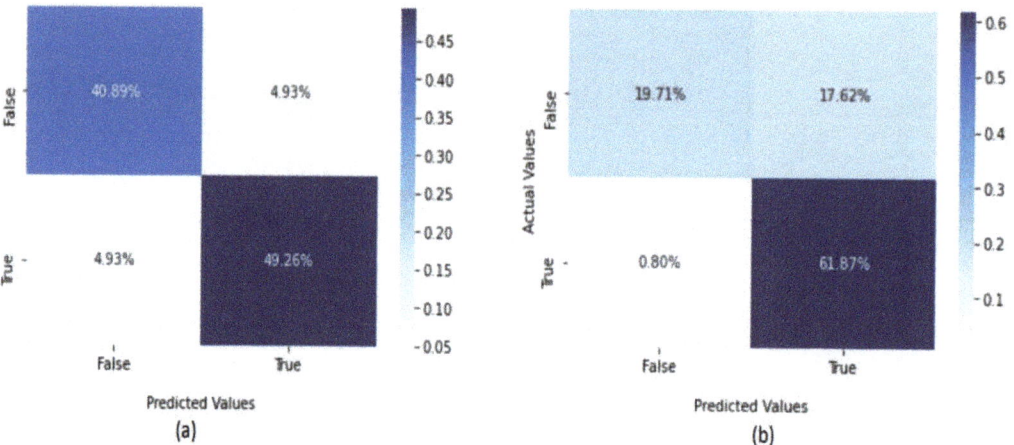

Figure 11. The confusion metrics of the KNN method using the (**a**) CICAndMal2017 and (**b**) Drebin datasets.

The results of the LDA method are presented in Table 7. Overall, the results were not adequate due to the complexity of the network dataset. The nonlinear algorithms are not appropriate for the analysis of network datasets. The accuracy of LDA was 45.32% in the CICAndMal201 dataset, a percentage that reached 81% in the case of the Drebin dataset.

Table 7. Results of the LDA method.

	CICAndMal201		
Metrics	Precision (%)	Recall (%)	F1-Score (%)
Normal	0.46	0.98	0.62
Attacks	0.33	0.01	0.02
Accuracy		45.32	
Weighted average	0.39	0.45	0.29
	Drebin Dataset		
Metrics	Precision (%)	Recall (%)	F1-score (%)
Normal	0.95	0.53	0.68
Attacks	0.78	0.98	0.87
Accuracy		81.35	
Weighted average	84	0.81	0.82

The confusion metrics of the LDA method are presented in Figure 12. The percentage of true positives was high (49%), whereas that of true negatives (classified as normal applications) was low (44.83%) in the CICAndMal2017 dataset. The percentage of false positives was high (53.69%), showing that the LDA model is not appropriate for this dataset. In the Drebin dataset, the confusion metrics showed that 19.15% were true negatives and 1.02% false positives, classifying normal applications as malware. Overall, the LDA method had good performance with the Drebin dataset.

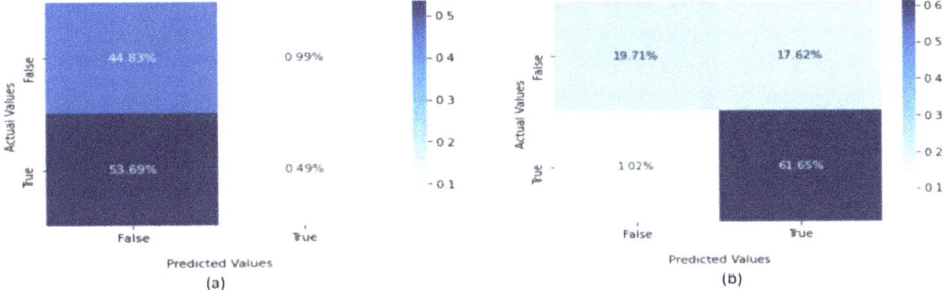

Figure 12. The confusion metrics for the (**a**) CICAndMal2017 and (**b**) Drebin datasets.

4.3.2. Performance of the Deep Learning Models

In this section, the results of the deep learning algorithms, namely LSTM, CNN-LSTM, and AE, are presented. The dataset was divided into 70% training and 30% test data. Table 8 show the results of the LSTM, CNN-LSTM, and AE models. The performance of the CNN-LSTM model achieved high accuracy (95.07%) compared with the LSTM and AE models in the CICAndMal2017 dataset.

Table 8. Results of the deep learning algorithms in the CICAndMal2017 dataset.

Models	Loss	Accuracy (%)	Precision (%)	Recall (%)	F1 Score (%)
LSTM	0.20	94.58	95.41	94.54	94.97
CNN-LSTM	0.16	95.07	97.16	93.63	95.53
AE	1.43	75.79	92.15	66.78	77.44

Figure 13 show the accuracy performance of the LSTM, CNN-LSTM, and AE algorithms using the CICAndMal2017 dataset. The performance plots show that the CNN-LSTM model achieved an accuracy of 99.9% in the training phase, and in the validation phase, the initial 75% accuracy reached 95.07%. The LSTM model achieved good performance in the training phase (99%) and the validation phase it reached 94.58%.

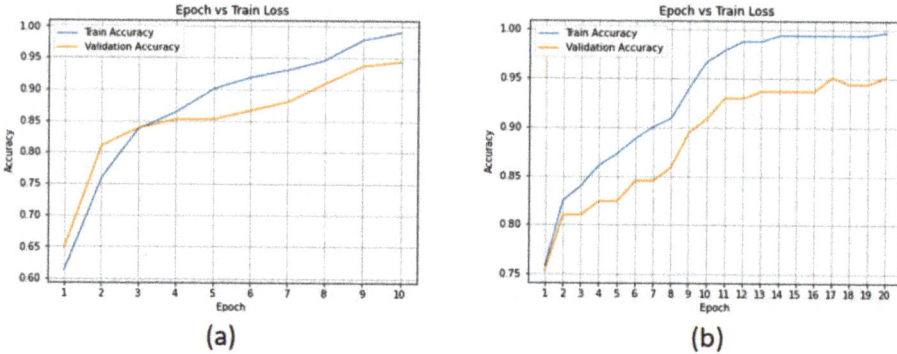

Figure 13. Performance of the deep learning models with the CICAndMal2017 dataset. (a) LSTM. (b) CNN-LSTM.

The binary_crossentropy method was used to calculate the accuracy loss in the training and testing phases. Figure 14 show the validation accuracy of the deep learning models. The accuracy loss of the LSTM model in the validation phase changed from 0.5 to 0.2, while in the case of the CNN-LSTM model, this changed from 0.6 to 0.2.

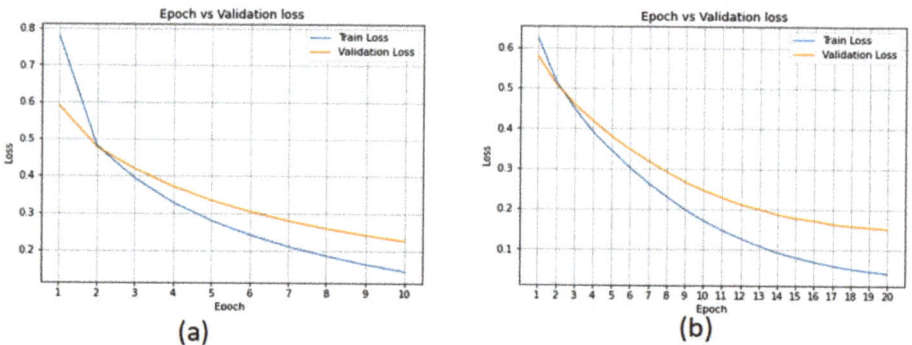

Figure 14. Accuracy loss of the deep learning models in the CICAndMal2017 dataset. (a) LSTM. (b) CNN-LSTM.

Table 9 show the results of the LSTM, CNN-LSTM, and AE models using the Drebin dataset. The LSTM model achieved high accuracy (99.40%). Furthermore, the CNN-LSTM model showed high accuracy of 97.20%, and the performance of the AE model was satisfying.

Table 9. Results of the deep learning models using the Drebin dataset.

Models	Loss	Accuracy (%)	Precision (%)	Recall (%)	F1 Score (%)
LSTM	0.05	99.40	99.32	99.74	99.53
AE	3.65	56.65	41.18	65.71	51.11
CNN-LSTM	0.09	97.20	97.72	97.92	97.82

Figure 15 show the accuracy performance of the deep learning models. The validation accuracy of the LSTM model started from 97% and reached 99.40% with 20 Epochs. The LSTM model in the training phase achieved an accuracy of 100%. The performance of the CNN-LSTM model was 97.20% in the validation phase.

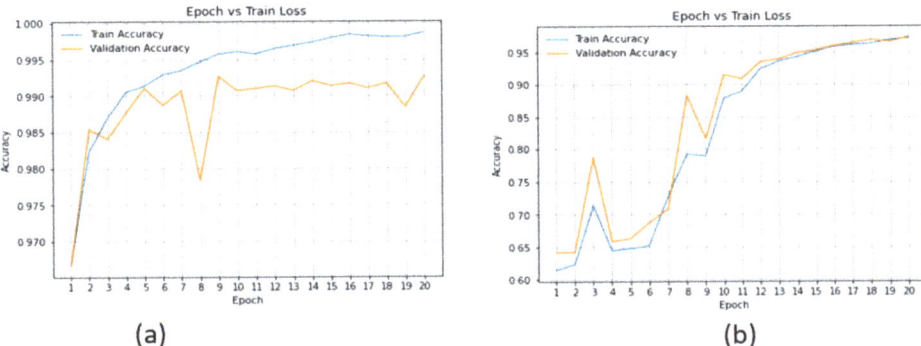

Figure 15. Performance of the deep learning models in the CICAndMal2017 dataset. (a) LSTM. (b) CNN-LSTM.

Figure 16 show the validation loss of the deep learning models. In the LSTM model, the validation loss changed from 0.10 to 0.7, whereas for the CNN-LSTM model, it changed from 0.7 to 0.1 with 20 Epoch.

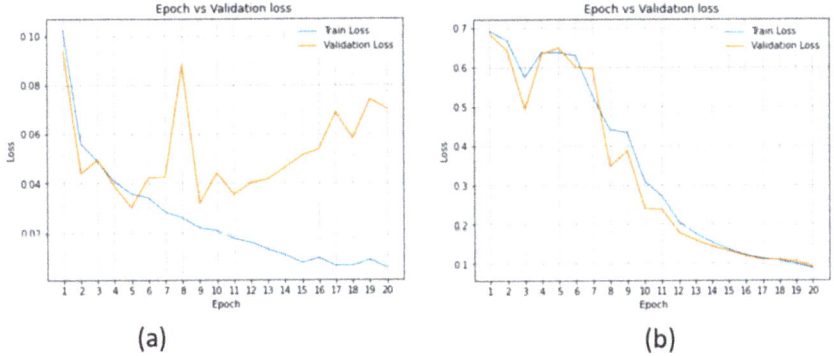

Figure 16. Accuracy loss of the deep learning models in the CICAndMal2017 dataset. (a) LSTM. (b) CNN-LSTM.

The accuracy performance of the AE model using the CICAndMal2017 and Drebin datasets is presented in Figure 17. The performance of AE was not satisfying, with the accuracy in the training phase being 79% and in the validation phase 75.79% for the CICAndMal2017 dataset. For the Drebin dataset, the accuracy in the validation phase was 56%. The accuracy percentage of the LSTM and CNN-LSTM models outperformed the AE model.

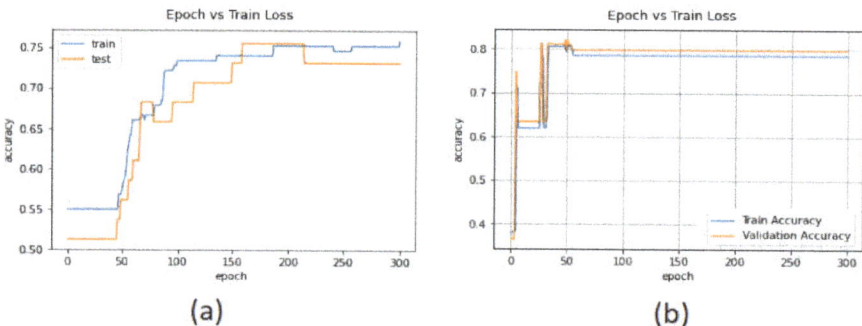

Figure 17. Accuracy of the AE model in the (**a**) CICAndMal2017 and (**b**) Drebin datasets.

Figure 18 display the accuracy loss of the AE model in both datasets. The accuracy loss was high (from 0.70 to 0.55) for the CICAndMal2017 dataset. Furthermore, the validation loss changed from 0.9 to 0.4 in the case of the Drebin dataset. Overall, the validation loss of the AE model was high; therefore, the AE model's performance is not appropriate for the detection of Android malicious attacks.

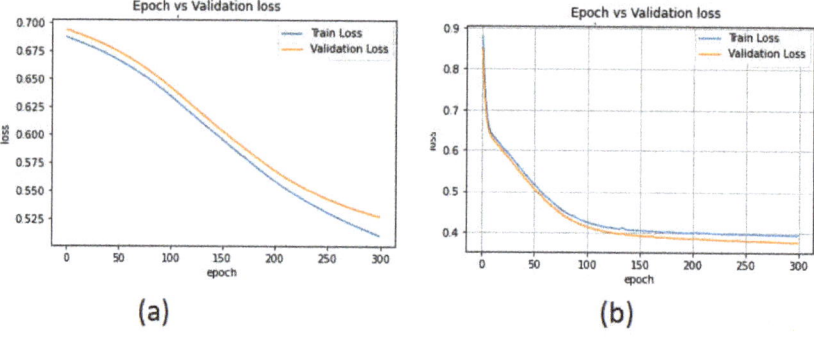

Figure 18. Accuracy loss of the autoencoder model in the (**a**) CICAndMal2017 and (**b**) Drebin datasets.

4.4. Sensitivity Analysis

Sensitivity analysis is an approach used to measure the influence of uncertainties of the input data variables. Analyzing the input data is very useful in extracting the patterns from the dataset. The Pearson's correlation coefficient was applied to find the correlation between the input features and the classes. Some features had significant relationships between the classes (normal and attacks) [70,71].

We selected the features that had a relationship >50% between the class. Figure 19 show the features that have a significant correlation with the classes variables in the CI-CAndMal2017 dataset. We considered four features with correlation >50%. The correlation coefficient results for the Drebin dataset are presented in Figure 20. It was observed that the Drebin dataset revealed a strong correlation between classes, while in the CICAndMal2017 dataset, they were <50%.

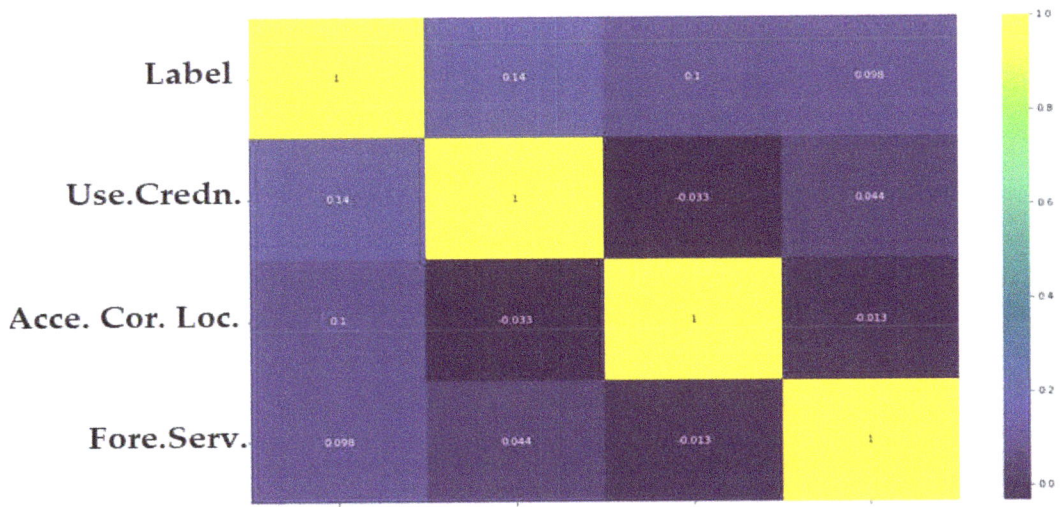

Figure 19. The correlation coefficient results using the CICAndMal2017 dataset.

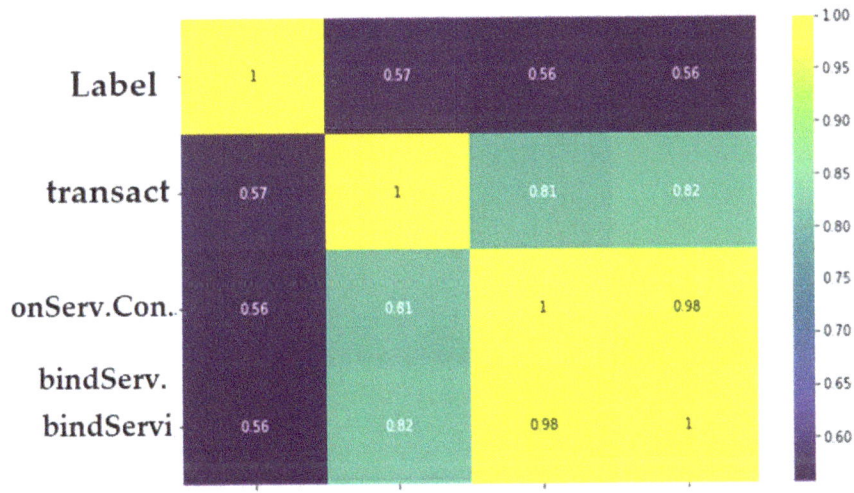

Figure 20. The correlation coefficient for the Drebin dataset.

We applied the statistical metrics mean absolute error (MAE), MSE, RMSE, and R^2 to identify the prediction error between the target class and the predicted values. The prediction error of the machine learning algorithms is presented in Table 10. The SVM algorithm displayed fewer prediction errors, and the R^2 between the predicted values and the target values was 100% for the CICAndMal2017dataset. The KNN method showed fewer prediction errors (MSE = 0.1842), and the relationship between the predicted and target values was 33.35%.

Table 10. Statistical analysis of the machine learning algorithms' results using the CICAndMal2017 dataset.

Models	MAE	MSE	RMSE	R^2 (%)
SVM	0.00	0.0	0.0	100
KNN	0.0985	0.09852	0.313	63.31
LDA	0.429	0.4189	0.647	53.68

Table 11 show the prediction potential of the SVM, KNN, and LDA methods. The prediction performance of the KNN method was R^2 = 33.35, achieving the best correlation between the predicted and target values in the Drebin dataset. Overall, the prediction results of the machine learning algorithms were satisfactory.

Table 11. Statistical analysis of the machine learning models using the Drebin dataset.

Models	MAE	MSE	RMSE	R^2 (%)
SVM	0.1915	0.1915	0.437	31.57
KNN	0.1842	0.1842	0.429	33.35
LDA	0.1864	0.1864	0.431	32.09
SVM	0.1915	0.1915	0.437	31.57

The prediction errors of the deep learning algorithms are summarized in Table 12. The LSTM model achieved lower prediction levels (MSE = 0.0054), and the correlation between the predicted and target values was 88.25% in the CICAndMal2017 dataset. In the Drebin dataset, the LSTM model showed lower prediction levels (MSE = 0.0059) and high correlation (R^2 = 97.39%). The prediction performance of LSTM was good in both datasets.

Table 12. Statistical analysis of the deep learning models.

CICAndMal2017 Dataset				
Models	MAE	MSE	RMSE	R^2 (%)
LSTM model	0.0054	0.0541	0.232	88.25
Autoencoder model	0.339	0.339	0.5830	31.74
CNN-LSTM	0.049	0.049	0.221	80.31
Drebin dataset				
Models	MAE	MSE	RMSE	R^2 (%)
LSTM model	0.0059	0.0059	0.077	97.39
Autoencoder model	0.2425	0.2279	0.177	17.91
CNN-LSTM	0.027	0.027	0.1671	87.84

5. Discussion

With rapidly developing technology, the use of smartphones with new features and associated Android applications has increased. Statista reported that 1.3 billion smartphones will be used by 2023. This also brings challenges for the researchers and developers of security mechanisms for these applications, originating in the new complexities and vulnerabilities of the Android applications that hackers can quickly exploit.

Considering that Android applications of digital e-commerce, e-business, savings, and online banking are associated with confidential and appreciated information communicated within the mobile network, it is important to evaluate the application data in terms of accomplishing proper security. Machine and deep learning algorithms are used to monitor the detection of malicious attacks against Android applications to ensure that security openings do not occur within this network. The present research contributes to the area of cybersecurity by developing a system based on machine learning and deep learning algorithms to detect anomalies in signature databases, thus, permitting the system to detect unknown attacks.

As we know, the network has a very complex format; in this study, nonlinear models were proposed to achieve high accuracy, whereas linear, namely LDA and KNN, models achieved slightly worse performance. The accuracy performance of LDA was 45.32% in the CICAndMal2017 dataset, and the accuracy performance improved to 81.35% using the Drebin dataset. It was observed that the KNN model achieved little accuracy, 81.57%, using the Drebin dataset. We observed that the LDA and KNN algorithms are not appropriate for detecting Android malware. In deep learning models, the AE mode results were not satisfactory for detecting the mobile attacks. The AE achieved 75.79% and 56.65% with respect to the CICAndMal2017 and Drebin datasets. The AE is composed of the encoder and decoder models; the encoder compresses the input data, whereas the decoder is used to recreate the input from the encoder. Overall, we observed that these models did not achieve good results due to the research datasets being binary data.

Furthermore, using the support vector machine, LSTM and CNN-LSTM algorithms achieved high accuracy performance for developing an appropriate system that can support the security of smartphones against malware. Two standard datasets were used. The SVM model achieved an accuracy of 100% using the CICAndMal2017 dataset and the LSTM algorithm achieved 99.40% using the Drebin dataset.

Our system was compared with existing systems of machine learning and deep learning models that detect malware for the security of Android applications. The mechanism of the proposed system is based on the pattern of dataset behavior for detecting the attacks. The LSTM model had an accuracy of 99.40% in the case of the Drebin dataset, indicating that it is a robust model to handle Android security vulnerabilities. Recently, by employing a CNN model on an Android platform, the system was found to achieve high accuracy; however, our system is more accurate against all systems. Table 13 show the results of our system against existing security systems using the same dataset. The graphic representation of our system and other existing systems' results with respect to the accuracy metrics is presented in Figure 21. Overall, the system we propose achieved the highest accuracy.

Table 13. Results of the proposed system against existing security systems using the Drebin dataset.

Reference	Year	Datasets	Model	Accuracy (%)
Ref. [72]	2021	Drebin	CNN	91
Ref. [73]	2018	Drebin	RF, J.48, NB, Simple Logistic, BayesNet TAN, BayesNet K2, SMO PolyKernel, IBK, SMO NPolyKernel	88–96
Ref. [74]	2021	Drebin	CBR, SVM, DT	95
Ref. [75]	2019	Drebin	Random forest tree	96.7
Ref. [76]	2018	Drebin	DT	97.7
Ref. [77]	2019	Drebin	RF with 1000 decision trees	98.7
Ref. [78]	2019	Drebin	SVM	93.7
Ref. [79]	2019	Drebin	Random forest tree	94
Ref. [80]	2019	Drebin	Random forest tree	96
Ref. [81]	2016	Drebin	Random forest tree	97
Ref. [82]	2021	Drebin	CNN	98.2
Proposed model	2022	Drebin	LSTM	99.40
			CNN-LSTM	97.82

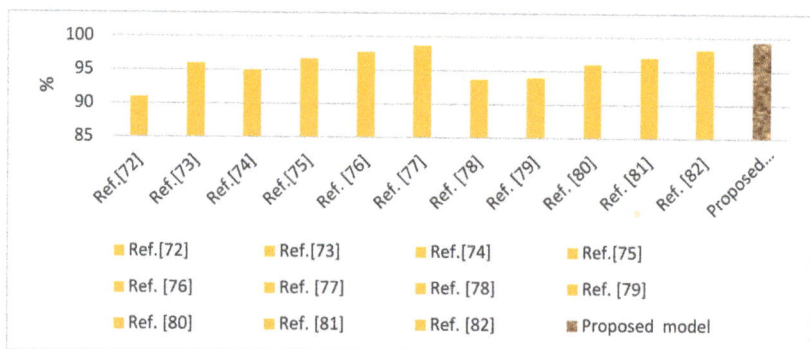

Figure 21. Comparative performance of the proposed system against existing systems in the detection of malware against Android applications using the Drebin dataset.

Table 14 display the results of the proposed system and other existing Android cybersecurity systems that use the machine and deep learning algorithms applied to different Android datasets. To confirm the results of the proposed system against other Android security systems, we compared recent systems' results with ours, with the latter achieving high accuracy. The graphic representation of these results is presented in Figure 22.

Table 14. Results of the proposed system against existing security systems using different Andriod datasets.

Reference	Year	Datasets	Model	Accuracy (%)
Ref. [83]	2019	MalGenome, Kaggle, Androguard	Random forest tree	93
Ref. [84]	2018	Google Play, VirusShare, MassVet	LSTM	97.4
Ref. [85]	2017	Genome, IntelSecurity, MacAfee, Google Play	Deep CNN	87
	2022	Drebin	LSTM	99.40
			CNN-LSTM	97.82

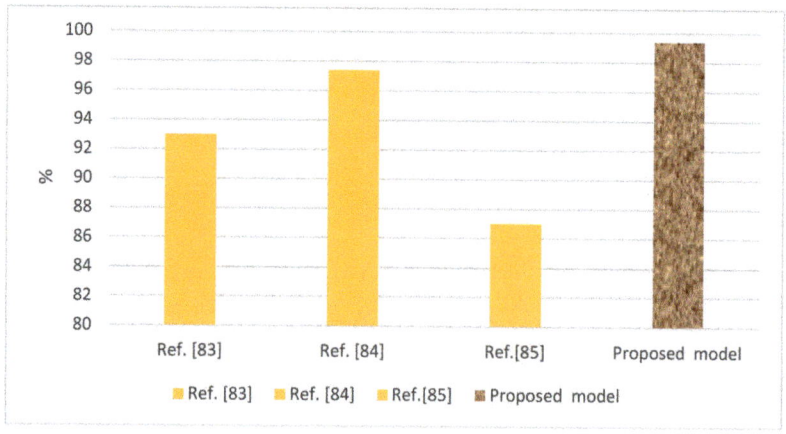

Figure 22. Comparative performance of the proposed system against existing systems in the detection of malware against Android applications using different datasets.

6. Conclusions

Smartphones are becoming more and more popular, constituting a profitable target for hackers due to their susceptibility to security breaches. Android is an open gate for attackers who exploit it with malicious applications, benefiting from the system's security flaws. An emerging method for signature-based malicious attack detection is the antivirus applications against new malware, created with AI, machine learning, and deep learning algorithms that predict malware. In this study, a security system was built and designed based on the support vector machine (SVM), k-nearest neighbors (KNN), linear discriminant analysis (LDA), long short-term memory (LSTM), convolution neural network-long short-term memory (CNN-LSTM), and autoencoder algorithms. According to the promising results of the present research, the following conclusions can be drawn:

The proposed system was evaluated and examined using two standard Android malware applications datasets: CICAndMal2017 and Drebin. The SVM, KNN, and LDA methods proved to be efficient machine learning algorithms and successfully detected malware, with SVM being the most effective. The LSTM and CNN-LSTM models are proposed to detect malicious applications, with the LSTM model being more efficient for developing Android security. Sensitive analysis examining the metrics MSE, RMSE, and R^2 revealed the errors between the predicted output and the target values in the validation phase. The LSTM and CNN-LSTM algorithms achieved fewer prediction errors in the Drebin dataset, while the SVM method was more effective in the case of the CICAndMal2017 dataset. The validation phase results of the machine learning and deep learning methods were satisfying, with the LSTM and SVM models achieving superior performance. The results of the present study were compared with recent research findings, confirming the robustness and effectiveness of our results. We implemented machine learning and deep learning algorithms and experimented with them to obtain optimal malware detection. Both of the proposed classifiers achieved good accuracy, but the LSTM accuracy was 99.40%, indicating it can outperform other state-of-the-art models.

Author Contributions: Conceptualization, T.H.H.A. and H.A.; methodology, T.H.H.A.; software, T.H.H.A.; validation, T.H.H.A. and H.A.; formal analysis, T.H.H.A. and H.A.; investigation, T.H.H.A. and H.A.; resources, T.H.H.A.; data curation, T.H.H.A. and H.A.; writing—original draft preparation, T.H.H.A. and H.A.; writing—review and editing, H.A.; visualization, T.H.H.A. and H.A.; supervision, T.H.H.A.; project administration, T.H.H.A. and H.A.; funding acquisition, T.H.H.A. and H.A. All authors have read and agreed to the published version of the manuscript.

Funding: This research and the APC were funded by the Deanship of Scientific Research at King Faisal University for financial support under grant No. NA00036.

Institutional Review Board Statement: Not applicable.

Informed Consent Statement: Not applicable.

Data Availability Statement: The data presented in this study are available here: https://www.kaggle.com/saurabhshahane/Android-permission-dataset; https://www.kaggle.com/shashwatwork/android-malware-dataset-for-machine-learning (accessed on 25 Novmber 2021).

Acknowledgments: The authors extend their appreciation to the Deanship of Scientific Research at King Faisal University for funding this research work through project number NA00038.

Conflicts of Interest: The authors declare that they have no conflict of interest.

References

1. McAfee Mobile Threat Report Q1. 2020. Available online: https://www.mcafee.com/en-us/consumer-support/2020-mobilethreat-report.html (accessed on 2 December 2021).
2. Yerima, S.Y.; Khan, S. Longitudinal Performance Analysis of Machine Learning based Android Malware Detectors. In Proceedings of the 2019 International Conference on Cyber Security and Protection of Digital Services (Cyber Security), Oxford, UK, 3–4 June 2019.
3. Grill, B.B.; Ruthven, M.; Zhao, X. "Detecting and Eliminating Chamois, a Fraud Botnet on Android" Android Developers Blog. March 2017. Available online: https://android-developers.googleblog.com/2017/03/detecting-and-eliminating-chamois-fraud.html (accessed on 12 December 2021).

4. Clarke, E.; Emerson, E.; Sistla, A. Automatic verification of finite-state concurrent systems using temporal logic specifications. *ACM Trans. Program. Lang. Syst.* **1986**, *8*, 244–263. [CrossRef]
5. Andersen, J.R.; Andersen, N.; Enevoldsen, S.; Hansen, M.M.; Larsen, K.G.; Olesen, S.R.; Srba, J.; Wortmann, J.K. CAAL: Concurrency workbench, Aalborg edition. In Proceedings of the Theoretical Aspects of Computing—ICTAC 2015—12th International Colloquium, Cali, Colombia, 29–31 October 2015; Springer: Cham, Switzerland, 2015; pp. 573–582.
6. Alothman, B.; Rattadilok, P. Android botnet detection: An integrated source code mining approach. In Proceedings of the 12th International Conference for Internet Technology and Secured Transactions (ICITST), Cambridge, UK, 11–14 December 2017; pp. 111–115.
7. Haystack. Mobile Issues. Available online: https://safeguarde.com/mobile-apps-stealing-your-information/ (accessed on 14 January 2022).
8. AV-TEST. Security Institute. Available online: https://www.av-test.org/en/statistics/malware/ (accessed on 14 January 2022).
9. Alzahrani, A.J.; Ghorbani, A.A. Real-Time Signature-Based Detection Approach For Sms Botnet. In Proceedings of the 2015 13th Annual Conference on Privacy, Security and Trust (PST), Izmir, Turkey, 21–23 July 2015; pp. 157–164.
10. Girei, D.A.; Shah, M.A.; Shahid, M.B. An Enhanced Botnet Detection Technique For Mobile Devices Using Log Analysis. In Proceedings of the 2016 22nd International Conference on Automation and Computing (ICAC), Colchester, UK, 7–8 September 2016; pp. 450–455.
11. Gilski, P.; Stefanski, J. Android OS: A Review. *Tem. J.* **2015**, *4*, 116. Available online: https://www.temjournal.com/content/41/14/temjournal4114.pdf (accessed on 19 May 2021).
12. Android Developers. Privacy in Android 11. Available online: https://developer.android.com/about/versions/11/privacy (accessed on 10 January 2022).
13. Syarif, A.R.; Gata, W. Intrusion Detection System Using Hybrid Binary PSO and K-Nearest Neighborhood Algorithm. In Proceedings of the 2017 11th International Conference on Information & Communication Technology and System (ICTS), Surabaya, Indonesia, 31 October 2017; pp. 181–186.
14. Huanran., W.; Hui, H.; Weizhe, Z. Demadroid: Object Reference Graph-Based Malware Detection in Android. *Secur. Commun. Netw.* **2018**, *2018*, 7064131.
15. LeCun, Y.; Bengio, Y.; Hinton, G. Deep learning. *Nature* **2015**, *521*, 436–444. [CrossRef] [PubMed]
16. Liu, K.; Xu, S.; Xu, G.; Zhang, M.; Sun, D.; Liu, H. A Review of Android Malware Detection Approaches Based on Machine Learning. *IEEE Access* **2020**, *8*, 124579–124607. [CrossRef]
17. Goeschel, K. Reducing False Positives In Intrusion Detection Systems Using Data-Mining Techniques Utilizing Support Vector Machines, Decision Trees, And Naive Bayes for Off-Line Analysis. In Proceedings of the SoutheastCon 2016, Norfolk, VA, USA, 30 March–3 April 2016; pp. 1–6.
18. Kuttranont, P.; Boonprakob, K.; Phaudphut, C.; Permpol, S.; Aimtongkhamand, P.; KoKaew, U.; Waikham, B.; So-In, C. Parallel KNN and Neighborhood Classification Implementations on GPU for Network Intrusion Detection. *J. Telecommun. Electron. Comput. Eng. (JTEC)* **2017**, *9*, 29–33.
19. Mehedi, S.T.; Anwar, A.; Rahman, Z.; Ahmed, K. Deep Transfer Learning Based Intrusion Detection System for Electric Vehicular Networks. *Sensors* **2021**, *21*, 4736. [CrossRef]
20. Kalash, M.; Rochan, M.; Mohammed, N.; Bruce, N.D.B.; Wang, Y.; Iqbal, F. Malware Classification with Deep Convolutional Neural Networks. In *Proceedings of the 2018 9th IFIP International Conference on New Technologies, Mobility and Security (NTMS), Paris, France, 26–28 February 2018*; Institute of Electrical and Electronics Engineers (IEEE): Piscataway, NJ, USA, 2018; pp. 1–5.
21. Diro, A.; Chilamkurti, N. Leveraging LSTM networks for attack detection in fog-to-things communications. *IEEE Commun. Mag.* **2018**, *56*, 124–130. [CrossRef]
22. Čeponis, D.; Goranin, N. Investigation of Dual-Flow Deep Learning Models LSTM-FCN and GRU-FCN Efficiency against Single-Flow CNN Models for the Host-Based Intrusion and Malware Detection Task on Univariate Times Series Data. *Appl. Sci.* **2020**, *10*, 2373. [CrossRef]
23. Alrawashdeh, K.; Purdy, C. Toward an Online Anomaly Intrusion Detection System Based On Deep Learning. In Proceedings of the 2016 15th IEEE International Conference on Machine Learning and Applications (ICMLA), Anaheim, CA, USA, 18–20 December 2016; pp. 195–200.
24. Hojjatinia, S.; Hamzenejadi, S.; Mohseni, H. Android Botnet Detection using Convolutional Neural Networks. In Proceedings of the 2020 28th Iranian Conference on Electrical Engineering (ICEE), Tabriz, Iran, 4–6 August 2020.
25. Farnaaz, N.; Jabbar, M. Random forest modeling for network intrusion detection system. *Procedia Comput. Sci.* **2016**, *89*, 213–217. [CrossRef]
26. Alkahtani, H.; Aldhyani, T.H.H. Botnet Attack Detection by Using CNN-LSTM Model for Internet of Things Applications. *Secur. Commun. Netw.* **2021**, *2021*, 3806459. [CrossRef]
27. Min, E.; Long, J.; Liu, Q.; Cui, J.; Chen, W. TR-IDS: Anomaly-based intrusion detection through text-convolutional neural network and random forest. *Secur. Commun. Netw.* **2018**, *2018*, 4943590. [CrossRef]
28. Zeng, Y.; Gu, H.; Wei, W.; Guo, Y. Deep—Full—Range: A Deep Learning Based Network Encrypted Traffic Classification and Intrusion Detection Framework. *IEEE Access* **2019**, *7*, 45182–45190. [CrossRef]
29. Alkahtani, H.; Aldhyani, T.; Al-Yaari, M. Adaptive anomaly detection framework model objects in cyberspace. *Appl. Bionics Biomech.* **2020**, *2020*, 6660489. [CrossRef] [PubMed]

30. Kadir, A.F.A.; Stakhanova, N.; Ghorbani, A.A. Android Botnets: What Urls Are Telling Us. In *Proceedings of the International Conference on Network and System Security, New York, NY, USA, 3 5 November 2015*; Springer: New York, NY, USA, 2015; pp. 78–91.
31. Anwar, S.; Zain, J.M.; Inayat, Z.; Haq, R.U.; Karim, A.; Jabir, A.N. A Static Approach Towards Mobile Botnet Detection. In Proceedings of the 2016 3rd International Conference on Electronic Design (ICED), Phuket, Thailand, 11–12 August 2016; pp. 563–567.
32. Alqatawna, J.F.; Faris, H. Toward a Detection Framework for Android Botnet. In Proceedings of the 2017 International Conference on New Trends in Computing Sciences (ICTCS), Amman, Jordan, 11–13 October 2017; pp. 197–202.
33. Abdullah, Z.; Saudi, M.M.; Anuar, N.B. ABC: Android botnet classification using feature selection and classification algorithms. *Adv. Sci. Lett.* **2017**, *23*, 4717–4720. [CrossRef]
34. Toldinas, J.; Venčkauskas, A.; Damaševičius, R.; Grigaliūnas, Š.; Morkevičius, N.; Baranauskas, E. A Novel Approach for Network Intrusion Detection Using Multistage Deep Learning Image Recognition. *Electronics* **2021**, *10*, 1854. [CrossRef]
35. Karim, A.; Rosli, S.; Syed, S. DeDroid: A Mobile Botnet Detection Approach Based on Static Analysis. In Proceedings of the 7th International Symposium on UbiCom Frontiers Innovative Research, Systems and Technologies, Beijing, China, 10–14 August 2015.
36. The Drebin Dataset. Available online: https://www.sec.cs.tu-bs.de/~||danarp/drebin/index.html (accessed on 28 December 2021).
37. Deng, L. A tutorial survey of architectures, algorithms, and applications for deep learning. *APSIPA Trans. Signal Inf. Process.* **2014**, *3*, e2. [CrossRef]
38. Berman, D.S.; Buczak, A.L.; Chavis, J.S.; Corbett, C.L. A survey of deep learning methods for cyber security. *Information* **2019**, *10*, 122. [CrossRef]
39. Yilmaz, S.; Sen, S. Early Detection of Botnet Activities Using Grammatical Evolution. In *Applications of Evolutionary Computation*; Springer International Publishing: Berlin/Heidelberg, Germany, 2019; pp. 395–404.
40. Yu, Y.; Long, J.; Liu, F.; Cai, Z. Machine Learning Combining with Visualization For Intrusion Detection: A survey. In Proceedings of the International Conference on Modeling Decisions for Artificial Intelligence, Sant Julià de Lòria, Andorra, 19–21 September 2016; pp. 239–249.
41. Ahmed, A.A.; Jabbar, W.A.; Sadiq, A.S.; Patel, H. Deep learning-based classification model for botnet attack detection. *J. Ambient. Intell. Humaniz. Comput.* **2020**, *2020*, 1–10. [CrossRef]
42. Alauthman, M.; Aslam, N.; Al-kasassbeh, M.; Khan, S.; Al-Qerem, A.; Raymond Choo, K. An efficient reinforcement learning-based Botnet detection approach. *J. Netw. Comput. Appl.* **2020**, *150*, 102479. [CrossRef]
43. Mazini, M.; Shirazi, B.; Mahdavi, I. Anomaly network-based intrusion detection system using a reliable hybrid artificial bee colony and AdaBoost algorithms. *J. King Saud Univ. Comput. Inf. Sci.* **2019**, *31*, 541–553. [CrossRef]
44. Asadi, M.; Jabraeil Jamali, M.A.; Parsa, S.; Majidnezhad, V. Detecting botnet by using particle swarm optimization algorithm based on voting system. *Future Gener. Comput. Syst.* **2020**, *107*, 95–111. [CrossRef]
45. Al Shorman, A.; Faris, H.; Aljarah, I. Unsupervised intelligent system based on one class support vector machine and Grey Wolf optimization for IoT botnet detection. *J. Ambient Intell. Humaniz. Comput.* **2020**, *11*, 2809–2825. [CrossRef]
46. Lin, K.C.; Chen, S.Y.; Hung, J.C. Botnet Detection Using Support Vector Machines with Artificial Fish Swarm Algorithm. *J. Appl. Math.* **2014**, *2014*, 986428. [CrossRef]
47. Chen, T.; Mao, Q.; Yang, Y.; Lv, M.; Zhu, J. TinyDroid: A lightweight and efficient model for Android malware detection and classification. *Mob. Inf. Syst.* **2018**, *2018*, 4157156. [CrossRef]
48. Nisa, M.; Shah, J.H.; Kanwal, S.; Raza, M.; Khan, M.A.; Damaševičius, R.; Blažauskas, T. Hybrid malware classification method using segmentation-based fractal texture analysis and deep convolution neural network features. *Appl. Sci.* **2020**, *10*, 4966. [CrossRef]
49. Amin, M.; Shah, B.; Sharif, A.; Ali, T.; Kim, K.I.; Anwar, S. Android malware detection through generative adversarial networks. *Trans. Emerg. Telecommun. Technol.* **2019**, *33*, e3675. [CrossRef]
50. Arp, D.; Spreitzenbarth, M.; Hubner, M.; Gascon, H.; Rieck, K.; Siemens, C. Drebin: Effective and Explainable Detection Of Android Malware In Your Pocket. In Proceedings of the 2014 Network and Distributed System Security Symposium, San Diego, CA, USA, 23–26 February 2014.
51. Google Play. Available online: https://play.google.com/ (accessed on 2 January 2022).
52. VirusShare. Available online: https://virusshare.com/ (accessed on 2 January 2022).
53. Intel Security/MacAfee. Available online: https://steppa.ca/portfolio-view/malware-threat-intel-datasets/ (accessed on 20 December 2021).
54. Wandoujia App Market. Available online: https://www.wandoujia.com/apps (accessed on 2 January 2022).
55. Google Playstore Appsin Kaggle. Available online: https://www.kaggle.com/gauthamp10/google-playstore-apps (accessed on 2 January 2022).
56. CICMaldroid Dataset. Available online: https://www.unb.ca/cic/datasets/maldroid-2020.html (accessed on 2 January 2022).
57. Alkahtani, H.; Aldhyani, T.H. Intrusion Detection System to Advance Internet of Things Infrastructure-Based Deep Learning Algorithma. *Complexity* **2021**, *2021*, 5579851. [CrossRef]
58. Odusami, M.; Abayomi-Alli, O.; Misra, S.; Shobayo, O.; Damasevicius, R.; Maskeliunas, R. Android Malware Detection: A Survey. In *Communications in Computer and Information Science, Proceedings of the International Conference on Applied Informatics, Bogota, Colombia, 1–3 November 2018*; Springer: Cham, Switzerland, 2018; pp. 255–266.

59. Yerima, S.Y.; Sezer, S. DroidFusion: A Novel Multilevel Classifier Fusion Approach for Android Malware Detection. *IEEE Trans. Cyber.* **2019**, *49*, 453–466. [CrossRef]
60. Liu, G.; Zhao, H.; Fan, F.; Liu, G.; Xu, Q.; Nazir, S. An Enhanced Intrusion Detection Model Based on Improved kNN in WSNs. *Sensors* **2022**, *22*, 1407. [CrossRef]
61. Aldallal, A.; Alisa, F. Effective Intrusion Detection System to Secure Data in Cloud Using Machine Learning. *Symmetry* **2021**, *13*, 2306. [CrossRef]
62. Zheng, D.; Hong, Z.; Wang, N.; Chen, P. An Improved LDA-Based ELM Classification for Intrusion Detection Algorithm in IoT Application. *Sensors* **2020**, *20*, 1706. [CrossRef] [PubMed]
63. Yann, L.; Yoshua, B. Convolutional Networks for Images, Speech, and Time-Series. *Handb. Brain Theory Neural Netw.* **1995**, *10*, 2571–2575.
64. Rawat, W.; Wang, Z. Deep Convolutional Neural Networks for Image Classification: A Comprehensive Review. *Neural Comput.* **2017**, *29*, 2352–2449. [CrossRef]
65. Hochreiter, S.; Schmidhuber, J. Long Short-Term Memory. *Neural Comput.* **1997**, *9*, 1735–1780. [CrossRef]
66. Aldhyani, T.H.H.; Alkahtani, H. Attacks to Automatous Vehicles: A Deep Learning Algorithm for Cybersecurity. *Sensors* **2022**, *22*, 360. [CrossRef]
67. Khan, M.A.; Khan, M.A.; Jan, S.U.; Ahmad, J.; Jamal, S.S.; Shah, A.A.; Pitropakis, N.; Buchanan, W.J. A Deep Learning-Based Intrusion Detection System for MQTT Enabled IoT. *Sensors* **2021**, *21*, 7016. [CrossRef]
68. Tang, C.; Luktarhan, N.; Zhao, Y. SAAE-DNN: Deep Learning Method on Intrusion Detection. *Symmetry* **2020**, *12*, 1695. [CrossRef]
69. Kunang, Y.N.; Nurmaini, S.; Stiawan, D.; Zarkasi, A.; Jasmir, F. Automatic Features Extraction Using Autoencoder in Intrusion Detection System. In Proceedings of the International Conference on Electrical Engineering and Computer Science (ICECOS), Pangkal Pinang, Indonesia, 2–4 October 2018; pp. 219–224.
70. Ginocchi, M.; Ponci, F.; Monti, A. Sensitivity Analysis and Power Systems: Can We Bridge the Gap? A Review and a Guide to Getting Started. *Energies* **2021**, *14*, 8274. [CrossRef]
71. Nasirzadehdizaji, R.; Balik Sanli, F.; Abdikan, S.; Cakir, Z.; Sekertekin, A.; Ustuner, M. Sensitivity Analysis of Multi-Temporal Sentinel-1 SAR Parameters to Crop Height and Canopy Coverage. *Appl. Sci.* **2019**, *9*, 655. [CrossRef]
72. Millar, S.; McLaughlin, N.; del Rincon, J.M.; Miller, P. Multi-view deep learning for zero-day Android malware detection. *J. Inf. Secur. Appl.* **2021**, *58*, 102718. [CrossRef]
73. Kapratwar, A.; Di Troia, F.; Stamp, M. *Static and Dynamic Analysis of Android Malware*; ICISSP: Porto, Portugal, 2017; pp. 653–662.
74. Qaisar, Z.H.; Li, R. Multimodal information fusion for android malware detection using lazy learning. *Multimed. Tools Appl.* **2021**, 1–15. [CrossRef]
75. Salehi, M.; Amini, M.; Crispo, B. Detecting Malicious Applications Using System Services Request Behavior. In Proceedings of the 16th EAI International Conference on Mobile and Ubiquitous Systems: Computing, Networking and Services, Houston, TX, USA, 12–14 November 2019; pp. 200–209.
76. Koli, J. RanDroid: Android Malware Detection Using Random Machine Learning Classifiers. In Proceedings of the 2018 Technologies for Smart-City Energy Security and Power (ICSESP), Bhubaneswar, India, 28–30 March 2018; pp. 1–6.
77. Kabakus, A.T. What static analysis can utmost offer for Android malware detection. *Inf. Technol. Control* **2019**, *48*, 235–249. [CrossRef]
78. Lou, S.; Cheng, S.; Huang, J.; Jiang, F. TFDroid: Android Malware Detection By Topics And Sensitive Data Flows Using Machine Learning Techniques. In Proceedings of the 2019 IEEE 2nd International Conference on Information and Computer Technologies (ICICT), Kahului, HI, USA, 14–17 March 2019; pp. 30–36.
79. Onwuzurike, L.; Mariconti, E.; Andriotis, P.; Cristofaro, E.D.; Ross, G.; Stringhini, G. MaMaDroid: Detecting Android malware by building Markov chains of behavioral models (extended version). *ACM Trans. Priv. Secur. (TOPS)* **2019**, *22*, 1–34. [CrossRef]
80. Zhang, H.; Luo, S.; Zhang, Y.; Pan, L. An efficient Android malware detection system based on method-level behavioral semantic analysis. *IEEE Access* **2019**, *7*, 69246–69256. [CrossRef]
81. Meng, G.; Xue, Y.; Xu, Z.; Liu, Y.; Zhang, J.; Narayanan, A. Semantic Modelling Of Android Malware For Effective Malware Comprehension, Detection, and Classification. In Proceedings of the 25th International Symposium on Software Testing and Analysis, Saarbrücken, Germany, 18–20 July 2016; pp. 306–317.
82. Vu, L.N.; Jung, S. AdMat: A CNN-on-Matrix Approach to Android Malware Detection and Classification. *IEEE Access* **2021**, *9*, 39680–39694. [CrossRef]
83. Jannat, U.S.; Hasnayeen, S.M.; Shuhan, M.K.B.; Ferdous, M.S. Analysis and Detection Of Malware in Android Applications Using Machine Learning. In Proceedings of the 2019 International Conference on Electrical, Computer and Communication Engineering (ECCE), Cox'sBazar, Bangladesh, 7–9 February 2019; pp. 1–7.
84. Xu, K.; Li, Y.; Deng, R.H.; Chen, K. Deeprefiner: Multi-Layer Android Malware Detection System Applying Deep Neural Networks. In Proceedings of the 2018 IEEE European Symposium on Security and Privacy (EuroS&P), London, UK, 24–26 April 2018; pp. 473–487.
85. McLaughlin, N.; Martinez del Rincon, J.; Kang, B.; Yerima, S.; Miller, P.; Sezer, S.; Safaei, Y.; Trickel, E.; Zhao, Z.; Doupé, A.; et al. Deep Android Malware Detection. In Proceedings of the Seventh ACM on Conference on Data and Application Security and Privacy, Scottsdale, AZ, USA, 22–24 March 2017; pp. 301–308.

Article

Authentication and Authorization in Microservices Architecture: A Systematic Literature Review

Murilo Góes de Almeida * and Edna Dias Canedo *

Department of Computer Science, University of Brasília (UnB), P.O. Box 4466, Brasilia 70910-900, Brazil
* Correspondence: murilo.almeida@aluno.unb.br (M.G.d.A.); ednacanedo@unb.br or edna.canedo@gmail.com (E.D.C.); Tel.: +55-61-98114-0478 (E.D.C.)

Abstract: The microservice architectural style splits an application into small services, which are implemented independently, with their own deployment unit. This architecture can bring benefits, nevertheless, it also poses challenges, especially about security aspects. In this case, there are several microservices within a single system, it represents an increase in the exposure of the safety surface, unlike the monolithic style, there are several applications running independently and must be secured individually. In this architecture, microservices communicate with each other, sometimes in a trust relationship. In this way, unauthorized access to a specific microservice could compromise an entire system. Therefore, it brings a need to explore knowledge about issues of security in microservices, especially in aspects of authentication and authorization. In this work, a Systematic Literature Review is carried out to answer questions on this subject, involving aspects of the challenges, mechanisms and technologies that deal with authentication and authorization in microservices. It was found that there are few studies dealing with the subject, especially in practical order, however, there is a consensus that communication between microservices, mainly due to its individual and trustworthy characteristics, is a concern to be considered. To face the problems, mechanisms such as OAuth 2.0, OpenID Connect, API Gateway and JWT are used. Finally, it was found that there are few open-source technologies that implement the researched mechanisms, with some mentions of the Spring Framework.

Keywords: microservice; authentication; authorization; security; SLR

Citation: de Almeida, M.G.; Canedo, E.D. Authentication and Authorization in Microservices Architecture: A Systematic Literature Review. *Appl. Sci.* **2022**, *12*, 3023. https://doi.org/10.3390/app12063023

Academic Editors: Leandros Maglaras, Helge Janicke and Mohamed Amine Ferrag

Received: 5 February 2022
Accepted: 8 March 2022
Published: 16 March 2022

Publisher's Note: MDPI stays neutral with regard to jurisdictional claims in published maps and institutional affiliations.

Copyright: © 2022 by the authors. Licensee MDPI, Basel, Switzerland. This article is an open access article distributed under the terms and conditions of the Creative Commons Attribution (CC BY) license (https://creativecommons.org/licenses/by/4.0/).

1. Introduction

The microservice architectural style is represented by an ecosystem of small services, each running in its own process and communicating through lightweight protocols, such as HTTP (Hypertext Transfer Protocol), built around business resources and deployed independently [1]. Breaking an application into microservices can bring some benefits, such as optimizing management, scalability, availability and reliability [2,3]. However, it may bring challenges in relation to security, because, in this case, an individual attention about it must be observed in each microservice developed, different from the monolithic style where security strategies are applied in a single application [3,4]. Furthermore, there are few practical demonstrations in the literature describing solutions to improve the security of [4] service-oriented architectures.

Regardless of the implemented architecture, the authentication and authorization aspects are relevant, considering them as key elements for the security mechanisms [5]. Authentication is the process of determining whether someone or something is, in fact, who they claim to be. Authorization is the process of giving someone or something permission to do or possess something [6]. There are protocols that deal with authorization and authentication issues, such as OAuth 2.0, the standard for delegated authorization, and OpenID Connect, the authentication layer on top of OAuth 2.0 [7]. It is important to note that there is a distinction between user authentication and service authentication. In the case of authentication between microservices, there are specific mechanisms for this, such as Mutual Transport Layer Security (MTLS) [7]. Using MTLS, each microservice will legitimately identify who it talks to, while also ensures data confidentiality and integrity in this communication [8].

According to some studies, microservices are usually designed in such way that there is a relationship of trust between them [3,9]. However, it is possible to find microservice architectures that use the "zero-trust" paradigm [10]. In this last case, there is a premise that trust is never granted implicitly but must be continually evaluated [11]. Thus, a lack of observation about authentication and authorization in a single microservice can affect the entire ecosystem. It is important that studies related to security issues in microservices emphasize aspects involving authentication and authorization. Therefore, in this paper, we carried out a Systematic Literature Review (SLR) to identify in the literature the studies that address authentication and authorization in microservice environments, what are their challenges, security mechanisms used to deal with these challenges and open-source technologies that implement the mechanisms identified in the review. The focus on open-source is to provide technologies that can reduce costs, free access to source code and customization [12]. There are advantages for use open-source in the public sector, such as avoiding monopoly dominance in the market [12]. Last, but not least, even software developed by commercial firms is being released under open-source licenses as well [13]. It is important to note that the adoption of open-source, although it has the advantage of free use, it will not necessarily bring an adequate cost/benefit for the organization [14]. Therefore, it is recommended that its adoption be based on metrics such as the Total Cost of Ownership (TCO), an instrument that assesses the cost of adapting, managing and maintaining the proposed software [14].

Our main findings reveal that authentication and authorization challenges involving microservices are mostly related to the communication between them and the complexity of implementing security in each microservice, generating a complexity both in the development and in the increase of the attack surface, since individual attention must be given to each microservice. The most mentioned mechanisms in the literature that address the challenges of authentication and authorization in microservices are OAuth 2.0, JWT, API Gateway and OpenID Connect, in addition to Single Sign-on strategy. These mechanisms can be implemented together, with their respective role in the security context. The API Gateway acts as an intermediary between the external client and the microservices, providing a private network environment that allows the exchange of data between them [15]. Single Sign On (SSO) allows users to authenticate only once and use all apps associated with their user accounts, without requiring them to enter their credentials each time they access a different app [16]. Finally, we identified that the Spring Framework is widely used in the context of open-source applications.

2. Systematic Literature Review

To achieve the research goal, we performed a Systematic Literature Review (RSL), in accordance with the guidelines proposed by Kitchenham and Charles [17] and the structuring applied by Kitchenham et al. [18]. According to the authors, an RSL is "a means of identifying, evaluating and interpreting all available research relevant to a specific research question, or topic area, or phenomenon of interest" [17]. In addition, we used the online tool Parsifal [19] to support the screening and analysis of the identified studies.

2.1. Research Questions

We conducted the SLR to answer the following research questions (RQ):

1. RQ.1. What are the challenges mentioned in the literature to perform authentication and authorization in the context of microservice architecture systems?
2. RQ.2. What mechanisms are used in the literature to deal with the challenges related to authentication and authorization in a microservices architecture?

3. RQ.3. What are the main open-source technology solutions that implement the authentication and authorization mechanisms identified in the literature?

2.2. Search Process

To identify studies in the literature, we performed an automatic search in the main digital databases in the field of Computer Science. The digital databases used in the systematic literature review were: DBLP (https://dblp.uni-trier.de, accessed on 4 February 2022), IEEE Digital Library (http://ieeexplore.ieee.org, accessed on 4 February 2022) and Scopus (http://www.scopus.com, accessed on 4 February 2022). The search string used in digital databases was defined according to the keywords that must appear in the search results. The search string used was:

("MICROSERVICE" OR "MICROSERVICES") AND ("SECURITY" AND "AUTHENTICATION" AND "AUTHORIZATION") AND ("CHALLENGE*" OR "PROBLEM*" OR "ISSUE*" OR "SOLUTION*" OR "PROTOCOL*" OR "MECHANISM*" "STRATEG*" OR "IMPLEMENTATION*" OR "OPENSOURCE" OR "OPEN-SOURCE" OR "OPEN SOURCE").

We also applied the "snowballing" process which aims to prevent relevant studies from being omitted [20]. In this process, references about the research object in each selected study are verified. Thus, we searched for papers where selected studies were cited.

2.3. Inclusion and Exclusion Criteria

The selection criteria for primary studies seek to identify papers that provide information about the research questions. Therefore, we defined the following inclusion and exclusion criteria, based on the research questions:

Inclusion Criteria

- **IC.1** Studies dealing with challenges involving authentication and authorization in microservices;
- **IC.2** Studies related to security mechanisms that deal with authentication and authorization challenges in microservices;
- **IC.3** Studies related to open-source technologies that implement security mechanisms.

Exclusion Criteria

- **EC.1** Studies that do not address the research object;
- **EC.2** Studies prior to 2010;
- **EC.3** Duplicate studies;
- **EC.4** Studies published as short paper.

2.4. Quality Assessment

To differentiate selected studies according to quality criteria we check in each selected study whether they answer the research questions. The criteria adopted were:

1. Is the research objective clearly described?
2. Do the authors describe the limitation of the study?
3. Does the study identify problems and/or challenges involving authentication and authorization in microservices architecture?
4. Does the study identify the mechanisms that mitigate the problems and/or challenges involving authentication and authorization in microservices architecture?
5. Does the study present solutions that implement security mechanisms using open-source technology?

The answer of each quality criterion question received a score, as follows:

1. Yes (1);
2. Partially (0.5);
3. No (0).

Although the primary studies were selected using specific criteria, there is an individual assessment of the quality for each study, to verify which of them are more aligned with the research questions that were defined.

2.5. Data Collection and Analysis

The following data were collected in the selected primary studies: (1) Authentication and/or authorization challenges found in microservices; (2) The mechanisms that deal with the authentication and/or authorization challenges found in microservices; (3) Open-source technologies that implement mechanisms which deal with authentication and authorization challenges in microservices.

The identified challenges, mechanisms and solutions were organized in a ranking to verify the most mentioned in the primary studies. This ranking aims to show what manuscripts have more answers about the research questions, this does not mean that the lowest rated manuscripts are worse than the first ones, it just means that the top-rated manuscripts have more information to answer our questions. Subsequently, the items most present in the studies were submitted to an individual analysis for a better understanding of their basic concepts. Finally, it was verified which specific mechanisms deal with the challenges found.

3. SLR Results

This section presents the results of performing the systematic literature review. Figure 1 shows the complete execution process of the proposed protocol to execute the SLR, with the respective quantity of studies identified in each step of the protocol. In the automatic search performed in the digital databases using the initial query, 22 papers were found. These studies were submitted to the snowballing process, resulting in 13 new selected papers. Of the 22 papers found initially, 11 were eliminated due to the exclusion criteria (5 studies that do not address the research object and 6 duplicate). Thus, 11 primary studies were selected from the digital databases and 13 studies on snowballing execution, totaling 24 primary studies. The selected primary studies are shown in Table 1. The filters applied during the SLR based on inclusion and inclusion criteria are demonstrated in Figure 2.

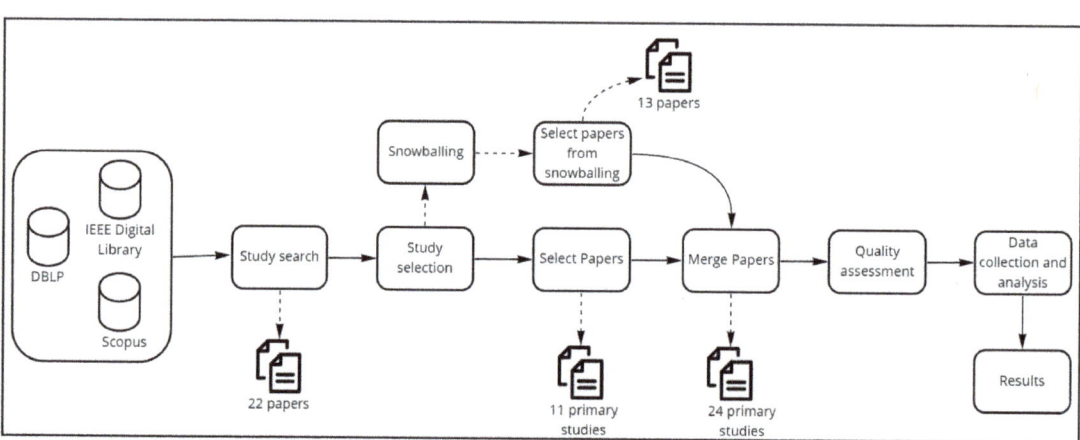

Figure 1. Protocol application process.

Table 1. Selected Studies.

ID	Year	Title	Ref
S1	2021	Security in microservice-based systems: A Multivocal literature review	[4]
S2	2021	Security in microservices architectures	[3]
S3	2020	Authentication and authorization in microservice-based systems: survey of architecture patterns	[8]
S4	2020	Information system development for restricting access to software tool built on microservice architecture	[21]
S5	2020	Research on Unified Authentication and Authorization in Microservice Architecture	[22]
S6	2020	Secure Edge Computing Management Based on Independent Microservices Providers for Gateway-Centric IoT Networks	[23]
S7	2019	Applying Spring Security Framework and OAuth 2.0 To Protect Microservice Architecture API	[24]
S8	2019	A survey on security issues in services communication of Microservices-enabled fog applications	[25]
S9	2019	Enhancing security to the MicroService (MS) architecture by implementing Authentication and Authorization (AA) service using Docker and Kubernetes	[26]
S10	2019	Implementing secure applications in smart city clouds using microservices	[16]
S11	2019	Microservice Security Agent Based On API Gateway in Edge Computing	[15]
S12	2019	Securing Microservices	[27]
S13	2019	Security Mechanisms Used in Microservices-Based Systems: A Systematic Mapping	[28]
S14	2018	Authentication and authorization orchestrator for microservice-based software architectures	[29]
S15	2018	Defense-in-depth and Role Authentication for Microservice Systems	[30]
S16	2018	Fine-Grained Access Control for Microservices	[31]
S17	2018	Overcoming Security Challenges in Microservice Architectures	[7]
S18	2018	Security considerations for microservice architectures	[32]
S19	2018	Unified account management for high performance computing as a service with microservice architecture	[33]
S20	2017	A Secure Microservice Framework for IoT	[34]
S21	2017	Access control with delegated authorization policy evaluation for data-driven microserviceworkflows	[35]
S22	2017	Authentication and Authorization of End User in Microservice Architecture	[36]
S23	2017	Integrating Continuous Security Assessments in Microservices and Cloud Native Applications	[37]
S24	2015	Security-as-a-Service for Microservices-Based Cloud Applications	[9]

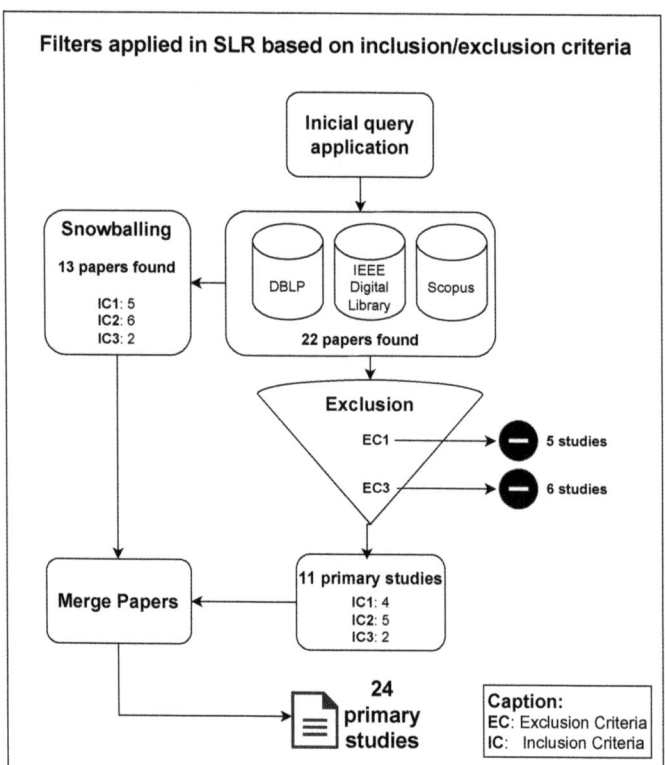

Figure 2. Filters applied in SLR process.

3.1. Quality Assessment of Reviews Carried out

According to the quality criteria, the selected studies were analyzed and scored, as shown in Table 2. All primary studies mentioned challenges involving authorization and authentication in microservices (AQ3), as well as mechanisms to mitigate such problems (AQ4), even if partially. However, there is a smaller amount of work (15) mentioning open-source technologies that implement the mechanism (AQ5). In general, the studies are clear about the objective (AQ1), but 11 of them do not describe its limitations (AQ2).

Table 2. Ranking of scores according to Quality Assessments.

ID	AQ1	AQ2	AQ3	AQ4	AQ5	Total
S1	1	1	1	1	1	5.0
S8	1	0.5	1	1	1	4.5
S16	1	1	1	1	0.5	4.5
S23	1	1	1	0.5	1	4.5
S17	1	1	1	1	0.5	4.5
S21	1	0.5	0.5	1	1	4.0
S5	1	0	1	1	1	4.0
S6	1	0.5	0.5	1	1	4.0
S13	1	1	1	1	0	4.0
S7	1	0.5	0.5	0.5	1	3.5
S3	1	0	1	1	0.5	3.5
S15	0.5	0	1	1	1	3.5
S10	1	0.5	1	1	0	3.5
S11	1	0	0.5	1	1	3.5
S20	1	1	0.5	0.5	0	3.0
S14	1	0	1	1	0	3.0
S12	1	0	1	1	0	3.0
S2	1	0	1	1	0	3.0
S19	1	0.5	0.5	0.5	0.5	3.0
S4	0.5	0	1	0.5	0.5	2.5
S24	0.5	0.5	1	0.5	0	2.5
S22	0.5	0	0.5	0.5	0.5	2.0
S18	0.5	0	0.5	0.5	0	1.5
S9	0	0	0.5	0.5	0	1.0

3.2. Quality Factors

We have done a verification to understand if there is any kind of relationship between the quality score and the year the study was published. Although it is possible to verify that the average score increased over the years, the standard deviation and the coefficient of variation show that the data are heterogeneous, and it is not possible to conclude that the quality has increased over the period, as shown in the Table 3. It is possible to verify in this situation that the standard deviation increases in the same proportion as the average, in addition to the coefficient of variation being in a high degree.

Table 3. Average study quality score by year.

	2015	2017	2018	2019	2020	2021
Number of Studies	1	4	6	7	4	2
Rating Average	2.5	3.38	3.33	3.29	3.50	4.00
Standard deviation	0	1.1087	1.1255	1.1127	0.7071	1.4142
Coefficient of variation	0	0.3285	0.3376	0.3386	0.2020	0.3536

We performed analyzes on data extracted from selected studies to answer the research questions.

3.3. RQ.1. What Are the Challenges Mentioned in the Literature to Perform Authentication and Authorization in the Context of Microservice Architecture Systems?

The challenges identified about authentication and authorization in the context of microservice architecture systems are presented in Table 4. Such challenges were presented according to the number of mentions in the selected studies, therefore, it does not mean that these are the most critical in terms of vulnerabilities or how much they occur in a microservices environment. The number of mentions of the challenges found in the studies does not necessarily reflect a level of priority in which they should be observed in a practical environment. Among the identified challenges, the five most mentioned in the literature were: "Communication between microservices" (13 mentions), "Trust between microservices compromised by unauthorized access" (12 mentions), "Individual concern for each microservice" (12 mentions), "Increased attack surface" (12 mentions), and "Microservice Access Control" (10 mentions).

Table 4. Challenges related to authentication and authorization in microservices

Pos	Challenge	ID	Number of Occurrences
1st	Communication between microservices	S1, S2, S3, S4, S7, S9, S10, S15, S16, S17, S18, S23, S24	13
2nd	Trust between microservices compromised by unauthorized access	S1, S2, S4, S6, S8, S12, S15, S16, S19, S21, S23, S24	12
3rd	Individual concern for each microservice	S1, S2, S5, S10, S12, S13, S15, S16, S21, S22, S23, S24	12
4th	Increased attack surface (compared to monolithic)	S1, S2, S3, S7, S8, S13, S14, S16, S17, S23	10
5th	Microservice access control	S5, S8, S10, S11, S14, S15, S19, S20, S21	9
6th	Authorization between services	S1, S7, S8, S15, S16, S17, S18, S21	8
7th	Lack of studies about microservices	S1, S2, S4, S8, S10, S13, S17	7
8th	Lack of security patterns in microservices	S1, S13, S15, S17, S20, S21	6
9th	Different teams working on different microservices must have the same understanding of security	S3, S15, S17, S20, S23	5
10th	Bypass on Api Gateway	S3, S4, S6, S12	4
11th	Intrusion detection/monitoring	S1, S12, S24	3
12th	Escalation of privileges	S2, S16, S24	3
13th	Lack of study demonstrating practical implementation of security in microservices	S7, S13, S23	3
14th	Coordinate authentication server with new microservices	S1, S22	2
15th	Lack of attention in attack reaction/recovery	S1, S13	2
16th	Token validation at each microservice request	S5, S6	2
17th	Public Images may be compromised	S1	1
18th	Many applications in commercial microservices without possibility to evaluate code	S4	1
19th	Use of authentication/authorization server that handles all microservices	S3	1
20th	Possibility of development in various technologies	S23	1

In general, the studies that mentioned the existing challenges made comparisons between monolithic and microservices architectures, explaining that in the monolithic model, there is only one surface to be protected, however, in the microservice environment, each autonomous service must be a point of concern regarding security, making it more complex to keep this entire ecosystem properly protected. Although each service needs particular attention, Yarygina and Bagge [7] alerted that manual security provisioning of hundreds or thousands of service instances is infeasible. Pereira-Vale et al. [4] compared monolithic with microservices using a KLOC metric (kilo Lines of Code), they say that in a monolithic application, every 100 kloc will have an average of 39 vulnerabilities. The same quantity of lines of code in a microservice application, will have an average of 180 vulnerabilities. They also alert about the decomposition of monolithic into microservices because security needs to be a global property, not the sum of local security defenses.

Nehme et al. [27] argue the importance of authentication and authorization in the context of microservice security. They mentioned that "Microservices should only be invoked after requesting authentication and, ideally, authorization if levels of privileges are available.". Pereira-Vale et al. [28] performed a systematic mapping about security mechanisms used in microservices and they discovered that the most reported security mechanisms are related to authorization, authentication and credentials. Banat et al. [29] also agreed that authentication and authorization need to be carefully observed in a microservice architecture, because this scenario presents many points of access for users and the other parts of the application. They argued that the data being especially sensitive, the crucial point of the development is the authentication and the authorization process. Cao et al. [33] proposed the implementation of a global authentication and authorization mechanism named Unified Account Management as a Service (UAMS). In this implementation, they used a RESTful API divided into several microservices. All sensitive data is transferred encrypted by the HTTPS protocol.

The studies also highlighted that microservices have the characteristic of communicating with each other, usually through the HTTP protocol, and this is a point of attention that differs from the traditional monolithic approach and should be properly analyzed and observed in terms of security. Regarding the communication issue, the authors mentioned the implementation of Transport Layer Security (TLS), used to protect the communication channels [30].

The challenges presented, in general, complement each other, or even act transversally. Mateus-Coelho et al. [3] stated that "Microservices are often designed to trust their peers and, if one of them is compromised and accessed improperly, it is possible that there is a great advantage for all others to be exploited". Dongjin et al. [25] agreed when they affirmed that "When a single service is controlled by an attacker, the service may maliciously influence other services". Nehme et al. [31] mentioned an access control problem that may be found in microservice architecture named "confused deputy attack", in their words, it is a privilege escalation attack in which a microservice that is trusted by other microservices is compromised. Sun et al. [9] brought the concern about trust between services, they affirmed that the "compromise of a single microservice may bring down the entire application". They also reported the challenge of monitoring and auditing the microservices interaction over the network and proposed a design of a security-as-service infrastructure for microservices-based cloud applications, that helps to monitor the network aiming to find possible non-expected behaviors in the communication between them.

Pereira-Vale et al. [4] stated that the communication between microservices is exposed through the network environment, which creates a potential attack surface. The authors also mentioned the problem of increasing the attack surface, as the decomposition of an application into several services increases the attack surface and the security of the application becomes more difficult to manage, because it becomes the sum of several independent defenses, rather than being managed in a global way, as was done in the monolithic approach.

Nguyen and Baker [24] warned that network communication between microservices can occur in the internet environment, which increases, in addition to exposure, the number of possible attackers. Kramer et al. [16] reinforce this concern to implement secure application in smart city clouds using microservices, because it will handle with a huge amount of data, including sensitive information about infrastructure and citizens. Xu et al. [15]

shared the same concerned, in this case, using microservice in IoT devices. Lu et al. [34] is also concerned about security in IoT devices using microservices architecture, mainly because of sensitive data that can be shared among services. They encourage the use of API gateways, that will remove all the concerns about microservices, because all interactions with the components will be performed with the API Gateway. Safaryan et al. [21] are also concerned about communication among different microservices because it is carried out through network interaction, being necessary to secure each of the service and the network. They also alert about the need of a pattern to be implemented. The lack of a correct pattern can compromise the network environment. Nguyen and Baker [24] agreed about the need to observe communication between the services. Banati et al. [29] argued that the network used in microservices communication can be secured using system administration tools such as VPN, Firewall and HTTPS.

Dongjin et al. [25] and Jander et al. [30] raised a concern that if a single service were controlled by an attacker, it could maliciously influence all other services. Concerns related to communication between microservices, increased surface exposure, access control and individual concern in each microservice, the API Gateway strategy, and use of mechanisms such as OAuth 2.0 and JWT were widely mentioned. They are also concerned about the industry, which are not fully aware about security issues involving microservices.

API Gateway helps to limit exposure between microservices as requests will be centered on it, and no longer on a microservice directly [3,8]. Jin et al. [23] mention that API gateway will secure a microservices environment because it will filter all requests. They also proposed an edge gateway to manage microservices. Nehme et al. [27] not recommended the access token validation in the gateway level, this role need to be performed in an authentication server. In contrast, Torkura et al. [37] proposed a security gateway used as security control for enforcing security policies. They also alerted about discoverability, that means, a gateway feature that allows a microservice to subscribe in it. If the discovery service can accept any subscription, vulnerable microservices could be pushed to production environments. OAuth 2.0 is a popular authorization protocol and could protect access to microservices from unauthorized access as access tokens are issued to trusted clients that could access certain services [25]. Nguyen and Baker [24] explained that OAuth 2.0 is not only used in web-based application but can be applied into backend services with no need of web browser or user interaction. ShuLin and JiePing [22] mentioned that the JWT is an open standard (RFC 7519) that defines a compact and independent way to securely transmit information between parties as a JSON object. This information can be verified and trusted because it is digitally signed.

Barabanov and Makrushin [8] warned that implementing authorization directly in the source code of each microservice can lead to future problems, especially in different teams working on independent microservices, because of new security updates must be performed in all projects, individually. He and Yang [36] followed in the same line, they alerted that imitate the way of monolithic structure in each microservice has several deficiencies, mainly when a new service join in the system, it will be necessary to implement the security function in this case. They proposed a solution creating a specific service focused on authentication and authorization, as a result, each service is focused on its own business, ensuring better scalability and decoupling the system.

Jander et al. [30] mentioned that different teams can implement their own internal security approach in the microservice which they have responsibility for, but this may require more specialized knowledge from the teams. Torkura et al. [37] brought the concern that development by different teams can bring microservices using not only different standards in development, but also different technologies, which must implement the same security standards.

Lu et al. [34] stated that the development of microservices by different teams and even different companies is completely possible, is there is an alignment on the security implementations in each microservice. Finally, it is important to mention that the lack of security patterns in microservices, added to the few studies on the subject, both theoretical and practical, can influence the management of the development of the architecture. Torkura et al. [37] warned that there are several literatures that highlight security problems in microservice architectures, however, none of them offer practical solutions to deal with these situations.

Pereira-Vale et al. [4] alerted about the use of an authentication and authorization service to be used in an architecture of microservices. The use of this server must be robust enough to authenticate the user and carry out the token validations that are made with each client request. The lack of concern about these challenges can cause a single point of failure of failure (SPOF), that is, if this part of system fails, will affect the entire application [38]. ShuLin and JiePing [22] stated that the authentication server may affect the performance of the entire system, mainly if there are several requests to this server.

Preuveneers and Joosen [35] alerted about the flow of the data among microservices. Even if each individual microservice is protected, it is important to note if the workflow is valid. In that work, they presented a workflow-oriented framework to avoid not expected communication between microservices.

Mateus-Coelho et al. [3] enumerated some examples of mechanisms which must be observed during the developing in a microservice architecture: complex passwords, authentication, web security flaws (what are the most flaws observed), people and processes. They also listed the most critical web application security risks: injection, broken authentication and session management, cross-site scripting, broken access control, security misconfiguration, sensitive data exposure, insufficient attack protection, cross-site request forgery, components with known vulnerabilities and under protected APIs. All of them may be exploited in a microservice environment. Nguyen and Baker [24] also pointed some web security risks and carried out some experiments using CSRF attack, XSS attack and Brute Force attack in an API endpoint protected by OAuth 2.0. In this case, all of tests were prevented by the configuration proposed in the Proof of Concept presented in that work.

3.4. RQ.2. What Mechanisms Are Used in the Literature to Deal with the Challenges Related to Authentication and Authorization in a Microservices Architecture?

The mechanisms identified in the literature used to deal with authentication and authorization challenges in microservices architecture are presented in Table 5. The OAuth 2.0 protocol was the most mentioned (16 mentions), followed by JWT (14 mentions), API Gateway (14 mentions), Single Sign-ON (8 mentions) and OpenID Connect (7 mentions). Figure 3 shows the number of occurrences of the mechanisms used in the selected studies. Some mechanisms were used in only one study. It is important to emphasize that, in general terms, the identified mechanisms do not need to be implemented in a unique way, that is, they can coexist in the same environment, each one acting with a specific purpose. We identified in the selected studies some implementations in which the mechanisms act together, to mitigate possible vulnerabilities involving authentication and authorization in the microservices environment. It important to observe that some studies only point the mechanisms without explain deeply or demonstrate a practical implementation of them, such as the work of Pereira-Vale et al. [28], which is more concerned in perform a systematic mapping of security mechanisms.

Table 5. Security mechanisms used in microservices architecture

Pos	Mechanism	ID	Number of Occurrences
1°	OAuth 2.0	S1, S3, S5, S6, S7, S8, S10, S11, S12, S13, S14, S15, S16, S17, S20, S21	16
2°	JWT	S1, S3, S4, S5, S6, S9, S11, S12, S13, S14, S15, S17, S21, S22	14
3°	API Gateway	S2, S3, S4, S5, S6, S11, S12, S13, S14, S16, S17, S19, S20, S22	14
4°	Single Sign-ON SSO	S1, S2, S9, S10, S14, S19, S21, S22	8
5°	OpenID Connect	S1, S3, S8, S12, S16, S17, S21	7
6°	HTTPS	S2, S10, S14, S15, S17, S19, S20	7
7°	RBAC	S1, S3, S5, S13, S14, S17, S21	7
8°	ABAC	S1, S8, S14, S17, S20, S21	6
9°	XACML	S1, S3, S13, S15, S16, S21	6
10°	HMAC	S2, S3, S5, S14, S21	5
11°	SAML	S1, S2, S13, S14, S21	5
12°	TLS	S1, S10, S14, S15, S16	5
13°	OAuth	S1, S5, S8, S16	4
14°	Multilevel Security	S1, S3, S13	3
15°	DAC	S14, S21	2
16°	IAM	S14, S21	2
17°	RSA	S5, S11	2
18°	SASL	S1, S13	2
19°	SSL	S2, S13	2
20°	MTLS	S1, S17	2
21°	OpenID	S2, S14	2
22°	API Keys	S2	1
23°	Captcha	S19	1
24°	CAS	S8	1
25°	X509 Certificates	S1	1
26°	EAS	S3	1
27°	ECDSA	S5	1
28°	GSI	S8	1
29°	IDS	S12	1
30°	LDAP	S8	1
31°	MFA	S19	1
32°	NGAC	S3	1
33°	PBAC	S1	1

We will present a brief description of the most mentioned mechanisms in the selected studies:

- **OAuth 2.0 (Open Authorization):** The OAuth 2.0 protocol is defined by RFC 6749 [39]. According to Banati et al. [29] OAuth 2.0 is an authorization framework that allows users to access different services without having to share their credentials. In a practical scenario, the user authenticates to an authorization server and receives an authorization code or an access token, which can be used to access resources, without the need to contact the authorization server again or have to inform the username and password [25]. Access tokens are validated on each request for some service [15,35]. This procedure poses a risk for affecting the performance of a distributed architecture because if there is a large number of requests, the authentication server may be affected [22]. OAuth 2.0 is one of the protocols most used by microservice architectures for access delegation [25,31] and can be applied both in web applications and in backend services, in addition to meeting both the authentication and authorization proposal [24,28]. It is important to mention that OAuth 2.0 is widely used as an authorization protocol to protect services that use the REST (Representational State

Transfer) [23,25,27], in addition to adopting the HTTPS protocol in data communication [25].

- **JWT (JSON Web Token)**: The JWT is defined by RFC 7519 [40]. It is an open standard that provides a compact and independent way to securely transmit information between applications using a JSON (Javascript Object Notation) object. This information is verifiable and trustworthy as it is digitally signed using a secret [22]. It has a format divided into three parts: Header, Payload and Signature. The header is separated into two parts, token type and the algorithm, that may be a HMAC, SHA256 or RSA [29]. The JWT has an advantage over traditional tokens, this verification can be done directly on the resource server, without connecting to the authentication server [22,36]. Using JWT it is possible to retrieve user information directly from the token [22,26]. In addition to user information, it is common to find in a JWT their permissions and expiration time of the token [36]. The JWT has adherence to "stateless" applications, that is, those that do not keep a session on the server side, but stay data with the client side, and must be used in each client request to the resource [26]. In a microservices environment, JWTs can be transferred during the communication between them [35]. Finally, it is important to know that JWT can be integrated with the OAuth 2.0 protocol [4,23].

- **API Gateway**: In the microservices environment, API Gateway acts as an intermediary between the client and the microservices, providing a private network environment that allows the exchange of private data [3,15], that is, clients do not communicate directly with services, but only with a Gateway, which is responsible for communicates with the requested service. It can be an input that performs the filtering of client requests, making the appropriate forwarding to the microservice [23], and it can check the user's credentials, to find out if he owns the proper authorization [7,37]. We realized that API Gateway is a technique to decrease microservice exposure. Nevertheless, it is important in a future work compares the communication in different scenarios. These scenarios could be using or not using the API gateway between the client and microservices. Consequently, it will be possible to collect the strengths and weakness of both approaches and verify the possibility of hybrid scenarios.
Lu et al. [34], stated that the API Gateway can aggregate multiple microservices in a single client interface, being an element that stands between the client and the requested service. Using the API Gateway helps to reduce the exposure of systems, then, the microservices are all protected behind the API Gateway [8]. Although several advantages for its implementation have been observed, its use may not prove advantageous when it becomes a single decision point, because, in case of failure in this element, the entire application may become inaccessible [8]. An API Gateway can use services such as JWT and OAuth 2.0 [7,8,15,23,25,34,36].

- **OpenID Connect**: OpenID Connect is an open authentication standard that ensures users have only one digital identity for multiple applications or services [29]. Dongjin et al. [25] stated that it is an authentication layer over the OAuth protocol, allowing services to read the user's basic information. Nehme et al. [27,31] observed that OpenID Connect is built on top of the OAuth protocol. Yarygina and Bage [7] reinforced that OpenID Connect provisions the user's identity. OpenID Connect can be used in conjunction with OAuth 2.0 [4,8,35]. There is a difference between OpenID and OpenID Connect (OIDC). According to the OpenID Foundation website, "OpenID Connect performs many of the same tasks as OpenID 2.0, but does so in a way that is API-friendly, and usable by native and mobile applications" [41]. They also explain that "OpenID Connect defines optional mechanisms for robust signing and encryption. Whereas integration of OAuth 1.0a and OpenID 2.0 required an extension, in OpenID Connect, OAuth 2.0 capabilities are integrated with the protocol itself" [41].

- **SSO (Single Sign-On)**: SSO allows a user to be authenticated only once when logging into a particular system, therefore, users can access all authorized resources and services on a system without needing another authentication [29,36]. According to

Banati et al. [29], the main purpose of this mechanism is the exchange of authorization credentials and not the authentication by itself. The authors also reinforced that the mechanism guarantees unified authentication in microservices and the implementation of this feature can improve the user experience [33]. In the same way as the API Gateway, the implementation of an SSO server may cause a "Single Point of Failure", that is, if there are problems in this system, every application can be compromised, as it centralizes all authentication of a system [36]. It is possible to implement a Single Sign-On system based on OAuth 2.0 [16,29,35].

- **HTTPS**: The Hyper Text Transfer Protocol Secure is defined by RFC 2818 [42]. It describes the use of HTTP over TLS (Transport Layer Security). Using the HTTPS protocol will ensure that the communication be encrypted [16]. It provides a channel between two hosts identified by certificates [30]. The use of HTTPS not just limited to encrypt data, but ensures that a given client is communication with whom he wants to [3].
- **RBAC**: Role-Based Access Control is used in authorization process [4]. It is a very know identity-based access control model [35]. The use of a role-based access control will increase the flexibility of the system because the role will define what access the client is allowed [22]. RBAC is user-centric access control model, it does not account for relationship between the requesting entity and the resource [7]. RBAC authorization roles can be incorporated into JWT tokens as an additional attribute [7].
- **ABAC**: Attribute-Based Access Control, based in the words of Preuveneers and Joosen [35] "grants access rights to subjects through the use of policies or rules that combine various types of attributes to facilitate user access to the right resources under the right conditions". They complemented that it offers more expressivity and flexibility compared to another access control models such as RBAC. The primary goal of ABAC in the words of Yu et al. [25] is "an access control model is to fulfill the requirements of highly heterogeneous environments such as multi-cloud environment". They also pointed the benefit of centralized security management and orchestration that will protect the application according to consistent policies. ABAC is recommended to be used when there is fine-grained authorization of resources, such as access to a specific API call [7].
- **XACML**: eXtensible Access Control Markup Language is defined by RFC 7061 [43]. According to this document, XACML "defines an architecture and a language for access control (authorization). The language consists of requests, responses, and policies". It is used to create access control policies and can be used with OAuth 2.0 protocol [31]. Nehme et al. [31] proposed a model using XACML along with OAuth 2.0. In this case, OAuth 2.0 acts as an authorization service and XACML with policy administration and decision points. Barabanov and Marushin [8] discourage the use of XAML because it use a complicated syntax, causing more work for developers, adding to the fact that there were not many open-source integrations.
- **HMAC**: Hash-based Message Authentication Code is defined by RFC 2104. It provides a way to check the integrity of an information transmitted in a medium [44]. In the words of Mateus-Coelho et al., HMAC consists in "hash-based messaging code to sign the request". According to the same authors, there are many examples that can be found in internet suggesting the use of HMAC over HTTP. HMAC algorithm can also be used to sign a JWT [22].
- **SAML**: The Security Assertion Markup Language (SAML) 2.0 is defined by RFC 7522 and is defined as an XML-based framework that allows identity and security information to be shared across security domains [45]. In a microservices environment, SAML is used to exchange user attributes stored at the identity provider [35]. Mateus-Coelho et al. [3] affirmed that "SAML and OpenID is perfect for Authentication and Authorization of someone's on a system but it's also great for service-to-service authentication as well". Nevertheless, they admitted that SAML is complex when it is compared to other technologies such as Api Keys.

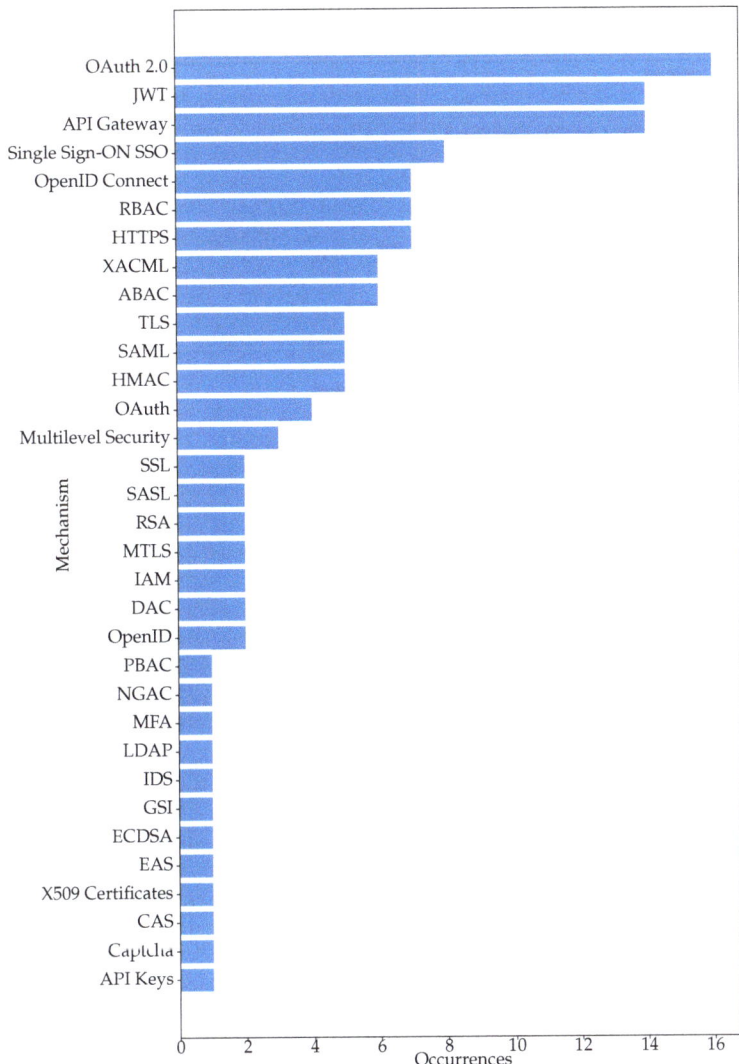

Figure 3. Number of occurrences versus security mechanisms found

3.5. RQ.3.What Are the Main Open-Source Technology Solutions That Implement the Authentication and Authorization Mechanisms Identified in the Literature?

The main open-source solutions that implement the authentication and authorization mechanisms identified in the literature are presented in Table 6. It is possible to notice that the Spring [46] ecosystem libraries are the most mentioned (Spring Security, Spring Cloud, Spring Boot, Gateway Zuul and Eureka Server), totaling 10 mentions. We realized that Spring Boot and Eureka are not focused on security, but they have specific security libraries that can be used together. For instance, implementing Spring Boot allows to implement Spring Security library, and Eureka helps to implement an API Gateway using Spring Cloud. Some of these studies demonstrated a practical implementation of Spring framework using security mechanisms [21,22,24,25].

Although the research found several references to the Spring ecosystem, which is built by Java programming language, it is important to mention that there are alternative frameworks based in other languages that implement the security mechanisms found into a microservice environment, such as GoKit (Golang) [47], Flask (Python) [48] and .NET Core (C#) [49].

The other open-source solution mentioned more than once is called Kong [50] (2 mentions), the others being mentioned only once. It is important to note that the Spring Framework uses the Java programming language and has several libraries for implementing security mechanisms in microservices, such as API Gateway, OAuth 2.0 and OpenID Connect [46]. The Kong application refers to the "Kong API Gateway", that is, among all the mechanisms raised, it supports the implementation of an API Gateway.

Table 6. Open-source technologies that implement security in microservices architecture.

Open-Source	ID
Spring Security	S1, S4, S5, S7, S8
Kong	S6, S11
Spring Boot	S6, S7
Gateway Zuul	S4, S5
Eureka Server	S5
Jadex	S15
Jarvis	S1
Lagom	S15
VertX	S15

4. Discussions

Given the challenges, mechanisms and open-source solutions presented, it was verified which of the mechanisms and solutions could be implemented to face the challenges, according to what was collected in this RSL. Table 7 presents the solutions that act on the related challenges. It was verified that part of the challenges does not have a direct link on the open-source mechanism and/or implementation. The mechanisms identified could be applied to face 09 challenges of the 20 listed in Table 4. Although it seems a low number, these challenges are the most mentioned by the authors.

The mechanisms were widely mentioned by the authors, most of them can face the challenges, with emphasis once again on the implementation of OAuth 2.0, JWT, API Gateway, OpenID Connect and Single Sign ON (SSO). Nevertheless, it does not mean that they are better or will solve any kind of security issues related to microservices architecture. Even the less mentioned mechanisms could be more appropriate, depending on the case. It is important to know what each mechanism is individually and what it does, for then, implement a good security architecture in a system.

Table 7. Linking Challenges, Mechanism and Open-Source Solutions.

Challenge	Mechanism	Open-Source
Increased attack surface (compared to monolithic)	API Gateway (S2, S3, S4, S6, S12, S13, S16, S17, S20, S22), OAuth 2.0 (S7, S12, S13), SSO (S14)	Spring (S4, S7), Kong (S6, S11)
Authorization between services	OAuth 2.0 (S1, S5, S6, S7, S8, S10, S12, S13, S14, S15, S16, S17, S21), SAML(S2), OpenID(S2, S14, S16, S17), JWT (S9, S12, S13, S14, S15, S17, S21)	Spring (S1, S5)
Bypass in Api Gateway	XACML (S3), NGAC (S3), JWT (S3, S6, S12), OpenID (S3, S12, S16), OAuth 2.0 (S3, S6, S12, S16), TLS (S3), IDS (S12)	Spring (S4), Kong (S6)
Communication between microservices	TLS (S1, S10, S14, S15), MTLS (S17), SSL (S2), HTTPS (S2, S10, S14, S15, S17, S20), SAML(S2, S14, S21), XACML (S16, S21), OpenID(S2, S3, S8, S16, S17), JWT (S4, S5, S6, S12, S13, S14,15, S16, S17, S21), OAuth 2.0 (S5, S6, S7, S8, S10, S13, S15, S17, S21), GSI (S8)	Spring (s5), Kong (S6)
Trust between microservices compromised by unauthorized access	JWT (S1, S3, S4, S5, S6, S9, S11, S12, S13, S14, S15, S17, S21, S22), OAuth 2.0 (S1, S3, S5, S6, S7, S8, S10, S11, S12, S13, S14, S15, S16, S17, S20, S21), OpenID Connect S1, S2, S3, S8, S12, S14, S16, S17, S21)	Spring (S4, S5)
Microservice Access Control	OAuth 2.0 (S1, S5, S8, S10, S11, S12, S13, S14, S15, S16, S20, S21), OpenID (S1, S2, S3, S8, S12, S14, S16), TLS(S1), MTLS(S1), SASL (S1), SSO (S1, S2), JWT (S1, S5, S11, S12, S13, S14, S15, S21), HMAC(S2, S21), ABAC (S8, S17, S20, S21), RBAC (S17, S21), CAS (S8), RSA (S11), XACML (S16), Captcha (S20), Multiple FA (S20), DAC (S21), IAM (S21)	Spring (S1, S5, S11)
Coordinate authentication server with new microservices	LDAP (S8), SSO (S1, S2, 10, S13, s14, S20, S21, S22), OAuth 2.0 (S12, S14, S16), OpenID (S12, S14, S16),	
Individual concern for each microservice	JWT (S1, S3, S4, S5, S6, S9, S11, S12, S13, S14, S15, S17, S21, S22), OAuth 2.0 (S1, S3, S5, S6, S7, S8, S10, S11, S12, S13, S14, S15, S16, S17, S20, S21), OpenID Connect (S1, S2, S3, S8, S12, S14, S16, S17, S21)	Spring (S4, S5)
Use of authentication/authorization server that handles all microservices	SSO (S1, S2, 10, S13, s14, S20, S21, S22), OAuth 2.0 (S12, S14, S16)	

Study Limitations

The study was performed with searches in DBLP, IEEE and Scopus databases. To prevent relevant works from being discarded, the snowballing process was applied. Even with this concern, it is possible that, increasing the number of databases for consultations, new studies may be found. However, as verified in some studies collected, there is currently a lack of studies on the subject [3,4,7,16,21,25,28]. It was also verified that there is a lack of studies demonstrating the practical implementation of security in microservices [24,28,37]. Hence, it is likely that over the next few years, if the research related to the subject in question increases, a new systematic review will be necessary, in order to complement the knowledge collected in this work. We cannot conclude that the mechanisms less mentioned in the studies are less used, therefore, it is important to explore all of them, that can be done in a future work. Finally, it is important to note that this study is more focused on identifying answers to the research questions, that is, it is possible that the answers to these questions may be the subject of further studies pointing out which challenges are most critical in terms of vulnerability, how much they occur in a practical environment, or even which of these challenges should be addressed with priority. The mechanisms can also be implemented and tested in order to find out in a practical environment which of the challenges are mitigated with the implemented mechanism.

5. Final Remarks

As verified during the execution of this work and demonstrated in Table 4, there is a lack of studies related to security in microservices architecture. The lack increases when the study is specific for authentication and authorization, especially in a practical approach. It is important that the subject be better explored, because, as verified in this work, within a microservice environment, it is necessary to be concerned with security aspects in each service, individually, as the adoption of this architecture can increase the attack surface and still generate attention points in the communication between them, in this way, the lack of attention in these questions can make the applications vulnerable to unauthorized accesses. Of all the points listed in Table 4, there are issues related to the implementation of technologies themselves, however, there are other aspects related to the subject, such as the organization of development teams working on different microservices within the same system, therefore, is a theme with vast field to be explored.

Several mechanisms were found that mitigate the main points of attention observed, all of them listed in Table 5, with OAuth 2.0 being the most mentioned, along with the Json Web Token (JWT) and the use of API Gateway. The correct implementation of these can reduce the possibility of any type of unauthorized access to one or more microservices, making the environment better protected. There are few studies on practical implementations, thus, a scenario for future work is foreseen, especially with proposals for specific patterns within this context.

Finally, it was found that the literature indicates few open-source solutions that implement the mechanisms found. In this case, a viable alternative expands the search into new sources, including gray literature, which is literature produced at all levels of government, academic, business and industrial, in print and electronic formats, but which is not controlled by commercial publishers, that is, where publication is not the primary activity of the producing body[51]. Such findings can be properly experimented with scientific rigor and identified as technical solutions that solve the challenges collected in this work.

Author Contributions: Writing—original draft preparation, M.G.d.A. and E.D.C.; writing—review and editing, M.G.d.A. and E.D.C.; visualization, M.G.d.A. and E.D.C. All authors have read and agreed to the published version of the manuscript.

Funding: This research received no external funding.

Institutional Review Board Statement: Not applicable.

Informed Consent Statement: Not applicable.

Data Availability Statement: Not applicable.

Conflicts of Interest: The authors declare no conflict of interest.

References

1. Lewis, J.; Fowler, M. *Microservices—A Definition of This New Archtectural Term*; EA PAD: Redwood City, CA, USA, 2014.
2. Merson, P. *Microservices beyond the Hype: What You Gain and What You Lose*; SEI Digital Library: San Diego, CA, USA, 2015.
3. Mateus-Coelho, N.; Cruz-Cunha, M.; Ferreira, L.G. Security in microservices architectures. *Procedia Comput. Sci.* **2021**, *181*, 1225–1236. doi: [CrossRef]
4. Pereira-Vale, A.; Fernandez, E.B.; Monge, R.; Astudillo, H.; Márquez, G. Security in microservice-based systems: A Multivocal literature review. *Comput. Secur.* **2021**, *103*, 102200. doi: [CrossRef]
5. Pippal, S.K.; Kumari, A.; Kushwaha, D.S. CTES based Secure approach for Authentication and Authorization of Resource and Service in Clouds. In Proceedings of the 2011 2nd International Conference on Computer and Communication Technology (ICCCT-2011), Allahabad, India, 15–17 September 2011; IEEE: Piscataway, NJ, USA, 2011; pp. 444–449.

6. Halonen, T. Authentication and authorization in mobile environment. In *Tik-110.501 Seminar on Network Security*; Citeseer: Princeton, NJ, USA, 2000.
7. Yarygina, T.; Bagge, A.H. Overcoming Security Challenges in Microservice Architectures. In Proceedings of the 12th IEEE International Symposium on Service-Oriented System Engineering, SOSE 2018 and 9th International Workshop on Joint Cloud Computing, JCC 2018, Bamberg, Germany, 26–29 March 2018; pp. 11–20. doi: [CrossRef]
8. Barabanov, A.; Makrushin, D. Authentication and authorization in microservice-based systems: Survey of architecture patterns. *arXiv* **2020**, arXiv:2009.02114.
9. Sun, Y.; Nanda, S.; Jaeger, T. Security-as-a-service for microservices-based cloud applications. In Proceedings of the IEEE 7th International Conference on Cloud Computing Technology and Science, CloudCom 2015, Vancouver, BC, Canada, 30 November–3 December 2015; pp. 50–57. doi: [CrossRef]
10. Mehraj, S.; Banday, M.T. Establishing a Zero Trust Strategy in Cloud Computing Environment. In Proceedings of the 2020 International Conference on Computer Communication and Informatics (ICCCI), Coimbatore, India, 22–24 January 2020; pp. 1–6. doi: [CrossRef]
11. Rose, S.; Borchert, O.; Mitchell, S.; Connelly, S. *Zero Trust Architecture*; Technical Report; National Institute of Standards and Technology: Gaithersburg, MA, USA, 2020.
12. Rossi, B.; Russo, B.; Succi, G. Adoption of free/libre open source software in public organizations: Factors of impact. *Inf. Technol. People* **2012**, *25*, 156–187. doi: [CrossRef]
13. Hippel, E.V.; Krogh, G.V. Open source software and the "private-collective" innovation model: Issues for organization science. *Organ. Sci.* **2003**, *14*, 209–223. doi: [CrossRef]
14. Lavazza, L. Beyond Total Cost of Ownership: Applying Balanced Scorecards to Open-Source Software. In Proceedings of the International Conference on Software Engineering Advances (ICSEA 2007), Cap Esterel, France, 25–31 August 2007; p. 74. doi: [CrossRef]
15. Xu, R.; Jin, W.; Kim, D. Microservice security agent based on API gateway in edge computing. *Sensors* **2019**, *19*, 4905. doi: [CrossRef] [PubMed]
16. Krämer, M.; Frese, S.; Kuijper, A. Implementing secure applications in smart city clouds using microservices. *Future Gener. Comput. Syst.* **2019**, *99*, 308–320. doi: [CrossRef]
17. Kitchenham, B.; Charters, S. *Guidelines for Performing Systematic Literature Reviews in Software Engineering*; EBSE: Goyang, Korea, 2007.
18. Kitchenham, B.; Brereton, O.P.; Budgen, D.; Turner, M.; Bailey, J.; Linkman, S. Systematic literature reviews in software engineering—A systematic literature review. *Inf. Softw. Technol.* **2009**, *51*, 7–15. doi: [CrossRef]
19. Freitas, V. Parsifal. 2021. Available online: https://parsif.al (accessed on 17 October 2021).
20. Wohlin, C. Guidelines for Snowballing in Systematic Literature Studies and a Replication in Software Engineering. In Proceedings of the 18th International Conference on Evaluation and Assessment in Software Engineering, London, UK, 13–14 May 2014; EASE '14; Association for Computing Machinery: New York, NY, USA, 2014. doi: [CrossRef]
21. Safaryan, O.; Pinevich, E.; Roshchina, E.; Cherckesova, L.; Kolennikova, N. Information system development for restricting access to software tool built on microservice architecture. *E3S Web Conf.* **2020**, *224*, 01041. doi: [CrossRef]
22. Shulin, Y.; Jieping, H. Research on Unified Authentication and Authorization in Microservice Architecture. In Proceedings of the International Conference on Communication Technology Proceedings, ICCT, Nanning, China, 28–31 October 2020; pp. 1169–1173. doi: [CrossRef]
23. Jin, W.; Xu, R.; You, T.; Hong, Y.G.; Kim, D. Secure edge computing management based on independent microservices providers for gateway-centric IoT networks. *IEEE Access* **2020**, *8*, 187975–187990. doi: [CrossRef]
24. Nguyen, Q.; Baker, O. Applying Spring Security Framework and OAuth2 To Protect Microservice Architecture API. *J. Softw.* **2019**, *14*, 257–264. doi: [CrossRef]
25. Yu, D.; Jin, Y.; Zhang, Y.; Zheng, X. A survey on security issues in services communication of Microservices-enabled fog applications. *Concurr. Comput. Pract. Exp.* **2019**, *31*, e4436. doi: [CrossRef]
26. Bhutada, S.; Jyothi, K.K. Enhancing Security to the Microservice (MS) Architecture By Implementing Authentication and Authorization Service using Docker and Kubernetes. *Int. J. Innov. Technol. Explor. Eng.* **2019**, *8*, 401–407.
27. Nehme, A.; Jesus, V.; Mahbub, K.; Abdallah, A. Securing Microservices. *IT Prof.* **2019**, *21*, 42–49. MITP.2018.2876987 [CrossRef]
28. Pereira-Vale, A.; Marquez, G.; Astudillo, H.; Fernandez, E.B. Security mechanisms used in microservices-based systems: A systematic mapping. In Proceedings of the 2019 45th Latin American Computing Conference, CLEI 2019, Panama City, Panama, 30 September–4 October 2019. doi: [CrossRef]
29. Banati, A.; Kail, E.; Karoczkai, K.; Kozlovszky, M. Authentication and authorization orchestrator for microservice-based software architectures; Authentication and authorization orchestrator for microservice-based software architectures. In Proceedings of the 41st International Convention on Information and Communication Technology, Electronics and Microelectronics (MIPRO), Opatija, Croatia, 21–25 May 2018.
30. Jander, K.; Braubach, L.; Pokahr, A. Defense-in-depth and Role Authentication for Microservice Systems. *Procedia Comput. Sci.* **2018**, *130*, 456–463. doi: [CrossRef]

31. Nehme, A.; Jesus, V.; Mahbub, K.; Abdallah, A. Fine-Grained Access Control for Microservices. *Lect. Notes Comput. Sci. Incl. Subser. Lect. Notes Artif. Intell. Lect. Notes Bioinform.* **2019**, *11358 LNCS*, 285–300. doi: [CrossRef]
32. Richter, D.; Neumann, T.; Polze, A. Security considerations for microservice architectures. In Proceedings of the CLOSER 2018-Proceedings of the 8th International Conference on Cloud Computing and Services Science, Funchal, Portugal, 19–21 March 2018; pp. 608–615. doi: [CrossRef]
33. Cao, R.; Lu, S.; Wang, X.; Xiao, H.; Chi, X. Unified Account Management for High Performance Computing as a Service with Microservice Architecture. In Proceedings of the Unified Account Management for High Performance Computing as a Service with Microservice Architecture, Taipei, Taiwan, 16–23 March 2018.
34. Lu, D.; Huang, D.; Walenstein, A.; Medhi, D. A Secure Microservice Framework for IoT. In Proceedings of the Proceedings-11th IEEE International Symposium on Service-Oriented System Engineering, SOSE 2017, San Francisco, CA, USA, 6–9 April 2017; pp. 9–18. doi: [CrossRef]
35. Preuveneers, D.; Joosen, W. Access Control with Delegated Authorization Policy Evaluation for Data-Driven Microservice Workflows. *Future Internet* **2017**, *9*, 58. doi: [CrossRef]
36. He, X.; Yang, X. Authentication and Authorization of End User in Microservice Architecture. *J. Phys. Conf. Ser.* **2017**, *910*, e012060. doi: [CrossRef]
37. Torkura, K.A.; Sukmana, M.I.; Meinel, C. Integrating continuous security assessments in microservices and cloud native applications. In Proceedings of the UCC 2017-Proceedings of the 10th International Conference on Utility and Cloud Computing, Austin, TX, USA, 5–8 December 2017; pp. 171–180. doi: [CrossRef]
38. Dooley, K. *Designing Large Scale Lans: Help for Network Designers*; O'Reilly Media: Sebastopol, CA, USA, 2001.
39. Hardt, D. Rfc 6749: The oauth 2.0 authorization framework. *Internet Eng. Task Force IETF* **2012**, *10*, 1–75.
40. Jones, M.; Bradley, J.; Sakimura, N. *Rfc 7519: Json Web Token (jwt)*; IETF: Fremont, CA, USA, 2015.
41. Foundation, O. How Is OpenID Connect Different than OpenID 2.0? 2022. Available online: https://openid.net/connect/ (accessed on 4 February 2022).
42. Rescorla, E. *Rfc 2818: HTTP over TLS*; Internet Engineering Task Force (IETF): Fremont, CA, USA, 2000.
43. Sinnema, R.; Wilde, E. *Rfc 7061: eXtensible Access Control Markup Language (XACML) XML Media Type*; Internet Engineering Task Force (IETF): Fremont, CA, USA, 2013.
44. Krawczyk, H.; Bellare, M.; Canetti, R. *MAC: Keyed-Hashing for Message Authenticatio*; Internet Engineering Task Force (IETF): Fremont, CA, USA, 1997.
45. Campbell, B.; Mortimore, C.; Jones, M. *Security Assertion Markup Language (SAML) 2.0 Profile for OAuth 2.0 Client Authentication and Authorization Grants*; Internet Engineering Task Force (IETF): Fremont, CA, USA, 2015.
46. Pivotal. Spring Cloud. 2021. Available online: https://spring.io/projects/spring-cloud (accessed on 14 October 2021).
47. Kit, G. Go Kit—A Toolkit for Microservices. 2022. Available online: https://gokit.io/ (accessed on 4 February 2022).
48. Projects, P. Flask Web Development. 2022. Available online: https://flask.palletsprojects.com/ (accessed on 4 February 2022).
49. Microsoft. NET Core. 2022. Available online: https://dotnet.microsoft.com/ (accessed on 4 February 2022).
50. Kong. Kong API Gateway. 2021. Available online: https://konghq.com/kong (accessed on 14 October 2021).
51. Garousi, V.; Felderer, M.; Mäntylä, M.V. Guidelines for including grey literature and conducting multivocal literature reviews in software engineering. *Inf. Softw. Technol.* **2019**, *106*, 101–121. [CrossRef]

Article

Less Is More: Robust and Novel Features for Malicious Domain Detection

Chen Hajaj [1,*], Nitay Hason [2] and Amit Dvir [2]

[1] Ariel Cyber Innovation Center, Data Science and Artificial Intelligence Research Center, Department of Industrial Engineering and Management, Ariel University, Ariel 4076414, Israel
[2] Ariel Cyber Innovation Center, Department of Computer Science, Ariel University, Ariel 4076414, Israel; nitay.has@gmail.com (N.H.); amitdv@g.ariel.ac.il (A.D.)
* Correspondence: chenha@ariel.ac.il; Tel.: +972-7472-33019

Abstract: Malicious domains are increasingly common and pose a severe cybersecurity threat. Specifically, many types of current cyber attacks use URLs for attack communications (e.g., C&C, phishing, and spear-phishing). Despite the continuous progress in detecting cyber attacks, there are still critical weak spots in the structure of defense mechanisms. Since machine learning has become one of the most prominent malware detection methods, a robust feature selection mechanism is proposed that results in malicious domain detection models that are resistant to evasion attacks. This mechanism exhibits a high performance based on empirical data. This paper makes two main contributions: First, it provides an analysis of robust feature selection based on widely used features in the literature. Note that even though the feature set dimensional space is cut by half, the performance of the classifier is still improved (an increase in the model's F1-score from 92.92% to 95.81%). Second, it introduces novel features that are robust with regard to the adversary's manipulation. Based on an extensive evaluation of the different feature sets and commonly used classification models, this paper shows that models based on robust features are resistant to malicious perturbations and concurrently are helpful in classifying non-manipulated data.

Keywords: malware detection; robust features; domain

1. Introduction

Cybersecurity attacks have become a significant issue for governments and civilians [1]. Many of these attacks are based on malicious web domains or URLs (see Figure 1 for an example of a URL structure). These domains are used for phishing [2–6] (e.g., spear phishing), Command and Control (C&C) [7] and a vast set of virus and malware [8] attacks. Therefore, the ability to identify a malicious domain in advance is a massive game-changer [9–26].

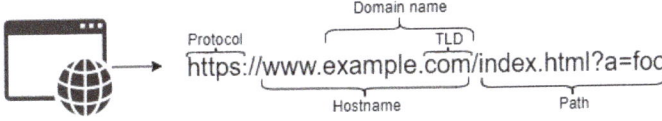

Figure 1. The URL structure.

A common way of identifying malicious/compromised domains is to collect information about the domain names (alphanumeric characters) and network information (such as DNS and passive DNS data). This information is then used to extract a set of features, according to which machine learning (ML) algorithms are trained based on a massive amount of data [11–15,17–22,24,26–28]. A mathematical approach can also be used in various ways [16,26], such as measuring the distance between a known malicious domain name and the analyzed domain (benign or malicious) [26]. Nonetheless, while ML-based solutions are widely used, many of them are not robust; an attacker can easily bypass these models with minimal feature perturbations (e.g., changing the domain's length or modifying network parameters such as Time To Live (TTL)) [29,30]. In this context, one of the main problems is how to train a robust malicious domain classifier, one that is immune to the presence of an intelligent adversary that can manipulate domain properties, to classify malicious domains as benign.

For this purpose, a feature selection process is executed to differentiate between robust and non-robust features. Given the robust feature set, the defender is still guaranteed to provide an efficient classifier, which is harder to manipulate. Even if the attacker has black-box access to the model, tampering with the domain properties or network parameters will have a negligible effect on the classifier's accuracy. In order to achieve this goal, we collected a broad set of both malicious and benign URLs. In addition, we reviewed related work and identified a set of features commonly used for the classification task. These features were then artificially manipulated to show that some, although widely used, are not robust in the face of adversarial perturbations. In a complementary manner, we engineered an original set of novel and robust features. Therefore, we created a hybrid set of features, combining the robust well-known features with our novel features. Finally, the different feature sets (e.g., common, robust common, and novel) were evaluated using common machine learning algorithms, with emphasis on the importance of feature selection and feature engineering processes.

The rest of the paper is organized as follows: Section 2 summarizes related work. Section 3 describes the methodology and the novel features. Section 4 presents the empirical analysis and evaluation. Finally, Section 5 concludes and summarizes this work.

2. Related Work

The issue of identifying malicious domains is a fundamental problem in cybersecurity. This section discusses recent results in identifying malicious domains, focusing on two significant methodologies, mathematical theory (MT) approaches and machine learning (ML)-based techniques.

The use of graph theory to identify malicious domains was more pervasive in the past [16,26,31–33]. Yadav et al. [26] presented a method for recognizing malicious domain names based on fast flux. Fast flux is a DNS technique used by botnets to hide phishing and malware delivery sites behind an ever-changing network of compromised hosts acting as proxies. They analyzed the DNS queries and responses to detect if and when domain names were being generated by a Domain Generation Algorithm (DGA). Their solution was based on computing the distribution of alphanumeric characters for groups of domains and by statistical metrics with the KL (Kullback Leibler) distance, Edit distance and Jaccard measure to identify these domains. For a fast-flux attack using the Jaccard Index, they achieved impressive results, with 100% detection and 0% false positives. However, for smaller numbers of generated domains for each TLD, their false-positive results were much higher, at 15% when 50 domains were generated for the TLD using the KL-divergence over unigrams, and 8% when 200 domains were generated for each TLD using the Edit distance.

Dolberg et al. [16] described a system called *Multi-dimensional Aggregation Monitoring (MAM)* that detects anomalies in DNS data by measuring and comparing a "steadiness" metric over time for domain names and IP addresses using a tree-based mechanism. The steadiness metric is based on a domain similar to IP resolution patterns when comparing DNS data over a sequence of consecutive time frames. The domain name to IP mappings were based on an aggregation scheme and measured steadiness. In terms of detecting malicious domains, the results showed that an average steadiness value of 0.45 could be used as a reasonable threshold value, with a 73% true positive rate and only a 0.3% false positive one. The steadiness values might not be considered a good indicator when fewer malicious activities are present (e.g., <10%).

However, the most common approach to identifying malicious domains is by means of machine learning (ML) and Deep Learning (DL) [11,14,20,23,24,27,28,34–42]. Researchers can train ML algorithms to label URLs as malicious or benign using a set of extracted features. Shi et al. [23] proposed a machine learning methodology to detect malicious domain names using the Extreme Learning Machine (ELM) [19], which is closest to the one employed here. ELM is a new neural network with a high accuracy and fast learning speed. The authors divided their features into four categories: construction-based, IP-based, TTL-based, and *WHOIS*-based categories. Their evaluation resulted in a high detection rate with an accuracy exceeding 95% and a fast learning speed. However, as shown below, a significant fraction of the features used in this work emerged as non-robust and ineffective in the presence of an intelligent adversary.

Sun et al. [24] presented a system called *HinDom*, which generates a heterogeneous graph (in contrast to homogeneous graphs created by Rahbarinia et al. [22] and Yadav et al. [26]) in order to robustly identify malicious attacks (e.g., spam, phishing, malware, and botnets). Even though HinDom collected DNS and pDNS data, it also has the ability to collect information from various clients inside networks (e.g., CERNET2 and TUNET); thus, its perspective is different from the perspective of this study (i.e., client perspective). Nevertheless, HinDom has achieved remarkable results using a transductive classifier and achieved a high accuracy and F1-scores of 99% and 97.5%, respectively.

Bilge et al. [13] created a system called *Exposure*, which is designed to detect malicious domain names. Their system uses passive DNS data collected over some time to extract features related to known malicious and benign domains. Passive DNS Replication [11,13,20,22,25,27,28] refers to the reconstruction of DNS zone data by recording and aggregating live DNS queries and responses. Passive DNS data can be collected without requiring the cooperation of zone administrators. The Exposure system is designed to detect malware- and spam-related domains. It can also detect malicious fast-flux and DGA-related domains based on their unique features. The system computes the following four sets of features from anonymized DNS records: (a) time-based features related to the periods and frequencies that a specific domain name was queried in; (b) DNS-answer-based features calculated according to the number of distinctive resolved IP addresses and domain names, the countries in which the IP addresses reside, and the ratio of the resolved IP addresses that can be matched with valid domain names and other services; (c) TTL-based features that are calculated based on a statistical analysis of the TTL over a given time series; and (d) domain name-based features that are extracted by computing the ratio of the numerical characters to the domain name string, and the ratio of the size of the longest meaningful substring in the domain name. Using a Decision Tree model, Exposure reported a total of 100,261 distinct domains as being malicious, which resulted in 19,742 unique IP addresses. The combination of features used to identify malicious domains led to the successful identification of several domains related to botnets, flux networks, and DGAs, with low false-positive and high detection rates. It may not be possible to generalize the detection rate results reported by the authors (98%) since they were highly dependent on comparisons with biased datasets. Despite the positive results, once an identification scheme is published, it is always possible for an attacker to evade detection by mimicking the behaviors of benign domains.

Rahbarinia et al. [22] presented a system called *Segugio*, which is an anomaly detection system based on passive DNS traffic to identify malware-controlled domain names based on their relationship to known malicious domains. The system detects malware-controlled domains by creating a machine domain bipartite graph representing the underlying relations between new domains and known benign/malicious domains. The system operates by calculating the following features: (a) machine behavior, based on the ratio of "known malicious" and "unknown" domains that query a given domain d over the total number of machines that query d. The larger the total number of queries and the fraction of malicious related queries, the higher the probability that d is a malware-controlled domain; (b) Domain activity, where given a time period, domain activity is computed by counting the total number of days in which a domain was actively queried; (c) IP abuse, where, given a set of IP addresses that the domain resolves to, this feature represents the fraction of those IP addresses that were previously targeted by known malware-controlled domains. Using a Random Forest model, Segugio was shown to produce high true positive and meager false positive rates (94% and 0.1%, respectively). It was also able to detect malicious domains earlier than commercial blacklisting websites. However, Segugio is a system that can only detect malware-related domains based on their relationship to previously known domains and therefore cannot detect new (unrelated to previous malicious domains) malicious domains. Additional information concerning malicious domain filtering and malicious URL detection can be found in [34,42].

Adversarial machine learning is a subfield of machine learning in which instances used to train the model and instances in the wild may be characterized by different distributions. For example, given perturbations on a malicious instance so that it will be falsely classified as benign. These manipulated instances are commonly called *adversarial examples (AE)* [43]. AE are samples that an attacker changes based on some model classification function knowledge. These examples are slightly different from correctly classified examples. Therefore, the model fails to classify them correctly. AE are widely used in the fields of spam filtering [44], network intrusion detection systems (IDS) [45], anti-virus signature tests [46] and biometric recognition [47].

Attackers commonly follow one of two models to generate adversarial examples: (1) white-box attacker [48–51], which has full knowledge of the classifier and the train/test data and (2) black-box attacker [48,52,53], which has access to the model's output for each given input. Various methods have emerged to tackle AE-based attacks and make ML models robust. The most promising are those based on game-theoretic approaches [54–56], robust optimization [48,49,57], and adversarial retraining [30,58,59]. These approaches mainly concern *feature-space models* of attacks where feature space models assume that the attacker changes the values of features directly. Note that these attacks may be an abstraction of reality as random modifications to feature values may not be realizable or avoid the manipulated instance functionality.

Note that the topic of robust feature selection has attracted an increasing number of researchers in recent years [30,60,61]. In the domain of PDF malware, Tong et al. [30] extracted a set of features termed "conserved features" that the adversary cannot unilaterally modify without compromising malicious functionality. In the domain of APK malware, Chen et al. [60] demonstrated the need for robust feature selection in their tool, Android HIV. This tool takes advantage of non-robust features to easily bypass state-of-the-art android malware classifiers.

3. Methodology

The structure of this section is as follows: Section 3.1 outlines the characteristics and methods of collection of the dataset. Section 3.2 presents our evaluation metrics. Section 3.3 defines each of the well-known features from the literature. Section 3.4 covers the evaluation of their robustness, and Section 3.5 presents novel features and evaluates their robustness.

3.1. Data Collection

The main ingredient of ML models is the data on which the models are trained. Data collection should be as heterogeneous as possible to model reality. The data collected for this work include both malicious and benign URLs: the benign URLs are based on the Alexa top 1 million [62], and the malicious domains were crawled from multiple sources [63,64] to allow diversity and due to the fact they are fairly rare.

According to [65], 25% of all URLs in 2020 were malicious, suspicious, or moderately risky. Therefore, to make a realistic dataset, all the evaluations include all 1356 malicious active unique URLs, and consequently, 5345 benign active unique URLs as well. For each instance, the URL and domain information properties were crawled from *Whois* and their

DNS records. *Whois* is a widely used Internet record listing that identifies who owns a domain, how to get in contact with them, the creation date, update dates, and expiration date of the domain. *Whois* records have been proven to be extremely useful and have developed into an essential resource for maintaining the integrity of the domain name registration and website ownership. Note that according to a study by ICANN (Internet Corporation for Assigned Names and Numbers) [66], many malicious attackers abuse the Whois system. Hence, only the information that could not be manipulated was used. A graphical representation of the data collection framework is illustrated in Figure 2.

Finally, based on these resources (*Whois* and DNS records), the following features were generated: the length of the domain, the number of consecutive characters, and the entropy of the domain from the URLs' datasets. Next, the lifetime of the domain and the active time of domain were calculated from the *Whois* data. Based on the DNS response dataset (a total of 263,223 DNS records), the number of IP addresses, distinct geo-locations of the IP addresses, average Time to Live (TTL) value, and the Standard deviation of the TTL were extracted. For extracting the novel features (Section 3.5), Virus Total (*VT*) [67] and *Urlscan* [68] were used, where *Urlscan* was used to extract parameters such as the IP address of the page element of the URL.

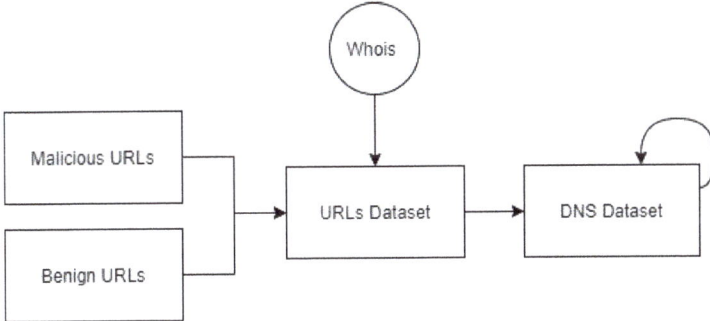

Figure 2. Data collection framework.

3.2. Evaluation Metrics

Machine Learning (ML) is a subfield of computer science aimed at causing computers to act and improve over time autonomously by feeding them data in the form of observations and real-world interactions. In contrast to traditional programming, where input and algorithms are provided to receive an output, with ML, a list of inputs and their associated outputs are provided to extract the algorithm that maps the two.

ML algorithms are often categorized as either supervised or unsupervised. In supervised learning, each example is a pair consisting of an input vector (also called data point) and the desired output value (class/label). Unsupervised learning learns from data that have not been labeled, classified, or categorized. Instead of responding to feedback, unsupervised learning identifies commonalities in the data and reacts based on the presence or absence of such commonalities in each new piece of data.

In order to evaluate how a supervised model is adapted to a problem, the dataset needs to be split into two, namely, a training set and testing set. The training set is used to train the model, and the testing set is used to evaluate how well the model "learned" (i.e., by comparing the model predictions with the known labels). Usually, the train/test distribution is around 75%/25% (depending on the problem and the amount of data). Standard evaluation criteria are as follows: recall, precision, accuracy, F1-score, and loss. All of these criteria can easily be extracted from the evaluation's confusion matrix.

A confusion matrix (Table 1) is commonly used to describe the performance of a classification model. Recall (Equation (2)) is defined as the number of correctly classified malicious examples out of all the malicious ones. Similarly, precision (Equation (3)) is the number of correctly classified malicious examples from all examples classified as malicious (both correctly and wrongly classified). Accuracy (Equation (1)) is used as a statistical measure of how well a classification test correctly identifies or excludes a condition. That is, the accuracy is the proportion of true results (both true positives and true negatives) among the total number of cases examined. Finally, the F1-score (Equation (4)) is a measure of a test's accuracy. It considers both the precision and the recall of the test to compute the score. The F1-score is the harmonic average of the precision and recall, where an F1-score reaches its best value at 1 (perfect precision and recall) and worst at 0. These criteria are used as the main evaluation metric.

The problem of identifying malicious web domains is a supervised classification problem, as the correct label (i.e., malicious or benign) can be extracted using a blacklist-based method, as we describe in the next section.

$$Accuracy = \frac{TP + TN}{TP + FP + TN + FN} = \frac{T}{P + N} \quad (1)$$

$$Recall = \frac{TP}{TP + FN} \quad (2)$$

$$Precision = \frac{TP}{TP + FP} = \frac{TP}{P} \quad (3)$$

$$F_1 - score = 2 \cdot \frac{Precision \cdot Recall}{Precision + Recall} \quad (4)$$

Table 1. Confusion matrix.

		Prediction Outcome		
		Positive	Negative	Total
Actual Value	Positive	True Positive	False Negative	TP + FN
	Negative	False Positive	True Negative	FP + TN
	Total	P	N	

3.3. Feature Engineering

Based on the previous works surveyed, a set of features that are commonly used for malicious domain classification [11,13,22,23,27,28,35,69,70] were extracted. Specifically, the following nine features were used as the baseline (note that the focus of this work is on the potential use of robust features and not on the specific features; thus, WLOG, we evaluated a set of nine commonly used features):

- **Length of domain**: The length of a domain is calculated by the domain name followed by the TLD (gTLD or ccTLD). Hence, the minimum length of a domain is four since the domain name needs to be at least one character (most domain names have at least three characters), and the TLD (gTLD or ccTLD) is composed of at least three characters (including the dot character) as well. For example, for the URL http://www.ariel-cyber.co.il; accessed on 20 March 2022, the length of the domain is 17 (the number of characters for the domain name—"ariel-cyber.co.il").
- **Number of consecutive characters**: This is the maximum number of consecutive repeated characters in the domain. This includes the domain name and the TLD (gTLD or ccTLD). For example, for the domain "caabbbccccd.com" the maximum number of consecutive repeated characters value is 4, due to the four consecutive "c" characters.
- **Entropy of the domain**: The entropy of a domain is defined as: $-\sum_{j=1}^{n_i} \frac{count(c_j^i)}{length(Domain_{(i)})} \cdot \log_2 \frac{count(c_j^i)}{length(Domain_{(i)})}$, where each $Domain_{(i)}$ consists of n_i distinct characters

$\{c_1^i, c_2^i, \ldots, c_{n_i}^i\}$. For example, for the domain "google.com", the entropy is $-(5 \cdot (\frac{1}{10} \log_2 \frac{1}{10}) + 2 (\frac{2}{10} \log_2 \frac{2}{10}) + 3(\frac{3}{10} \log_2 \frac{3}{10})) = 1.25$ The domain has 5 characters that appear once ("l", "e", ".", "c", and "m"), one character that appears twice ("g") and one character that appears three times ("o").

- **Number of IP addresses**: This is the number of distinct IP addresses in the domain's DNS record. For example, for the list ["1.1.1.1", "1.1.1.1", and "2.2.2.2"], the number of distinct IP addresses is 2.
- **Distinct geo-locations of the IP addresses**: For each IP address in the DNS record, the countries for each IP were listed and the number of different countries was counted. For example, for the list of IP addresses ["1.1.1.1", "1.1.1.1", and "2.2.2.2"] the list of countries is ["Australia", "Australia", and "France"] and the number of distinct countries is 2. Note that this feature relates to the number of different countries and not the country itself.
- **Mean TTL value**: For all the DNS records of the domain in the DNS dataset, the TTL values were averaged. For example, if a domain's DNS records were checked 30 times, and in 20 of them the TTL value was "60" and in 10 the TTL value was "1200", the mean is $\frac{20 \cdot 60 + 10 \cdot 1200}{30} = 440$.
- **Standard deviation of the TTL**: The standard deviations of the TTL values for all the DNS records of the domain in the DNS dataset were calculated. For the "Mean TTL value" example above, the standard deviation of the TTL values is 537.401.
- **Lifetime of domain**: This is the interval between a domain's expiration date and creation date in years. For example, the domain "ariel-cyber.co.il", according to *Whois* information, which was updated on 4 June 2018, was created on 14 May 2015 and expires on 14 May 2022. Therefore, the lifetime of the domain is the number of years from 14 May 2015 to 14 May 2022, i.e., 8.
- **Active time of domain**: Similar to the lifetime of a domain, the active time of a domain is calculated as the interval between a domain's updated date and creation date in years. Using the same example as in the "Lifetime of domain", the active time of the "ariel-cyber.co.il" domain is the number of years between 14 May 2015 and 14 May 2021, i.e., 6.

3.4. Robust Feature Selection

Next, the robustness of the set of features described above was evaluated to filter those that could significantly harm the classification process due to the adversary's manipulations. Table 2 lists the common features along with the mean value and standard deviation (note that the std in some cases (e.g., mean TTL value) is higher due to fact that these features have a positive value by definition.) For malicious and benign URLs based on our dataset, note that some features have similar mean values for both benign and malicious instances while they are commonly used. Furthermore, whereas "Standard deviation of the TTL" has distinct values for benign and malicious domains, we later show that an intelligent adversary can easily manipulate this feature, leading to a benign classification of malicious domains.

In order to understand the malicious abilities of an adversary, the base features were manipulated over a wide range of possible values, one feature at a time. This analysis considers an intelligent adversary with black-box access to the model (i.e., a set of features or output for a given input). The robustness analysis is based on an ANN model that classifies the manipulated samples, where the train set is the empirically crawled data, and the test set includes the manipulated malicious samples. Figure 3 depicts the possible adversary manipulations over any of the features. We chose recall for the evaluation metric, representing the average detection rate after modifications.

Table 2. Classic features and statistical properties (*—robust features).

Feature	Benign Mean (std)	Malicious Mean (std)
Length of domain	14.38 (4.06)	15.54 (4.09)
Number of consecutive characters *	1.29 (0.46)	1.46 (0.5)
Entropy of the domain	4.85 (1.18)	5.16 (1.34)
Number of IP addresses	2.09 (1.25)	1.94 (0.94)
Distinct geo-locations of the IP addresses	1.00 (0.17)	1.02 (0.31)
Mean TTL value *	7578.13 (17,781.47)	8039.92 (15,466.29)
Standard deviation of the TTL	2971.65 (8777.26)	2531.38 (7456.62)
Lifetime of domain *	10.98 (7.46)	6.75 (5.77)
Active time of domain *	8.40 (6.79)	4.64 (5.66)

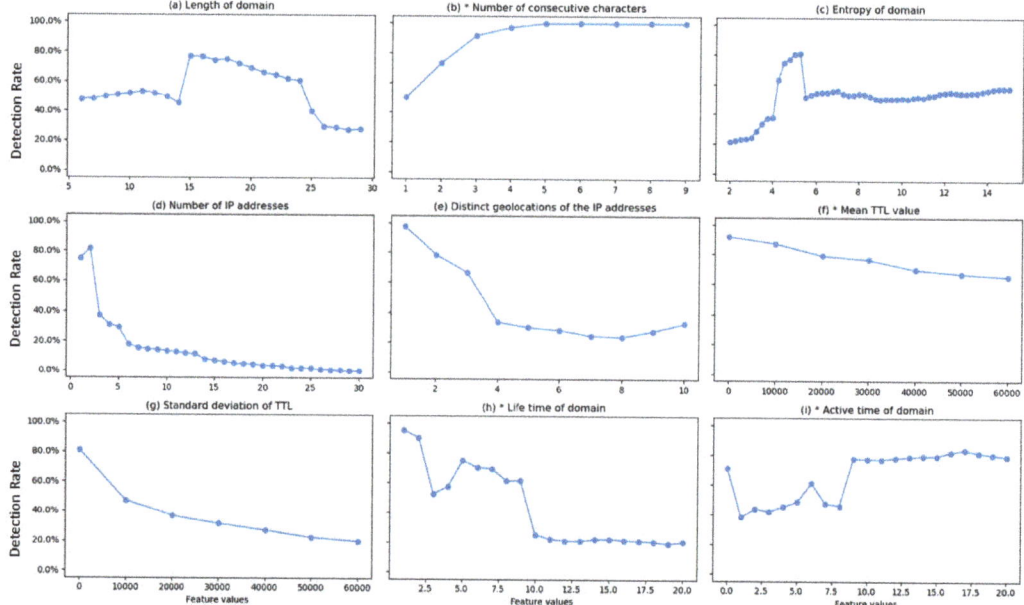

Figure 3. Base feature manipulation graphs (*—robust features).

The well-known features were divided into three groups: robust features, robust features that seemed non-robust (defined as semi-robust), and non-robust features. Next, it it is shown how an attacker can manipulate the classifier for each feature and define its robustness:

1. **"Length of domain"**: an adversary can easily purchase a short or long domain to result in a benign classification for a malicious domain; hence, this feature was classified as *non-robust*.
2. **"Number of consecutive characters"**: as depicted in Figure 3, manipulating the "Number of consecutive characters" feature can significantly lower the prediction percentage (e.g., move from three consecutive characters to one or two). Still, as depicted in Table 2, on average, there were 1.46 consecutive characters in malicious

domains (with a low standard deviation). Therefore, as this feature's minimal value is 1, it is considered to be a *robust feature*.

3. **"Entropy of the domain"**: in order to manipulate the "Entropy of the domain" feature as a benign domain entropy, the adversary can create a domain name with an entropy of less than 4. For example, the domain "ddcd.cc" is available for purchase. The entropy for this domain is 1.44. This value falls precisely in the entropy area of the benign domains defined by the trained model. This example breaks the model and causes a malicious domain to look like a benign URL. Hence, this feature was classified as *non-robust*.

4. **"Number of IP addresses"**: note that an adversary can add many A records to the DNS zone file of its domain to imitate a benign domain. Thus, to manipulate the number of IP addresses, an intelligent adversary only needs to have several different IP addresses and add them to the zone file. This fact causes this feature to be classified as *non-robust*.

5. **"Distinct Geo-locations of the IP addresses"**: in order to be able to circumvent the model with the "Distinct Geolocations of the IP addresses" feature, the adversary needs to use several IP addresses from different geo-locations. Suppose the adversary can determine how many different countries are sufficient to mimic the number of distinct countries of benign domains. In that case, he will be able to append this number of IP addresses (a different IP address from each geo-location) in the DNS zone file. Moreover, because this feature counts the number of the countries, the attacker can choose a set of countries to meet the desired number. Thus, this feature was also classified as *non-robust* (this assumption gave us the motivation for one of our novel features which is based on the rank of the countries and not only the number of the countries).

6. **"Mean TTL value" and "Standard deviation of the TTL"**: there is a clear correlation between the "Mean TTL value" and the "Standard deviation of the TTL" features since the value manipulated by the adversary is the TTL itself. Thus, it makes no difference if the adversary cannot manipulate the "Mean TTL value" feature if the model uses both. In order to robustify the model, it is better to use the "Mean TTL value" feature without the "Standard deviation of the TTL". Solely in terms of the "Mean TTL value" feature, Figure 3 shows that manipulation will not result in a false classification since the prediction percentage does not drop dramatically, even when this feature is drastically manipulated. Therefore, this feature ("Mean TTL value") is considered to be *robust*.

    An adversary can set the DNS TTL values to [0,120,000] (according to the RFC 2181 [71] the TTL value range is from 0 to $2^{31} - 1$). Figure 3 shows that even manipulating the value of this feature to 60,000 will deceive the model and cause a malicious domain to be wrongly classified as a benign URL. Therefore, the "Standard deviation of the TTL" is considered a *non-robust* feature.

7. **"Lifetime of domain"**: As for the lifetime of domains, based on Shi et al. [23], we know that a benign domain's lifetime is typically much longer than a malicious domain's lifetime. In order to deceive the model by manipulating the "Lifetime of domain" feature, the adversary must buy an old domain that is available on the market. Even though it is possible to buy an appropriate domain, it is expensive (if feasible). Hence, we considered this to be a *robust* feature.

8. **"Active time of domain"**: Similar to the previous feature, in order to overcome the "Active time of domain" feature, an adversary has to find a domain with a particular active time, which is much more tricky. It is complex, expensive, and perhaps unfeasible. Therefore we considered it to be a *robust* feature.

Based on the analysis above, the *robust* features presented in Table 2 were selected, and the *non-robust* ones were dropped. Using this subset, the model was trained and achieved an accuracy of 95.71% with an F1-score of 88.78%, compared to an accuracy of 97.2% and an F1-score of 90.23% when using all the features (i.e., including the robust ones). Therefore, we extended our analysis and searched for new features that would meet the robustness requirements to build a robust model with a higher F1-score.

3.5. Novel Features

We aim to validate that manipulating the features in order to result in the misclassification of malicious instances will require a disproportionate effort that will deter the attacker from doing so. The four novel features were designed according to this paradigm based on two communication information properties, passive DNS changes, and the expiration time of the SSL certificate. For each IP, we used *Urlscan* [68] to extract the geo-location, which in turn was appended to a communication country list. The communication Autonomous System Numbers (ASNs) is a list of ASNs, extracted using *Urlscan*, each IP address, and appended the ASNs list. Benign-malicious ratio tables for communication countries, and communication ASNs (Figures 4 and 5) were created using the URL dataset and the *Urlscan* service. The ratio tables were calculated for each element E (country—for the communication countries ratio table; ASN—for the communication ASNs ratio table). Each table represents the probability that a URL associated with a country (ASN) is malicious. In order to extract the probabilities, the number of malicious URLs associated with E was divided by the total URLs associated with E. Initially, due to the heterogeneity of the dataset (i.e., there exist some elements that appear only a few times), the ratio tables appeared to be biased. To overcome this challenge, an initial threshold was set as an insertion criterion which is later detailed in Algorithm 1.

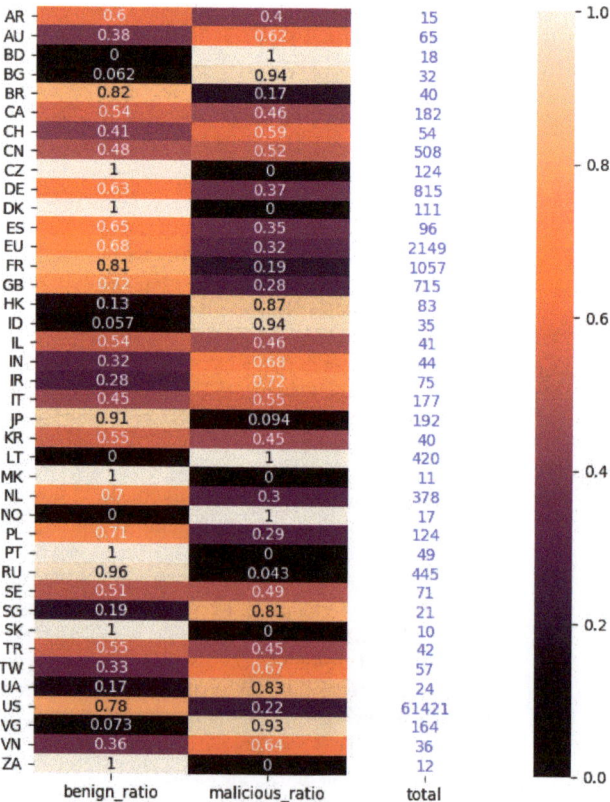

Figure 4. Communication countries ratio.

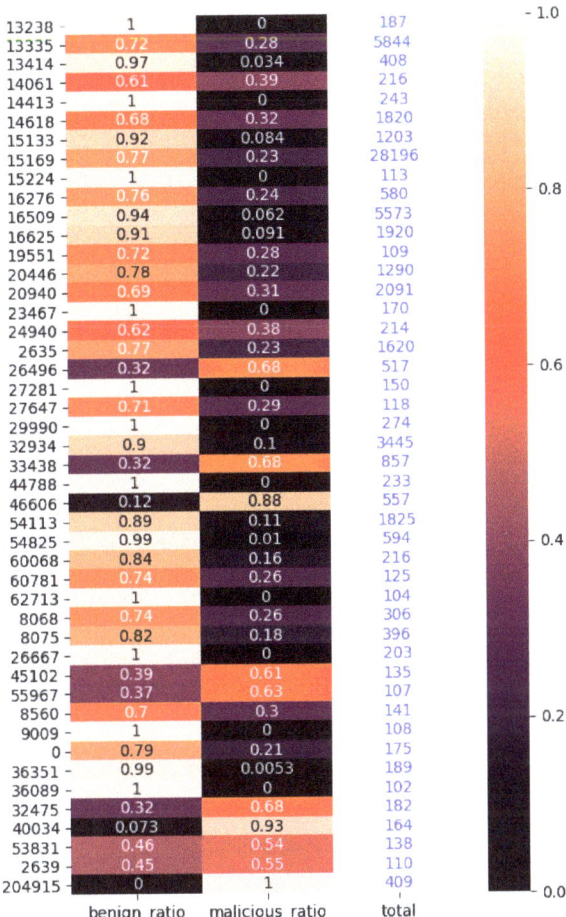

Figure 5. Communication ASNs ratio.

The following is a detailed summary of the novel features:

- **Communication Countries Rank (CCR)**: This feature looks at the communication countries with respect to the communication IPs, and uses the countries ratio table to rank a specific URL. The motivation is to gain a broader perspective.
- **Communication ASNs Rank (CAR)**: Similarly, this feature analyzes the communication ASNs with respect to the communication IPs, and uses the ASNs ratio table to rank a specific URL. While there is some correlation between the ASNs and the countries, the second feature examines each Autonomous System (AS) within each country to gain a broader perspective.
- **Number of passive DNS changes**: When inspecting the passive DNS records, benign domains emerged as having much more significant DNS changes that the sensors (of the company that collects the DNS records) could identify, unlike malicious domains (i.e., 26.4 vs. 8.01, as reported in Table 3). The number of DNS record changes was counted for the "Number of passive DNS changes", which is somewhat similar to other features described in other works [11,25]. Nonetheless, these features require much more elaborated information, which is not publicly available. On the other hand,

this feature can be extracted from passive DNS records obtained from VirusTotal, which are scarce (in terms of record types).
- **Expiration time of SSL certificate**: When installing an SSL certificate, a Certificate Authority (CA) conducts a validation process. Depending on the type of certificate, the CA verifies the organization's identity before issuing the certificate. When analyzing our data, it was noted that most malicious domains do not use valid SSL certificates and those that only use one for a short period. Therefore, this feature was engineered in order to represent the time the SSL certificate remains valid. The "Expiration time of SSL certificate", in contrast to the binary feature version used by Ranganayakulu et al. [69], extends the scope and represents both the existence of an SSL certificate and the remaining time until the SSL certificate expires.

Algorithm 1 Communication Rank

Input: URL, Threshold, Type
Output: Rank (CCR or CAR)

  if Type = Countries **then**
    ItemsList = communication countries list of the URL
  else
    ItemsList = ASNs list of the URL
  end if
  $Rank = 0$
  for Item in ItemsList **do**
    $Ratio = 0.75$ {Init value}
    $Total\_norm = 1$ {Init value}
    if $TotalOccurrences(Item) >= Threshold$ **then**
      $Total\_norm = Normalize(Item)$
      $Ratio = BenignRatio(Item)$
    end if
    $Rank\mathrel{+}= (\log_{0.5}(Ratio + \epsilon))/Total\_norm$
  end for

Table 3. Novel features and statistical properties.

Feature	Benign Mean (std)	Malicious Mean (std)
Communication Countries Rank (CCR)	31.31 (91.16)	59.40 (215.15)
Communication ASNs Rank (CAR)	935.59 (12,258.99)	12,979.38 (46,384.86)
Number of passive DNS changes	26.40 (111.99)	8.01 (16.63)
Expiration time of SSL certificate	1.547×10^7 (2.304×10^7)	4.365×10^6 (1.545×10^7)

Algorithm 1 receives a URL as an input and returns its communication country rate or the ASN communication rate (based on the type of the input in the algorithm). For each item (i.e., country or ASN), first the algorithm initializes the value of the ratio variable to 0.75 (according to [65], 25% of all URLs in 2020 were malicious, suspicious, or moderately risky). It then normalizes an item's total occurrences (Total_norm) to be 1. Next, in Step 9, if an item's total number of occurrences is $\geq$ to the threshold, the algorithm replaces the ratio. It normalizes occurrences to the correct values according to the ratio tables given in Figures 4 and 5. Finally, the algorithm sums the rank with a log base of 0.5 of the ratio (ϵ is a very small value that was added for the special case where $Ratio = 0$) and divides this value by the normalized total occurrences.

Figure 6 depicts the detection rate as a function of the novel features' values for each feature in Table 3. This evaluation proves that manipulating our novel features does not affect the robust model (i.e., the detection rate remains steady). The negative correlation between "Expiration time of SSL certificate" feature and the detection rate may raise concern. Nevertheless, it is noteworthy that the average value for malicious domains is three times higher than the benign ones. While, theoretically, the adversary can lower this value, the implications of such an action mean acquiring (or attaining for free) an SSL certificate. Since there is a validation process involved in the acquisition of an SSL certificate, doing so will cause the adversary to lose its anonymity and disclose its identity.

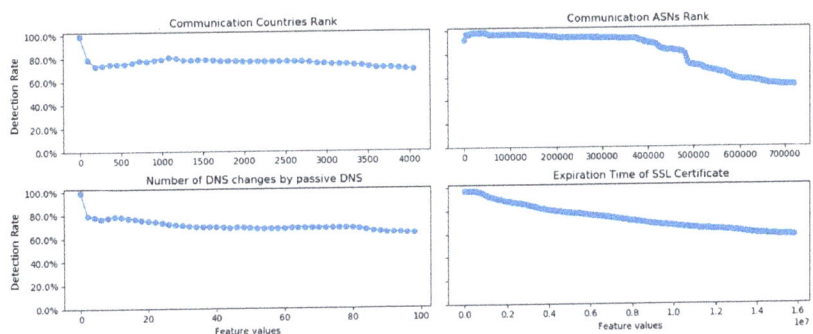

Figure 6. Novel robust feature manipulation graphs.

4. Empirical Analysis and Evaluation

This section describes the testbed used to evaluate models based on the types of features (both robust and not). General settings are provided for each of the models (e.g., the division of the data into training and test sets), as well as the parameters used to configure each of the models, and the efficiency of each model. (our code is publicly available at https://github.com/nitayhas/robust-malicious-url-detection; accessed on 20 March 2022).

4.1. Experimental Design

In addition to intelligently choosing the model parameters, one should verify that the data used for the learning phase accurately represent the domain malware's real-world distribution. Hence, the dataset was constructed such that 75% were benign domains, and the remaining 25% were malicious domains (~5000 benign URLs and ~1350 malicious domains, respectively) [65].

There are many ways to define the efficiency of a model. A broad set of metrics was extracted to account for most of them, including accuracy, recall, F1-score, and training time. Note that for each model, the dataset was split into train and test sets where 75% of the data (both benign and malicious) were assigned to the train set, and the remaining domains were assigned to the test set. Note that the entire dataset included 75% benign samples. Later, when we trained a model, we used 75% of the dataset for the training process and 25% for the evaluation (i.e., test set).

The evaluation measured the efficiency of the different models while varying the robustness of the features included in the model. Specifically, four classical models (i.e., Logistic Regression, SVM, ELM, and ANN) were trained using the following feature sets:

- Base (*B*)—the set of commonly used features in previous works (see Table 2 for more details).
- Base Robust (*BR*)—the subset of robust base features (marked with a * in Figure 3).
- "TCP" (*TCP*)—the four novel features: Time of SSL certificate, Communication ranks (CCR and CAR) and PassiveDNS changes (see Table 3).
- Base Robust + "TCP" (*BRTCP*)—the combination (union) of *BR* and *TCP*, the robust subset of all features.
- Base + "TCP" (*BTCP*)—the union of *B* and *TCP*.

4.2. Experimental Results

Four commonly used classification models were considered: Logistic Regression (LR), Support Vector Machines (SVM), Extreme Learning Machine (ELM), and Artificial Neural Networks (ANN). All the models were trained and evaluated on a Dell XPS 8920 computer, Windows 10 64Bit OS with 3.60GHz Intel Core i7-7700 CPU, 16GB of RAM, and NVIDIA GeForce GTX 1060 6GB. In the following paragraphs, we describe the experimental results for each model, followed by a short discussion of the findings and their implications.

4.2.1. Logistic Regression

As a baseline for the evaluation process, and before using the nonlinear models, the LR classification model was used. The LR model with the five feature sets (Base, Robust Base, TCP, BRTCP, BTCP) was trained. Table 4 shows that the different feature sets resulted in similar accuracy rates. However, the accuracy rate measures how well the model predicts (i.e., TP + TN) with respect to all the predictions (i.e., TP + TN + FP + FN). Thus, given the unbalanced dataset (75% of the dataset are benign and 25% are malicious domains), ~90% accuracy is not necessarily a sufficient result for malware detection. For example, the *TCP* feature set has high accuracy and, at the same time, a very poor F1-Score, due to the high precision rate and poor recall rate. As the recall is low for all features sets, the accuracy rate is not a good measure in this domain. Consequently, we focused on the F1-score measure, the harmonic mean of the precision, and the recall measures.

4.2.2. Support Vector Machine (SVM)

Compared to the results of the LR model (Table 4), the results of the SVM model (Table 5) show a significant improvement in the recall and F1-score measures; e.g., for *Base*, the recall and the F1-score measures were both above 90%. It should be noted that the model that trained on the *Base* feature set resulted in a higher recall (and F1-score) compared to the one trained on the *Robust Base* feature set. Nonetheless, it is also noteworthy that the *Robust Base* feature set is robust to adversarial manipulation and uses less than half of the features provided in the training phase with the *Base* feature set. This discussion also applies to the *BRTCP* and *BTCP* feature sets. Another advantage of including the novel features is that models converge much faster. The results are based on the analysis of a non-manipulated dataset. As stated above, the *Base* feature set includes some non-robust features. Hence, an intelligent adversary can manipulate the values of these features, resulting in the wrong classification of malicious instances (to the extreme of 0% recall). However, an intelligent adversary will need to invest much more effort with a model that was trained using the *Robust Base* or *TCP* features since each was specifically chosen to avoid such manipulations. In order to find models that were also efficient on the non-manipulated dataset, the two sophisticated models were examined in the analysis, the ELM model Shi et al. [23] provided and the ANN model.

Table 4. Model performance—logistic Regression.

Feature Set	Accuracy	Recall	F1-Score
Base	89.99%	38.82%	53.21%
Robust Base	88.33%	38.87%	49.42%
TCP	86.20%	8.30%	14.99%
BRTCP	88.82%	52.46%	65.57%
BTCP	92.86%	64.14%	72.48%

Table 5. Model performance—SVM.

Feature Set	Accuracy	Recall	F1-Score
Base	96.49%	91.20%	91.36%
Robust Base	90.14%	56.51%	69.93%
TCP	83.10%	60.21%	54.21%
BRTCP	96.78%	91.37%	92.02%
BTCP	97.95%	90.73%	92.83%

4.2.3. ELM

The architecture of the ELM is the one previously used [23]: one input layer, one hidden layer, and one output layer. Activation function: first layer—ReLU; hidden layer—Sigmoid. Overall, the ELM model resulted (see Table 6) in a high accuracy and higher recall rates compared to Table 4, for any feature set. When compared to the SVM models, the *Base* model resulted in a lower recall rate (though a higher F1-score was achieved with the ELM model). On the other hand, the *Robust Base* resulted in a higher recall rate with the ELM model compared to the SVM model. Even though the *Robust Base* feature set had a low dimensional space, the three rates (i.e., accuracy, recall, and F1-score) were higher than those of the *Base* feature set. Using the sets that include the novel features increased these metrics while improving the robustness of the model at the same time.

Table 6. Model performance—ELM.

Feature Set	Accuracy	Recall	F1-Score
Base	98.17%	88.81%	92.92%
Robust Base	98.83%	92.24%	95.81%
TCP	98.88%	94.64%	96.84%
BRTCP	98.86%	95.82%	97.07%
BTCP	98.19%	93.09%	95.34%

4.2.4. ANN

The architecture of the neural network was as follows: one input layer, three hidden layers, and one output layer. Activation function: first layer—ReLU; first hidden layer—RELU; second hidden layer—LeakyReLU; third hidden layer—Sigmoid. Batch size: -150, with a learning rate of 0.01; solver: Adam with $\beta_1 = 0.9$ and $\beta_2 = 0.999$. Similar to the ELM results, the ANN results (Table 7) show high performance with all feature sets. For the "basic" feature sets (i.e., *Base* and *Robust Base*), the ELM models resulted in higher recall and F1-score. Nevertheless, the main focus was in the *BTCP* feature set and, more specifically, on the *BRTCP* variant, where the ANN models resulted in a higher recall and F1-score.

Table 7. Model performance—ANN.

Feature Set	Accuracy	Recall	F1-Score
Base	97.20%	88.03%	90.23%
Robust Base	95.71%	83.63%	88.78%
TCP	98.03%	96.83%	95.24%
BRTCP	99.36%	98.77%	98.42%
BTCP	99.82%	99.47%	99.56%

Our analysis concludes with Figure 7, which depicts the F1-scores of the feature sets for all the models.

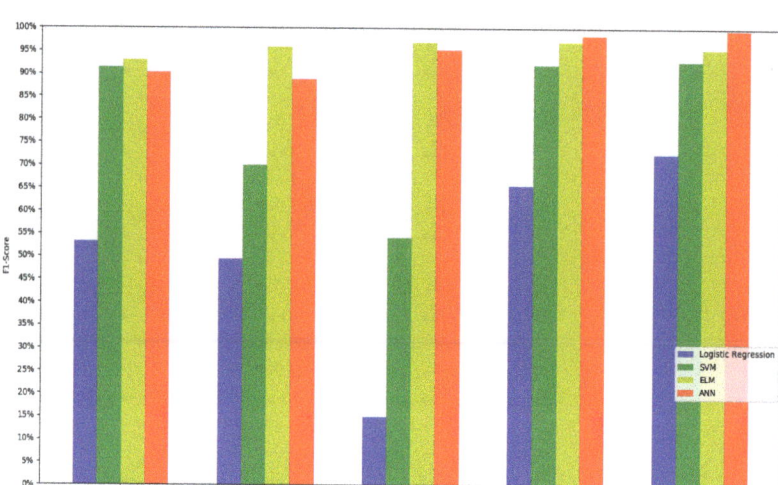

Figure 7. The F1-Score by feature sets and models.

All the results provided in this article are based on clean data (i.e., with no adversarial manipulation). Naturally, given an adversarial environment where the attacker can manipulate the values of the features, models which are based on the *Robust Base* or *TCP* feature sets will dominate models that are trained using the *Base* dataset. Thus, by showing that the *Robust Base* feature set does not dramatically decrease the performance of the classifier using clean data and that adding the novel feature improves the model's performance as well as its robustness, it leads to the conclusion that malicious domain classifiers should use this feature set for robust malicious domain detection.

5. Conclusions

Numerous attempts have been made to tackle the problem of identifying malicious domains. However, many fail to successfully classify malware in realistic environments where an adversary can manipulate the features in order to make the model wrongly classify malicious domains. Specifically, this research used a large empirical dataset that was crawled over a significant amount of time at different hours of the day, and captures traffic generated in various countries and continents. Based on this rich dataset, this paper tackled the case where an attacker has access to the model (i.e., a set of features or output for a given input) and tampers with the domain properties. This tampering has a catastrophic effect on the model's efficiency. As a countermeasure, we propose two feature-based mechanisms: (I) an intelligent feature selection procedure that is robust to adversarial manipulation. We evaluated the robustness of each feature, taking into account both the hardness of changing its value and the effects of such manipulations on the classifier; (II) a novel and robust feature engineering process. Based on the domains' properties, we engineered a set of four features which are robust to adversarial manipulation and, together with the common features, improve the classifiers' performance.

We empirically evaluated the common feature set as well as our novel ones using a large dataset, which took into account both malicious and benign models. To extend our evaluation, we picked a broad set of well-known machine learning algorithms. Our evaluation showed that models trained using the robust features are more precise in terms of manipulated data while maintaining good results on clean data as well.

From the industry perspective, our solution can be easily adopted either in any organization's DPI center solution, Firewall, Load Balancer, behavioral analytic or as a client agent that will query a cloud-service dataset. Further research is needed to create models that classify malicious domains into malicious attack types, either in terms of a more extensive list of models or by sampling data in a stratified way, validating the amount of data for any feature value. Another promising direction would be to cluster a set of malicious domains into one cyber campaign.

Author Contributions: Conceptualization, C.H., N.H. and A.D.; Data Curation, N.H.; Formal Analysis, C.H. and N.H.; Funding Acquisition, C.H.; Investigation, N.H.; Methodology, C.H. and N.H.; Project Administration, C.H.; Software, N.H.; Supervision, C.H. and A.D.; Validation, C.H. and A.D.; Visualization, N.H. and A.D.; Writing—Original Draft Preparation, C.H., N.H. and A.D.; Writing—Review and Editing, C.H. and A.D. All authors have read and agreed to the published version of the manuscript.

Funding: This research was funded by Ariel University and Holon Institute of Technology (RA1900000614).

Institutional Review Board Statement: Not applicable.

Informed Consent Statement: Not applicable.

Data Availability Statement: Publicly available datasets were analyzed in this study. This data can be found here: https://github.com/nitayhas/robust-malicious-url-detection; accessed on 20 March 2022.

Acknowledgments: This work was supported by the Ariel Cyber Innovation Center in conjunction with the Israel National Cyber directorate of the Prime Minister's Office. The authors express special thanks to Nissim Harel of Holon Institute of Technology and Asaf Nadler of Akamai Technologies for the fruitful discussions and their insights.

Conflicts of Interest: The authors declare no conflict of interest.

References

1. Vincent, N.E.; Pinsker, R. IT risk management: interrelationships based on strategy implementation. *Int. J. Account. Inf. Manag.* **2020**, *28*, 553–575. [CrossRef]
2. Blum, A.; Wardman, B.; Solorio, T.; Warner, G. Lexical feature based phishing URL detection using online learning. In Proceedings of the Workshop on Artificial Intelligence and Security, Krakow, Poland, 15–18 February 2010; pp. 54–60.
3. Khonji, M.; Iraqi, Y.; Jones, A. Phishing detection: A literature survey. *IEEE Commun. Surv. Tutor.* **2013**, *15*, 2091–2121. [CrossRef]
4. Le, A.; Markopoulou, A.; Faloutsos, M. Phishdef: Url Names Say It All. In Proceedings of the 2011 IEEE INFOCOM, Shanghai, China, 10–15 April 2011; pp. 191–195.
5. Prakash, P.; Kumar, M.; Kompella, R.R.; Gupta, M. Phishnet: Predictive Blacklisting to Detect Phishing Attacks. In Proceedings of the 2010 IEEE INFOCOM, San Diego, CA, USA, 14–19 March 2010; pp. 1–5.
6. Sheng, S.; Wardman, B.; Warner, G.; Cranor, L.F.; Hong, J.; Zhang, C. An empirical analysis of phishing blacklists. In Proceedings of the Conference on Email and Anti-Spam, Mountain View, CA, USA, 16–17 July 2009.
7. Sandell, N.; Varaiya, P.; Athans, M.; Safonov, M. Survey of decentralized control methods for large scale systems. *IEEE Trans. Autom. Control* **1978**, *23*, 108–128. [CrossRef]
8. Canali, D.; Cova, M.; Vigna, G.; Kruegel, C. Prophiler: A fast filter for the large-scale detection of malicious web pages. In Proceedings of the International Conference on World Wide Web, Hyderabad, India, 28 March–1 April 2011; pp. 197–206.
9. Hason, N.; Dvir, A.; Hajaj, C. Robust Malicious Domain Detection. In *Cyber Security Cryptography and Machine Learning*; Dolev, S., Kolesnikov, V., Lodha, S., Weiss, G., Eds.; Springer: Cham, Switzerland, 2020; pp. 45–61.
10. Ahmed, M.; Khan, A.; Saleem, O.; Haris, M. A Fault Tolerant Approach for Malicious URL Filtering. In Proceedings of the International Symposium on Networks, Computers and Communications, Rome, Italy, 19–21 June 2018; pp. 1–6.
11. Antonakakis, M.; Perdisci, R.; Dagon, D.; Lee, W.; Feamster, N. Building a Dynamic Reputation System for DNS. In Proceedings of the 19th USENIX conference on Security, Washington, DC, USA, 11–13 August 2010; pp. 273–290.
12. Berger, H.; Dvir, A.Z.; Geva, M. A wrinkle in time: A case study in DNS poisoning. *Int. J. Inf. Secur.* **2021**, *20*, 313–329. [CrossRef]
13. Bilge, L.; Sen, S.; Balzarotti, D.; Kirda, E.; Kruegel, C. Exposure: A Passive DNS Analysis Service to Detect and Report Malicious Domains. *Trans. Inf. Syst. Secur.* **2014**, *16*, 1–28. [CrossRef]
14. Caglayan, A.; Toothaker, M.; Drapeau, D.; Burke, D.; Eaton, G. Real-time detection of fast flux service networks. In Proceedings of the Conference For Homeland Security, Cybersecurity Applications and Technology, Washington, DC, USA, 3–4 March 2009; pp. 285–292.
15. Choi, H.; Zhu, B.B.; Lee, H. Detecting Malicious Web Links and Identifying Their Attack Types. *WebApps* **2011**, *11*, 218.
16. Dolberg, L.; François, J.; Engel, T. Efficient Multidimensional Aggregation for Large Scale Monitoring. In Proceedings of the 26th Large Installation System Administration Conference, Washington, DC, USA, 3–8 November 2013; pp. 163–180.
17. Harel, N.; Dvir, A.; Dubin, R.; Barkan, R.; Shalala, R.; Hadar, O. MiSAL-A minimal quality representation switch logic for adaptive streaming. *Multimed. Tools Appl.* **2019**, *78*, 1–26.

18. Hu, Z.; Chiong, R.; Pranata, I.; Susilo, W.; Bao, Y. Identifying malicious web domains using machine learning techniques with online credibility and performance data. In Proceedings of the Congress on Evolutionary Computation (CEC), Vancouver, BC, Canada, 24–29 July 2016; pp. 5186–5194.
19. Huang, G.B.; Zhu, Q.Y.; Siew, C.K. Extreme learning machine: Theory and applications. *Neurocomputing* **2006**, *70*, 489–501. [CrossRef]
20. Nelms, T.; Perdisci, R.; Ahamad, M. ExecScent: Mining for New C&C Domains in Live Networks with Adaptive Control Protocol Templates. In Proceedings of the 22nd USENIX Security Symposium, Washington, DC, USA, 14–16 August 2013; pp. 589–604.
21. Peng, T.; Harris, I.; Sawa, Y. Detecting phishing attacks using natural language processing and machine learning. In Proceedings of the International Conference on Semantic Computing, Laguna Hills, CA, USA, 31 January–2 Februay 2018; pp. 300–301.
22. Rahbarinia, B.; Perdisci, R.; Antonakakis, M. Efficient and accurate behavior-based tracking of malware-control domains in large ISP networks. *ACM Trans. Priv. Secur.* **2016**, *19*, 4. [CrossRef]
23. Shi, Y.; Chen, G.; Li, J. Malicious Domain Name Detection Based on Extreme Machine Learning. *Neural Process. Lett.* **2017**, *48*, 1–11. [CrossRef]
24. Sun, X.; Tong, M.; Yang, J.; Xinran, L.; Heng, L. HinDom: A Robust Malicious Domain Detection System based on Heterogeneous Information Network with Transductive Classification. In Proceedings of the International Symposium on Research in Attacks, Intrusions and Defenses, Beijing, China, 23–25 September 2019; pp. 399–412.
25. Torabi, S.; Boukhtouta, A.; Assi, C.; Debbabi, M. Detecting Internet Abuse by Analyzing Passive DNS Traffic: A Survey of Implemented Systems. *Commun. Surv. Tutor.* **2018**, *20*, 3389–3415. [CrossRef]
26. Yadav, S.; Reddy, A.K.K.; Reddy, A.L.N.; Ranjan, S. Detecting Algorithmically Generated Domain-flux Attacks with DNS Traffic Analysis. *Trans. Netw.* **2012**, *20*, 1663–1677. [CrossRef]
27. Antonakakis, M.; Perdisci, R.; Lee, W.; Vasiloglou, N.; Dagon, D. Detecting Malware Domains at the Upper DNS Hierarchy. In Proceedings of the 20th USENIX Security Symposium, San Francisco, CA, USA, 8–12 August 2011; Volume 11, pp. 1–16.
28. Perdisci, R.; Corona, I.; Giacinto, G. Early detection of malicious flux networks via large-scale passive DNS traffic analysis. *IEEE Trans. Dependable Secur. Comput.* **2012**, *9*, 714–726. [CrossRef]
29. Papernot, N.; McDaniel, P.; Wu, X.; Jha, S. Distillation as a Defense to Adversarial Perturbations against Deep Neural Networks. In Proceedings of the IEEE Symposium on Security and Privacy, San Jose, CA, USA, 22–26 May 2016.
30. Tong, L.; Li, B.; Hajaj, C.; Xiao, C.; Zhang, N.; Vorobeychik, Y. Improving Robustness of ML Classifiers against Realizable Evasion Attacks Using Conserved Features. In Proceedings of the 28th USENIX Security Symposium, Santa Clara, CA, USA, 14–16 August 2019.
31. Jung, J.; Sit, E. An empirical study of spam traffic and the use of DNS black lists. In Proceedings of the SIGCOMM Conference on Internet Measurement, Taormina Sicily, Italy, 25–27 October 2004; pp. 370–375.
32. Mishsky, I.; Gal-Oz, N.; Gudes, E. A topology based flow model for computing domain reputation. In Proceedings of the IFIP Annual Conference on Data and Applications Security and Privacy, Fairfax, VA, USA, 13–15 July 2015; pp. 277–292.
33. Othman, H.; Gudes, E.; Gal-Oz, N. Advanced Flow Models for Computing the Reputation of Internet Domains. In Proceedings of the IFIP International Conference on Trust Management, Toronto, ON, Canada, 9–13 July 2017; pp. 119–134.
34. Dey, S.; Jain, E.; Das, A. Machine Learning Features for Malicious URL Filtering—The Survey. *arXiv* **2019**, arXiv:2019.0621.
35. Sahoo, D.; Liu, C.; Hoi, S.C. Malicious URL detection using machine learning: A survey. *arXiv* **2017**, arXiv:1701.07179.
36. Shahzad, H.; Sattar, A.R.; Skandaraniyam, J. From Real Malicious Domains to Possible False Positives in DGA Domain Detection. In Proceedings of the 2021 IEEE 13th International Conference on Computer Research and Development (ICCRD), Beijing, China, 5–7 January 2021; pp. 6–10. [CrossRef]
37. Zhang, S.; Zhou, Z.; Li, D.; Zhong, Y.; Liu, Q.; Yang, W.; Li, S. Attributed Heterogeneous Graph Neural Network for Malicious Domain Detection. In Proceedings of the 2021 IEEE 24th International Conference on Computer Supported Cooperative Work in Design (CSCWD), Dalian, China, 5–7 May 2021; pp. 397–403. [CrossRef]
38. Iwahana, K.; Takemura, T.; Cheng, J.C.; Ashizawa, N.; Umeda, N.; Sato, K.; Kawakami, R.; Shimizu, R.; Chinen, Y.; Yanai, N. MADMAX: Browser-Based Malicious Domain Detection Through Extreme Learning Machine. *IEEE Access* **2021**, *9*, 78293–78314. [CrossRef]
39. Kumi, S.; Lim, C.; Lee, S.G. Malicious url detection based on associative classification. *Entropy* **2021**, *23*, 182. [CrossRef]
40. Janet, B.; Kumar, R.J.A. Malicious URL Detection: A Comparative Study. In Proceedings of the 2021 International Conference on Artificial Intelligence and Smart Systems (ICAIS), Coimbatore, India, 25–27 March 2021; pp. 1147–1151.
41. Srinivasan, S.; Vinayakumar, R.; Arunachalam, A.; Alazab, M.; Soman, K. DURLD: Malicious URL detection using deep learning-based character level representations. In *Malware Analysis Using Artificial Intelligence and Deep Learning*; Springer: Berlin/Heidelberg, Germany, 2021; pp. 535–554.
42. Cyprienna, R.A.; Zo Lalaina Yannick, R.; Randria, I.; Raft, R.N. URL Classification based on Active Learning Approach. In Proceedings of the 2021 3rd International Cyber Resilience Conference (CRC), Langkawi Island, Malaysia, 29–31 January 2021; pp. 1–6. [CrossRef]
43. Goodfellow, I.J.; Shlens, J.; Szegedy, C. Explaining and Harnessing Adversarial Examples; In Proceedings of the 3rd International Conference on Learning Representations, San Diego, CA, USA, 7–9 May 2015.
44. Nelson, B.; Barreno, M.; Chi, F.J.; Joseph, A.D.; Rubinstein, B.I.; Saini, U.; Sutton, C.A.; Tygar, J.D.; Xia, K. Exploiting Machine Learning to Subvert Your Spam Filter. *LEET* **2008**, *8*, 1–9.

45. Fogla, P.; Sharif, M.I.; Perdisci, R.; Kolesnikov, O.M.; Lee, W. Polymorphic Blending Attacks. In Proceedings of the 15th USENIX Security Symposium, Austin, TX, USA, 10–12 August 2006; pp. 241–256.
46. Newsome, J.; Karp, B.; Song, D. Paragraph: Thwarting signature learning by training maliciously. In Proceedings of the International Workshop on Recent Advances in Intrusion Detection, Hamburg, Germany, 20–22 September 2006; pp. 81–105.
47. Rodrigues, R.N.; Ling, L.L.; Govindaraju, V. Robustness of multimodal biometric fusion methods against spoof attacks. *J. Vis. Lang. Comput.* **2009**, *20*, 169–179. [CrossRef]
48. Madry, A.; Makelov, A.; Schmidt, L.; Tsipras, D.; Vladu, A. Towards Deep Learning Models Resistant to Adversarial Attacks. In Proceedings of the Sixth International Conference on Learning Representations, Vancouver, BC, Canada, 30 April–3 May 2018.
49. Raghunathan, A.; Steinhardt, J.; Liang, P. Certified Defenses against Adversarial Examples. In Proceedings of the Sixth International Conference on Learning Representations, Vancouver, BC, Canada, 30 April–3 May 2018.
50. Song, Y.; Kim, T.; Nowozin, S.; Ermon, S.; Kushman, N. Pixeldefend: Leveraging Generative Models to Understand and Defend against Adversarial Examples. In Proceedings of the Sixth International Conference on Learning Representations, Vancouver, BC, Canada, 30 April–3 May 2018.
51. Berger, H.; Hajaj, C.; Mariconti, E.; Dvir, A. Crystal Ball: From Innovative Attacks to Attack Effectiveness Classifier. *IEEE Access* **2022**, *10*, 1317–1333. [CrossRef]
52. Papernot, N.; McDaniel, P.; Goodfellow, I.; Jha, S.; Celik, Z.B.; Swami, A. Practical black-box attacks against machine learning. In Proceedings of the Asia Conference on Computer and Communications Security, Abu Dhabi, United Arab Emirates, 2–6 April 2017; pp. 506–519.
53. Shahpasand, M.; Hamey, L.; Vatsalan, D.; Xue, M. Adversarial Attacks on Mobile Malware Detection. In Proceedings of the International Workshop on Artificial Intelligence for Mobile, Hangzhou, China, 24–24 February 2019; pp. 17–20.
54. Br'uckner, M.; Scheffer, T. Stackelberg games for adversarial prediction problems. In Proceedings of the International Conference on Knowledge Discovery and Data Mining, San Diego, CA, USA, 21–24 August 2011; pp. 547–555.
55. Singh, A.; Lakhotia, A. Game-theoretic design of an information exchange model for detecting packed malware. In Proceedings of the International Conference on Malicious and Unwanted Software, Fajardo, PR, USA, 18–19 October 2011; pp. 1–7.
56. Zolotukhin, M.; H'am'al'ainen, T. Support vector machine integrated with game-theoretic approach and genetic algorithm for the detection and classification of malware. In Proceedings of the Globecom Workshops, Atlanta, GA, USA, 9–13 December 2013; pp. 211–216.
57. Xu, H.; Caramanis, C.; Mannor, S. Robustness and regularization of support vector machines. *J. Mach. Learn. D* **2009**, *10*, 1485–1510.
58. Li, B.; Vorobeychik, Y. Evasion-robust classification on binary domains. *Trans. Knowl. Discov. Data* **2018**, *12*, 50. [CrossRef]
59. Nissim, N.; Moskovitch, R.; BarAd, O.; Rokach, L.; Elovici, Y. ALDROID: Efficient update of Android anti-virus software using designated active learning methods. *Knowl. Inf. Syst.* **2016**, *49*, 795–833. [CrossRef]
60. Chen, X.; Li, C.; Wang, D.; Wen, S.; Zhang, J.; Nepal, S.; Xiang, Y.; Ren, K. Android HIV: A study of repackaging malware for evading machine-learning detection. *IEEE Trans. Inf. Forensics Secur.* **2019**, *15*, 987–1001. [CrossRef]
61. Fidel, G.; Bitton, R.; Katzir, Z.; Shabtai, A. Adversarial robustness via stochastic regularization of neural activation sensitivity. *arXiv* **2020**, arXiv:2009.11349.
62. Alexa. Available online: https://www.alexa.com (accessed on 1 February 2022).
63. PhishTank. Available online: https://www.phishtank.com (accessed on 1 February 2022).
64. ScumWare. Available online: https://www.scumware.org (accessed on 1 February 2022).
65. WEBROOT. Available online: https://mypage.webroot.com/rs/557-FSI-195/images/2020%20Webroot%20Threat%20Report_US_FINAL.pdf (accessed on 1 February 2022).
66. A Study of Whois Privacy and Proxy Service Abuse. Available online: https://gnso.icann.org/sites/default/files/filefield_41831/pp-abuse-study-20sep13-en.pdf (accessed on 1 February 2022).
67. VirusTotal. Available online: https://www.virustotal.com (accessed on 1 February 2022).
68. urlscan.io. Available online: https://www.urlscan.io (accessed on 1 February 2022).
69. Ranganayakulu, D.; Chellappan, C. Detecting malicious URLs in E-mail–An implementation. *AASRI* **2013**, *4*, 125–131. [CrossRef]
70. Xiang, G.; Hong, J.; Rose, C.P.; Cranor, L. Cantina+: A feature-rich machine learning framework for detecting phishing web sites. *Trans. Inf. Syst. Secur.* **2011**, *14*, 21. [CrossRef]
71. Clarifications to the DNS Specification. Available online: https://tools.ietf.org/html/rfc2181 (accessed on 1 February 2022).

Article

Evaluation of Contextual and Game-Based Training for Phishing Detection

Joakim Kävrestad [1,*], Allex Hagberg [2], Marcus Nohlberg [1], Jana Rambusch [1], Robert Roos [2] and Steven Furnell [3]

1. School of Informatics, University of Skövde, 541 28 Skövde, Sweden; marcus.nohlberg@his.se (M.N.); jana.rambusch@his.se (J.R.)
2. Xenolith AB, 541 34 Skövde, Sweden; allex@xenolith.se (A.H.); robert@xenolith.se (R.R.)
3. School of Computer Science, University of Nottingham, Nottingham NG7 2RD, UK; steven.furnell@nottingham.ac.uk
* Correspondence: joakim.kavrestad@his.se

Abstract: Cybersecurity is a pressing matter, and a lot of the responsibility for cybersecurity is put on the individual user. The individual user is expected to engage in secure behavior by selecting good passwords, identifying malicious emails, and more. Typical support for users comes from Information Security Awareness Training (ISAT), which makes the effectiveness of ISAT a key cybersecurity issue. This paper presents an evaluation of how two promising methods for ISAT support users in acheiving secure behavior using a simulated experiment with 41 participants. The methods were game-based training, where users learn by playing a game, and Context-Based Micro-Training (CBMT), where users are presented with short information in a situation where the information is of direct relevance. Participants were asked to identify phishing emails while their behavior was monitored using eye-tracking technique. The research shows that both training methods can support users towards secure behavior and that CBMT does so to a higher degree than game-based training. The research further shows that most participants were susceptible to phishing, even after training, which suggests that training alone is insufficient to make users behave securely. Consequently, future research ideas, where training is combined with other support systems, are proposed.

Keywords: usable security; cybersecurity training; ISAT; SETA; phishing; user awareness; security behavior

Citation: Kävrestad, J.; Hagberg, A.; Nohlberg, M.; Rambusch, J.; Roos, R.; Furnell, S. Evaluation of Contextual and Game-Based Training for Phishing Detection. *Future Internet* 2022, 14, 104. https://doi.org/10.3390/fi14040104

Academic Editors: Leandros Maglaras, Helge Janicke and Mohamed Amine Ferrag

Received: 7 March 2022
Accepted: 22 March 2022
Published: 25 March 2022

Publisher's Note: MDPI stays neutral with regard to jurisdictional claims in published maps and institutional affiliations.

Copyright: © 2022 by the authors. Licensee MDPI, Basel, Switzerland. This article is an open access article distributed under the terms and conditions of the Creative Commons Attribution (CC BY) license (https://creativecommons.org/licenses/by/4.0/).

1. Introduction

The world is continuing a journey towards an increasingly digital state [1]. The use of computers and online services has been a natural component of the lives of most people in developed countries for decades and adoption in developing regions is on the rise [2]. Furthermore, populations that previously demonstrated low adoption rates are now adopting and using digital services at a rapid pace [3,4]. This development is positive. On a national level, Internet adoption has been shown to positively impact financial development [2]. On the individual level, the use of digital services makes it easier for the individual to access information, healthcare, and more, while enabling social contact in situations where meeting physically is challenging or even impossible [5,6].

However, digitalization is not without risk. The move to more digital work, leisure and more also means a move to more digital crime and threats [7]. Digital threats expose users and organizations to risks daily, and the need for cybersecurity to protect against those risks is undeniable. The threat landscape is multi-faceted and includes various types of threats that can be broadly classified as technological or human [8]. Technological threats include, for instance, malware or hacking where the attacker is using technological means to destroy or gain access to devices or services. Human threats involve exploiting user behavior, typically for the same purpose. A common type of human threat is phishing, where an attacker sends an email to the target victim and attempts to persuade the victim into behaving in an insecure way by, for instance, downloading an attachment or clicking a link and then submitting login credentials to some service. Phishing is continuously reported as the most common threat to both organizations and individuals, and therefore the topic of this paper [9–11].

At its core, phishing is when an attacker attempts to trick a user into insecure behavior. Insecure behavior typically includes downloading a malicious attachment, clicking a link or giving up sensitive information in reply to the email [12]. Phishing has traditionally been easy to spot as generic messages which are often poorly formatted with poor spelling and grammar [13]. While that is still true for some of today's phishing campaigns, now many phishing emails are well-written and use various techniques to invoke trust [12]. Furthermore, attackers employ targeted attacks where they tailor emails to a specific recipient, a technique known as spear-phishing [9]. In such an attack, the attacker may steal the email address of a friend or coworker of the target victim and make the email appear to come from that known sender. The attacker may also research the victim and ensure that the content of the malicious email is content that the victim would, given the victim's job position or interest, expect to receive [14].

Techniques used by attackers and techniques used to defend against phishing both include technical and human aspects [15]. An attacker will exploit human behavior to invoke trust and persuade the victim into insecure behavior. As part of the attack, the attacker may also exploit technical weaknesses in the email protocols to pose as a trusted sender or use another technical weakness to take control of the victim's system once the victim opens a malicious attachment [12]. Likewise, several organizations employ technical measures, such as automatic filters, to defend against phishing. However, educating users on detecting phishing emails remains the most commonly suggested defense mechanism. While both technical and human aspects of phishing are important, the primary focus of this paper is on the human side, particularly on user behavior and how it can be understood and improved.

As explained by the knowledge, attitude, and behavior (KAB) model, behavior is influenced by knowledge, and attitude [16]. KAB describes that increased knowledge about an expected behavior will lead to increased awareness and, finally, a change in behavior. This relationship has been evaluated in the security domain and found to hold [17].

Information Security Awareness Training (ISAT) is commonly suggested as the way to improve user awareness [18–20]. There are several different ways to train users presented in the literature. These include providing lectures, text-based warnings, video instructions sent out via email at regular intervals, instructive games and training automatically provided to users in risky situations [21–25]. There are, however, several publications suggesting that many training efforts fail to support users towards secure behavior to a high enough degree [26,27]. Suggested reasons include that it is hard to make users participate in on-demand training, that acquired knowledge is not retained for long enough, and that knowledge does not necessarily translate to correct behavior [20,28]. Some research even suggests that training methods are not empirically evaluated to a high enough extent [29,30].

This paper seeks to evaluate the effectiveness of two promising methods for ISAT; game-based training and Context-Based Micro-Training (CBMT). Game-based training means that users are presented with an educative game and is argued to increase user participation rates and provide a more realistic training environment compared to lectures, videos, or similar [31]. CBMT means that users are presented with condensed information in situations where the training is of direct relevance. In the context of phishing, a user will receive training when opening a mailbox. CBMT is argued to increase users' awareness and has been evaluated in the context of password security with positive results [32]. The research question addressed in this paper is:

> To what extent can the two methods, game-based training and CBMT, support users to accurately differentiate between phishing and legitimate email?

The research was carried out as a simulated experiment with 41 participants. The participants were asked to identify phishing emails while their behavior was monitored using an eye-tracking technique. The results show that both training methods can support users towards secure behavior and that CBMT does so to a higher degree than game-based training, which makes the first contribution of this research. The research further shows that most participants were susceptible to phishing, even after training which suggests that training alone is not enough to make users behave securely. The upcoming section will elaborate on ISAT and justify the selection of CBMT and game-based training as the focus of this research. The rest of this paper will, in turn, present the research methodology results, and discuss those results and their limitations.

2. Information Security Awareness Training

ISAT has been discussed in the scientific literature for several decades, and the importance of providing ISAT as a means of improving user behavior is widely acknowledged [33–35]. ISAT intends to increase user knowledge and awareness through training. There are many and diverse, options for ISAT, and recent publications [35–37] categorize ISAT methods differently. In general terms, ISAT methods can be described as seen in Table 1. Table 1 is based on the classifications by [36,37].

Table 1. Overview of ISAT methods.

Method	Description
Classroom training	Typically provided on-site as a lecture attended as a specific point in time.
Broadcasted online training	Typically, brief training delivered as broadcast to large user groups using e-mail or social networks.
E-learning	ISAT typically delivered using an online platform that is accessible to users on-demand.
Simulated or contextual training	Training delivered to users during a real or simulated event.
Gamified training	Gamified training is described as using gamification to develop ISAT material.

While ISAT has been long discussed in scientific literature and used in practice, several publications suggest that many ISAT methods fail to adequately support users towards secure behavior [26,27]. This notion is emphasized by the continuous reports of incidents where human behavior is a key component [38,39]. Three core reasons for why ISAT does not always provide its intended effect can be found in recent research:

- Knowledge acquired during training deteriorates over time [21].
- It is challenging to get users to participate in training delivered on-demand [28].
- Users are provided with knowledge, but not motivated to act in accordance to that knowledge [20].

The ISAT methods included in this research are game-based training and Context-Based Micro-Training (CBMT). Gamified training means that game concepts are applied to ISAT, with the intent to better motivate users to actively participate [28]. It is considered in this research since it is argued to better motivate and engage users when compared to other ISAT alternatives. There are several examples of gamified ISAT. The landscape includes multi-player competitive games, story-based single-player games, board games, role-playing games, quizzes, and more [28,40].

CBMT is an example of contextual training. ISAT using the CBMT method is delivered to users in short sequences and in situations where the training is of direct relevance. Phishing training is, for instance, delivered to users that are in a situation with an elevated risk of being exposed to phishing. It is argued to counter the knowledge retention and user participation problems by automatically appearing in those relevant situations [32]. It is also argued to motivate users towards secure behavior by providing them with training that directly relates to the users' current situation.

3. Materials and Methods

The purpose of this study was to evaluate user behavior when assessing if emails are malicious or not. To that end, a controlled experiment where the participants were exposed to an inbox and asked to classify the email contained in that inbox was conducted. The participants were scored based on how accurately they classified the emails. Furthermore, the participants' behavior was monitored during the experiment by an eye tracker that recorded where the participants were looking on screen. Before the experiments, the participants were randomised into three groups; game-based training, CBMT-based training or control. A between-group analysis was performed to identify differences between training methods and answer the research question posed. As detailed at the end of paper statements, data supporting this paper is available as open data (https://doi.org/10.5878/g6d9-7210 (accessed on 6 March 2022)). Furthermore, the study did not require ethical review, but all participants signed a written informed consent form detailing how the study was executed and how data were handled. An overview of the research process is presented in Figure 1. The rest of this section provides a detailed description of the experiment environment, data collection procedures, collected variables, and data analysis procedures.

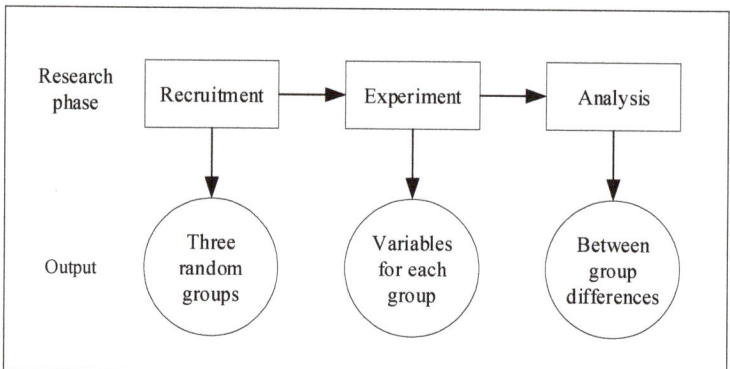

Figure 1. Research process overview.

3.1. Experiment Environment

An experiment environment containing an email system was set up on Ubuntu Linux using the email server and management platform Modoboa (https://modoboa.org/en/ (accessed on 6 March 2022)). Both Ubuntu Linux and Modoboa were installed with default settings. Modoboa allowed for the creation of unlimited email domains and addresses and provided a webmail interface. Several email domains were configured so that different types of emails could be created:

- Legitimate emails from service providers such as Google and banks.
- Phishing emails that imitated phishing emails from hijacked sender accounts.
- Phishing emails from domains made up to look similar to real domains, for instance, lundstro.mse instead of lundstrom.se.

The fictitious company Lundström AB, and the character Jenny Andersson were developed. The company was given the domain lundstrom.se and the character was given the email address jenny@lundstrom.se. A persona was developed for Jenny Andersson. The experiment participants were asked to assume Jenny´s persona and classify the email in her inbox. The persona was expressed as follows:

Jenny is 34 years old and works as an accountant at a small company (Lundström AB), and her manager is Arne Lundtröm. She lives with her husband and kids in a small town in Sweden. Your email address is jenny@lundstrom.se. You use the banks SBAB and Swedbank and is interested in investing in Bitcoin. You are about to remodel your home and have applied for loans at several banks to finance that. You shop a fair bit online and are registered at several e-stores without really remembering where. You are currently about to remodel your bathroom. Ask the experiment supervisor if you need additional information about Jenny or the workplace during the experiment.

Jenny's inbox was populated with 11 emails where five were legitimate, and six were phishing. The legitimate emails were crafted as reasonable questions from her manager or communications from banks and craftsmen. The communications from banks and craftsmen were based on real emails taken from one of the researcher's inboxes. The six phishing emails were crafted to include different phishing identifiers. Five different phishing identifiers were included in the experiment. They are commonly mentioned in scientific and popular literature and were the following [41–44]:

1. Incorrect sender address where the attacker may use an arbitrary incorrect sender address, attempt to create an address that resembles that of the true sender, or use a sender name to hide the actual sender address.
2. Malicious attachments where the attacker will attempt to make the recipient download an attachment with malicious content. A modified file extension may disguise the attachment.
3. Malicious links that are commonly disguised so that the user needs to hover over them to see the true link target.
4. Persuasive tone where an attacker attempts to pressure the victim to act rapidly.
5. Poor spelling and grammar that may indicate that a text is machine translated or not written professionally.

The included phishing emails are described as follows:

1. The first phishing email came from the manager's real address and mimicked a spear-phishing attempt, including a malicious attachment and hijacked sender address. The attachment was a zip file with the filename "annons.jpeg.zip (English: advertisement.jpeg.zip)". The text body prompted the recipient to open the attached file. In addition to a suspicious file extension, the mail signatures differed from the signature in other emails sent by the manager.
2. The second phishing email came from Jenny's own address and prompted the recipient to click a link that supposedly led to information about Bitcoin. The email could be identified as phishing by the strange addressing and the fact that the tone in the email was very persuasive.
3. The third phishing email appeared to be a request from the bank SBAB. It prompted the user to reply with her bank account number and deposit a small sum of money into another account before a loan request could be processed. It could be identified by improper grammar, an incorrect sender address (that was masked behind a reasonable sender name), and the request itself.
4. The fourth phishing email was designed to appear from Jenny´s manager. It prompted Jenny to quickly deposit a large sum of money into a bank account. It could be identified by the request itself and because the sender address was arne@lundstro.mse instead of arne@lundstrom.se.
5. The fifth phishing email mimicked a request from Google Drive. It could be identified by examining the target of the included links that lead to the address xcom.se instead of google.
6. The sixth phishing email appeared to be from the bank Swedbank and requested the recipient to go to a web page and log in to prove the ownership of an account. It could

be identified as phishing by examining the link target, the sender address, which was hidden behind a sender name, and the fact that it contained several spelling errors.

The experiment was set up so that most phishing emails had similar legitimate counterparts. The legitimate emails included where:

1. The first legitimate email was a request from Jenny's manager Arne. The request prompted Jenny to review a file on a shared folder.
2. The second legitimate email was a notification from a Swedish bank. It prompted Jenny to go to the bank website and log in. It did not contain any link.
3. The third legitimate email was an offering from a plumber. While containing some spelling errors, it did not prompt Jenny to make any potentially harmful actions.
4. The fourth legitimate email is a request for a meeting from Jenny's manager Arne.
5. The fifth email is a notification from a Swedish bank. This notification prompts the user to go to the bank website and log in. It does not contain any greeting or signature with address.

The webmail interface is demonstrated in Figure 2. Figure 2 displays the layout of the included emails and is annotated to show the ordering of the emails. Legitimate emails are denoted Ln, in green, and phishing emails are denoted Pn in red.

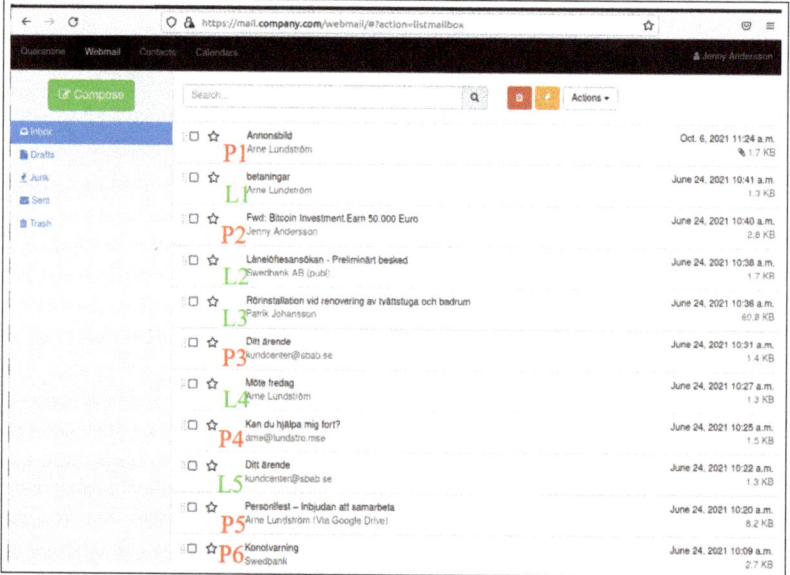

Figure 2. Webmail interface used in the experiment.

3.2. Participant Recruitment

Participants were recruited using a convenience sampling approach where students and employees from the University of Skövde were recruited. Participants with education or work experience in cybersecurity were excluded from the study. All participants were invited with a direct email that they were asked to reply to in order to participate. Upon registration, participants were randomly assigned to one of the three groups and provided with a description of the experiment, a description of the persona, and an informed consent form. The three groups were the following:

- Game: Participants in this group were prompted to play an educational game before arriving for the experiment. The game is called Jigsaw (https://phishingquiz.withgoogle.com/) (accessed on 6 March 2022) and is developed by Google. It is an example of game-based training that is implemented as a quiz and was selected for use

in this research because it is readily available for users. It also covers all the identifiers of phishing previously described. Jigsaw takes about five minutes to complete.
- CBMT: Participants in this group received computerized training developed by the research team according to the specifications of CBMT. It was written information that appeared to the participants when they opened Jenny's inbox, as demonstrated in Figure 3. The participants were presented with a few tips and prompted to participate in further training, which led the participants to a text-based slide show in a separate window. The training takes about five minutes to complete.
- CONTROL: This group completed the experiment without any intervention.

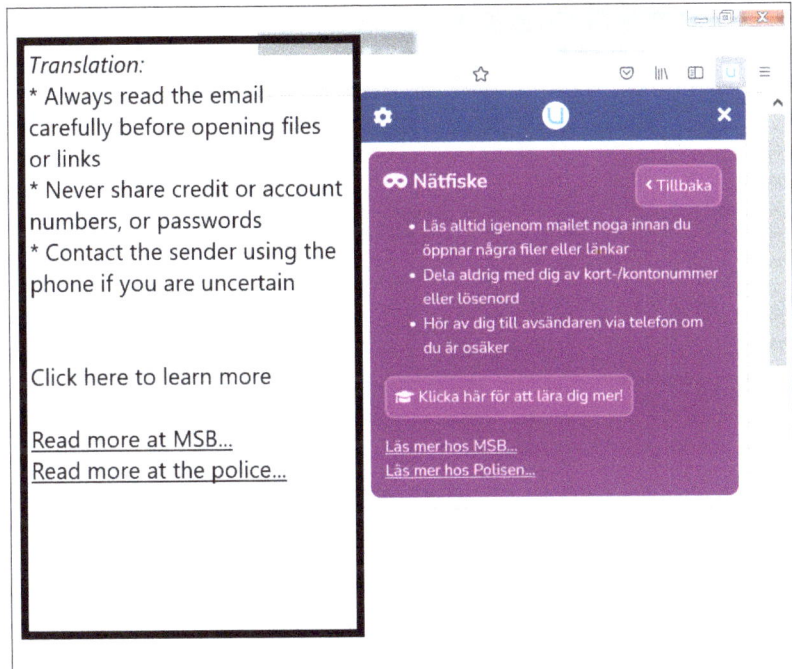

Figure 3. Demonstration of CBMT-based training.

3.3. Experiment Procedure

On arriving for the experiment, the participant was seated in a regular office in front of a 24″ computer monitor that displayed the experiment environment. The monitor was equipped with a Gazepoint GP3 HD eye tracker (https://www.gazept.com/product/gp3 hd/) (accessed on 6 March 2022). The participant was asked to read the informed consent form and given the opportunity to ask questions about the experiment and study before signing it. The participant was then asked to respond to a survey with demographic questions and asked to take a seat in front of the monitor. The eye tracker was calibrated using the manufacturer's built-in calibration sequence with nine points [45]. The calibration was considered successful when the control software deemed all nine points valid. In cases where the eye tracker could not be successfully calibrated, eye-tracking data were disregarded for that participant. This happened for three participants.

The participant was then reminded of Jenny's persona and asked to classify the email in Jenny's inbox. The participant was instructed to delete all phishing emails and keep all legitimate emails. The participant was asked to think aloud during the experiment, especially about how decisions to delete emails were made. The participant was also told that at least one of the emails was phishing and that a score was to be calculated based on the participants' performance. The intent was to make the participant as aware of phishing as possible. The rationale was that mere inclusion in the experiment would increase the participant's awareness level, and by priming all participants to high awareness would make the awareness levels of the participants comparable. Consequently, the gathered data reflects the participants' best ability to delete phishing rather than the ability they can be assumed to have during their daily work. Gazepoint analysis UX Edition (https://www.gazept.com/product/gazepoint-analysis-ux-edition-software/) (accessed on 6 March 2022) was used to monitor the participant's performance in real time on an adjacent screen and for post-experiment analysis of the collected eye-tracking data. Following the experiment, the participants in the game group were asked if they had played the game before the experiment as instructed. The experiment process, from the participant's point of view, is visualized in Figure 4.

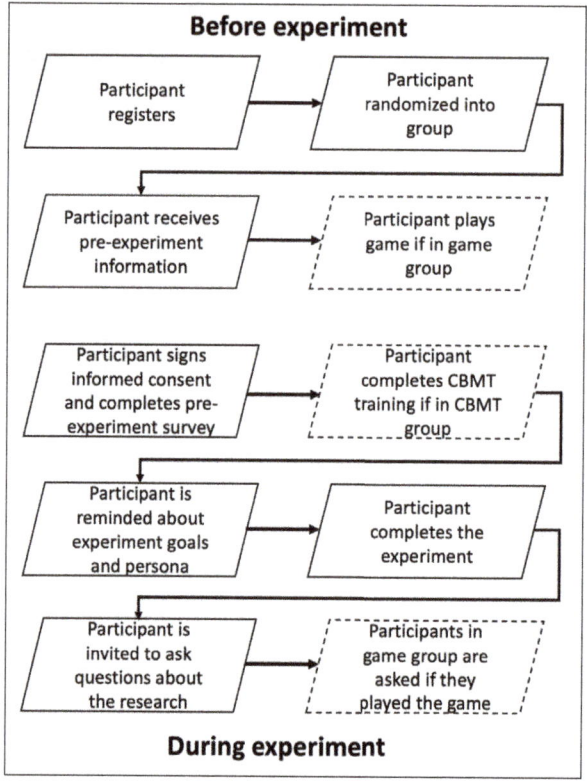

Figure 4. Visualization of experiment procedure. Dashed boxes only applied to some groups.

3.4. Collected Variables

Variables reflecting the participants' demographic background, score and behavior were captured during the experiment. The demographic variables were collected to enable a descriptive presentation of the sample's demographic attributes. The score variables reflected the total number of correct classifications the participants made. The behavior variables described how the participants acted during the experiment by counting how many of the previously described phishing identifiers the participants used. Two behavior variables were collected. The first was collected manually during the experiment (behavior_manual). It was based on real-time monitoring, and the participants expressed thoughts. It reflected how many of the following actions the participant performed at least once:

1. Evaluated the sender address by hovering over the displayed name to see the real sender address.
2. Evaluated attachments by acknowledging their existence and describing it as suspicious or legitimate.
3. Evaluated links by hovering over them or in some other way verified the link destination.
4. Evaluated if the tone in the email was suspiciously persuasive.
5. Evaluated if spelling and grammar made the email suspicious.

Please note that the variables reflect what identifiers the participants used but not if they accurately interpreted the information provided by the identifier. A participant who, for instance, incorrectly evaluated a sender address as legitimate would still get the point for evaluating the sender address. The second behavior variable, behavior_tracked, was computed automatically by defining Areas of Interest in Gazepoint analysis UX Edition and counting how many times the participant gazed in those areas. Areas of Interest are defined screen areas that allow for collecting the number of times the participants gaze in those particular areas. The following three Areas of Interest were defined.

- Address, which covered the area holding the sender and recipient addresses.
- Attachment covering the area where email attachments are visible.
- Link covering the area where the true link destination appears.

The Areas of Interest were only active when they included the intended information. For instance, the Attachment area was only active when an attachment was visible on the screen. The Areas of Interest are demonstrated in Figure 5 which also shows how red dots denote where the participant is currently looking.

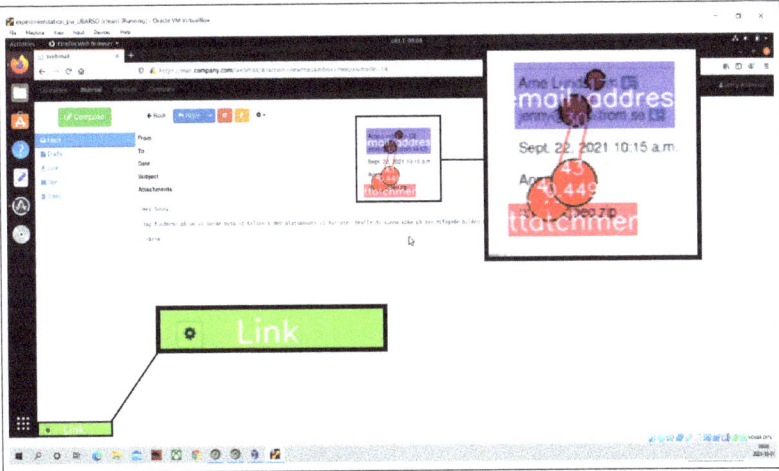

Figure 5. Demonstration of how areas of interest were defined, with AOI definitions enlarged. Please note that the Link area contains the target address of a link that is hovered over.

3.5. Data Analysis

The data were analyzed using SPSS version 25. The demographic properties of the sample were first described followed by a descriptive overview of the three variables SCORE, behavior_manual, and behavior_tracked. The proportion of participants that received perfect scores was then reported. A perfect score means that a participant identified all 11 emails correctly, or used all phishing identifiers assessed by the variables behavior_manual and behavior_tracked, respectively.

Next, Kruskal–Wallis H tests were used, with pairwise Mann–Whitney U test with Bonferroni correction as post hoc procedure, to identify significant between-group differences. Kruskal–Wallis H test performed on three or more samples will return a significant result if at least one sample is different from the others. In such a case, the Mann–Whitney U test with Bonferroni correction is used between all pairs in the sample to analyze what individual samples that are different from each other. Kruskal–Wallis H test was used over ANOVA because the data must show a normal distribution for ANOVA to be robust, and most samples did not in this case [46]. The conventional significance level of 0.05 is used throughout this paper.

4. Results

This section outlines the results of the study. It is divided into two sections were the first section outlines a descriptive overview of the data. The second section outlines the results in relation to the research question. It should be noted that three participants in the Game group reported that they did not play the provided game. This is to be expected given previous works suggesting that it is challenging to get users to participate in training [28]. All statistical procedures have been performed with and without those three participants. Results concerning the Game group are reported as $n(m)$ were n is the result when the complete group is considered and m is the result when participants that did not play the game are omitted.

4.1. Data Overview

Data was collected over a period of about two months and included 41 participants. Two participants were removed from the data set since they reported having formal training in cybersecurity. The data collection period was intended to be longer, but data collections stopped after a security incident where the IT department warned all students and staff at the university about phishing involving attachments. Continued data collection would have risked the validity of the data set. The mean participant age was 37. Twenty-three participants identified themselves as female and 16 as male. Twenty-three participants reported being employees and 16 reported being students. An overview of the mean and median values for the collected variables and the distribution form of the variables is presented in Table 2. Please note that eye-tracking failed for three participants and the participants included for the variable behavior_tracked is therefore only 36.

Table 2. Data overview.

Variable	Group	Mean	Median	Normal Distribution
SCORE out of 11	Control (n = 11)	8.82	9	YES
	CBMT (n = 14)	10	10	NO
	Game (n = 14)	8.86 (9.09)	9 (9)	NO
	Total (n = 39)	9.26	9	NO
behavior_manual out of 5	Control (n = 11)	3	3	NO
	CBMT (n = 14)	4.57	5	NO
	Game (n = 14)	3.64 (3.82)	3.5 (4)	NO
	Total (n = 39)	3.79	4	NO
behavior_tracked out of 3	Control (n = 10)	1.9	2	NO
	CBMT (n = 12)	2.5	3	NO
	Game (n = 14)	2.29 (2.55)	2 (3)	NO
	Total (n = 36)	2.25	2	NO

4.2. The Effect of Training

The effect of training was assessed by first examining the proportion of participants that received perfect scores. A perfect score means that the participants used all phishing identifiers or identified all emails correctly. The proportions of perfect scores are presented in Table 3.

Table 3 suggests that participants who received training performed better than participants in the control group for the behavior variables and that the participants in the CBMT group outperformed the other groups for the variable SCORE. The same tendency is seen in Table 2 where mean and median results for the different sample groups are presented. Table 2 suggests that participants in the group game performed slightly better than the control group while the participants in the group CBMT outperformed the other groups with a bigger margin. The exception is for the variable behavior_tracked where the groups CBMT and game performed equally when participants who reported not playing the game were omitted from the game group.

Table 3. Proportions of perfect scores.

Variable	Group	Perfect Scores
SCORE	Control (n = 11)	0%
	CBMT (n = 14)	21.4%
	Game (n = 14)	0% (0%)
	Total (n = 39)	7.7% (8.3%)
behavior_manual	Control (n = 11)	0%
	CBMT (n = 14)	64.3%
	Game (n = 14)	14.3% (18.2%)
	Total (n = 39)	28.2% (30.6%)
behavior_tracked	Control (n = 10)	9.1%
	CBMT (n = 12)	57.1%
	Game (n = 14)	42.9% (54.5%)
	Total (n = 36)	38.5% (45.5%)

Kruskal–Wallis H test was used to identify variables with statistically significant between-group differences. The results are presented in Table 4.

Table 4. Kruskal–Wallis H tests.

Variable	Kruskal–Wallis H	p-Value
SCORE	13.965 (12.531)	0.001 (0.002)
behavior_manual	16.270 (15.434)	0.000 (0.000)
behavior_tracked	5.569 (7.332)	0.062 (0.026)

The Kruskal–Wallis H tests suggest that at least one sample is different from the others when $p < 0.05$, as is the case for the variables SCORE and behavior_manual. The same is also true for the variable behavior_tracked when users who did not play the game are omitted. Pairwise Mann–Whitney U tests with Bonferroni correction was used to test what variables that were significantly different from each other. The results are presented in Table 5.

Table 5. Pairwise post hoc tests. Please note that post hoc tests for the variable behavior_tracked were only computed in the case when participants in the group Game, who did not play the game was omitted because the corresponding Kurskal-Wallis H tests was only significant in that case.

Variable	Groups	p-Value
SCORE	Control-Game	1.000 (1.000)
	Control-CBMT	0.005 (0.003)
	Game-CBMT	0.003 (0.023)
behavior_manual	Control-Game	0.502 (0.277)
	Control-CBMT	0.000 (0.000)
	Game-CBMT	0.021 (0.102)
behavior_tracked	Control-Game	X (0.083)
	Control-CBMT	X (0.036)
	Game-CBMT	X (1.000)

In this case, the difference between two variables is statistically significant when $p < 0.05$. Table 5 shows that CBMT is separated from the groups game and control for the variables SCORE and behavior_manual while control and game cannot be separated. For behavior_tracked, game and CBMT cannot be separated but are both separated from control.

5. Discussion

This research explores how effectively Information Security Awareness Training (ISAT) can support users to accurately identify phishing emails. The research evaluated two methods that were discussed as being promising in recent literature, namely game-based training and training based on CBMT. The research was conducted as a simulated experiment that measured how the participants behaved when assessing whether emails were phishing or not, and how accurately they classified email. The statistical analysis shows that participants in the CBMT group had higher scores than users in the game or control group. In terms of behavior, participants in the CBMT group performed better than the game and control group for the manually collected variable. However, the CBMT and game groups were equally strong for the variable computed based on eye-tracking data. In conclusion, both game-based training and CBMT are shown to improve user behavior in relation to phishing while only CBMT can be shown to improve users' ability to accurately classify phishing emails.

One reason could be that CBMT provides an awareness increasing mechanism in addition to training while game-based training does not. The game-based training is delivered to participants on a regular basis and was mimicked in the experiment by letting the participants take the training prior to arriving for the experiment. CBMT is, by design, presented to users when they are entering a risky situation and that was mimicked by presenting the CBMT training to participants just before starting the experiment. The difference in how the training was delivered could account for the difference in results between the two groups. In fact, the effect of awareness increasing mechanisms have been evaluated in prior research with good results [47,48]. This research extends those results by suggesting that awareness increasing mechanisms combined with training are likely to have a positive effect on users' ability to accurately identify phishing emails.

While training was proven to improve participants' ability to identify phishing, it can be noted that less than 10% of the participants were able to identify all emails correctly. Furthermore, less than 50% of the participants evaluated all of the phishing identifiers and even if the participants in the CBMT group received training just before starting the experiment, 35.7% of those participants missed one or more phishing identifiers. Yet, most organizations explicitly or implicitly expect users to correctly identify all phishing emails all the time. The present research shows that even if users are provided with training just before being tasked with identifying phishing, and instructed to actively search for phishing, very few users are able to fulfill the expectations of that security model. The implication of this result is that the security model or the feasibility of using training alone to reach it must be questioned. One could, for instance, question if we should follow a paradigm where users are expected to change according to how computers work. A more useful paradigm could be to modify the way that computers work to match the abilities of the users. A similar viewpoint is presented by [49] who questions why the responsibility for cybersecurity is individualized through the notion of the "stupid user". Instead, ref. [49] suggest that user-oriented threats should be managed by security professionals, and managers, at a collective level. Likewise, ref. [50] calls for a more holistic approach to anti-phishing methods.

5.1. Limitations

A given limitation of this study comes from participation bias. Participation bias is known to impact simulated experiments in cybersecurity awareness [22]. The expected effect in this study is that participants are more aware than they would be in a naturally occurring situation. Thus, the scores are expected to reflect the participants' best ability rather than their average performance. Using a between-group design, we still argue that differences between ISAT methods identified in this research are valid. However, it is likely that the actual performance of the included methods will be lower in a natural environment. On a similar note, the method cannot account for organizational factors such as leadership support and social pressure, which are know to impact cybersecurity behavior [51].

A second limitation concerns sampling where this research included participants studying, or working at, a university. As such, the results are representative of that population and any inference beyond that population should be avoided. On this topic, recent research argues that there are indeed demographic differences in the ability to detect phishing [52]. The number of participants is a further limitation and a higher participant number would have been preferable. In this case, data collection was stopped following a cybersecurity incident that prompted the IT department to broadcast a phishing warning. Participants performing the experiment after that event would have been exposed to information not presented to other participants and that would have introduced bias into the dataset.

A third possible discussion under the umbrella of limitations is how the different types of training were presented to the participants. The participants placed in the game group were asked to play a game before arriving for the experiment while participants in the CBMT group were subjected to training on arrival. There is, therefore, a chance that participants in the game group forgot some of the training, or forgot to play the game entirely. The design is argued to mimic the natural behavior of the two training types and both retention and failure to play are two previously discussed obstacles with game-based training delivered in a format that requires active participation [28]. Consequently, any effect of the experimental design mimics an expected effect in a natural environment.

5.2. Future Work

While training can undoubtedly support users to identify phishing emails, this study suggests that training alone is not enough and that opens up several future research directions. First, future studies could focus on combining training with modifying the way emails are presented to users. One could imagine that finding ways to make it easier for users to find and interpret phishing identifiers could improve users' ability to identify malicious emails. A possible example could be to rewrite links in the text body of emails to always show the full link address, which is unclickable, instead of allowing clickable hyperlinks with arbitrary display names. A similar possible direction is to further research predicting user susceptibility to phishing using artificial intelligence [53]. That could identify a user in need to training and then provide tailored training. A second direction for future work could be to replicate this study with a different population. That would allow for identification of differences and similarities between, for instance, technical and non-technical users, male and female users, and users of different age.

A more theoretical direction for future work could be to evaluate the strength of the relationships in the KAB model and to evaluate the relationship between behavior and actual outcomes of that behavior. In certain situations, including phishing, applying a correct behavior is not enough, since a user also has to interpret the result of that behavior. For instance, a correct behavior would make a user control the real target of a link, and to make a decision about the email the user needs to interpret the trustworthiness of the link target. Furthermore, one could assess the possible effect of usability on the relationship between the constructs in the KAB model. One can imagine that knowledge about a certain behavior is more likely to result in that behavior if the effort to comply is low.

Author Contributions: Conceptualization, J.K., M.N. and J.R.; methodology, J.K, M.N. and J.R.; software, R.R. and A.H.; validation, All.; formal analysis, J.K.; investigation, J.K.; resources, J.K.; data curation, J.K.; writing—original draft preparation, J.K.; writing—review and editing, M.N. and S.F.; supervision, S.F.; project administration, J.K.; funding acquisition, J.K, M.N., J.R., R.R. and A.H. All authors have read and agreed to the published version of the manuscript.

Funding: This research was funded by The Swedish Post and Telecom Authority grant number 19-10617.

Institutional Review Board Statement: Ethical review and approval were waived for this study, due to fact that it does not require ethical clearance under the Swedish Ethical Review Act. Ethical Review Act dictates that research including sensitive personal data, physical interventions on living or deceased persons, methods that aim to affect persons physically och mentally, methods that can harm persons physically or mentally, or biological material from living of deceased persons [54]. Since this research does not fall under any of those criteria, ethical clearance has not been applied for. The study has been discussed with the chairperson of the council of research ethics at the University of Skövde.

Informed Consent Statement: Informed consent was obtained from all subjects involved in the study.

Data Availability Statement: Data supporting this research can be found at: https://doi.org/10.5878/g6d9-7210 (accessed on 6 March 2022).

Conflicts of Interest: The authors declare no conflict of interest. The funders had no role in the design of the study; in the collection, analyses, or interpretation of data; in the writing of the manuscript, or in the decision to publish the results.

Abbreviations

Abbreviation

The following abbreviations are used in this manuscript:

ISAT	Information Security Awareness Training
CBMT	Context-Based Micro-Training
SETA	Security Education, Training, and Awareness
KAB	Knowledge, Attitude, and Behaviour
SPSS	Statistical Package for the Social Sciences
ANOVA	Analysis of variance

References

1. OECD. *Hows Life in the Digital Age?* OECD Publishing: Paris, France, 2019; p. 172.
2. Owusu-Agyei, S.; Okafor, G.; Chijoke-Mgbame, A.M.; Ohalehi, P.; Hasan, F. Internet adoption and financial development in sub-Saharan Africa. *Technol. Forecast. Soc. Chang.* **2020**, *161*, 120293. [CrossRef]
3. Anderson, M.; Perrin, A. *Technology Use among Seniors*; Pew Research Center for Internet & Technology: Washington, DC, USA, 2017.
4. Bergström, A. Digital equality and the uptake of digital applications among seniors of different age. *Nord. Rev.* **2017**, *38*, 79. [CrossRef]
5. Milana, M.; Hodge, S.; Holford, J.; Waller, R.; Webb, S. A Year of COVID-19 Pandemic: Exposing the Fragility of Education and Digital in/Equalities. 2021. Available online: https://www.tandfonline.com/doi/full/10.1080/02601370.2021.1912946 (accessed on 6 March 2022)
6. Watts, G. COVID-19 and the digital divide in the UK. *Lancet Digit. Health* **2020**, *2*, e395–e396. [PubMed] [CrossRef]
7. Joseph, D.P.; Norman, J. An analysis of digital forensics in cyber security. In *First International Conference on Artificial Intelligence and Cognitive Computing*; Springer: Berlin/Heidelberg, Germany, 2019; pp. 701–708.
8. Sfakianakis, A.; Douligeris, C.; Marinos, L.; Lourenço, M.; Raghimi, O. *ENISA Threat Landscape Report 2018: 15 Top Cyberthreats and Trends*; ENISA: Athens, Greece, 2019.
9. Bhardwaj, A.; Sapra, V.; Kumar, A.; Kumar, N.; Arthi, S. Why is phishing still successful? *Comput. Fraud. Secur.* **2020**, *2020*, 15–19. [CrossRef]
10. Dark Reading. Phishing Remains the Most Common Cause of Data Breaches, Survey Says. Available online: https://www.darkreading.com/edge-threat-monitor/phishing-remains-the-most-common-cause-of-data-breaches-survey-says (accessed on 1 December 2021).
11. Butnaru, A.; Mylonas, A.; Pitropakis, N. Towards lightweight url-based phishing detection. *Future Internet* **2021**, *13*, 154. [CrossRef]
12. Gupta, B.B.; Arachchilage, N.A.; Psannis, K.E. Defending against phishing attacks: taxonomy of methods, current issues and future directions. *Telecommun. Syst.* **2018**, *67*, 247–267. [CrossRef]
13. Vishwanath, A.; Herath, T.; Chen, R.; Wang, J.; Rao, H.R. Why do people get phished? Testing individual differences in phishing vulnerability within an integrated, information processing model. *Decis. Support Syst.* **2011**, *51*, 576–586. [CrossRef]
14. Steer, J. Defending against spear-phishing. *Comput. Fraud. Secur.* **2017**, *2017*, 18–20. [CrossRef]

15. Lacey, D.; Salmon, P.; Glancy, P. Taking the bait: a systems analysis of phishing attacks. *Procedia Manuf.* **2015**, *3*, 1109–1116. [CrossRef]
16. Khan, B.; Alghathbar, K.S.; Nabi, S.I.; Khan, M.K. Effectiveness of information security awareness methods based on psychological theories. *Afr. J. Bus. Manag.* **2011**, *5*, 10862–10868.
17. Parsons, K.; McCormac, A.; Butavicius, M.; Pattinson, M.; Jerram, C. Determining employee awareness using the human aspects of information security questionnaire (HAIS-Q). *Comput. Secur.* **2014**, *42*, 165–176. [CrossRef]
18. Puhakainen, P.; Siponen, M. Improving employees' compliance through information systems security training: an action research study. *MIS Q.* **2010**, *34*, 757–778. [CrossRef]
19. Bin Othman Mustafa, M.S.; Kabir, M.N.; Ernawan, F.; Jing, W. An enhanced model for increasing awareness of vocational students against phishing attacks. In Proceedings of the 2019 IEEE International Conference on Automatic Control and Intelligent Systems (I2CACIS), Selangor, Malaysia, 29 June 2019; pp. 10–14.
20. Bada, M.; Sasse, A.M.; Nurse, J.R. Cyber security awareness campaigns: Why do they fail to change behaviour? *arXiv* **2019**, arXiv:1901.02672.
21. Reinheimer, B.; Aldag, L.; Mayer, P.; Mossano, M.; Duezguen, R.; Lofthouse, B.; von Landesberger, T.; Volkamer, M. An investigation of phishing awareness and education over time: When and how to best remind users. In Proceedings of the Sixteenth Symposium on Usable Privacy and Security (SOUPS 2020), Santa Clara, CA, USA, 7–11 August 2020; pp. 259–284.
22. Lastdrager, E.; Gallardo, I.C.; Hartel, P.; Junger, M. How Effective is Anti-Phishing Training for Children? In Proceedings of the Thirteenth Symposium on Usable Privacy and Security (SOUPS 2017), Santa Clara, CA, USA, 12–14 July 2017; pp. 229–239.
23. Junglemap. Nanolearning. Available online: https://junglemap.com/nanolearning (accessed on 7 January 2021).
24. Gokul, C.J.; Pandit, S.; Vaddepalli, S.; Tupsamudre, H.; Banahatti, V.; Lodha, S. PHISHY—A Serious Game to Train Enterprise Users on Phishing Awareness. In Proceedings of the 2018 Annual Symposium on Computer-Human Interaction in Play Companion Extended Abstracts, Melbourne, Australia, 28–31 October 2018; pp. 169–181.
25. Lim, I.K.; Park, Y.G.; Lee, J.K. Design of Security Training System for Individual Users. *Wirel. Pers. Commun.* **2016**, *90*, 1105–1120. [CrossRef]
26. Hatfield, J.M. Social engineering in cybersecurity: The evolution of a concept. *Comput. Secur.* **2018**, *73*, 102–113. [CrossRef]
27. Renaud, K.; Zimmermann, V. Ethical guidelines for nudging in information security & privacy. *Int. J. Hum.-Comput. Stud.* **2018**, *120*, 22–35.
28. Gjertsen, E.G.B.; Gjaere, E.A.; Bartnes, M.; Flores, W.R. Gamification of Information Security Awareness and Training. In Proceedings of the 3rd International Conference on Information Systems Security and Privacy, SiTePress, Setúbal, Portugal, 19–21 February 2017; pp. 59–70.
29. Abraham, S.; Chengalur-Smith, I. Evaluating the effectiveness of learner controlled information security training. *Comput. Secur.* **2019**, *87*, 101586. [CrossRef]
30. Siponen, M.; Baskerville, R.L. Intervention effect rates as a path to research relevance: information systems security example. *J. Assoc. Inf. Syst.* **2018**, *19*. [CrossRef]
31. Wen, Z.A.; Lin, Z.; Chen, R.; Andersen, E. What. hack: Engaging anti-phishing training through a role-playing phishing simulation game. In Proceedings of the 2019 CHI Conference on Human Factors in Computing Systems, Glasgow, UK, 4–9 May 2019; pp. 1–12.
32. Kävrestad, J.; Nohlberg, M. Assisting Users to Create Stronger Passwords Using ContextBased MicroTraining. In *IFIP International Conference on ICT Systems Security and Privacy Protection*; Springer: Berlin/Heidelberg, Germany, 2020; pp. 95–108.
33. Siponen, M.T. A conceptual foundation for organizational information security awareness. *Inf. Manag. Comput. Secur.* **2000**, *8*, 31–41. [CrossRef]
34. Bulgurcu, B.; Cavusoglu, H.; Benbasat, I. Information security policy compliance: an empirical study of rationality-based beliefs and information security awareness. *MIS Q.* **2010**, *34*, 523–548. [CrossRef]
35. Hu, S.; Hsu, C.; Zhou, Z. Security education, training, and awareness programs: Literature review. *J. Comput. Inf. Syst.* **2021**, 1–13. [CrossRef]
36. Aldawood, H.; Skinner, G. An academic review of current industrial and commercial cyber security social engineering solutions. In Proceedings of the 3rd International Conference on Cryptography, Security and Privacy, Kuala Lumpur, Malaysia, 19–21 January 2019; pp. 110–115.
37. Al-Daeef, M.M.; Basir, N.; Saudi, M.M. Security awareness training: A review. *Proc. World Congr. Eng.* **2017**, *1*, 5–7.
38. EC-Council. The Top Types of Cybersecurity Attacks of 2019, Till Date, 2019. Available online: https://blog.eccouncil.org/the-top-types-of-cybersecurity-attacks-of-2019-till-date/ (accessed on 31 May 2021).
39. Cybint. 15 Alarming Cyber Security Facts and Stats. 2020. Available online: https://www.cybintsolutions.com/cyber-security-facts-stats/ (accessed on 20 March 2022).
40. Sharif, K.H.; Ameen, S.Y. A review of security awareness approaches with special emphasis on gamification. In Proceedings of the 2020 International Conference on Advanced Science and Engineering (ICOASE), Duhok, Iraq, 23–24 December 2020; pp. 151–156.
41. Williams, E.J.; Hinds, J.; Joinson, A.N. Exploring susceptibility to phishing in the workplace. *Int. J.-Hum.-Comput. Stud.* **2018**, *120*, 1–13. [CrossRef]

42. Chiew, K.L.; Yong, K.S.C.; Tan, C.L. A survey of phishing attacks: Their types, vectors and technical approaches. *Expert Syst. Appl.* **2018**, *106*, 1–20. [CrossRef]
43. Microsoft. Protect Yourself from Phishing. Available online: https://support.microsoft.com/en-us/windows/protect-yourself-from-phishing-0c7ea947-ba98-3bd9-7184-430e1f860a44 (accessed on 30 December 2021).
44. Imperva. Phishing Attacks. Available online: https://www.imperva.com/learn/application-security/phishing-attack-scam/ (accessed on 30 December 2021).
45. Cuve, H.C.; Stojanov, J.; Roberts-Gaal, X.; Catmur, C.; Bird, G. Validation of Gazepoint low-cost eye-tracking and psychophysiology bundle. *Behav. Res. Methods* **2021**, 1–23. [CrossRef] [CrossRef]
46. MacFarland, T.W.; Yates, J.M. Kruskal–Wallis H-test for oneway analysis of variance (ANOVA) by ranks. In *Introduction to Nonparametric Statistics for the Biological Sciences Using R*; Springer: Berlin/Heidelberg, Germany, 2016; pp. 177–211.
47. Zimmermann, V.; Renaud, K. The nudge puzzle: matching nudge interventions to cybersecurity decisions. *ACM Trans. Comput.-Hum. Interact. (TOCHI)* **2021**, *28*, 1–45. [CrossRef]
48. Van Bavel, R.; Rodríguez-Priego, N.; Vila, J.; Briggs, P. Using protection motivation theory in the design of nudges to improve online security behavior. *Int. J. Hum.-Comput. Stud.* **2019**, *123*, 29–39. [CrossRef]
49. Klimburg-Witjes, N.; Wentland, A. Hacking humans? Social Engineering and the construction of the "deficient user" in cybersecurity discourses. *Sci. Technol. Hum. Values* **2021**, *46*, 1316–1339. [CrossRef]
50. Alabdan, R. Phishing attacks survey: Types, vectors, and technical approaches. *Future Internet* **2020**, *12*, 168. [CrossRef]
51. Mashiane, T.; Kritzinger, E. Identifying behavioral constructs in relation to user cybersecurity behavior. *Eurasian J. Soc. Sci.* **2021**, *9*, 98–122. [CrossRef]
52. Das, S.; Nippert-Eng, C.; Camp, L.J. Evaluating user susceptibility to phishing attacks. *Inf. Comput. Secur.* **2022**, *309*, 1–18. [CrossRef]
53. Yang, R.; Zheng, K.; Wu, B.; Li, D.; Wang, Z.; Wang, X. Predicting User Susceptibility to Phishing Based on Multidimensional Features. *Comput. Intell. Neurosci.* **2022**, *2022*, 7058972. [CrossRef] [PubMed]
54. Swedish Research Council. Good Research Practice. Available online: https://www.vr.se/english/analysis/reports/our-reports/2017-08-31-good-research-practice.html (accessed on 30 December 2021).

Article

Adaptative Perturbation Patterns: Realistic Adversarial Learning for Robust Intrusion Detection

João Vitorino *, Nuno Oliveira and Isabel Praça *

Research Group on Intelligent Engineering and Computing for Advanced Innovation and Development (GECAD), School of Engineering, Polytechnic of Porto (ISEP/IPP), 4249-015 Porto, Portugal; nunal@isep.ipp.pt

* Correspondence: jpmvo@isep.ipp.pt (J.V.); icp@isep.ipp.pt (I.P.)

Abstract: Adversarial attacks pose a major threat to machine learning and to the systems that rely on it. In the cybersecurity domain, adversarial cyber-attack examples capable of evading detection are especially concerning. Nonetheless, an example generated for a domain with tabular data must be realistic within that domain. This work establishes the fundamental constraint levels required to achieve realism and introduces the adaptative perturbation pattern method (A2PM) to fulfill these constraints in a gray-box setting. A2PM relies on pattern sequences that are independently adapted to the characteristics of each class to create valid and coherent data perturbations. The proposed method was evaluated in a cybersecurity case study with two scenarios: Enterprise and Internet of Things (IoT) networks. Multilayer perceptron (MLP) and random forest (RF) classifiers were created with regular and adversarial training, using the CIC-IDS2017 and IoT-23 datasets. In each scenario, targeted and untargeted attacks were performed against the classifiers, and the generated examples were compared with the original network traffic flows to assess their realism. The obtained results demonstrate that A2PM provides a scalable generation of realistic adversarial examples, which can be advantageous for both adversarial training and attacks.

Keywords: realistic adversarial examples; adversarial attacks; adversarial robustness; machine learning; tabular data; intrusion detection

Citation: Vitorino, J.; Oliveira, N.; Praça, I. Adaptative Perturbation Patterns: Realistic Adversarial Learning for Robust Intrusion Detection. *Future Internet* 2022, 14, 108. https://doi.org/10.3390/fi14040108

Academic Editors: Leandros Maglaras, Helge Janicke and Mohamed Amine Ferrag

Received: 8 March 2022
Accepted: 27 March 2022
Published: 29 March 2022

Publisher's Note: MDPI stays neutral with regard to jurisdictional claims in published maps and institutional affiliations.

Copyright: © 2022 by the authors. Licensee MDPI, Basel, Switzerland. This article is an open access article distributed under the terms and conditions of the Creative Commons Attribution (CC BY) license (https://creativecommons.org/licenses/by/4.0/).

1. Introduction

Machine learning is transforming the way modern organizations operate. It can be used to automate and improve various business processes, ranging from the recognition of patterns and correlations to complex regression and classification tasks. However, adversarial attacks pose a major threat to machine learning models and to the systems that rely on them. A model can be deceived into predicting incorrect results by slightly modifying original data, which creates an adversarial example. This is especially concerning for the cybersecurity domain because adversarial cyber-attack examples capable of evading detection can cause significant damage to an organization [1,2].

Depending on the utilized method, the data perturbations that result in an adversarial example can be created in one of three settings: black-, gray- and white-box. The first solely queries a model's predictions, whereas the second may also require knowledge of its structure or the utilized feature set, and the latter needs full access to its internal parameters. Even though machine learning is inherently susceptible to these examples, a model's robustness can be improved by various defense strategies. A standard approach is performing adversarial training, a process where the training data is augmented with examples generated by one or more attack methods [3,4].

Nonetheless, a method can only be applied to a given domain if the examples it generates are realistic within that domain. In cybersecurity, a domain with tabular data, if an adversarial example does not resemble real network traffic, a network-based intrusion detection system (NIDS) will never actually encounter it because it cannot be transmitted through a computer network. Furthermore, if an example can be transmitted but is incompatible with its intended malicious purpose, evading detection will be futile because no damage can be caused. Consequently, training machine learning models with unrealistic cyber-attack examples only deteriorates their generalization to real computer networks and attack scenarios. Therefore, the generation of realistic adversarial examples for domains with tabular data is a pertinent research topic.

This work addressed the challenge of generating realistic examples, with a focus on network-based intrusion detection. The main contributions are the establishment of the fundamental constraint levels required to achieve realism and the introduction of the adaptative perturbation pattern method (A2PM) to fulfil these constraints in a gray-box setting. The capabilities of the proposed method were evaluated in a cybersecurity case study with two scenarios: Enterprise and Internet of Things (IoT) networks. It generated adversarial network traffic flows for multi-class classification by creating data perturbations in the original flows of the CIC-IDS2017 and IoT-23 datasets.

Due to the noticeably different internal mechanics of an artificial neural network (ANN) and a tree-based algorithm, the study analyzed the susceptibility of both types of models to the examples created by A2PM. A total of four multilayer perceptron (MLP) and four random forest (RF) classifiers were created with regular and adversarial training, and both targeted and untargeted attacks were performed against them. To provide a thorough analysis, example realism and time consumption were assessed by comparing the generated examples with the corresponding original flows and recording the time required for each A2PM iteration.

The present article is organized into multiple sections. Section 2 defines the fundamental constraint levels and provides a survey of previous work on adversarial examples. Section 3 describes the proposed method and the key concepts it relies on. Section 4 presents the case study and an analysis of the obtained results. Finally, Section 5 addresses the main conclusions and future work.

2. Related Work

In recent years, adversarial examples have drawn attention from a research perspective. However, since the focus has been the image classification domain, the generation of realistic examples for domains with tabular data remains a relatively unexplored topic. The common adversarial approach is to exploit the internal gradients of an ANN in a white-box setting, creating unconstrained data perturbations [5–7]. Consequently, most state-of-the-art methods do not support other types of machine learning models nor other settings, which severely limits their applicability to other domains. This is a pertinent aspect of the cybersecurity domain, where white-box is a highly unlikely setting. Considering that a NIDS is developed in a secure context, an attacker will commonly face a black-box setting, or occasionally gray-box [8,9].

The applicability of a method for adversarial training is significantly impacted by the models it can attack. Despite an adversarially robust generalization still being a challenge, significant progress has been made in ANN robustness research [10–14]. However, various other types of algorithms can be used for a classification task. This is the case of network-based intrusion detection, where tree-based algorithms, such as RF, are remarkably well-established [15,16]. They can achieve a reliable performance on regular network traffic, but their susceptibility to adversarial examples must not be disregarded. Hence, these algorithms can benefit from adversarial training and several defense strategies have been developed to intrinsically improve their robustness [17–20].

In addition to the setting and the supported models, the realism of the examples generated by a method must also be considered. Martins et al. [21] performed a systematic review of recent developments in adversarial attacks and defenses for cybersecurity and observed that none of the reviewed articles evaluated the applicability of the generated examples to a real intrusion detection scenario. Therefore, it is imperative to establish the fundamental constraints an example must comply with to be applicable to a real scenario on a domain with tabular data. We define two constraint levels:

1. Domain constraints—Specify the inherent structure of a domain.
2. Class-specific constraints—Specify the characteristics of a class.

To be valid on a given domain, an example can solely reach the first level. Nonetheless, full realism is only achieved when it is also coherent with the distinct characteristics of its class, reaching the second. In a real scenario, each level will contain concrete constraints for the utilized data features. These can be divided into two types:

- Intra-feature constraints—Restrict the value of a single feature.
- Inter-feature constraints—Restrict the values of one or more features according to the values present in other features.

In a real computer network, an example must fulfil the domain constraints of the utilized communication protocols and the class-specific constraints of each type of cyber-attack. Apruzzese et al. [8] proposed a taxonomy to evaluate the feasibility of an adversarial attack against a NIDS, based on access to the training data, knowledge of the model and feature set, reverse engineering and manipulation capabilities. It can provide valuable guidance to establish the concrete constraints of each level for a specific system.

Even though some methods attempt to fulfil a few constraints, many exhibit a clear lack of realism. Table 1 summarizes the characteristics of the most relevant methods of the current literature, including the constraint levels they attempt to address. The keyword 'CP' corresponds to any model that can output class probabilities for each data sample, instead of a single class label.

Table 1. Summary of relevant methods and addressed constraint levels.

Method	Setting	Supported Models	Domain Constraints	Class-Specific Constraints
FGSM [3]	White-box	ANN	✗	✗
C&W [22]	White-box	ANN	✗	✗
DeepFool [23]	White-box	ANN	✗	✗
Houdini [24]	White-box	ANN	✗	✗
StrAttack [25]	White-box	ANN	✗	✗
ZOO [26]	White-box	ANN	✗	✗
JSMA [27]	White-box	ANN	✓	✗
Polymorphic [28]	Gray-box	ANN	✗	✓
Reconstruction [29]	Gray-box	ANN	✗	✗
OnePixel [30]	Black-box	CP	✓	✗
RL-S2V [31]	Black-box	CP	✗	✗
BMI-FGSM [32]	Black-box	Any	✗	✗
GAN [33]	Black-box	Any	✗	✗
WGAN [34]	Black-box	Any	✗	✗
Boundary [35]	Black-box	Any	✗	✗
Query-Efficient [36]	Black-box	Any	✗	✗

Regarding the Polymorphic attack [28], it addresses the preservation of original class characteristics. Chauhan et al. developed it for the cybersecurity domain, to generate examples compatible with a cyber-attack's purpose. The authors start by applying a feature selection algorithm to obtain the most relevant features for the distinction between benign network traffic and each cyber-attack. Then, the values of the remaining features, which are considered irrelevant for the classification, are perturbed by a Wasserstein generative adversarial network (WGAN) [34]. On the condition that there are no class-specific constraints for the remaining features, this approach could improve the coherence of an example with its class. Nonetheless, the unconstrained perturbations created by WGAN disregard the domain structure, which inevitably leads to invalid examples.

On the other hand, both the Jacobian-based saliency map attack (JSMA) [27] and the OnePixel attack [30] could potentially preserve a domain structure. The former was developed to minimize the number of modified pixels in an image, requiring full access to the internal gradients of an ANN, whereas the latter only modifies a single pixel, based on the class probabilities predicted by a model. These methods perturb the most appropriate features without affecting the remaining features, which could be beneficial for tabular data. However, neither validity nor coherence can be ensured because they do not account for any constraint when creating the perturbations.

To the best of our knowledge, no previous work has introduced a method capable of complying with the fundamental constraints of domains with tabular data, which hinders the development of realistic attack and defense strategies. This is the gap in the current literature addressed by the proposed method.

3. Proposed Method

A2PM was developed with the objective of generating adversarial examples that fulfil both domain and class-specific constraints. It benefits from a modular architecture to assign an independent sequence of adaptative perturbation patterns to each class, which analyze specific feature subsets to create valid and coherent data perturbations. Even though it can be applied in a black-box setting, the most realistic examples are obtained in gray-box, with only knowledge of the feature set. To fully adjust it to a domain, A2PM only requires a simple base configuration for the creation of a pattern sequence. Afterwards, realistic examples can be generated from original data to perform adversarial training or to directly attack a classifier in an iterative process (Figure 1).

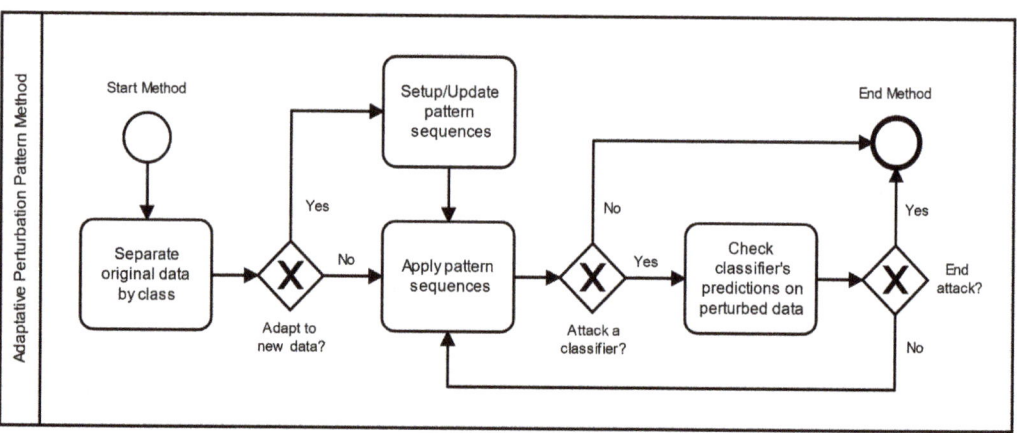

Figure 1. Adaptative perturbation pattern method (business process model and notation).

The generated examples can be untargeted, to cause any misclassification, or targeted, seeking to reach a specific class. New data perturbations could be generated indefinitely, but it would be computationally expensive. Hence, early stopping is employed to end the attack when the latest iterations could not cause any further misclassifications. Besides static scenarios where the full data is available, the proposed method is also suitable for scenarios where it is provided over time. After the pattern sequences are created for an initial batch of data, these can be incrementally adapted to the characteristics of subsequent batches. If novel classes are provided, the base configuration is used to autonomously create their respective patterns.

The performed feature analysis relies on two key concepts: value intervals and value combinations. The following subsections detail the perturbation patterns built upon these concepts, as well as the advantages of applying them in sequential order.

3.1. Interval Pattern

To perturb uncorrelated numerical variables, the main aspect to be considered is the interval of values each one can assume. This is an intra-feature constraint that can be fulfilled by enforcing minimum and maximum values.

The interval pattern encapsulates a mechanism that records the valid intervals to create perturbations tailored to the characteristics of each feature (Figure 2). It has a configurable 'probability to be applied', in the (0, 1] interval, which is used to randomly determine if an individual feature will be perturbed or not. Additionally, it is also possible to specify only integer perturbations for specific features.

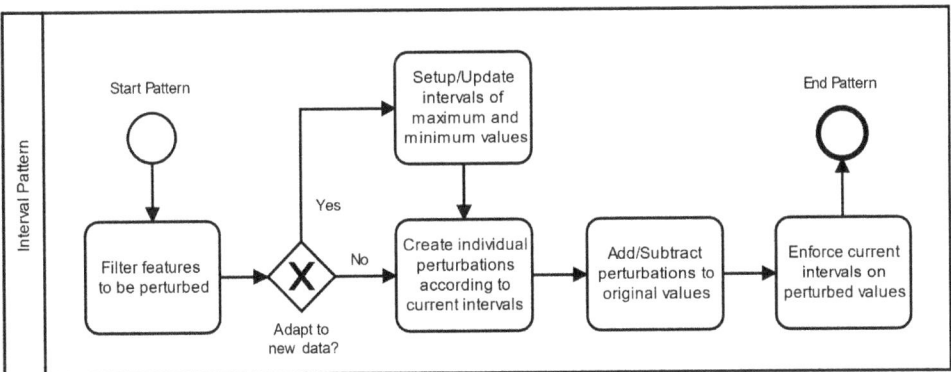

Figure 2. Interval pattern (business process model and notation).

Instead of a static interval, moving intervals can be utilized after the first batch to enable an incremental adaptation to new data, according to a configured momentum. For a given feature and a momentum $k \in [0, 1]$, the updated minimum m_i and maximum M_i of a batch i are mathematically defined as:

$$m_i = m_{i-1} * k + \min(x_i) * (1-k) \quad (1)$$

$$M_i = M_{i-1} * k + \max(x_i) * (1-k) \quad (2)$$

where $\min(x_i)$ and $\max(x_i)$ are the actual minimum and maximum values of the samples x_i of batch i.

Each perturbation is computed according to a randomly generated number and is affected by the current interval, which can be either static or moving. The random number $\varepsilon \in (0, 1]$ acts as a ratio to scale the interval. To restrict its possible values, it is generated within the standard range of [0.1, 0.3], although other ranges can be configured. For a given feature, a perturbation P_i of a batch i can be represented as:

$$P_i = (M_i - m_i) * \varepsilon \quad (3)$$

After a perturbation is created, it is randomly added or subtracted to the original value. Exceptionally, if the original value is less or equal to the current minimum, it is always increased, and vice-versa. The resulting value is capped at the current interval to ensure it remains within the valid minimum and maximum values of that feature.

3.2. Combination Pattern

Regarding uncorrelated categorical variables, enforcing their limited set of qualitative values is the main intra-feature constraint. Therefore, the interval approach cannot be replicated even if they are encoded to a numerical form, and a straightforward solution can be recording each value a feature can assume. Nonetheless, the most pertinent aspect of perturbing tabular data is the correlation between multiple variables. Since the value present in a variable may influence the values used for other variables, there can be several inter-feature constraints. To improve beyond the previous solution and fulfil both types of constraints, several features can be combined into a single common record.

The combination pattern records the valid combinations to perform a simultaneous and coherent perturbation of multiple features (Figure 3). It can be configured with locked features, whose values are used to find combinations for other features without being modified. Due to the simultaneous perturbations, its 'probability to be applied', in the (0, 1] interval, can affect several features.

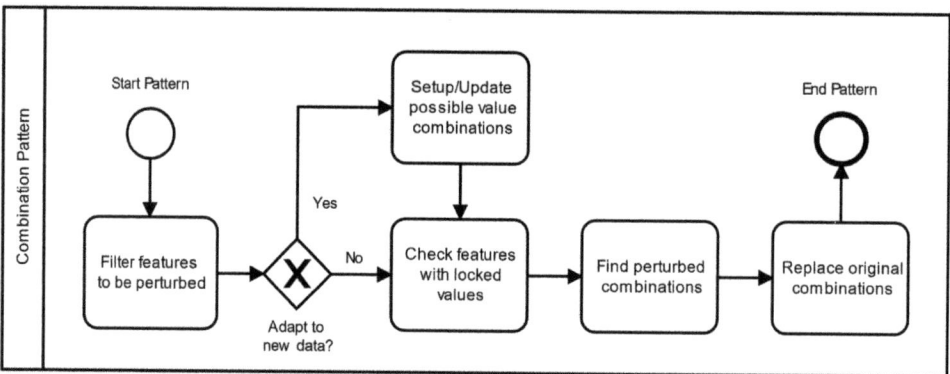

Figure 3. Combination pattern (business process model and notation).

Besides the initially recorded combinations, new data can provide additional possibilities. These can be merged with the previous or used as gradual updates. For a given feature and a momentum $k \in [0, 1]$, the number of updated combinations C_i of a batch i is mathematically expressed as:

$$C_i = C_{i-1} * k + \text{unique}(x_i) \qquad (4)$$

where $\text{unique}(x_i)$ is the number of unique combinations of the samples x_i of batch i.

Each perturbation created by this pattern consists of a combination randomly selected from the current possibilities, considering the locked features. It directly replaces the original values, ensuring that the features remain coherent.

3.3. Pattern Sequences

Domains with diverse constraints may require an aggregation of several interval and combination patterns, which can be performed by pattern sequences. Furthermore, the main advantage of applying multiple patterns in a sequential order is that it enables the fulfilment of countless inter-feature constraints of greater complexity. It is pertinent to note that all patterns in a sequence are independently adapted to the original data, to prevent any bias when recording its characteristics. Afterwards, the sequential order is enforced to create cumulative perturbations on that data.

To exemplify the benefits of using these sequences, a small, but relatively complex, domain will be established. It contains three nominal features, F0, F1 and F2, and two integer features, F3 and F4. For an adversarial example to be realistic within this domain, it must comply with the following constraints:

- F0 must always keep its original value,
- F1 and F4 can be modified but must have class-specific values,
- F2 and F3 can be modified but must have class-specific values, which are influenced by F0 and F1.

The base configuration corresponding to these constraints specifies the feature subsets that each pattern will analyze and perturb:
1. Combination pattern—Modify {F1};
2. Combination pattern—Modify {F2, F3}, Lock {F0, F1};
3. Interval pattern—Modify {F3, F4}, Integer {F3, F4}.

A2PM will then assign each class to its own pattern sequence. For this example, the 'probability to be applied' will be 1.0 for all patterns, to demonstrate all three cumulative perturbations (Figure 4). The first perturbation created for each class is replacing F1 with another valid qualitative value, from 'B' to 'C'. Then, without modifying the original F0 nor the new F1, a valid combination is found for F0, F1, F2 and F3. Since the original F2 and F3 were only suitable for 'A' and 'B', new values are found to match 'A' and 'C'. Finally, the integer features F3 and F4 are perturbed according to their valid intervals. Regarding F3, to ensure it remains coherent with F0 and F1, the perturbation is created on the value of the new combination.

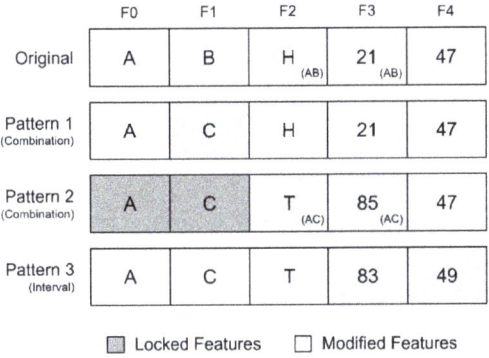

Figure 4. Exemplification of a perturbation pattern sequence.

4. Experimental Evaluation

A case study was conducted to evaluate the capabilities of the proposed method, as well as its suitability for multi-class classification on the cybersecurity domain. Assessments of example realism and time consumption were performed by comparing the examples generated by A2PM with the original data and recording the time required for each iteration. To thoroughly analyze example realism, the assessment included examples generated by the potential alternatives of the current literature: JSMA and OnePixel.

Since the internal mechanics of an ANN and a tree-based algorithm are noticeably different, the susceptibility of both types of models to A2PM was analyzed by performing targeted and untargeted attacks against MLP and RF classifiers. Two scenarios were considered: Enterprise and IoT networks. For these scenarios, adversarial network traffic flows were generated using the original flows of the CIC-IDS2017 and the IoT-23 datasets, respectively. In addition to evaluating the robustness of models created with regular training, the effects of performing adversarial training with A2PM were also analyzed.

The study was conducted on relatively common hardware: a machine with 16 gigabytes of random-access memory, an 8-core central processing unit, and a 6-gigabyte graphics processing unit. The implementation relied on the Python 3 programming language and several libraries: *Numpy* and *Pandas* for data preprocessing and manipulation, *Tensorflow* for the MLP models, *Scikit-learn* for the RF models, and *Adversarial-Robustness-Toolbox* for the alternative methods. The following subsections describe the most relevant aspects of the case study and present an analysis of the obtained results.

4.1. Datasets and Data Preprocessing

Both CIC-IDS2017 and IoT-23 are public datasets that contain multiple labeled captures of benign and malicious network flows. The recorded data is extremely valuable for intrusion detection because it includes various types of common cyber-attacks and manifests real network traffic patterns.

CIC-IDS2017 [37] consists of seven captures of cyber-attacks performed on a standard enterprise computer network with 25 interacting users. It includes denial-of-service and brute-force attacks, which were recorded in July 2017 and are available at the Canadian Institute for Cybersecurity. In contrast, IoT-23 [38] is directed at the emerging IoT networks, with wireless communications between interconnected devices. It contains network traffic created by malware attacks targeting IoT devices between 2018 and 2019, divided into 23 captures and available at the Stratosphere Research Laboratory.

From each dataset, two captures were selected and merged, to be utilized for the corresponding scenario. Table 2 provides an overview of their characteristics, including the class proportions and the label of each class, either 'Benign' or a specific type of cyber-attack. The 'PartOfAHorizontalPortScan' label was shortened to 'POAHPS'.

Table 2. Main characteristics of utilized datasets.

Scenario	Dataset (Captures)	Total Samples	Class Samples	Class Label
Enterprise Network	CIC-IDS2017 (Tuesday and Wednesday)	1,138,612	873,066	Benign
			230,124	Hulk
			10,293	GoldenEye
			7926	FTP-Patator
			5897	SSH-Patator
			5796	Slowloris
			5499	Slowhttptest
			11	Heartbleed
IoT Network	IoT-23 (1-1 and 34-1)	1,031,893	539,587	POAHPS
			471,198	Benign
			14,394	DDoS
			6714	C&C

Before their data was usable, both datasets required a similar preprocessing stage. First, the features that did not provide any valuable information about a flow's benign or malicious purpose, such as timestamps and IP addresses, were discarded. Then, the categorical features were converted to numeric values by performing one-hot encoding. Due to the high cardinality of these features, the very low frequency categories were aggregated into a single category designated as 'Other', to avoid encoding qualitative values that were present in almost no samples and therefore had a small relevance.

Finally, the holdout method was applied to randomly split the data into training and evaluation sets with 70% and 30% of the samples. To ensure that the original class proportions were preserved, the split was performed with stratification. The resulting CIC-IDS2017 sets were comprised of eight imbalanced classes and 83 features, 58 numerical and 25 categorical, whereas the IoT-23 sets contained four imbalanced classes and approximately half the structural size, with 42 features, 8 numerical and 34 categorical.

4.2. Base Configurations

After the data preprocessing stage, the distinct characteristics of the datasets were analyzed to identify their concrete constraints and establish the base configurations for A2PM. Regarding CIC-IDS2017, some numerical features had discrete values that could only have integer perturbations. Due to the correlation between the encoded categorical features, they required combined perturbations to be compatible with a valid flow. Additionally, to guarantee the coherence of a generated flow with its type of cyber-attack, the encoded features representing the utilized communication protocol and endpoint, designated as port, could not be modified. Hence, the following configuration was used for the Enterprise scenario, after it was converted to the respective subset of feature indices:

1. Interval pattern—Modify {numerical features}, Integer {discrete features};
2. Combination pattern—Modify {categorical features}, Lock {port, protocol}.

Despite the different features of IoT-23, it presented similar constraints. The main difference was that, in addition to the communication protocol, a generated flow had to be coherent with the application protocol as well, which was designated as service. The base configuration utilized for the IoT scenario was:

1. Interval pattern—Modify {numerical features}, Integer {discrete features};
2. Combination pattern—Modify {categorical features}, Lock {port, protocol, service}.

It is pertinent to note that, for the 'Benign' class, A2PM would only generate benign network traffic that could be misclassified as a cyber-attack. Therefore, the configurations were only applied to the malicious classes, to generate examples compatible with their malicious purposes. Furthermore, since the examples should resemble the original flows as much as possible, the 'probability to be applied' was 0.6 and 0.4 for the interval and combination patterns, respectively. These values were established to slightly prioritize the small-scale modifications of individual numerical features over the more significant modifications of combined categorical features.

4.3. Models and Fine-Tuning

A total of four MLP and four RF classifiers were created, one per scenario and training approach: regular or adversarial training. The first approach used the original training sets, whereas the latter augmented the data with one adversarial example per malicious flow. To prevent any bias, the examples were generated by adapting A2PM solely to the training data. The models and their fine-tuning process are described below.

An MLP [39] is a feedforward ANN consisting of an input layer, an output layer and one or more hidden layers in between. Each layer can contain multiple nodes with forward connections to the nodes of the next layer. When utilized as a classifier, the number of input and output nodes correspond to the number of features and classes, respectively, and a prediction is performed according to the activations of the output nodes.

Due to the high computational cost of training an MLP, it was fine-tuned using a Bayesian optimization technique [40]. A validation set was created with 20% of a training set, which corresponded to 14% of the original samples. Since an MLP accounts for the loss of the training data, the optimization sought to minimize the loss of the validation data. To prevent overfitting, early stopping was employed to end the training when this loss stabilized. Additionally, due to the class imbalance present in both datasets, the assigned class weights were inversely proportional to their frequency.

The fine-tuning led to a four-layered architecture with a decreasing number of nodes for both training approaches. The hidden layers relied on the computationally efficient rectified linear unit (ReLU) activation function and the dropout technique, which inherently prevents overfitting by randomly ignoring a certain percentage of the nodes during training. To address multi-class classification, the Softmax activation function was used to normalize the outputs to a class probability distribution. The MLP architecture for the Enterprise scenario was:

1. Input layer—83 nodes, 512 batch size;
2. Hidden layer—64 nodes, ReLU activation, 10% dropout;
3. Hidden layer—32 nodes, ReLU activation, 10% dropout;
4. Output layer—8 nodes, Softmax activation.

A similar architecture was utilized for the IoT scenario, although it presented a decreased batch size and an increased dropout:

1. Input layer—42 nodes, 128 batch size;
2. Hidden layer—32 nodes, ReLU activation, 20% dropout;

3. Hidden layer—16 nodes, ReLU activation, 20% dropout;
4. Output layer—4 nodes, Softmax activation.

The remaining parameters were common to both scenarios because of their equivalent classification tasks. Table 3 summarizes the MLP configuration.

Table 3. Summary of multilayer perceptron configuration.

Parameter	Value
Objective Loss	Categorical Cross-Entropy
Optimizer	Adam Algorithm
Learning Rate	0.001
Maximum Epochs	50
Class Weights	Balanced

On the other hand, an RF [41] is an ensemble of decision trees, where each individual tree performs a prediction according to a different feature subset, and the most voted class is chosen. It is based on the wisdom of the crowd, the idea that a multitude of classifiers will collectively make better decisions than just one.

Since training an RF has a significantly lower computational cost, a five-fold cross-validated grid search was performed with well-established hyperparameter combinations. In this process, five stratified subsets were created, each with 20% of a training set. Then, five distinct iterations were performed, each training a model with four subsets and evaluating it with the remaining one. Hence, the MLP validation approach was replicated five times per combination. The macro-averaged F1-Score, which will be described in the next subsection, was selected as the metric to be maximized. Table 4 summarizes the optimized RF configuration, common to both scenarios and training approaches.

Table 4. Summary of random forest configuration.

Parameter	Value
Splitting Criteria	Gini Impurity
Number of Trees	100
Maximum Depth of a Tree	32
Minimum Samples in a Leaf	2
Maximum Features	$\sqrt{\text{Number of Features}}$
Class Weights	Balanced

4.4. Attacks and Evaluation Metrics

A2PM was applied to perform adversarial attacks against the fine-tuned models for a maximum of 50 iterations, by adapting to the data of the holdout evaluation sets. The attacks were untargeted, causing any misclassification of malicious flows to different classes, as well as targeted, seeking to misclassify malicious flows as the 'Benign' class. To perform a trustworthy evaluation of the impact of the generated examples on a model's performance, it was essential to select appropriate metrics. The considered metrics and their interpretation are briefly described below [42,43].

Accuracy measures the proportion of correctly classified samples. Even though it is the standard metric for classification tasks, its bias towards the majority classes must not be disregarded when the minority classes are particularly relevant to a classification task, which is the case of network-based intrusion detection [44]. For instance, in the Enterprise scenario, 77% of the samples have the 'Benign' class label. Since A2PM was configured to not generate examples for that class, even if an adversarial attack was successful and all generated flows evaded detection, an accuracy score as high as 77% could still be achieved. Therefore, to correctly exhibit the misclassifications caused by the performed attacks, the accuracy of a model was calculated using the network flows of all classes except 'Benign'. This metric can be expressed as:

$$Accuracy = \frac{TP + TN}{TP + TN + FP + FN} \quad (5)$$

where TP and TN are the number of true positives and negatives, correct classifications, and FP and FN are the number of false positives and negatives, misclassifications.

Despite the reliability of accuracy for targeted attacks, it does not entirely reflect the impact of the performed untargeted attacks. Due to their attempt to cause any misclassification, their impact across all the different classes must also be measured. The F1-Score calculates the harmonic mean of precision and recall, considering both false positives and false negatives. To account for class imbalance, it can be macro-averaged, which gives all classes the same relevance. This is a reliable evaluation metric because a score of 100% indicates that all cyber-attacks are being correctly detected and there are no false alarms. Additionally, due to the multiple imbalanced classes present in both datasets, it is also the most suitable validation metric for the employed fine-tuning approach. The macro-averaged F1-Score is mathematically defined as:

$$\text{Macro-averaged F1-Score} = \frac{1}{C} * \sum_{i=1}^{C} \frac{2 * P_i * R_i}{P_i + R_i} \quad (6)$$

where P_i and R_i are the precision and recall of class i, and C is the number of classes.

4.5. Enterprise Scenario Results

In the Enterprise network scenario, adversarial cyber-attack examples were generated using the original flows of the CIC-IDS2017 dataset. The results obtained for the targeted and untargeted attacks were analyzed, and assessments of example realism and time consumption were performed. To assess the realism of the generated examples, these were analyzed and compared with the corresponding original flows, considering the intricacies and malicious purposes of the cyber-attacks. In addition to A2PM, the assessment included its potential alternatives: JSMA and OnePixel. To prevent any bias, a randomly generated number was used to select one example, detailed below.

The selected flow had the 'Slowloris' class label, corresponding to a denial-of-service attack that attempts to overwhelm a web server by opening multiple connections and maintaining them as long as possible [45]. The data perturbations created by A2PM increased the total flow duration and the packet inter-arrival time (IAT), while reducing the number of packets transmitted per second and their size. These modifications were mostly focused on enhancing time-related aspects of the cyber-attack, to prevent its detection. Hence, in addition to being valid network traffic that can be transmitted through a computer network, the adversarial example also remained coherent with its class.

On the other hand, JSMA could not generate a realistic example for the selected flow. It created a major inconsistency in the encoded categorical features by assigning a single network flow to two distinct communication endpoints: destination ports number 80 (P80) and 88 (P88). Due to the unconstrained perturbations, the value of the feature representing P88 was increased without accounting for its correlation with P80, which led to an invalid example. In addition to the original Push flag (PSH) to keep the connection open, the method also assigned the Finished flag (FIN), which signals for connection termination and therefore contradicts the cyber-attack's purpose. Even though two numerical features were also slightly modified, the adversarial example could only evade detection by using categorical features incompatible with real network traffic.

Similarly, OnePixel also generated an example that contradicted the 'Slowloris' class. The feature selected to be perturbed represented the Reset flag (RST), which also causes termination. Since the method intended to perform solely one modification, it increased the value of a feature that no model learnt to detect because it is incoherent with that cyber-attack. Consequently, neither JSMA nor OnePixel are adequate alternatives to A2PM for tabular data. Table 5 provides an overview of the modified features. The '–' character indicates that the original value was not perturbed.

Table 5. Modified features of an adversarial 'Slowloris' example.

Feature	Original Value	A2PM Value	JSMA Value	OnePixel Value
Flow duration	109,034,141	119,046,064	109,034,140	–
Mean flow IAT	13,600,000	19,374,259	–	–
Flow packets per second	0.0825	0.0429	0.0824	–
Mean forward packet length	49.4	48.1	–	–
Minimum forward segment size	40	36	–	–
Connection flags	'PSH'	–	'PSH' + 'FIN'	'PSH' + 'RST'
Destination port	'P80'	–	'P80' + 'P88'	–

Regarding the targeted attacks performed by A2PM, the models created with regular training exhibited significant performance declines. Even though both MLP and RF achieved over 99% accuracy on the original evaluation set, a single iteration lowered their scores by approximately 15% and 33%. In the subsequent iterations, more malicious flows gradually evaded MLP detection, whereas RF was quickly exploited. After 50 iterations, their very low accuracy evidenced their inherent susceptibility to adversarial examples. In contrast, the models created with adversarial training kept significantly higher scores, with fewer flows being misclassified as benign. By training with one generated example per malicious flow, both classifiers successfully learned to detect most cyber-attack variations. RF stood out for preserving the 99.91% it obtained on the original data throughout the entire attack, which highlighted its excellent generalization (Figure 5).

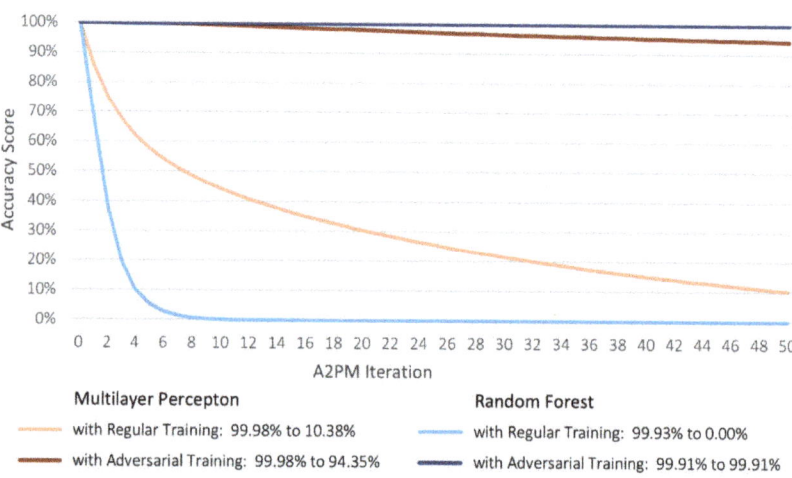

Figure 5. Targeted attack accuracy of Enterprise network scenario.

The untargeted attacks significantly lowered both evaluation metrics. The accuracy and macro-averaged F1-Score declines of the regularly trained models were approximately 99% and 79%, although RF was more affected in the initial iterations. The inability of both classifiers to distinguish between the different classes corroborated their high susceptibility to adversarial examples. Nonetheless, when adversarial training was performed, the models preserved considerably higher scores, with a gradual decrease of less than 2% per iteration. Despite some examples still deceiving them into predicting incorrect classes, both models were able to learn the intricacies of each type of cyber-attack, which mitigated the impact of the created data perturbations. The adversarially trained RF consistently reached higher scores than MLP in both targeted and untargeted attacks, indicating a better robustness (Figures 6 and 7).

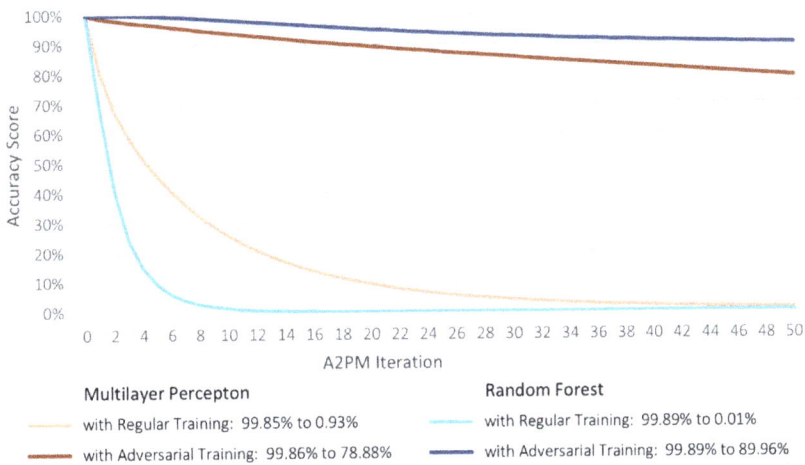

Figure 6. Untargeted attack accuracy of Enterprise network scenario.

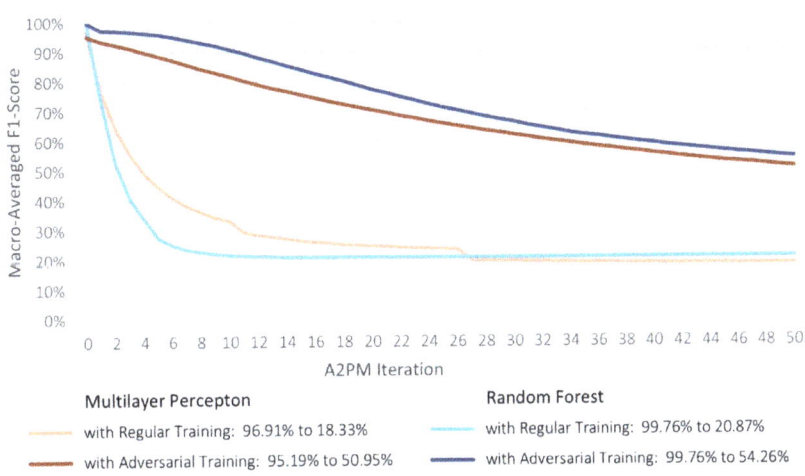

Figure 7. Untargeted attack F1-Score of Enterprise network scenario.

To analyze the time consumption of A2PM, the number of milliseconds required for each iteration was recorded and averaged, accounting for the decreasing quantity of new examples generated as an attack progressed. The generation was performed at a rate of 10 examples per 1.7 milliseconds on the utilized hardware, which evidenced the fast execution and scalability of the proposed method when applied to adversarial training and attacks in enterprise computer networks.

4.6. IoT Scenario Results

In the IoT network scenario, the adversarial cyber-attack examples were generated using the original flows of the IoT-23 dataset. The analysis performed for the previous scenario was replicated to provide similar assessments, including the potential alternatives of the current literature: JSMA and OnePixel.

The randomly selected flow for the assessment of example realism had the 'DDoS' class label, which corresponds to a distributed denial-of-service attack performed by the malwares recorded in the IoT-23 dataset. A2PM replaced the encoded categorical features of the connection state and history with another valid combination, already used by other original flows of the 'DDoS' class. Instead of an incomplete connection (OTH) with a bad packet checksum (BC), it became a connection attempt (S0) with a Synchronization flag (SYN). Hence, the generated network flow example remained valid and compatible with its intended malicious purpose, achieving realism.

As in the previous scenario, both JSMA and OnePixel generated unrealistic examples. Besides the original OTH, both methods also increased the value of the feature representing an established connection with a termination attempt (S3). Since a flow with simultaneous OTH and S3 states is neither valid nor coherent with the cyber-attack's purpose, the methods remain inadequate alternatives to A2PM for tabular data. In addition to the states, JSMA also assigned a single flow to two distinct communication protocols, transmission control protocol (TCP) and Internet control message protocol (ICMP), which further evidenced the inconsistency of the created data perturbations. Table 6 provides an overview of the modified features, with '–' indicating an unperturbed value.

Table 6. Modified features of an adversarial 'DDoS' example.

Feature	Original Value	A2PM Value	JSMA Value	OnePixel Value
Connection state	'OTH'	'S0'	'OTH' + 'S3'	'OTH' + 'S3'
Connection history	'BC'	'SYN'	–	–
Communication protocol	'TCP'	–	'TCP' + 'ICMP'	–

Regarding the targeted attacks, A2PM caused much slower declines than in the previous scenario. The accuracy of the regularly trained MLP only started being lower than 50% at iteration 43, and RF stabilized with approximately 86%. These scores evidenced the decreased susceptibility of both classifiers, especially RF, to adversarial examples targeting the 'Benign' class. Furthermore, with adversarial training, the models were able to preserve even higher rates during an attack. Even though many examples still evaded MLP detection, the number of malicious flows predicted to be benign by RF was significantly lowered, which enabled it to keep its accuracy above 99%. Hence, the latter successful detected most cyber-attack variations (Figure 8).

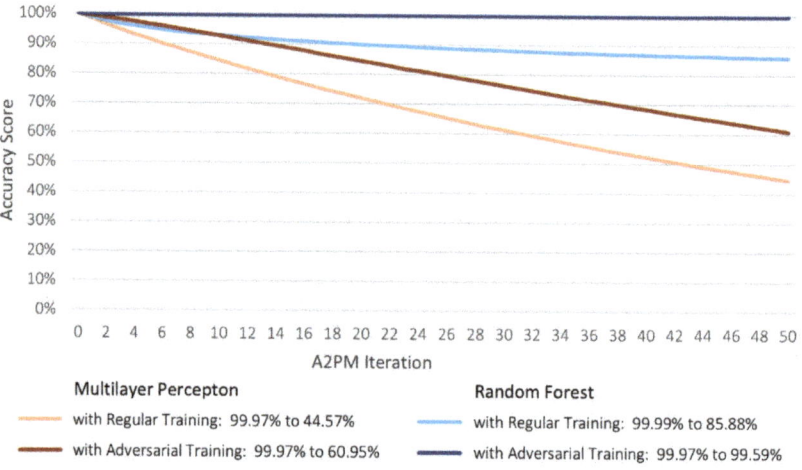

Figure 8. Targeted attack accuracy of IoT network scenario.

The untargeted attacks iteratively caused small decreases of both metrics. Despite RF starting to stabilize from the fifth iteration forward, MLP continued its decline for an additional 48% of accuracy and 17% of macro-averaged F1-Score. This difference in both targeted and untargeted attacks suggests that RF, and possibly tree-based algorithms in general, have a better inherent robustness to adversarial examples of IoT network traffic. Unlike in the previous scenario, adversarial training did not provide considerable improvements. Nonetheless, the augmented training data still contributed to the creation of more adversarially robust models because they exhibited fewer incorrect class predictions throughout the attack (Figures 9 and 10).

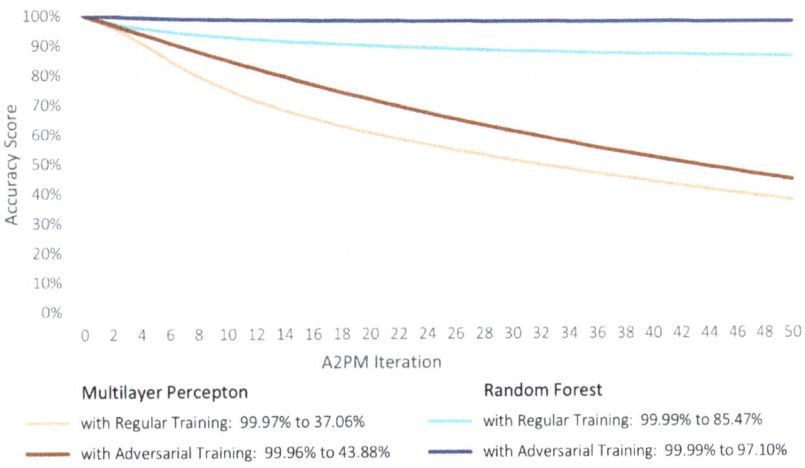

Figure 9. Untargeted attack accuracy of IoT network scenario.

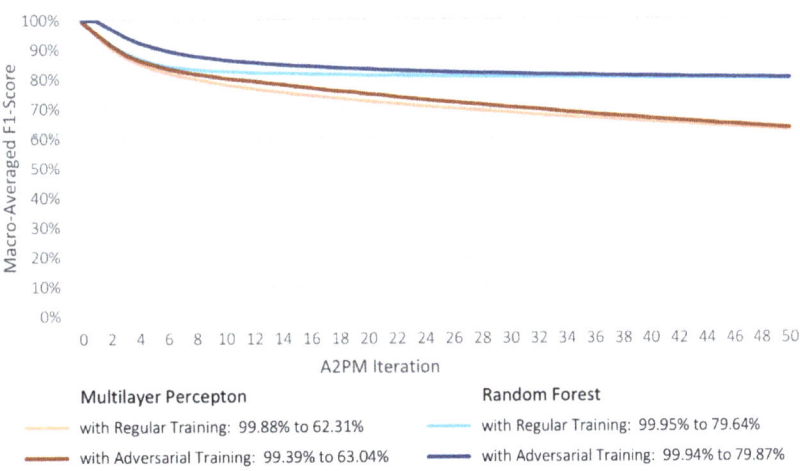

Figure 10. Untargeted attack F1-Score of IoT network scenario.

A time consumption analysis was also performed, to further analyze the scalability of A2PM on relatively common hardware. The number of milliseconds required for each iteration was recorded and averaged, resulting in a rate of 10 examples per 2.4 milliseconds. By comparing the rate obtained in both scenarios, it can be observed that it was 41% higher for IoT-23 than for CIC-IDS2017. Even though the former dataset had approximately half the structural size, a greater number of locked categorical features were provided to the combination pattern. Therefore, the increased rate suggests that the more complex inter-feature constraints are specified, the more time will be required to apply A2PM. Nonetheless, the time consumption was still reasonably low, which further evidenced the fast execution and scalability of the proposed method.

5. Conclusions

This work established the domain and class-specific constraint levels, which an adversarial example must comply with to achieve realism on tabular data and introduced A2PM to fulfil these constraints in a gray-box setting, with only knowledge of the feature set. The capabilities of the proposed method were evaluated in a cybersecurity case study with two scenarios: Enterprise and IoT networks. MLP and RF classifiers were created with regular and adversarial training, using the network flows of the CIC-IDS2017 and IoT-23 datasets, and targeted and untargeted attacks were performed against them. For each scenario, the impact of the attacks was analyzed, and assessments of example realism and time consumption were performed.

The modular architecture of A2PM enabled the creation of pattern sequences adapted to each type of cyber-attack, according to the concrete constraints of the utilized datasets. Both targeted and untargeted attacks successfully decreased the performance of all MLP and RF models, with significantly higher declines exhibited in the Enterprise scenario. Nonetheless, the inherent susceptibility of these models to adversarial examples was mitigated by augmenting their training data with one generated example per malicious flow. Overall, the obtained results demonstrate that A2PM provides a scalable generation of valid and coherent examples for network-based intrusion detection. Therefore, the proposed method can be advantageous for adversarial attacks, to iteratively cause misclassifications, and adversarial training, to increase the robustness of a model.

In the future, the patterns can be improved to enable the configuration of more complex intra and inter-feature constraints. Since it is currently necessary to use both interval and combination patterns to perturb correlated numerical features, a new pattern can be developed to address their required constraints. It is also imperative to analyze other datasets and other domains to contribute to robustness research. Future case studies can further reduce the knowledge required to create realistic examples.

Author Contributions: Conceptualization, J.V., N.O. and I.P.; methodology, J.V. and N.O.; software, J.V.; validation, N.O. and I.P.; investigation, J.V. and I.P.; writing, J.V. and I.P.; supervision, I.P.; project administration, I.P.; funding acquisition, I.P. All authors have read and agreed to the published version of the manuscript.

Funding: The present work has received funding from the European Union's Horizon 2020 research and innovation program, under project SeCoIIA (grant agreement no. 871967). This work has also received funding from UIDP/00760/2020.

Data Availability Statement: Publicly available datasets were analyzed in this work. The data can be found at: CIC-IDS2017 (https://www.unb.ca/cic/datasets/ids-2017.html, accessed on 7 March 2022), IoT-23 (https://www.stratosphereips.org/datasets-iot23, accessed on 7 March 2022). A novel method was developed in this work. An implementation in the Python 3 programming language can be found at: A2PM (https://github.com/vitorinojoao/a2pm, accessed on 7 March 2022).

Conflicts of Interest: The authors declare no conflict of interest. The funders had no role in the design of the study; in the collection, analyses, or interpretation of data; in the writing of the manuscript, or in the decision to publish the results.

References

1. Szegedy, C. Intriguing properties of neural networks. In Proceedings of the 2nd International Conference on Learning Representations, ICLR 2014, Banff, AB, Canada, 14–16 April 2014; Conference Track Proceedings. pp. 1–10.
2. European Union Agency for Cybersecurity; Malatras, A.; Dede, G. AI Cybersecurity Challenges: Threat Landscape for Artificial Intelligence. 2020. Available online: https://op.europa.eu/en/publication-detail/-/publication/e52bf2d7-4017-11eb-b27b-01aa75ed71a1/language-en (accessed on 7 March 2022). [CrossRef]

3. Goodfellow, I.J.; Shlens, J.; Szegedy, C. Explaining and harnessing adversarial examples. In Proceedings of the 3rd International Conference on Learning Representations, ICLR 2015, San Diego, CA, USA, 7–9 May 2015; Conference Track Proceedings. pp. 1–11.
4. European Union Agency for Cybersecurity; Malatras, A.; Agrafiotis, I.; Adamczyk, M. Securing Machine Learning Algorithms. 2022. Available online: https://op.europa.eu/en/publication-detail/-/publication/c7c844fd-7f1e-11ec-8c40-01aa75ed71a1/language-en (accessed on 7 March 2022). [CrossRef]
5. Yuan, X.; He, P.; Zhu, Q.; Li, X. Adversarial Examples: Attacks and Defenses for Deep Learning. *IEEE Trans. Neural Netw. Learn. Syst.* **2019**, *30*, 2805–2824. [CrossRef] [PubMed]
6. Pitropakis, N.; Panaousis, E.; Giannetsos, T.; Anastasiadis, E.; Loukas, G. A taxonomy and survey of attacks against machine learning. *Comput. Sci. Rev.* **2019**, *34*, 100199. [CrossRef]
7. Qiu, S.; Liu, Q.; Zhou, S.; Wu, C. Review of Artificial Intelligence Adversarial Attack and Defense Technologies. *Appl. Sci.* **2019**, *9*, 909. [CrossRef]
8. Apruzzese, G.; Andreolini, M.; Ferretti, L.; Marchetti, M.; Colajanni, M. Modeling Realistic Adversarial Attacks against Network Intrusion Detection Systems. *Digit. Threat. Res. Pract.* **2021**, *1*. [CrossRef]
9. Corona, I.; Giacinto, G.; Roli, F. Adversarial attacks against intrusion detection systems: Taxonomy, solutions and open issues. *Inf. Sci.* **2013**, *239*, 201–225. [CrossRef]
10. Madry, A.; Makelov, A.; Schmidt, L.; Tsipras, D.; Vladu, A. Towards deep learning models resistant to adversarial attacks. In Proceedings of the 6th International Conference on Learning Representations, ICLR 2018, Vancouver, BC, Canada, April 30–3 May 2018; Conference Track Proceedings. pp. 1–28.
11. Schmidt, L.; Talwar, K.; Santurkar, S.; Tsipras, D.; Madry, A. Adversarially robust generalization requires more data. *Adv. Neural Inf. Process. Syst.* **2018**, *31*, 5014–5026.
12. Ganin, Y.; Ustinova, E.; Ajakan, H.; Germain, P.; Larochelle, H.; Laviolette, F.; Marchand, M.; Lempitsky, V. *Domain-Adversarial Training of Neural Networks*; Advances in Computer Vision and Pattern Recognition Book Series; 2017; pp. 189–209. Available online: https://www.jmlr.org/papers/volume17/15-239/15-239.pdf (accessed on 7 March 2022).
13. Ullah, S.; Khan, M.A.; Ahmad, J.; Jamal, S.S.; e Huma, Z.; Hassan, M.T.; Pitropakis, N.; Arshad; Buchanan, W.J. HDL-IDS: A Hybrid Deep Learning Architecture for Intrusion Detection in the Internet of Vehicles. *Sensors* **2022**, *22*, 1340. [CrossRef] [PubMed]
14. Tramèr, F.; Kurakin, A.; Papernot, N.; Goodfellow, I.; Boneh, D.; McDaniel, P. Ensemble adversarial training: Attacks and defenses. In Proceedings of the 6th International Conference on Learning Representations, ICLR 2018, Vancouver, BC, Canada, 30 April–3 May 2018; Conference Track Proceedings. pp. 1–22.
15. Belavagi, M.C.; Muniyal, B. Performance Evaluation of Supervised Machine Learning Algorithms for Intrusion Detection. *Procedia Comput. Sci.* **2016**, *89*, 117–123. [CrossRef]
16. Primartha, R.; Tama, B.A. Anomaly detection using random forest: A performance revisited. In Proceedings of the 2017 International Conference Data Software Engineering, Palembang, Indonesia, 1–2 November 2017; pp. 1–6. [CrossRef]
17. Kantchelian, A.; Tygar, J.D.; Joseph, A.D. Evasion and hardening of tree ensemble classifiers. In Proceedings of the 33rd International Conference on Machine Learning, New York, NY, USA, 20–22 June 2016; Volume 5, pp. 3562–3573.
18. Chen, H.; Zhang, H.; Boning, D.; Hsieh, C.J. Robust decision trees against adversarial examples. In Proceedings of the 36th International Conference on Machine Learning, ICML 2019, Long Beach, CA, USA, 9–15 June 2019; Volume 97, pp. 1911–1926.
19. Vos, D.; Verwer, S. Efficient Training of Robust Decision Trees Against Adversarial Examples. In Proceedings of the 38th International Conference on Machine Learning, Online, 18–24 July 2021; Volume 139, pp. 10586–10595. Available online: https://proceedings.mlr.press/v139/vos21a.html (accessed on 7 March 2022).
20. Chen, Y.; Wang, S.; Jiang, W.; Cidon, A.; Jana, S. Cost-aware robust tree ensembles for security applications. In Proceedings of the 30th USENIX Security Symposium, Online, 11–13 August 2021; pp. 2291–2308. Available online: https://www.usenix.org/conference/usenixsecurity21/presentation/chen-yizheng (accessed on 7 March 2022).
21. Martins, N.; Cruz, J.M.; Cruz, T.; Abreu, P.H. Adversarial Machine Learning Applied to Intrusion and Malware Scenarios: A Systematic Review. *IEEE Access* **2020**, *8*, 35403–35419. [CrossRef]
22. Carlini, N.; Wagner, D. Towards Evaluating the Robustness of Neural Networks. In Proceedings of the 2017 IEEE Symposium on Security and Privacy (SP), San Jose, CA, USA, 22–26 May 2017; pp. 39–57. [CrossRef]
23. Moosavi-Dezfooli, S.M.; Fawzi, A.; Frossard, P. DeepFool: A Simple and Accurate Method to Fool Deep Neural Networks. In Proceedings of the IEEE Conference on Computer Vision and Pattern Recognition (CVPR), Las Vegas, NV, USA, 27–30 June 2016; pp. 2574–2582. [CrossRef]
24. Cisse, M.; Adi, Y.; Neverova, N.; Keshet, J. Houdini: Fooling Deep Structured Prediction Models. 2017. Available online: http://arxiv.org/abs/1707.05373 (accessed on 7 March 2022).
25. Xu, K. Structured adversarial attack: Towards general implementation and better interpretability. In Proceedings of the 7th International Conference on Learning Representations, ICLR 2019, New Orleans, LA, USA, 6–9 May 2019.
26. Chen, P.Y.; Zhang, H.; Sharma, Y.; Yi, J.; Hsieh, C.J. ZOO: Zeroth order optimization based black-box atacks to deep neural networks without training substitute models. In Proceedings of the 10th International Workshop on Artificial Intelligence and Security (AISec 2017), Dallas, TX, USA, 3 November 2017; pp. 15–26. [CrossRef]
27. Papernot, N.; McDaniel, P.; Jha, S.; Fredrikson, M.; Celik, Z.B.; Swami, A. The Limitations of Deep Learning in Adversarial Settings. In Proceedings of the 2016 IEEE European Symposium on Security and Privacy, Saarbruecken, Germany, 21–24 March 2016; pp. 372–387. [CrossRef]

28. Chauhan, R.; Sabeel, U.; Izaddoost, A.; Heydari, S.S. Polymorphic Adversarial Cyberattacks Using WGAN. *J. Cybersecur. Priv.* **2021**, *1*, 767–792. [CrossRef]
29. Xu, Y.; Zhong, X.; Yepes, A.J.; Lau, J.H. Grey-box Adversarial Attack And Defence For Sentiment Classification. In Proceedings of the 2021 Conference of the North American Chapter of the Association for Computational Linguistics: Human Language Technologies, Online, 6–11 June 2021; pp. 4078–4087. [CrossRef]
30. Su, J.; Vargas, D.V.; Sakurai, K. One Pixel Attack for Fooling Deep Neural Networks. *IEEE Trans. Evol. Comput.* **2019**, *23*, 828–841. [CrossRef]
31. Dai, H.; Li, H.; Tian, T.; Huang, X.; Wang, L.; Zhu, J.; Song, L. Adversarial Attack on Graph Structured Data. In Proceedings of the 35th International Conference on Machine Learning, Stockholm, Sweden, 10–15 July 2018; Volume 80, pp. 1115–1124. Available online: https://proceedings.mlr.press/v80/dai18b.html (accessed on 7 March 2022).
32. Lin, J.; Xu, L.; Liu, Y.; Zhang, X. Black-box adversarial sample generation based on differential evolution. *J. Syst. Softw.* **2020**, *170*, 110767. [CrossRef]
33. Goodfellow, I. Generative Adversarial Nets. In Proceedings of the Advances in Neural Information Processing Systems, Montreal, QC, Canada, 8–13 December 2014; Volume 27.
34. Arjovsky, M.; Chintala, S.; Bottou, L. Wasserstein Generative Adversarial Networks. In Proceedings of the 34th International Conference on Machine Learning, Sydney, Australia, 6–11 August 2017; pp. 1–44.
35. Brendel, W.; Rauber, J.; Bethge, M. Decision-based adversarial attacks: Reliable attacks against black-box machine learning models. In Proceedings of the 6th International Conference on Learning. Representations, ICLR 2018, Vancouver, BC, Canada, 30 April–3 May 2018; Conference Track Proceedings. pp. 1–12.
36. Cheng, M.; Zhang, H.; Hsieh, C.J.; Le, T.; Chen, P.Y.; Yi, J. Query-efficient hard-label black-box attack: An optimization-based approach. In Proceedings of the 7th International Conference on Learning Representations, ICLR 2019, New Orleans, LA, USA, 6–9 May 2019; pp. 1–12.
37. Sharafaldin, I.; Lashkari, A.H.; Ghorbani, A.A. Toward generating a new intrusion detection dataset and intrusion traffic characterization. In Proceedings of the 4th International Conference on Information Systems Security and Privacy, Funchal, Portugal, 22–24 January 2018; pp. 108–116. [CrossRef]
38. Garcia, S.; Parmisano, A.; Erquiaga, M.J. IoT-23: A labeled dataset with malicious and benign IoT network traffic. *Zenodo* **2020**. [CrossRef]
39. Murtagh, F. Multilayer perceptrons for classification and regression. *Neurocomputing* **1991**, *2*, 183–197. [CrossRef]
40. Snoek, J.; Larochelle, H.; Adams, R.P. Practical Bayesian Optimization of Machine Learning Algorithms. In Proceedings of the Advances in Neural Information Processing Systems, Lake Tahoe, NV, USA, 3–6 December 2012; Volume 25.
41. Breiman, L. Random forests. *Mach. Learn.* **2001**, *5*, 5–32. [CrossRef]
42. Powers, D.M.W. Evaluation: From precision, recall and F-measure to ROC, informedness, markedness and correlation. *arXiv* **2020**, arXiv:2010.16061. Available online: http://arxiv.org/abs/2010.16061 (accessed on 7 March 2022).
43. Hossin, M.; Sulaiman, M.N. A review on evaluation metrics for data classification evaluations. *Int. J. Data Min. Knowl. Manag. Process* **2015**, *5*, 1.
44. Oliveira, N.; Praça, I.; Maia, E.; Sousa, O. Intelligent cyber attack detection and classification for network-based intrusion detection systems. *Appl. Sci.* **2021**, *11*, 1674. [CrossRef]
45. Shorey, T.; Subbaiah, D.; Goyal, A.; Sakxena, A.; Mishra, A.K. performance comparison and analysis of slowloris, goldeneye and xerxes ddos attack tools. In Proceedings of the 2018 International Conference on Advances in Computing, Communications and Informatics, ICACCI 2018, Bangalore, India, 19–22 September 2018; pp. 318–322. [CrossRef]

Article

A Detection Method for Social Network Images with Spam, Based on Deep Neural Network and Frequency Domain Pre-Processing

Hua Shen [1,2,3], Xinyue Liu [2] and Xianchao Zhang [2,*]

[1] Faculty of Electronic Information and Electrical Engineering, Dalian University of Technology, Dalian 116024, China; huashen.cn@gmail.com
[2] School of Software, Dalian University of Technology, Dalian 116620, China; xyliu@dlut.edu.cn
[3] College of Mathematics and Information Science, Anshan Normal University, Anshan 114007, China
* Correspondence: xczhang@dlut.edu.cn

Abstract: As a result of the rapid development of internet technology, images are widely used on various social networks, such as WeChat, Twitter or Facebook. It follows that images with spam can also be freely transmitted on social networks. Most of the traditional methods can only detect spam in the form of links and texts; there are few studies on detecting images with spam. To this end, a novel detection method for identifying social images with spam, based on deep neural network and frequency domain pre-processing, is proposed in this paper. Firstly, we collected several images with embedded spam and combined the DIV2K2017 dataset to build an image dataset for training the proposed detection model. Then, the specific components of the spam in the images were determined through experiments and the pre-processing module was specially designed. Low-frequency domain regions with less spam are discarded through *Haar* wavelet transform analysis. In addition, a feature extraction module with special convolutional layers was designed, and an appropriate number of modules was selected to maximize the extraction of three different high-frequency feature regions. Finally, the different high-frequency features are spliced along the channel dimension to obtain the final classification result. Our extensive experimental results indicate that the spam element mainly exists in the images as high-frequency information components; they also prove that the proposed model is superior to the state-of-the-art detection models in terms of detection accuracy and detection efficiency.

Keywords: social networks; images with spam; *Haar* wavelet transform; feature extraction module

1. Introduction

Digital images are widely utilized in various social networks such as WeChat or Facebook, due to their convenience, fast acquisition, and abundance of redundant information [1–6]. While digital images bring convenience to people's lives, some security risks also follow. To receive free advertising and for other more harmful purposes, some criminals paste links, text, and additional pictures on images that seriously disrupt the order and security of social networks. Therefore, finding ways to accurately and quickly detect images containing spam is a huge challenge for researchers [7–10]. This research field is also of great significance for purifying social networks and improving the security of the social network environment.

In the past few decades, most research has focused on how to detect target objects, such as links, emails, texts, etc., and research on detecting images that include spam is still very rare. Zhu et al. [11] proposed a supervised matrix factorization method with social regularization (SMFSR) for spammer detection in social networks. Their method realized the detection task by combining the user's social behavior and social relationships, detecting some data from Renren.com and obtaining relatively good detection results. Hu et al. [12] focused on studying how to use network and content information together in Weibo to perform effective social spam detection. In addition, an optimization formula is designed to combine social network and content information for optimizing the model. The experimental results also show that their model can achieve good detection results on Twitter. Wu et al. [13] proposed a unified detection method for the collaborative combination of social spammers and spam messages on Weibo. Their approach combines social spam detection with spam detection exploiting the publishing relationship between the users and the message. Furthermore, an optimization schedule is introduced to improve the capability of their model, and an acceleration strategy is also proposed to improve the detection efficiency of the model. Chen et al. [14] analyzed the vulnerabilities of current detection methods from the perspective of three aspects: data, features, and models. Traditional machine learning technology is introduced to extract features for accomplishing binary classification tasks. In addition, the detection performance of the proposed method was evaluated in terms of the different aspects of the factors. Masood et al. [15] proposed a detection classification method for Twitter spam. The proposed method compared techniques based on several features, such as user characteristics, content characteristics, graphic characteristics, structural characteristics, temporal characteristics, etc. In addition, this paper also expounded on the future development direction of this field and offered solutions for some of the issues. Ahmed et al. [16] analyzed the advantages and challenges of machine learning in the field of spam detection and performed detailed comparative experiments to illustrate the scalability of machine learning in this field. In the same year, Sokhangoee et al. [17] proposed a new method for spam detection based on association-rule mining and genetic algorithm theory. The premise of this method effectively improved the detection accuracy for spam because more refined features can be extracted by combining a genetic algorithm and association rules. According to the above research, it can be seen that the current detection methods for links and text content are very mature; however, the detection methods for images that include spam are rarely studied, which shows that this field regarding images with spam is still in the initial stages.

In recent years, with the rapid development of computer hardware and network bandwidth, the field of artificial intelligence and deep learning has attracted extensive interest from researchers. So far, deep learning and CNN (convolutional neural networks) have provided many good solutions in various fields, such as image recognition [18,19], speech recognition, and natural language processing [20]. Therefore, in this era of deep learning, CNN provides an opportunity for the detection of images with spam. Xie et al. [21] proposed a detection method for pornographic images based on global classification and local sensitive information classification. CNN was introduced to extract image features such as color and texture, and an attention mechanism was utilized as the backbone of the network. Finally, discriminant results were obtained via the Softmax activation function. The experimental results show that their method can detect pornographic images efficiently from a specific dataset. Zhang et al. [22] proposed an image classification method for bad images, based on deep learning model integration, which achieved semantic complementarity by utilizing the image representation capabilities of multiple different deep networks and fused all the obtained features to improve the classification performance of the proposed model. Compared with traditional classification methods, their model has greatly improved upon previous accuracy rates. Cai et al. [23] proposed a method for detecting spam on the Internet, based on the BERT (Bidirectional Encoder Representation from Transformers) model, where the processing object comprises text information. Firstly, a bidirectional transformer structure was used to extract the contextual relationship information of the text content, then the trained BERT model was directly used to encode the sentences of the new task. Then, sentences of any length were encoded into fixed-length vectors to detect and analyze spam websites. From research in recent years, it can be seen that deep learning has made some progress in the field of spam detection, but most of the models focus on the detection of target objects, such as links and text, and research on spam detection in the context of images is still sparse.

To tackle the existing problems of detecting images containing spam, this paper proposes a detection method for social network images with spam based on deep neural network and frequency domain pre-processing. For this paper, first, we collected some images that included spam and combined the DIV2K2017 dataset to build a dataset for training the detection model. (Please note: the DIV2K dataset is a popular single-image super-resolution dataset that contains 1000 images of different scenes. In addition, this dataset contains low-resolution images with different types of degradations, which conform to all kinds of images that are common in everyday life; therefore, the dataset was suitable for training the proposed model). In the pre-processing stage, *Haar* wavelet transform analysis was utilized to extract different frequency domain information from the input image. Meanwhile, the low-frequency information of the image was discarded and the high-frequency information of three different frequency components was used as the input of the feature extraction stage, to improve the efficiency of the model. In the feature extraction stage, a feature extraction module with the designated convolution layers was designed, and an appropriate number of modules was selected through experiments to extract the vertical, horizontal, and diagonal high-frequency features of the input image, so as to maximize the extraction of the defective image characteristics of the information. The obtained different frequency domain features were subjected to the concat operation to obtain the final target feature, then the classification result was obtained. In addition, it has been verified through experiments that most spam exists in the image as high-frequency components, which provides a theoretical and experimental basis for the frame design of the model. The detection model also verified that it is completely feasible to apply deep learning to the field of spam detection.

Section 1 of this paper summarizes the background and research development of the social spam research field. Section 2 presents the proposed model framework in detail. Section 3 analyzes and summarizes the experimental results. Finally, a preliminary discussion is presented on the research significance of this paper and future research directions that are worthy of attention.

2. The Proposed Methods

According to the component of the spam existing in the image (please note: the experiments in Section 3.2 have verified that spam mainly exists in the image with high-frequency components, so the proposed detection model was designed based on experimental validation), the special detection model was designed to improve detection accuracy. The detection model can be divided into three stages to accomplish the detection task, which can be described as the pre-processing stage, the feature extraction stage, and the classification prediction stage. In the pre-processing stage, the input image is first decomposed by *Haar* wavelet analysis to obtain the low-frequency information, horizontal high-frequency information, vertical high-frequency information, and diagonal high-frequency information of the image. The experimental results show that most of the spam existed in the image as high-frequency information (see Section 4 for the experimental analysis). Therefore, in the feature extraction stage, a special feature extraction module and an appropriate number of modules are selected to extract the frequency feature. In the classification prediction stage, the obtained frequency domain features are subjected to the concat operation to obtain the final target feature, then the classification result is obtained. The overall architecture of the detection model is shown in Figure 1.

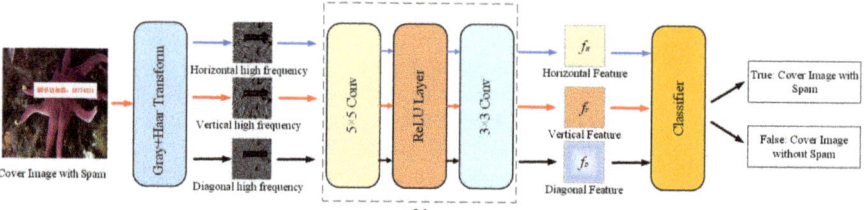

Figure 1. Visualization of the architecture for our proposed detection model.

2.1. The Stage of Pre-Processing

The primary focus of this paper was to verify that the spam mainly existed in the image in the form of high-frequency components, which also indicated that the low-frequency features of the input image have little effect on improving the accuracy of the detection model. To this end, in the pre-processing stage, the input image I_C was first subjected to wavelet transform analysis to obtain the corresponding low-frequency and high-frequency information; the operation is calculated as follows:

$$I_L, (I_H, I_V, I_D) = Haar(I_C) \tag{1}$$

where *Haar* represents the *Haar* wavelet transform, I_L is the corresponding low-frequency image after wavelet transform, I_H, I_V and I_D represent the horizontal high-frequency image, vertical high-frequency image, and diagonal high-frequency image after wavelet decomposition. At this stage, the low-frequency image containing few instances of spam information was discarded, and the three types of high-frequency images were reserved as the input information for the next stage.

2.2. The Stage of Feature Extraction

The task of the feature extraction stage is to extract representative features to determine whether the input image carries spam. The input of this stage is the horizontal high-frequency image I_H, the vertical high-frequency image I_V and the diagonal high-frequency image I_D after wavelet decomposition. The three high-frequency images enter the feature extraction block F with a fixed number of blocks with the same convolutional layer. The corresponding target feature can be obtained as follows:

$$f_H = nF(I_H) \tag{2}$$

$$f_V = nF(I_V) \tag{3}$$

$$f_D = nF(I_D) \tag{4}$$

where n represents the number of feature extraction blocks, F represents the feature extraction block with the designed convolutional layers, and the relationship between n and F is not a product operation. I_H, I_V and I_D are used as the input of F to get the feature vectors f_H, f_V and f_D, which represent the high-frequency features obtained in the feature extraction stage, respectively. During this stage, f_H, f_V and f_D represent the feature vectors for different high-frequency components. By selecting an appropriate number of feature extraction blocks, feature information that has spam in the images can be further extracted from high-frequency images, thereby improving the detection efficiency of the proposed model.

2.3. The Stage of Classification Prediction

In our model, unlike other current detection models, three feature components are obtained in the classification prediction stage, namely, the horizontal high-frequency feature, vertical high-frequency feature, and diagonal high-frequency feature, respectively. Therefore, the obtained high-frequency features are first concatenated by dimension; that is:

$$f = concat(f_H, f_V, f_D) \tag{5}$$

The final target feature f is obtained by splicing the high-frequency features, which contains most of the spam in the images, then the target features are operated as follows:

$$Result_{prediction} = Sigmoid(FC(f)) \tag{6}$$

As shown in Equation (6), the final target feature f is first sent to the fully connected layers, *FC*. Fully connected layers are able to map the learned distributed feature representation f to the sample label space. In this paper, FC layers consist of the input layer, hidden layer, and ReLU non-linear layer. The final target f is utilized as the input of the input layer. The ReLU layer is also used to enhance the nonlinear fitting ability of the model. The output of the FC layer is used as the input of the *Sigmoid* function. Finally, a prediction result is obtained through the *Sigmoid* function.

3. Experimental Results and Analysis

3.1. Dataset and Setup

In the process of our experiments, a PC with a GPU NVIDIA GeForce Tesla V100 16G was used, and the experimental environments Pytorch 1.1 and Python 3.7 were adopted. We built up our dataset to train the model proposed in this paper. To observe the detection effect of the proposed model, we collected some images with spam and combined the DIV2K2017 dataset to build a dataset for training the detection model (please note: the created dataset contained normal images without spam); some of the training images can be seen in Figure 2. The number of training images was 4000 and the size of the training images was cropped to 256 × 256; the number of test images for the test subset was set to 500. The image data in the training subset did not appear in the test subset. In addition, the architecture of the proposed detection model borrows from the idea of the VGG16 network; many experiments have been carried out on the setting of hyper-parameters, and the optimal parameter combination was selected. (In the training process, the batch size for the image dataset is set to 4, the number of training epochs was set to 350, and the learning rate was set to 0.005.)

Figure 2. Example training images from the collected and created image dataset.

3.2. The Elements of Spam in the Images

The specific components of the spam in the image determine the structural design of the proposed detection model. If the spam exists in the image in the form of high-frequency components, the feature extraction module of the proposed detection model can use the deep architecture to extract the high-frequency information of the image, to better detect the image with spam. Similarly, if the spam in the image comprises low-frequency components, the architecture of the detection model can appropriately reduce the number of network layers. Therefore, the components of the spam are first analyzed, and *Haar* wavelet transform analysis is utilized to decompose the image with the spam in the first-order frequency domain. The low-frequency information and the horizontal, vertical, and diagonal high-frequency information for images that include spam are obtained, respectively, as shown in Figure 3.

Original Image   *Low Frequency Component*   *Horizontal High Frequency Component*   *Vertical High Frequency Component*   *Diagonal High Frequency Component*

Figure 3. The corresponding frequency domain images after *Haar* wavelet decomposition of an image with spam.

From the experimental results in Figure 3, it is clear that spam mainly exists in the image in the form of high-frequency information, while the background occupies most of the low-frequency region of the image. In order to further verify that the spam exists in a specific region of the image, we also performed a *Haar* wavelet transform analysis on the original image without spam to obtain images corresponding to the different frequency domains. Then, we replaced the corresponding frequency domain of the image containing spam with the frequency domain of the original image for inverse *Haar* wavelet transform analysis. The reconstructed experimental results are shown in Figure 4.

Reconstructed Image Without Low Frequency Information   *Reconstructed Image Without Horizontal High Frequency Information*   *Reconstructed Image Without Vertical High Frequency Information*   *Reconstructed Image Without Diagonal High Frequency Information*   *Reconstructed Image With Only Low Frequency Information*

Figure 4. Reconstructed images lacking different frequency domain information.

From the analysis of the experimental results in Figure 4, when only the low-frequency components of spam are replaced, there is less loss of spam in the reconstructed image, and only the background of the image is not perfectly reconstructed; when the high-frequency components are replaced, it is clear that the reconstruction effect of spam is very poor and only the reconstructed background information is more prominent. When the image is reconstructed using only the low-frequency components, we can see that the background of the image is almost the only part to be reconstructed. Therefore, we can conclude that the spam mainly exists in the high-frequency components in the image; that is to say, as long as the detection model can extract most of the high-frequency features of the image containing spam, the detection accuracy for the model can be improved. From the analysis of the experimental results in Table 1, when the inputs of the model are only the low-frequency components, the detection accuracy can only reach 36.5%; when the inputs of the model are the high-frequency components, the detection accuracy can is as high as 86%; when the input of the model is the whole image, the detection accuracy drops to only 74.5%. The experimental results in Table 1 also indicate that the spam mainly exists in the high-frequency components in the image.

Table 1. The influence of the detection model under a combination of different frequency domain components.

	Detection Accuracy
Only low-frequency components	36.5%
Only high-frequency components	86%
Low-frequency components + high-frequency components	74.5%

3.3. The Architecture Depth of the Proposed Model

In order to verify the influence of the network architecture depth on the detection model in terms of its detection ability, we conducted an experimental comparison with different numbers used for the feature extraction block; that is, feature extraction blocks with different numbers (3, 7, 11, 15, 21, 25), and 400 images including spam (not included in the model training dataset) were randomly selected for testing. The experimental results are shown in Table 2.

Table 2. The influence on the detection ability of the detection model under different numbers of feature extraction blocks.

Number of the Feature Extraction Block	3	7	11	15	21	25
Detection accuracy	15%	36.5%	56.5%	82%	91%	84.5%

It can be seen from the experimental results in Table 2 that when the feature extraction block was set at 3, the model could only obtain a detection accuracy of 15%. As the number of the feature extraction block increased, its detection capability increased accordingly; when the feature extraction block number increased from 21 to 25, the detection accuracy dropped by 6.5%, which indicates that when the network architecture of the model reaches a certain level, its feature extraction ability will be affected. From the whole of the experimental results, the detection ability of the detection model with shallow layers is low; conversely, the detection ability of the model based on a deep architecture is stronger, which also verifies the conclusion drawn in Section 2.1: the spam mainly exists in the high-frequency components in the image.

3.4. The Influence of Pre-Processing Module

The main task of the proposed model was to detect the spam contained in the image. We know that most of the spam information existed in the image as high-frequency information. In order to improve the detection accuracy of the proposed model, an image pre-processing module was designed. Firstly, the input image was decomposed using *Haar* wavelet analysis to obtain low-frequency information and horizontal, vertical, and diagonal high-frequency information. Then the low-frequency information was discarded, and the horizontal, vertical, and diagonal high-frequency information was used as the input of the model. Finally, a classification result was obtained. In the experiment, we used the same image dataset to train the detection model with and without the image pre-processing module and randomly selected 200 images (not present in the training dataset) to test the trained detection model. Table 3 shows the experimental comparison results obtained by the models trained with and without the image pre-processing module.

Table 3. The comparison results obtained by the models trained with and without the image pre-processing module.

	Detection Accuracy	Training Time (min)
With Pre-processing Module	84.5%	657.3
Without Pre-processing Module	77%	771.4

From the experimental results in Table 3, it can be seen that the image pre-processing module is equivalent to performing a feature extraction operation on the image in advance; it takes less time to train this model than the model without an image pre-processing module. At the same time, the input of the model with the image pre-processing module comprises high-frequency information that focuses on the region where the spam exists and achieves a better detection accuracy. Compared to the model without the image pre-processing module, the detection accuracy was improved by nearly 8%.

3.5. Comparison with State-of-the-Arts

Table 3 compares the detection results between the proposed model and the current popular detection models. These comparison detection models include AlexNet [24], VGG13 [25], VGG16, VGG19, GoogleNet [26], and ResNet50 [18]. The same image dataset and hyper-parameters (learning rate, number of iterations, etc.) were used to train different detection models and 200 test images were randomly selected for testing. Regarding the other detection models, since the detection task was not aimed at detecting spam in the images, during the training process the input and output of other detection models were adjusted to suit the comparison task in this paper. In order to observe the performance of different detection models more intuitively, we compared the detection accuracy and training time, respectively. The detection accuracy can provide a visual indication of the performance of the new detection model, while the length of training time can indicate the ability of the model to extract features. The comparison results obtained are shown in Table 4.

Table 4. The comparison results between the proposed model and the current popular detection models.

	Detection Accuracy	Training Time (min)
AlexNet	32.5%	1412.5
VGG13	35.5%	1355
VGG16	44%	1156.3
VGG19	54%	968.5
GoogleNet	66.5%	1045.4
ResNet50	77%	825
The Proposed Method	91%	657.3

From the experimental results in Table 4, compared with the current popular detection models, the proposed method is superior in terms of detection accuracy (please note: the input and output of other detection models have been modified to meet the requirements of the detection task). In addition, from the perspective of the detection accuracy of VGG13, VGG16, and VGG19, VGG19 shows the best performance in terms of detection accuracy, because VGG19 has the deepest network architecture for extracting the detailed information (high-frequency information) in the input image. This also shows that the spam mainly exists in the high-frequency components in the image. In addition, from the perspective of training time, the proposed method can achieve a balanced state with the shortest time and number of iterations, which indicates that the proposed algorithm is superior to the other current detection models in terms of computational cost. From another point of view, the shorter the training time of the detection model, the stronger its ability to extract features. Therefore, it can be seen from the experimental results in Table 3 that the proposed model also has advantages in terms of feature extraction.

4. Conclusions

In this paper, a detection method is proposed for identifying social media images containing spam, based on a deep neural network and frequency domain pre-processing. Our research contributions can be summarized as follows:

(1) An image dataset including spam was collected and created; to the best of our knowledge, in the field of social network spam detection, this is the first time that an image-level training dataset has been proposed.
(2) It has been verified that the spam mainly existed in the high-frequency components in the images. On this basis, *Haar* wavelet transform analysis was introduced as the pre-processing module of the model, and the high-frequency information of the image is extracted as the input of the feature extraction module.
(3) In the feature extraction stage, a special feature extraction block is designed and an appropriate number is selected, according to our experiment and the spam component, which improves the accuracy and efficiency of the detection model.

Unlike the current detection models, this paper first verifies the specific components of spam in the image and then designs a more targeted detection framework, which can enhance the detection efficiency and accuracy of the proposed model. In future work, we will further expand the created image dataset and improve the recognition ability and efficiency of the proposed model. In addition, although the proposed model demonstrates good detection performance on fixed image datasets, it lacks breadth, which will be addressed. Improving the applicability of the model is another future research focus.

Author Contributions: H.S.: conceptualization, methodology, data preprocessing; data analysis, writing-original draft preparation; X.L.: data collection, writing-review and editing, visualization; X.Z.: conceptualization, supervision, project administration. All authors have read and agreed to the published version of the manuscript.

Funding: This research was funded by the National Natural Science Foundation of China (61272374, 61300190), the Key Project of the Chinese Ministry of Education (313011), and the Foundation of the Department of Education of Liaoning Province (L2015001).

Acknowledgments: We thank the anonymous reviewers for their careful reading of our manuscript and their many insightful comments and suggestions.

Conflicts of Interest: The authors declare no conflict of interest.

References

1. Chen, H.; He, X.; Qing, L.; Wu, Y.; Ren, C.; Sheriff, R.E.; Zhu, C. Real-world single image super-resolution: A brief review. *Inf. Fusion* **2022**, *79*, 124–145. [CrossRef]
2. Javed, I.T.; Toumi, K.; Alharbi, F.; Margaria, T.; Crespi, N. Detecting nuisance calls over internet telephony using caller reputation. *Electronics* **2021**, *10*, 353. [CrossRef]
3. Li, Q.; Wang, X.; Ma, B.; Wang, X.; Wang, C.; Gao, S.; Shi, Y. Concealed Attack for Robust Watermarking Based on Generative Model and Perceptual Loss. *IEEE Trans. Circuits Syst. Video Technol.* **2021**. [CrossRef]
4. Wang, Y.; Bashir, S.M.A.; Khan, M.; Ullah, Q.; Wang, R.; Song, Y.; Guo, Z.; Niu, Y. Remote sensing image super-resolution and object detection: Benchmark and state of the art. *Expert Syst. Appl.* **2022**, *197*, 116793. [CrossRef]
5. Minaee, S.; Boykov, Y.Y.; Porikli, F.; Plaza, A.J.; Kehtarnavaz, N.; Terzopoulos, D. Image segmentation using deep learning: A survey. *IEEE Trans. Pattern Anal. Mach. Intell.* **2021**. [CrossRef]
6. Tov, O.; Alaluf, Y.; Nitzan, Y.; Patashnik, O.; Cohen-Or, D. Designing an encoder for stylegan image manipulation. *ACM Trans. Graph.* **2021**, *40*, 1–14. [CrossRef]
7. Zhang, Z.; Hou, R.; Yang, J. Detection of social network spam based on improved extreme learning machine. *IEEE Access* **2020**, *8*, 112003–112014. [CrossRef]
8. Yang, C.; Harkreader, R.; Gu, G. Empirical evaluation and new design for fighting evolving twitter spammers. *IEEE Trans. Inf. Forensics Secur.* **2013**, *8*, 1280–1293. [CrossRef]
9. Jiang, M.; Cui, P.; Faloutsos, C. Suspicious behavior detection: Current trends and future directions. *IEEE Intell. Syst.* **2016**, *31*, 31–39. [CrossRef]
10. Rao, S.; Verma, A.K.; Bhatia, T. A review on social spam detection: Challenges, open issues, and future directions. *Expert Syst. Appl.* **2021**, *186*, 115742. [CrossRef]
11. Zhu, Y.; Wang, X.; Zhong, E.; Liu, N.; Li, H.; Yang, Q. Discovering spammers in social networks. In Proceedings of the AAAI Conference on Artificial Intelligence, Toronto, ON, Canada, 22–26 July 2012; Volume 26, pp. 171–177.
12. Hu, X.; Tang, J.; Zhang, Y.; Liu, H. Social spammer detection in microblogging. In Proceedings of the Twenty-Third International Joint Conference on Artificial Intelligence, Beijing, China, 3–9 August 2013.
13. Wu, F.; Shu, J.; Huang, Y.; Yuan, Z. Co-detecting social spammers and spam messages in microblogging via exploiting social contexts. *Neurocomputing* **2016**, *201*, 51–65. [CrossRef]

14. Chen, C.; Zhang, J.; Xie, Y.; Xiang, Y.; Zhou, W.; Hassan, M.M.; AlElaiwi, A.; Alrubaian, M. A performance evaluation of machine learning-based streaming spam tweets detection. *IEEE Trans. Comput. Soc. Syst.* **2015**, *2*, 65–76. [CrossRef]
15. Masood, F.; Almogren, A.; Abbas, A.; Khattak, H.A.; Din, I.U.; Guizani, M.; Zuair, M. Spammer detection and fake user identification on social networks. *IEEE Access* **2019**, *7*, 68140–68152. [CrossRef]
16. Ahmed, N.; Amin, R.; Aldabbas, H.; Koundal, D.; Alouffi, B.; Shah, T. Machine Learning Techniques for Spam Detection in Email and IoT Platforms: Analysis and Research Challenges. *Secur. Commun. Netw.* **2022**, *2022*, 1862888. [CrossRef]
17. Sokhangoee, Z.F.; Rezapour, A. A novel approach for spam detection based on association rule mining and genetic algorithm. *Comput. Electr. Eng.* **2022**, *97*, 107655. [CrossRef]
18. He, K.; Zhang, X.; Ren, S.; Sun, J. Deep residual learning for image recognition. In Proceedings of the IEEE Conference on Computer Vision and Pattern Recognition, Las Vegas, NV, USA, 27–30 June 2016; pp. 770–778.
19. Ma, W.; Tu, X.; Luo, B.; Wang, G. Semantic clustering based deduction learning for image recognition and classification. *Pattern Recognit.* **2022**, *124*, 108440. [CrossRef]
20. Kormilitzin, A.; Vaci, N.; Liu, Q.; Nevado-Holgado, A. Med7: A transferable clinical natural language processing model for electronic health records. *Artif. Intell. Med.* **2021**, *118*, 102086. [CrossRef]
21. Xie, X.; Niu, W.; Zhang, X.; Ren, Z.; Luo, Y.; Li, J. Co-Clustering Host-Domain Graphs to Discover Malware Infection. In Proceedings of the 2019 International Conference on Artificial Intelligence and Advanced Manufacturing, Dublin, Ireland, 17–19 October 2019; pp. 1–6.
22. Zhang, C.; Du, G.; Du, X. Illegal Image Classification Based on Ensemble Deep Model. *J. Beijing Jiaotong Univ.* **2017**, *41*, 21–26.
23. Cai, X. Internet bad information detection based on Bert model. *Telecommun. Sci.* **2020**, *36*, 121–126.
24. Krizhevsky, A.; Sutskever, I.; Hinton, G.E. Imagenet classification with deep convolutional neural networks. *Adv. Neural Inf. Processing Syst.* **2012**, *25*, 226–237. [CrossRef]
25. Simonyan, K.; Zisserman, A. Very deep convolutional networks for large-scale image recognition. *arXiv* **2014**, arXiv:1409.1556.
26. Szegedy, C.; Liu, W.; Jia, Y.; Sermanet, P.; Reed, S.; Anguelov, D.; Erhan, D.; Vanhoucke, V.; Rabinovich, A. Going deeper with convolutions. In Proceedings of the IEEE Conference on Computer Vision and Pattern Recognition, Boston, MA, USA, 7–12 June 2015; pp. 1–9.

Article

Security Ontology Structure for Formalization of Security Document Knowledge

Simona Ramanauskaitė [1,*], Anatoly Shein [2], Antanas Čenys [3] and Justinas Rastenis [3]

1. Department of Information Technologies, Vilnius Gediminas Technical University, LT-10223 Vilnius, Lithuania
2. Orchestra Group, Tel Aviv-Yafo 6688314, Israel; shein@orchestra.group
3. Department of Information Systems, Vilnius Gediminas Technical University, LT-10223 Vilnius, Lithuania; antanas.cenys@vilniustech.lt (A.Č.); justinas.rastenis@vilniustech.lt (J.R.)
* Correspondence: simona.ramanauskaite@vilniustech.lt

Abstract: Cybersecurity solutions are highly based on data analysis. Currently, it is not enough to make an automated decision; it also has to be explainable. The decision-making logic traceability should be provided in addition to justification by referencing different data sources and evidence. However, the existing security ontologies, used for the implementation of expert systems and serving as a knowledge base, lack interconnectivity between different data sources and computer-readable linking to the data source. Therefore, this paper aims to increase the possibilities of ontology-based cyber intelligence solutions, by presenting a security ontology structure for data storage to the ontology from different text-based data sources, supporting the knowledge traceability and relationship estimation between different security documents. The proposed ontology structure is tested by storing data of three text-based data sources, and its application possibilities are provided. The study shows that the structure is adaptable for different text data sources and provides an additional value related to security area extension.

Keywords: security; ontology; structure; formalization

Citation: Ramanauskaitė, S.; Shein, A.; Čenys, A.; Rastenis, J. Security Ontology Structure for Formalization of Security Document Knowledge. *Electronics* **2022**, *11*, 1103. https://doi.org/10.3390/electronics11071103

Academic Editor: Vijayakumar Varadarajan

Received: 13 March 2022
Accepted: 29 March 2022
Published: 31 March 2022

Publisher's Note: MDPI stays neutral with regard to jurisdictional claims in published maps and institutional affiliations.

Copyright: © 2022 by the authors. Licensee MDPI, Basel, Switzerland. This article is an open access article distributed under the terms and conditions of the Creative Commons Attribution (CC BY) license (https://creativecommons.org/licenses/by/4.0/).

1. Introduction

The development of modern information and communications technologies (ICTs) brings new possibilities for users and organizations, whereby the user is not strictly attached to physical data storage, can access their data anytime and anywhere, use different methods and services for data processing and sharing instantly, etc. Together with the ITC possibilities, the variety of cyberattack vectors has also increased. This is expected because of the complexity of modern technologies, as well as orientation to user experience (UX). Therefore, the spending on security and risk management increases every year, reaching 155 billion USD worldwide in 2021 [1].

The growth of spending on security and risk management is affected by multiple factors [2]: transition to remote or mixed working; cloud, SaaS security assurance; the rise of new threat landscapes. A solution to fight the current spending needs on security and risk management is cyber intelligence. In cyber intelligence, artificial intelligence (AI) solutions are used to automate the process, while providing additional benefits to specific security and risk management areas [3,4].

The development of cyber intelligence is affected by a lack of data for data analysis and decision support. While supervised learning AI solutions are mostly oriented on some specific tasks (data classification, anomaly detection), ontologies as a knowledge base for process automation might have a wider application (semantic modeling, extraction of needed knowledge, etc.) [5].

The ontology structure defines the simplicity of knowledge extraction, while the real value of the ontology relies upon the data it stores. The biggest portion of security knowledge at the moment is not structured; it is presented as text data and is, therefore, currently limited for application in cyber intelligence solutions. It is important to have a mechanism, assuring a wide range of up-to-date and qualitative data from different sources it. Manually updating security ontology is not practical because of the wide variety of data sources, potential impact of data interpretation, lack of resources, etc. Some methods for text transformation to ontology exist [6]; however, they concentrate on the estimation of concepts, instances, hypernyms, and hyponyms, with no relationship between the data source and concept. When adopting ontology knowledge application and decision justification by mapping knowledge to appropriate data sources, the ontology structure has to be suitably designed.

This paper aims to increase the possibilities of ontology-based cyber intelligence solutions by presenting a security ontology structure for data storage to the ontology from different text-based data sources, supporting the knowledge traceability and relationship estimation between different security documents. Therefore, the main contribution of the paper is answering the research question regarding the main principles of text-based security document formalization to the ontology for gathered data usability and generation of new knowledge.

The paper reviews related works on security ontology and text transformation to ontologies. On the basis of the review results, a new security ontology structure is proposed to provide a linking of the concepts to original data sources. The proposed structure is validated by presenting some numerical results of its application and directions of usage of such an ontology structure.

2. Related Works

"An ontology is a formal and explicit specification of a shared conceptualization" [7]. It is a basis of semantic modeling and allows the storage of different concepts, as well as their properties and relationships. Therefore, ontologies are known as knowledge bases rather than databases. Because of the properties of ontologies, they represent one of the solutions for cyber intelligence and a future research direction [8]. The potential of ontologies can be seen in different application areas, such as digital evidence review [9], software requirement and security issue detection [10], modeling of Internet of things design [11], security alert management [12], and as a standard for cyber threat sharing [13].

Ontologies are mostly created by area experts. The expert designs the ontology by formalizing its knowledge using different data sources. Ontologies based only on expert knowledge mostly present the landscape of an area, while additional tools and transformations are used to incorporate existing knowledge into the structure of the designed ontology. Ontologies, with formalized knowledge of different sources, have a higher value, as they present not only the general concepts of the area but consolidate knowledge of different data sources and serve as a knowledge base. However, the transformation from different data sources to ontology might be complicated because of different data formats and types. One of the most complex data types for formalization is text-based data. The same knowledge can be presented in very different texts, and word-to-word matching might not be enough for knowledge matching. Therefore, it is important to find the best solution for text-written knowledge extraction and transformation to ontology.

The next two sections are dedicated to analyzing the existence of security ontology, as well as presenting knowledge of different security area documents and existing solutions to transform text-written knowledge to ontology.

2.1. Security-Related Ontologies

The number of publications on security ontology-related topics in the Web of Science Core Collection has increased every year. Analyzing the publication number in the period between 2000–2021, the distribution of publications containing the terms "cyber ontology" or "security ontology" and publications containing the term "security" was very similar year by year (see Figure 1 below). Despite the number of publications on the general term "security" being more than 400 times higher (572,311 records for "security", compared to

2663 records for "cyber ontology" or "security ontology"), the tendencies were the same, i.e., the popularity of the topic is increasing in scientific publications. This indicates that the security ontology topic has been analyzed in scientific papers with the same growth as security in general.

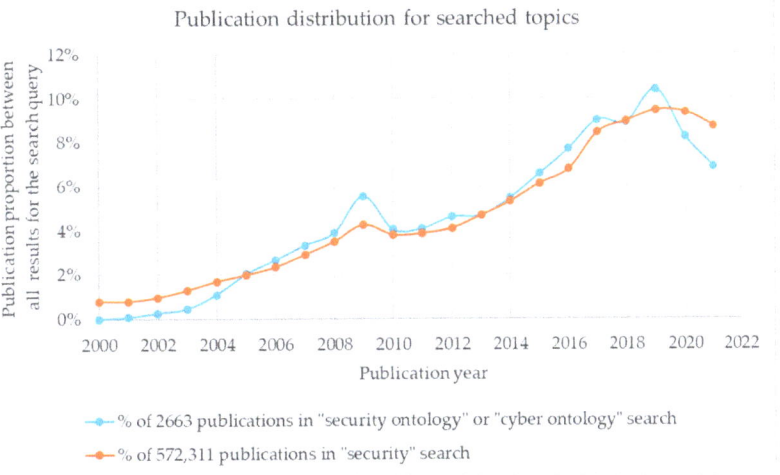

Figure 1. Growth of security ontology and security-related scientific papers.

Some of the analyzed papers proposed a new security ontology, others presented solutions based on the ontology or simply reviewed the current situation in the landscape of existing ontologies. One of the first attempts to present a general-purpose security ontology was by Schumacher [14]. The author presented nine concepts and relations between them. The same basic structure of concepts was applied by Tsoumas and Gritzalis [15] to present an idea of security management, based on security ontology. Since this time, the variety of security ontologies has increased and been directed to more specific areas of security, such as for annotating resources [16], for corporate assets and their threats [17], for incident analysis [18], for security requirement elicitation [19], for cloud security [20], for Internet of things security [21], and for the cross-site scripting attack [22].

One of the ways to increase the content of the ontology is to incorporate the data and content of some existing security-related systems. Example of such data sources are the CVE (Common Vulnerabilities and Exposures), CWE (Common Weakness Enumeration), CPE (Common Platform Enumeration), and CAPEC (Common Attack Pattern Enumeration and Classification) [23,24]. These sources have a clear structure and discrete values for specified attributes. Therefore, the transformation of the data to security ontology does not require intelligent solutions.

A big portion of security knowledge is presented as text in security standards and best practices. These security-related documents are also incorporated into security ontologies. One of the first cases to reflect security document data was presented by Parkin et al. [25]. These authors incorporated the ISO27002 standard structure (chapter, section, guideline, guideline step) to the ontology, by mapping it to the asset. Several other ontologies were also based on the ISO27002 standard [26,27]; however, all the ontologies were based on manual human work, where the security standard is analyzed, interpreted, and presented in the ontology by a human expert. Furthermore, in most cases, the requirements or guidelines of the security standards were not expressed in very basic and general concepts; they had a higher level of detail and were, thus, not fully adapted for fully automated content extraction. Therefore, solutions for text-based document analysis and transformation to ontology are needed.

2.2. Text Transformations to Ontology

Manually designing the ontology is not an option when wide and complex domains are presented and automated tools are needed to simplify the task [28]. Meanwhile, possible solutions for the automated ontology construction from text documents are implemented in different ways. Moreno and Perez [29] relied on statistics when multiple data sources were analyzed to extract the most frequent terms and incorporate them into the ontology. The principle of multiple data sources was used to extract the main knowledge in [30]. This approach is limited as it extracts just the terms identified in multiple sources. Therefore, more specific terms can be missed or ignored. At the same time, the detection of synonyms is very important to prevent ignorance or rarer terms and their synonyms. To solve this problem, some reference sources can be used. For example, in the tourism domain, the named entities are extracted as the main knowledge on the tourism domain ontology, mostly including locations, organizations, and persons [31]. Another option is to use natural language processing (NLP) solutions [32]. In most cases, the part of speech (POS) is estimated, where the nouns are identified as key concepts [33]. The concepts are additionally processed by using synonym tables [33]; however, this can be applied to narrow domain areas, as a detailed list of synonyms can be an issue for more complex domains. In such a case, clustering might be used to organize the concepts, find synonyms, and indicate relations between the concepts [34–36].

The ontology construction can be executed on very different levels to define terms, synonyms, concepts, concept hierarchies, relations, or rules [37]. A more detailed (including all levels of concepts) ontology increases its application possibilities, but also increases its construction complexity. Therefore, research on relation estimation between concepts is an important aspect of ontology construction. Semantic patterns can be extracted to identify relations between concepts [38,39], while grammar-based transformation [40] and supervised learning can also be applied [41,42]. For relation extraction, the semantic lexicon, syntactic structure analysis, and dependency analysis are mostly used [43].

Despite the variety of existing solutions for ontology learning from unstructured text, the performance of the transformations lacks accuracy, better results can be achieved when some specific domain is analyzed [44]. The transformation of unstructured text to ontology according to the domain allows adding some specific rules or solutions, enabling a more detailed presentation of the knowledge [39,45].

In cybersecurity, research on knowledge extraction from text exists [46,47]; however, automated ontology building or enrichment is mostly achieved using different data sources rather than unstructured text [48–50]. In the field of security ontology, Gillani and Ko [51,52] proposed a ProMine solution to enhance or maintain ontologies by using text-mining technologies. These solutions are based on the application of an existing ontology and external data sources (for synonym estimation) to indicate additional concepts from unstructured text. These solutions illustrate the need for a reference ontology to which the extracted data will be added. The ontology main structure is needed to define the main rules for the presentation of the extracted terms and relations. However, security ontology structures and transformations, dedicated to transforming security standards and best practices for data construction of security ontology, are still missing.

2.3. Summary of Related Work

The analysis of existing security ontologies and data source transformation methods as knowledge bases revealed that most knowledge source generation is based on manual work involving experts (see yellow blocks in Figure 2). Fully automated solutions [51,52] (presented as a green block in Figure 2) are oriented toward the presentation of security concepts without the presentation of the data source at different aggregation levels. Security metadata and aggregated data sources exist; however, full integration between formalized security concepts and data sources is missing in the existing security knowledge data sources (see X-axis in Figure 2).

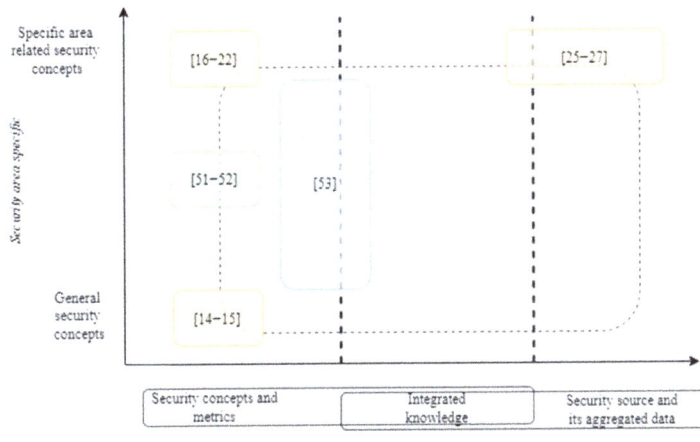

Figure 2. Summary of analyzed related words, based on presented security area and concept presentation level.

UCF Mapper [53] has a semi-automatic solution (presented in blue in Figure 2) when initial security text analysis is executed automatically, while human work is used to adjust the knowledge. This solution defines security document controls, linked together by using security concept similarity. The integration between security concepts and security documents exists but is implemented at a very abstract level only.

Further examples of partly integrated knowledge between security concepts and documents are security ontologies for mapping of security standards [25–27]. Those ontologies are mostly oriented toward security documents and aggregated knowledge, with just some links to formalized security concepts. This complicates their usage by automated systems; therefore, to realize the full potential of security document formalization, knowledge of different abstraction levels should be presented with its interconnections.

3. Security Ontology Structure for Text-Based Security Source Formalization

A security ontology for text-based security source formalization should define the structure and principles for knowledge extraction from different sources. This would allow automated composition of the security knowledge base using multiple data sources, rather than the perception of the ontology developer. Such a knowledge base presented as an ontology might serve as a base for different security intelligence tasks.

One of the requirements in modern knowledge and decision support systems is a justification of the decision. To do so, some relations between the data source and extracted terms, as well as the associated concepts, should be implemented. At the same time, the data source can give additional value and clarity for the decision traceability. Therefore, the structure of the proposed security ontology has three main layers (see Figure 3): data sources, including structure and content (documents); security concepts, as well as their properties and relations between concepts (knowledge); relations between data source and security knowledge, expressed as atomic sentences with links to concepts (mapping).

To present the security source, the document structure is important. Division into sections, subsection, controls, description, and other components is standard for a well-written security document; therefore, it should also be reflected in security ontology. However, different security documents might have different structural elements. This complicates the alignment of several data sources. Therefore, we ensure the adaptability of the security ontology structure to different data sources by applying class inheritance. The reference structure for data sources is composed of main classes and properties. Figure 4 illustrates the main structure of the ontology, where blue notated elements define main classes and gray elements denote properties, associated with an appropriate class. Each document is presented as an instance of the "security document" class, while its structure is presented as a hierarchical structure of chapters (presenting a tree of iterative chapters and subchapters). The content of the document should be defined on the basis of the type of content: control, testing procedure, attack description, and definition. Each text element has a "text" property and allows the presentation of the full, not formalized content. For the formalization, each text is divided into atomic sentences, as a link from the document structure to the security concepts.

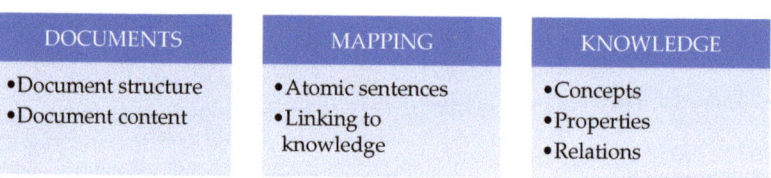

Figure 3. The main layers of the proposed ontology structure.

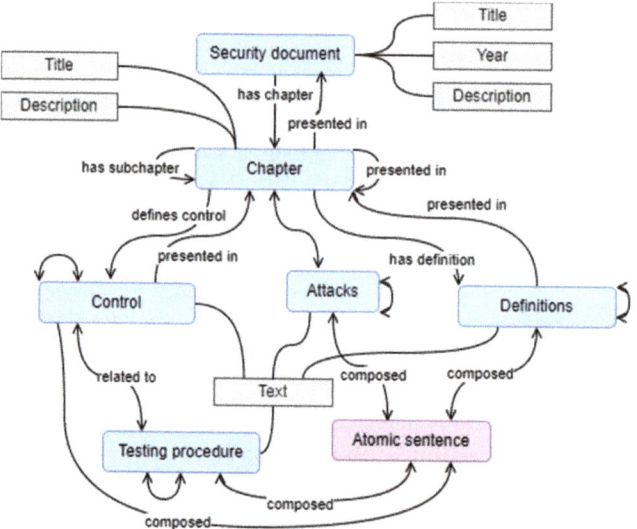

Figure 4. The main structure of security ontology data source layer: blue elements—classes; gray elements—data properties of the class; purple elements—class of mapping layer; arrows—data properties, connecting the instances of separate classes.

When a new data source is added to the ontology, new classes should be created for each of the actual (having an analogue component in it) classes in the security ontology data source reference structure. This will allow using reference classes for the selection of data in all inserted data sources. On the other hand, for more specific, defined data sources, the child classes can be used.

Close to the adding of inherited classes, the security document should be presented in the ontology, by creating instances of the classes. Instances reflect the object and data source rather than their structural elements. The naming of the instance can reflect the data source title for a faster search. One instance of security document class is created, in addition to one for each chapter and other elements of the data source.

Classes of the instance define the data source structure and go from abstract instances to more detail, where not only is the title presented but also the text defining the control, testing procedure, or concept description. The text is difficult to analyze as it contains multiple sentences, whereby one sentence might include different concepts on security. Therefore, each text should be divided into sentences. This can be achieved by using such solutions as finite state machines, part-of-speech tags, conceptual graphs, domain ontology and dependency trees, etc. The main functionality of sentence structure analysis and identification of the subsentence can be completed using widely available programming toolkits, such as NLTK. This simplifies the text division into sentences and later into atomic sentences, to be presented in the mapping layer.

In the mapping layer, each sentence should be divided into atomic sentences. In Figure 5, the sentence is presented in white, while atomic sentences are presented in red. Atomic sentences should present only one idea, without any side sentences. In one sentence, several atomic sentences might be presented and linked by some keywords. The keyword should also be used to link the atomic sentences in the ontology (in Figure 5, the link "leads to" defines the link between two atomic sentences of the same composite sentence). Consequently, appropriate object properties should be used, or an event should be newly added to the ontology if the analyzed sentence has a different keyword than presented in the ontology. However, to assure the ontology data's adaptability to machine usage, keyword processing should be used to eliminate nonmeaningful terms and term conversion to standard form. Therefore, "as a means of" could be converted to the term "leads to" or similar.

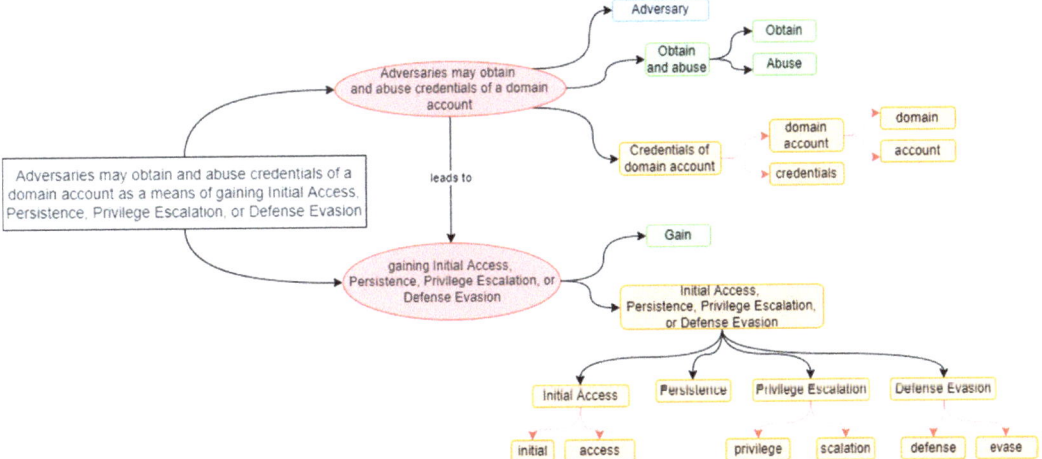

Figure 5. An example of sentence presentation as multiple atomic sentences, where each sentence is divided into subject, action, and object, as well as segmented to the lowest-granularity elements.

While atomic sentences present separate ideas and might indicate the logical sequence of concepts, text-based expression is not effective for machine usage. Therefore, each atomic sentence is divided into smaller elements, identifying the subject (blue in Figure 5), action (green in Figure 5), and object (yellow in Figure 5) in it. Such a separation allows an estimation of who is acting, what they are doing, and what object they are using for it. This division might be executed by a human or by natural language processing (NLP). Human-based transformation might be more accurate, as security experts might understand the meaning of the atomic sentence and express some terms in more popular synonyms (for a better match with other data sources). However, this is very time-consuming and, in some cases, requires not general, but very specific security knowledge and situational understanding. Therefore, NLP solutions for subject–object–verb extraction can be adapted for content extraction automation.

To make the content usable for machine systems and concept matching between several data sources, each subject, action, or object is divided into the lowest-level part-of-speech elements. The hierarchical structure of security concepts is adapted to present the idea, whereby a combination of several concepts might be differently interpreted in comparison to the sum of separate concepts. For example, "firewall and router configuration" might not be identical to the sum of "firewall configuration" and "router configuration", as the interdependencies between those two might also be considered. At the same time, the division into lower-level part-of-speech elements allows an estimation of concept similarity, with not only a full, but also a partial match.

The division of subject, action, and object into smaller elements covers the security knowledge layer. The terms for this layer are added by incorporating new data sources and identifying new, non-existing instances, which are needed to reflect the atomic sentence. At the same time, the object properties are important in this layer. The composite term is divided into lower-granularity terms according to NLP principles. Therefore, the object properties between concepts might indicate the logical operator (and, or, not), property of the elements, etc.

An example of text-based data division into atomic sentences and smaller components is presented in Figure 6. It presents the first two sentences of MITRE ATT&CK technique T1003.001, stored as four atomic sentences (two for each of the sentences). Additionally, the subject of the attack is presented in the purple background to help the identification of relations between system and attack behavior. This situation indicates that the storage of credential material in LSASS process memory is sensitive and directly related to attack actions, thus deserving attention for security assurance. The situation illustrates the formalization principle when two different sentences can be associated by analyzing the linking to the same security concepts. In the same manner, more sentences, chapters, or even security sources can be linked.

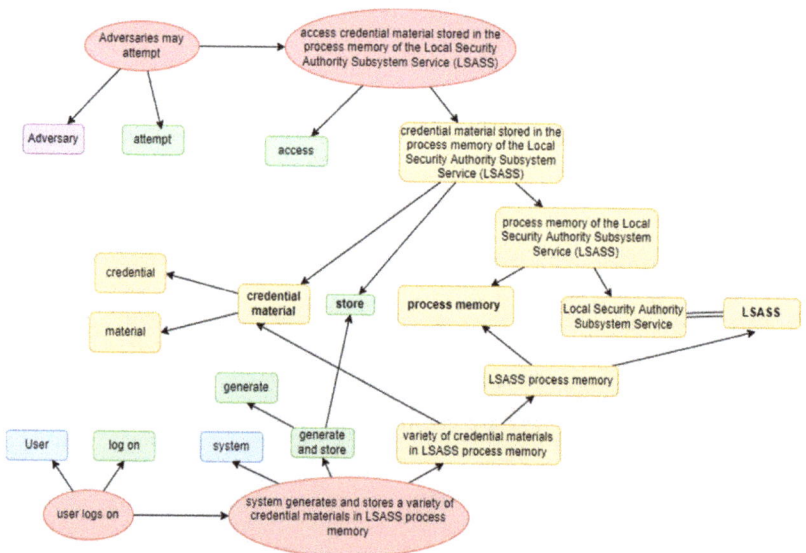

Figure 6. An example of attack technique description formalization, by indicating matching concepts between two sentences and the attack relation to system behavior.

The traceability to elements with a higher abstraction level helps identification of the flow, and this can be used for estimation of the distance between different elements. At the same time, such a structure is not optimized in the sense of data storage; it stores full text and duplicates its parts in lower-level elements. This solution is more oriented toward data usability rather than storage optimality. However, to solve the issue, the ontology data can be transformed, by filtering out unnecessary elements, i.e., leaving only the elements of the needed level (security document structure, atomic sentences, lowest-granularity elements, etc.).

4. Application of the Security Ontology and Its Data

This paper presented a security ontology structure and principles of how this ontology should be supported with data from different data sources. It is difficult to compare it to existing ontologies. This paper presented the reference structure for different security document presentations, while other existing security ontologies were mostly dedicated to knowledge presentation. Therefore, several approaches were applied to analyze the security ontology structure suitability to store knowledge of different text-based data sources, as well as its applicability.

4.1. Numeric Results of Sample Data Presentation in the Security Ontology

Structure suitability can be estimated by applying it for the formalization of different security data sources. In the current state, human-based sample data from ISO 27,002 (five chapters, three subchapters, and seven controls), PCI DSS (six chapters, four subchapters, two general requirements with four detailed requirements, and three testing procedures) standards, and MITRE ATT&CK enterprise techniques (descriptions of two techniques and two sub-techniques with 10 sentences in total) were added to validate the security ontology structure suitability for different text-based security data sources.

The formalization process does not require the adjustment of the reference security ontology structure. However, not all classes were used for instance creation in different data sources; MITRE ATT&CK techniques did not require chapter presentation, while security standard requirements and testing procedures were mostly presented, not attack techniques.

To review the specifics of the mapping layer, the results revealed (see Table 1) 1.88 atomic sentences on average for each analyzed sentence (requirement, testing procedure, control, technique descriptions). This illustrates the complex structure of the texts, presenting several interconnected concepts.

Table 1. Summary of the number of instances in sample data of the analyzed security documents.

Measurement	Value		
	ISO 27002	PCI DSS	MITRE ATT&CK Technique
Number of instances in document structure	8	10	0
Number of atomic sentence instances	13	12	24
Number of instances of initial terms (subjects, actions, objects)	30	29	55
Number of lower granularity term instances	64	55	147

At the same time, 2.33 lower-granularity term instances on average (not taking into account the match between different security documents) were generated from first-level composite term instances. This does not accurately reflect the situation as the majority of subjects and actions were presented as one term, while objects were mostly presented as complex structures, requiring hierarchical presentation to lower-level granularity term instances.

The experiment illustrates that text-based data (technique descriptions) were written in a more complex manner (usage of complex sentences and terms), but the same term was more often used (probably because the technique description was longer or contained more sentences) in comparison to analyzed security standard requirements and testing procedures.

4.2. Analysis of Knowledge Extraction Possibilities Using the Proposed Security Ontology

Suitability to store different text-based security data sources is not enough if there are no use cases of the presented data. Therefore, we present some use cases where the security ontology, with data integrated from different data sources, can add value in comparison to existing solutions. The list is not limited to these examples; however, it presents the most relevant, easily implementable application use cases.

4.2.1. Summary of Data Source Coverage or Security Landscape

To understand what concepts one or several selected data sources cover up, the data sources should be read and summarized. Using the security ontology, a list of mentioned terms can be easily obtained. The list can be reduced by adding requirements to provide only the most popular terms. At the same time, the data can be used to understand which area of the security landscape is covered by the data source in comparison to the full landscape of cybersecurity.

Different data sources analyze different aspects of the security area; therefore, the integration of different security data sources enables a wide security knowledge base. With the help of hierarchical term division into lower-granularity (simple words) terms and knowledge of different data sources, the link between different concepts can be established. Therefore, the term graph can be used as a presentation of a wide range of security areas, thereby forming the full landscape of cybersecurity. Such a data source can be used for learning purposes, as well as security area concept interdependencies analysis.

4.2.2. Mapping of Security Documents

Mapping of security documents allows a better understanding of what is common between multiple security documents, as well as their uniqueness and specifics. Some solutions to map different security standards using a reference ontology exist [26]. However, they are based on a very detailed security ontology, and the mapping of the security document to the security ontology must be done by a security expert. An automated solution is used by UCF [53], where text analysis is applied to extract the main terms. The mapping between the security documents is mostly implemented by the proportion of matching terms. However, the solution relies on the relational database rather than the ontology for knowledge storage; therefore, opposite statements such as "password is required" and "password is not required" lead to a high similarity because the proportion between matching terms is high.

Using the proposed security ontology structure and the hierarchical division of complex terms can allow more accurate mapping of security documents. The manual labeling of the most appropriate versions of the term used in UCF Mapper can be replaced by automated matching of terms where the relations are established by incorporating different security documents. Furthermore, the links between different granularity terms will allow the identification of opposite meanings or terms, enabling more accurate mapping of security documents.

4.2.3. Cybersecurity Threat Modeling

Security threat modeling tools exist; however, the data for the modeling must be manually transformed from different sources to a specified language or model. Xiong et al. [54] used the MITRE ACC&CK matrix as a data source and transformed it into an enterprise system. This allowed security threat modeling, but a manual expert-based transformation of the knowledge had to be implemented. Using the proposed security document formalization, the subject, action, object, and properties, as well as links between them, can be estimated. This might represent a basis for modeling different security situations. For example, the pre-conditions and post-conditions can be easily identified by analyzing the relations between atomic sentences. This can allow the generation of attack graphs or trees. Subject classification can be used to define attack and mitigation relations for security risk evaluation, while attack subject identification can be used to identify attack flows. This would align with the MITRE ENGENUITY attack flow project [55] as the formalization at different levels would enable flow automated identification based on the relations between concepts, while linking to the data source would allow relationship aggregation to the technique level.

5. Discussion and Conclusions

Security-related research and ontology applications are experiencing constant growth. However, the absence of fully functioning semantic web- or text-based security data source formalization solutions limits the exploitation of existing data sources in the cyber intelligence area. This paper goes one step further to solve the problem and provides an ontology structure, dedicated to linking the ontology content with a text-based data source.

The division of the proposed ontology structure into three layers allows a separation of the security area content, security document structure with texts, and mapping between the two. Therefore, the data can be easily filtered to use the security area content only, while additional layers can provide additional values, related to links between different data sources, automated mapping between them, etc.

While the ontology structure is suitable for human-based security document formalization, as shown in Section 4.1, the automated solution should be provided for simplification of text data transformation. The current solutions for text transformation to ontology are ideal for estimating concepts and their relations. For application to this security ontology structure, additional adaptation is needed, as document structure and sentence relation analysis must be incorporated.

Author Contributions: Conceptualization, S.R. and A.S.; methodology, S.R. and A.Č.; software, S.R.; validation, S.R., A.S. and J.R.; formal analysis, S.R. and A.Č.; resources, S.R. and A.S.; data curation, S.R.; writing—original draft preparation, S.R.; writing—review and editing, A.S., A.Č. and J.R.; visualization, S.R.; supervision, A.S. All authors have read and agreed to the published version of the manuscript.

Funding: This research received no external funding.

Data Availability Statement: The data presented in this study are available on request from the corresponding author. The data are not publicly available due to copyright requirements of the text data sources (security standards).

Conflicts of Interest: The authors declare no conflict of interest.

References

1. Spectrum News NY1. Available online: https://www.ny1.com/nyc/all-boroughs/ap-online/2022/03/08/beefing-up-security-google-buys-mandiant-for-54-billion (accessed on 13 March 2022).
2. Six Degrees. Available online: https://www.6dg.co.uk/blog/cyber-security-budget-trends/ (accessed on 13 March 2022).
3. Borum, R.; Felker, J.; Kern, S.; Dennesen, K.; Feyes, T. Strategic cyber intelligence. *Inf. Comput. Secur.* **2015**, *23*, 317–332. [CrossRef]
4. Bonfanti, M.E. Cyber Intelligence: In pursuit of a better understanding for an emerging practice. *Cyber Intell. Secur.* **2018**, *2*, 105–121.
5. Kinyua, J.; Awuah, L. AI/ML in Security Orchestration, Automation and Response: Future Research Directions. *Intell. Autom. Soft Comput* **2021**, *28*, 527–545. [CrossRef]
6. Sanagavarapu, L.M.; Iyer, V.; Reddy, R. A Deep Learning Approach for Ontology Enrichment from Unstructured Text. In *Cybersecurity & High-Performance Computing Environments: Integrated Innovations, Practices, and Applications*, 1st ed.; Li, K.C., Sukhija, N., Bautista, E., Gaudiot, J.L., Eds.; Taylor and Francis: New York, NY, USA, 2022.
7. Studer, R.; Benjamins, V.R.; Fensel, D. Knowledge engineering: Principles and methods. *Data Knowl. Eng.* **1998**, *25*, 161–197. [CrossRef]
8. Menges, F.; Sperl, C.; Pernul, G. Unifying cyber threat intelligence. In Proceedings of the International Conference on Trust and Privacy in Digital Business, Linz, Austria, 26–29 August 2019.
9. Wang, N. A Knowledge Model of Digital Evidence Review Elements Based on Ontology. *Digit. Forensics Forensic Investig.* **2017**, *9*, 281–290. [CrossRef]
10. Peldszus, S.; Bürger, J.; Kehrer, T.; Jürjens, J. Ontology-driven evolution of software security. *Data Knowl. Eng.* **2021**, *134*, 101907. [CrossRef]
11. Dwivedi, A.K.; Satapathy, S.M. Ontology-Based Modelling of IoT Design Patterns. *J. Inf. Knowl. Manag.* **2021**, *20* (Suppl. 1), 2140003. [CrossRef]
12. Kenaza, T. An ontology-based modelling and reasoning for alerts correlation. *Int. J. Data Min. Model. Manag.* **2021**, *13*, 65–80.
13. Asgarli, E.; Burger, E. Semantic ontologies for cyber threat sharing standards. In Proceedings of the 2016 IEEE Symposium on Technologies for Homeland Security (HST), Waltham, MA, USA, 10–11 May 2016.
14. Schumacher, M. 6. toward a security core ontology. In *Security Engineering with Patterns*; Goos, G., Hartmanis, J., Leeuwen, J., Eds.; Springer: Berlin/Heidelberg, Germany, 2003; pp. 87–96.

15. Tsoumas, B.; Gritzalis, D. Towards an ontology-based security management. In Proceedings of the 20th International Conference on Advanced Information Networking and Applications-Volume 1 (AINA'06), Vienna, Austria, 18–20 April 2006.
16. Kim, A.; Luo, J.; Kang, M. Security ontology for annotating resources. In Proceedings of the OTM Confederated International Conferences on the Move to Meaningful Internet Systems, Agia Napa, Cyprus, 31 October–4 November 2005.
17. Ekelhart, A.; Fenz, S.; Klemen, M.D.; Weippl, E.R. Security ontology: Simulating threats to corporate assets. In Proceedings of the International Conference on Information Systems Security, Samos Island, Greece, 30 August–2 September 2006.
18. Blackwell, C. A security ontology for incident analysis. In Proceedings of the Sixth Annual Workshop on Cyber Security and Information Intelligence Research, Oak Ridge, TN, USA, 21–23 April 2010.
19. Souag, A.; Salinesi, C.; Mazo, R.; Comyn-Wattiau, I. A security ontology for security requirements elicitation. In Proceedings of the International Symposium on Engineering Secure Software and Systems, Milan, Italy, 4–6 March 2015.
20. Singh, V.; Pandey, S.K. Cloud Security Ontology (CSO). In *Cloud Computing for Geospatial Big Data Analytics*; Das, H., Barik, R.K., Dubey, H., Roy, D.S., Eds.; Springer: Cham, Switzerland, 2019; pp. 81–109.
21. Gonzalez-Gil, P.; Martinez, J.A.; Skarmeta, A.F. Lightweight data-security ontology for IoT. *Sensors* **2020**, *20*, 801. [CrossRef]
22. Dora, J.R.; Nemoga, K. Ontology for Cross-Site-Scripting (XSS) Attack in Cybersecurity. *J. Cybersecur. Priv.* **2021**, *1*, 319–339. [CrossRef]
23. Guo, M.; Wang, J.A. An ontology-based approach to model common vulnerabilities and exposures in information security. In Proceedings of the ASEE Southest Section Conference, Marietta, GA, USA, 5–7 April 2009.
24. Zhu, L.; Zhang, Z.; Xia, G.; Jiang, C. Research on vulnerability ontology model. In Proceedings of the 2019 IEEE 8th Joint International Information Technology and Artificial Intelligence Conference (ITAIC), Chongqing, China, 24–26 May 2019.
25. Parkin, S.E.; van Moorsel, A.; Coles, R. An information security ontology incorporating human-behavioural implications. In Proceedings of the 2nd International Conference on Security of Information and Networks, Famagusta, Cyprus, 6–10 October 2009.
26. Ramanauskaitė, S.; Olifer, D.; Goranin, N.; Čenys, A. Security ontology for adaptive mapping of security standards. *Int. J. Comput. Commun. Control (IJCCC)* **2013**, *8*, 813–825. [CrossRef]
27. Fenz, S.; Plieschnegger, S.; Hobel, H. Mapping information security standard ISO 27002 to an ontological structure. *Inf. Comput. Secur.* **2016**, *25*, 452–473. [CrossRef]
28. Missikoff, M.; Velardi, P.; Fabriani, P. Text mining techniques to automatically enrich a domain ontology. *Appl. Intell.* **2003**, *18*, 323–340. [CrossRef]
29. Moreno, A.; Perez, C. From text to ontology: Extraction and representation of conceptual information. In Proceedings of the Conference on TIA, Nancy, France, 3–5 May 2001.
30. Buitelaar, P.; Olejnik, D.; Sintek, M. A protégé plug-in for ontology extraction from text based on linguistic analysis. In Proceedings of the European Semantic Web Symposium, Heraklion, Greece, 10–12 May 2004.
31. Velardi, P.; Fabriani, P.; Missikoff, M. Using text processing techniques to automatically enrich a domain ontology. In Proceedings of the International Conference on Formal Ontology in Information Systems, Ogunquit, ME, USA, 17–19 October 2001.
32. Witte, R.; Khamis, N.; Rilling, J. Flexible Ontology Population from Text: The OwlExporter. In Proceedings of the International Conference on Language Resources and Evaluation, LREC 2010, Valletta, Malta, 17–23 May 2010.
33. Kang, Y.B.; Haghighi, P.D.; Burstein, F. CFinder: An intelligent key concept finder from text for ontology development. *Expert Syst. Appl.* **2014**, *41*, 4494–4504. [CrossRef]
34. Biemann, C. Ontology learning from text: A survey of methods. *LDV Forum* **2005**, *20*, 75–93.
35. Poon, H.; Domingos, P. Unsupervised ontology induction from text. In Proceedings of the 48th Annual Meeting of the Association for Computational Linguistics, Uppsala, Sweden, 11–16 July 2010.
36. Lee, C.S.; Kao, Y.F.; Kuo, Y.H.; Wang, M.H. Automated ontology construction for unstructured text documents. *Data Knowl. Eng.* **2007**, *60*, 547–566. [CrossRef]
37. Buitelaar, P.; Cimiano, P.; Magnini, B. Ontology learning from text: An overview. *Ontol. Learn. Text Methods Eval. Appl.* **2005**, *123*, 3–12.
38. Dahab, M.Y.; Hassan, H.A.; Rafea, A. TextOntoEx: Automatic ontology construction from natural English text. *Expert Syst. Appl.* **2008**, *34*, 1474–1480. [CrossRef]
39. Kaushik, N.; Chatterjee, N. Automatic relationship extraction from agricultural text for ontology construction. *Inf. Processing Agric.* **2018**, *5*, 60–73. [CrossRef]
40. Mathews, K.A.; Kumar, P.S. Extracting ontological knowledge from textual descriptions through grammar-based transformation. In Proceedings of the Knowledge Capture Conference, Austin, TX, USA, 4–6 December 2017.
41. Celjuska, D.; Vargas-Vera, M. Ontosophie: A semi-automatic system for ontology population from text. In Proceedings of the International Conference on Natural Language Processing (ICON), Hyderabad, India, 19–22 December 2004.
42. Wang, J.; Liu, J.; Kong, L. Ontology construction based on deep learning. In Proceedings of the International Conference on Ubiquitous Information Technologies and Applications (CUTE 2016), Bangkok, Thailand, 19–21 December 2016.
43. Wong, W.; Liu, W.; Bennamoun, M. Ontology learning from text: A look back and into the future. *ACM Comput. Surv. (CSUR)* **2012**, *44*, 1–36. [CrossRef]
44. Al-Aswadi, F.N.; Chan, H.Y.; Gan, K.H. Automatic ontology construction from text: A review from shallow to deep learning trend. *Artif. Intell. Rev.* **2020**, *53*, 3901–3928. [CrossRef]

45. Couto, F.M.; Silva, M.J.; Coutinho, P.M. Finding genomic ontology terms in text using evidence content. *BMC Bioinform.* **2005**, *6*, 1–6. [CrossRef] [PubMed]
46. Mulwad, V.; Li, W.; Joshi, A.; Finin, T.; Viswanathan, K. Extracting information about security vulnerabilities from web text. In Proceedings of the 2011 IEEE/WIC/ACM International Conferences on Web Intelligence and Intelligent Agent Technology, Lyon, France, 22–27 August 2011.
47. Joshi, A.; Lal, R.; Finin, T.; Joshi, A. Extracting cybersecurity related linked data from text. In Proceedings of the 2013 IEEE Seventh International Conference on Semantic Computing, Washington, DC, USA, 16–18 September 2013.
48. Wali, A.; Chun, S.A.; Geller, J. A bootstrapping approach for developing a cyber-security ontology using textbook index terms. In Proceedings of the 2013 International Conference on Availability, Reliability and Security, Washington, DC, USA, 2–6 September 2013.
49. Geller, J.; Chun, S.A.; Wali, A. A Hybrid Approach to Developing a Cyber Security Ontology. In Proceedings of the 3rd International Conference on Data Management Technologies and Applications, Vienna, Austria, 29–31 August 2014.
50. Aksu, M.U.; Bicakci, K.; Dilek, M.H.; Ozbayoglu, A.M.; Tatli, E.I. Automated generation of attack graphs using NVD. In Proceedings of the Eighth ACM Conference on Data and Application Security and Privacy, Tempe, AZ, USA, 19–21 March 2018.
51. Gillani, S.; Ko, A. Incremental ontology population and enrichment through semantic-based text mining: An application for it audit domain. *Int. J. Semant. Web Inf. Syst. (IJSWIS)* **2015**, *11*, 44–66. [CrossRef]
52. Ko, A.; Gillani, S. Ontology maintenance through semantic text mining: An application for it governance domain. In *Innovations, Developments, and Applications of Semantic Web and Information Systems*; Lytras, M.D., Aljohani, N., Damiani, E., Chui, K.T., Eds.; IGI Global: Hershey, PN, USA, 2018; pp. 350–371.
53. UCF Mapper. Available online: https://www.ucfmapper.com/overview/mapping-approach/modern/ (accessed on 13 March 2022).
54. Xiong, W.; Legrand, E.; Åberg, O.; Lagerström, R. Cyber security threat modeling based on the MITRE Enterprise ATT&CK Matrix. *Softw. Syst. Modeling* **2021**, *21*, 1–21.
55. Attack Flow—Beyond Atomic Behaviors. Available online: https://medium.com/mitre-engenuity/attack-flow-beyond-atomic-behaviors-c646675cc793 (accessed on 26 March 2022).

Article

Ransomware-Resilient Self-Healing XML Documents

Mahmoud Al-Dwairi [1,†], Ahmed S. Shatnawi [2,*,†], Osama Al-Khaleel [1,†] and Basheer Al-Duwairi [3,†]

1. Department of Computer Engineering, Jordan University of Science and Technology, P.O. Box 3030, Irbid 22110, Jordan; mndwairi14@cit.just.edu.jo (M.A.-D.); oda@just.edu.jo (O.A.-K.)
2. Department of Software Engineering, Jordan University of Science and Technology, P.O. Box 3030, Irbid 22110, Jordan
3. Depatment of Network Engineering & Security, Jordan University of Science and Technology, P.O. Box 3030, Irbid 22110, Jordan; basheer@just.edu.jo
* Correspondence: ahmedshatnawi@just.edu.jo; Tel.: +962-7910-803-57
† These authors contributed equally to this work.

Abstract: In recent years, various platforms have witnessed an unprecedented increase in the number of ransomware attacks targeting hospitals, governments, enterprises, and end-users. The purpose of this is to maliciously encrypt documents and files on infected machines, depriving victims of access to their data, whereupon attackers would seek some sort of a ransom in return for restoring access to the legitimate owners; hence the name. This cybersecurity threat would inherently cause substantial financial losses and time wastage for affected organizations and users. A great deal of research has taken place across academia and around the industry to combat this threat and mitigate its danger. These ongoing endeavors have resulted in several detection and prevention schemas. Nonetheless, these approaches do not cover all possible risks of losing data. In this paper, we address this facet and provide an efficient solution that would ensure an efficient recovery of XML documents from ransomware attacks. This paper proposes a self-healing version-aware ransomware recovery (SH-VARR) framework for XML documents. The proposed framework is based on the novel idea of using the link concept to maintain file versions in a distributed manner while applying access-control mechanisms to protect these versions from being encrypted or deleted. The proposed SH-VARR framework is experimentally evaluated in terms of storage overhead, time requirement, CPU utilization, and memory usage. Results show that the snapshot size increases proportionately with the original size; the time required is less than 120 ms for files that are less than 1 MB in size; and the highest CPU utilization occurs when using the bzip2. Moreover, when the zip and gzip are used, the memory usage is almost fixed (around 6.8 KBs). In contrast, it increases to around 28 KBs when the bzip2 is used.

Keywords: ransomware; XML documents; secure document engineering self-healing

Citation: Al-Dwairi, M.; Shatnawi, A.S.; Al-Khaleel, O.; Al-Duwairi, B. Ransomware-Resilient Self-Healing XML Documents. *Future Internet* **2022**, *14*, 115. https://doi.org/10.3390/fi14040115

Academic Editor: Leandros Maglaras

Received: 12 March 2022
Accepted: 5 April 2022
Published: 7 April 2022

Publisher's Note: MDPI stays neutral with regard to jurisdictional claims in published maps and institutional affiliations.

Copyright: © 2022 by the authors. Licensee MDPI, Basel, Switzerland. This article is an open access article distributed under the terms and conditions of the Creative Commons Attribution (CC BY) license (https://creativecommons.org/licenses/by/4.0/).

1. Introduction

The progression of cybercrime and the development and adoption of new techniques to jeopardize sensitive information and impart damage across the Internet present an alarming threat to businesses, governments, and nations. Recent cybersecurity research (e.g., the works in [1–6]) confirms cybercriminals' determination to develop newer techniques for achieving their malicious objectives. Ransomware is just one of the methods that have been used recently by cybercriminals to achieve financial gains in return for releasing ransomware-encrypted files to their rightful owners. Ransomware attacks represent a real security threat to users' data files and various network resources that would contain backup files. Amongst others, a conservative estimate is that ransomware criminals received USD 412 million in payments in 2020 [7]. Ransomware attacks impact individuals and organizations in the public and private sectors, including, amongst many, the health sector, e-commerce, educational institutions, government agencies, and the business sectors, in a manner that leads to economic and moral loss. In 2017, the WannaCry Ransomware [8], a recent massive Ransomware attack, impacted up to 300,000 users in 150 countries worldwide, preventing them from accessing their devices and demanding Bitcoin payments in exchange for unlocking the files involved.

With an ever-increasing rate of storing and sharing data, document security is becoming one of the biggest challenges that faces both individuals and organizations. Here, digital documents are represented in many formats, one of the most popular of which includes the Extensible Markup Language (XML). When Ransomware attacks victims' machines, it will seek to lock or encrypt users' crucial files and documents, including XML-based documents such as ".docx" and ".odt" file types.

Since 2010, the rate of infection by Ransomware has increased significantly. This growing threat has received significant attention from both academia and industry. Many research studies have intensely served to analyze Ransomware and develop new techniques to detect it, as long as it considers backup. However, a significant portion of all proposed detection techniques claims to have a high detection success rate. Nonetheless, most detection and protection systems in use have several limitations.

In this study, we address the problem of recovering XML documents once a ransomware attack has taken place. We propose a self-healing version-aware XML recovery framework to combat Ransomware to achieve this goal. The proposed framework takes advantage of the structure of XML documents and combines link-based version control with well-known access-control mechanisms.

The Version-Control System (VCS) manages all the changes made to documents, including tracking and storing versioning data. In this paper, VCS will be tapped into by presenting a novel approach directed at recovering ransomware-infected XML-based files and documents. Version-Aware XML-based documents are part of a distributed version-control system that does not rely on a central repository but refers to the document file itself in tracking each subsequent version of a document.

The work presented in this paper focuses mainly on protecting XML-based documents such as ".docx" and ".odt" files from being encrypted by Ransomware. The proposed framework integrates decentralized version control that utilizes file links with access-control mechanisms to prevent Ransomware from tampering with the protected file version. Therefore, It ensures complete recovery of protected XML-based documents from ransomware infection. To that end, the main contributions of this work are as follows:

- A self-healing version-aware ransomware recovery framework for XML-based documents is identified.
- The proposed framework is evaluated according to different performance metrics, including storage overhead, CPU utilization, and memory requirements for about 500 XML-based documents of various sizes, ranging from a few kilobytes to 30 Megabytes.

The rest of this paper is organized as follows: Section 2 provides background information on information security, Ransomware, and version-control systems. Section 3 reviews some pieces of related work. Section 4 presents the proposed system. The performance evaluation part is presented in Section 5. Finally, we conclude in Section 6.

2. Background

The field of Information Security is one of the most critical fields in the IT world. Ensuring the protection of information assets is a top priority for users and organizations because the data stored on a computer are certainly worth more than the computer itself. Cybersecurity's critical goal is to protect data transferred over the network and its connected resources against any security threat. There are three main objectives for information security that are deemed primary pillars of cybersecurity. These pillars are Confidentiality, Integrity, and Availability; otherwise referred to as the Security Requirements Triad [9] or the CIA triangle. These three objectives are highly recognized across the security-concerned communities. Confidentiality means that the information is accessed only by authorized parties with sufficient privileges. It guarantees privacy, meaning that the individuals control what information is related to them, who can collect such information, and to whom a set of given data can be revealed. Integrity guarantees that the data stored on computers and other resources are correct and that either unauthorized people or malware do not manipulate pieces of data. It is more critical than availability and confidentiality. On the other hand, availability ensures connectivity for authorized users of network resources.

Two additional objectives are sometimes added to these pillars: Authenticity and Accountability. The extended model is known as the CIA+ model, as elaborated in [10]. Authenticity ensures that the message received is the same as the one sent without alteration or tampering; it ensures that it was sent from trusted sources; something that warrants truthfulness of origins. Accountability is related to the individual or organization's responsibility to trace the actions performed on their systems and perform preventive and defensive measures to counter these threats. This includes taking backup for essential data, instating fault isolation, ensuring proper intrusion detection and prevention, conducting after-action recovery, and taking legal action.

2.1. Ransomware

Ransomware is defined as a form of malware that prevents users from accessing their resources and files either by encryption or blockage until a ransom is rendered to restore access to infected files. It provides a means for money-based extortion that affects both individuals and organizations [11]. It is a piece of software designed and implemented by cybercriminals to gain access to legitimate users without their knowledge and to perform malicious activities such stealing sensitive data and asking for a ransom. Due to a lack of proper technical background with little knowledge of how to preserve their data, short of making necessary file backups, some users, especially naive ones, end up paying ransom to restore access to their files. This ultimately leads cybercriminals and attackers to gain more significant revenues and helps to make this an opportunity for thriving businesses [12].

In 1989, the first ransomware attack was reported when infected floppy disks with AIDS Trojan were distributed amongst biologists. The malware encrypted all the victims' system files with a ransom of USD 189 to undo the damage. The earliest variants of Ransomware were developed in 1980 [13]. Ransom was paid via postal mail. Today, ransomware authors order that payment is rendered via credit cards or cryptocurrency such as bitcoin [14].

In recent years there has been an increasing proliferation rate of different types of ransomware families that are spread like a worm, which involve advanced recovery-prevention schemes. This impacts home users, organizations, and the infrastructures of vital governmental establishments around the world [11].

WannaCry and Petaya [8] are examples of recent Ransomware which spreads through insecure compromised websites, exploiting weaknesses inherent in Microsoft Windows. On 12 May 2017, WannaCry was first observed as part of massive attacks over multiple countries [15]. These attacks affected many vital sectors, including government organizations and the healthcare and telecommunications sectors. WannaCry is an example of crypto Ransomware that is based on public-key cryptography; something that is rather challenging to mitigate or recover from, as the encryption keys are stored on a remote command and control server (C&C). In the following subsections, we explain the ransomware lifecycle and main ransomware categories:

2.1.1. Ransomware Lifecycle

The authors of [16] analyzed 25 ransomware families and found that they all possess similar dynamics. They differ somewhat, however, according to the ransomware versions in place, but exhibit a similar overall high-level pattern. In general, the ransomware lifecycle spans the following six steps [16]:

- Ransomware distribution: Like other malicious software programs, Ransomware uses social-engineering strategies to seduce victims to click links that lead to ridiculous content or download a malicious dropper or payload that causes infection.

- Infection: The malicious code is downloaded at this stage, and the execution of the code begins. At this stage, a victim's machine will have been compromised by Ransomware, with the underlying files still not yet encrypted. Encryption is a reversible process, involving highly intensive CPU calculations operations. Encryption does not readily happen in a typical ransomware attack as it requires time for data evaluation by the malware and the scope for data encryption. Once this stage becomes active, all the automatic detection systems will have stopped. The firewall, proxy, antivirus, and intrusion detection programs will have been compromised to allow all malicious communications to take place, ultimately putting the ransomware in total control.
- C2 Communications: The malicious code continues to maintain access to its command-and-control server (C2) at this stage. Here, an attacker manages a C2 server and begins to send commands to the compromised system. The primary C2 communications objective with Ransomware entails the acquisition of an encryption key. Once that is complete, different systems are changed, and persistence is determined.
- File search-scanning: This is when things start to slow down a bit. The malware searches the computer to find files to encrypt first. It also scans for cloud data that are synced through folders and shown as local data. Then it starts searching for file shares. This may take time, depending on how much activity there is across the network. The goal is to examine the available information and determine the victim's level of permissions (e.g., list, published, delete).
- Encryption: The encryption starts once all data have been inventoried. Local file encryption may take minutes, but it may take several hours to encrypt a network file; this is because data on network file shares are locally copied and encrypted in most ransomware attacks. Then this is followed by uploading the encrypted files and removing the original ones. This phase takes a bit of extra time.
- Ransom demand: At this stage, a victim will receive a ransom message instructing them to render ransom; the Ransomware message is issued immediately once encryption has taken place. The Ransomware shows a screen that instructs its victim to pay before criminals delete the key to decrypt the files. The last function usually performed by Ransomware is to end and uninstall itself from a victim's machine. At this point, the hackers are ready to receive the ransom to their Bitcoin wallet.

2.1.2. Ransomware Categories

Ransomware falls under three main categories ranging from severe to damaging: Scareware, Locker Ransomware, and crypto Ransomware. Table 1 summarizes these categories. Scareware is a form of malicious software that overwhelms users' screens with warnings and pop-ups claiming that issues are detected on the users' PC and it requires money to fix them. If the victim falls in for this trick and installs the malware on their machines, the cybercriminal/s would use this malware to access their files, send out fake emails in their names, and/or track their online activity. Locker Ransomware is malicious software that infects the operating system and prevents users from accessing their files and data. It hijacks one or more of the victim's system services, such as desktops, smartphones, and applications, depriving users of those tools from accessing them [11]. This attack usually takes the form of a locking computer interface asking the user to pay a ransom for re-access. Often, infected computers are left with limited capabilities to allow the user to communicate with ransomware and conduct-related activities to pay the requested ransom. For example, W32. Rasith is a worm that locks the victim's desktop, making the system unusable [17]. This type is not limited to PCs or servers alone, but it also affects mobile devices. Android.Lockdroid.H is an example of a trojan that locks the screen of mobile devices and displays a ransom message [17]. Since Locker ransomware is designed to prevent access to the device's interface, the underlying system and files are left untouched. It is possible to restore the computer to a state close to its original condition. Thus, Locker ransomware is less effective at eliciting ransom payments.

Although cryptography is regarded as a critical defense mechanism in computer and network applications [18], it can also be used to perform crypto crimes. The work in [19] is one of the earliest research studies on fraudulent cryptographic use. What distinguishes Ransomware from conventional malware is that it utilizes cryptography techniques, including symmetric and asymmetric key-based encryption, against victims, as discussed in [20]. This type is the most common type of Ransomware. It is the most harmful type and can cause a great deal of damage, thereby extorting vast amounts of money. This type of Ransomware is considered the most dangerous because once the attacker gets hold of the files, there is no way to restore them until a ransom is rendered for file restoration. Here, WannaCry [8] is one famous example.

Crypto ransomware encrypts victims' files, file contents, and file names without notification by utilizing different cryptographic methods and notifies victims that their data have been encrypted, forcing them to pay a ransom to decrypt files [12]. Since 2016, crypto Ransomware attacks have increased dramatically. According to a report by [21], 58.43% of ransomware attacks are conducted by a crypto Ransomware strain called TeslaCrypt. CTB-Locker was considered one of the primary ransomware attacks in 2016. CTB-Locker can attack multiple victims at the same time. Thus, during the same attack, it can extort several victims. This infects web servers by encrypting webroot, causing web servers, host applications, and websites to become paralyzed [21].

Table 1. Ransomware Categories.

Category	Symptoms	Example
Locker	prevents users from accessing their files and data	W32. Rasith Data
Crypto	Encrypts victims' files, file contents, and file names without notification by utilizing different cryptographic methods and notifies victims that their data have been encrypted, forcing them to pay a ransom to decrypt files.	WannaCry
Double extortion	Encrypts files and asks victims to pay a ransom. Attackers threaten to publicize stolen data if their demands are not met.	Maze
RaaS	Involves perpetrators leasing access to ransomware from the ransomware author, who delivers it as a paid service.	Locky

2.2. Version-Control System (VCS)

Version-control systems (VCS) are used to manage all changes made to documents, including tracking and storing version data. In this paper, VCS will be tapped into by presenting a novel approach to recovering XML documents affected when Ransomware attacks victims' machines, causing locking of file encryption. Version-Aware XML-based documents is a distributed version-control system that does not rely on a central repository but refers to the document file to utilize the changes between different versions of the same document. version-control is a system used for tracking all files or file set changes over time to allow for the subsequent release of a specific version of the file so that you can obtain a specific version of the file later. As VCS became popular, new techniques continued to evolve. It uses two main techniques to store versions of data. The first one is to keep a copy of each new version of the file, while the second one would keep only the deltas, which are the data differences between the two versions of the file. There are two major version-control types: centralized and distributed. A centralized version-control system is based on client–server architecture where a central repository is used to store the document versions. Centralized VCS must be used online as it requires the end-user (client) to be connected to the system (central repository) at all times. Using this approach makes it possible to elicit single points of failure [22].

A distributed version-control system, also known as Version-Aware XML document (used in our approach) was first introduced in [22]. In contrast to centralized VCS, version–aware VCS does not depend on a central repository to store versions data. It utilizes reverse deltas stored inside the document file itself, which are the data differences between the two versions of a file, rather than storing the whole document every time. By using Version-Aware XML document technology, users are not worried about the need to use a repository or network connection to remote servers. LibreOffice documents (ODT) are XML schemas that store files, styles, and settings. The authors of [23] created a Custom Microsoft Word plugin to support Version-Aware XML documents technology. Revisions of the document content are stored as a separate copy (snapshot) in a sub-directory inside the document. Shatnawi et al. [24,25] proposed a secure framework for XML documents that improves security for XML documents and their provenance and provides persistent integrity, detects tampering, and provides tools for performing forensics by utilizing version-aware XML document technology. Their approach provides an extensive document history with author signatures at each step, which also enhances the performance when applying security policies applied to documents.

3. Related Work

Cybersecurity researchers have extensively investigated malware attacks over the last few years. In particular, Ransomware has received significant attention among existing research works. Many researchers have studied Ransomware, analyzed its characteristics and properties, and explored how it affects impacted victims. Meanwhile, they have conducted their research work by proposing different approaches to detect and recover from ransomware attacks.

3.1. Ransomware Analysis

To recover from a ransomware attack and mitigate its impacts, we should understand how Ransomware is staged and, in the process, analyze what takes place. Analysis can be achieved by looking at the structure of Ransomware and what it does by invoking a reverse-engineering approach for multiple occurrences. The authors of [26] used reverse engineering to study ransomware samples based on code quality, functionality, and cryptographic primitives, if any. In their study, they concluded that the code is relatively basic for the most part, with high-level languages used in most instances. Both symmetric and asymmetric cryptography were employed. The analyzed samples were mainly purposed to masses, with no specific objects being targeted. While reverse engineering provides an in-depth look inside the structure of Ransomware, it is not considered a cost-effective alternative to performing reverse engineering for every ransomware sample to find a way to prevent attacks due to the complications and overheads involved.

The work in [27] performed a long-term ransomware attack analysis and reports the results of examining over 1300 samples collected between 2006 and 2014 belonging to 15 separate Ransomware families. They show that monitoring the activities in the file system would help with Ransomware detection. They concluded that families of Ransomware share very similar features in their core part, though their implementation differs. The author of [28] conducted their study on malware samples, which is readily valid for Ransomware. They proposed TTAnalyze, which can analyze the behavior of malware that comes as a Windows-executable file process on a virtual processor under an isolated environment. Other researchers were involved in studying the behavior of ransomware families on the network rather than on the local machine. The authors of [29] have, in particular, sought to analyze the network behavior of the CryptoWall Ransomware family. Here, they used HoneyPot technology, which is based on dynamic analysis concepts and an automatic run-time malware analytical system. They completed their study with the conclusion that they could identify infected machines in a dedicated environment and understand ransomware samples' network behavior. Malicious parties commonly associate Ransomware with a particular type of server called Command and Control (C&C)

servers. These are used to automatically control Ransomware and anonymously instruct it on what to do to infect other machines on the network. An approach is presented in [30] to detect communication activities between infected hosts and Command and Control servers by finding communication aggregates from multiple internal hosts that share common characteristics. The authors concluded that three aggregation functions could detect communication based on the hosts' destination, payload, and platform.

Another research effort was conducted in [31] to study how Command and Control servers operate. Instead of detecting communication activities to these servers, the authors proposed a way to make automata that can reveal the hidden specification of closed-type protocols. The solution they created does not require any information upfront, such as source code or specifications about the implementation, and was found to be able to successfully develop automata for FTP traces. The same principle could be applied to C&C servers, which are closed-type protocol automata that send replies to ransomware requests. The work in [32] presents the analysis of 14 strains of ransomware families that infect Windows platforms. This study compares the baseline of standard operating-system behavior operations, and Windows Application Programming Interface (API) calls made through Ransomware processes. This study reports notable features of Ransomware, as indicated by the frequency of API calls, without identifying code signatures within the ransomware code in order to provide a better understanding of what a particular Ransomware does to the system in API calls. The work in [33] applies data-mining techniques to connect components of multi-level code to find unique association rules to classify ransomware families through implementing static or dynamic reverse-engineering processes. The authors carried out this study using 450 ransomware samples in which they were able to identify the strong connection between the different code components that emerged from the experiments.

In [34], the authors examined ransomware attacks in a healthcare setting, duties, and the costs related to such infections as they would affect the healthcare business in general. They also discussed risk-impacts mitigation. They suggested that healthcare facilities should have a disaster plan with appropriate data backups and recovery plans and increase employees' awareness.

3.2. Ransomware Detection

In this section, we discuss the main research efforts for ransomware detection, mitigation and prevention. Detection methods rely on ransomware attack behaviors that affect computer systems such as files or network systems. They give an alarming signal to the end-users to prompt responses towards their files and important data. A SDN-based system that can improve protection against Ransomware by observing the ransomware attack is presented in [29]. By analyzing the behavior of two popular Ransomware, Cryp-toWall and Locky, they could be leveraged to detect Ransomware based on HTTPS messaging sequences and content size based on network-communication observations.

The authors of [35] proposed a Paybreak recovery solution to recover corrupted files on a victim's machine by extracting the encryption keys used to decrypt infected files following a Ransomware attack. PayBreak effectively implements a key escrow mechanism to store session keys in a key vault that can be encoded with a public user key; thus, the user may decrypt the key vault with his private key following ransomware attack. In another research work, Continella proposed ShieldFS in [36]. In this approach, the proposed scheme acts upon the operating and file system levels and serves as a shield to detect and correct any suspicious activities.

Kharraz in [27] carried out a long-term study of ransomware attacks and presents results leveraging analysis of more than 1300 samples collected between 2006 and 2014 that belong to 15 different Ransomware families. Further, the study showed that monitoring activities in the file system would ultimately help with Ransomware detection. R-locker, a general technique intended to prevent crypto Ransomware action, was first introduced by [37]. The researchers used the honeyfile technique to prevent a ransom once it accessed a trap file. Therefore, the honeyfile technique helps to preserve the data on the system. Moreover, while the ransom is blocked, a countermeasure to eliminate the issue would be beneficial to eradicate the environment's problem.

The study presented in [29] came with the ultimate objective of detecting the underlying Ransomware and mitigating its impact on the systems. The work in [38] provides a signature-based detection approach by observing the original semantics of the dataset of malware. Here, semantics are required to be as effective as malware. However, the authors conclude that malware could be detected commensurate with these signatures at higher error rates with broad classes such as Trojans. In [39], the authors introduced CryptoDrop; an early warning system for ransomware attacks to notify users during any unusual file operation. Based on popular ransomware behavior criteria, the proposed solution tracks victim data and identified Ransomware in the process. Their study conducted experiments on 492 real-world samples of Ransomware, representing 14 families, and was able to achieve high detection rates with low false positives. Ransomware designers continually keep improving their techniques to spread their attacks, especially for Ransomware types that are not easily detected. They use encryption algorithms to hide malicious code within benign code to be executed later.

Shafiqq, Khayam, and Farooq [40] proposed a detection scheme to detect embedded malware, malicious code that is hidden within benign files, using statistical abnormal detection. Yfuksel, den Hartog, and Etalle [41] described a protocol-aware anomaly detection framework that aims to monitor a network from embedded malware access by scanning a network for SBM and Microsoft Remote Procedure Call (RPC) messages. The work presented in [42] studies the whole life cycle of Ransomware creation, design, and implementation using Dynamic Data Exchange (DDE) in Python scripting language and REST APIs in PHP, with the back-end being a MySQL database. Their study aimed to prove that even though many security measures and several top-quality antivirus programs are currently in use, ransomware authors continue to develop and write dangerous malicious codes that can be distributed easily through connected devices. Meanwhile, various research endeavors have widely explored analysis and detection of Ransomware based on its characteristics, leveraging machine learning techniques. In [43], Lim and Ramli applied machine learning techniques to classify extracted static and behavioral analysis, and they developed an efficient malware analysis framework based on the mentioned analysis features addressed thus far.

An approach to efficiently detect Ransomware was presented in [44]. The authors incorporated feature-generation engines and machine learning in a reverse-engineering framework. The purpose of malware code segments is to achieve better examination and interpretation in the proposed framework by performing multilevel analyses such as raw binaries, libraries, function calls, and assembly language. Binaries are decoded to assembly level instructions and DLL libraries using the object-code dump tool (Linux) and portable executable (PE) parser. The experiments were conducted using supervised ML techniques on both Ransomware and normal binaries. Seven of the eight ML classifiers that were tested had a detection rate of at least 90%.

In [45], G. Cusack, O. Michel, and E. Keller proposed a solution using programmable data-transmission from the network-traffic-monitoring engines between the infected computer and command and control server. They derived high-level flow features from this traffic and used this dataset to detect Ransomware. A detection rate of around 0.86 was achieved in this classification model.

While Ransomware is commonly found to infect personal computers rather more frequently, the rapid spread and increased usage of mobile devices and smartphones have often led Ransomware writers and hackers to pay particular attention to this evolving market. Although mobile applications are subject to specific standards by stores before they are made available to end-users, users can still find and download infected applications from these stores. Andronio, Zanero, and Maggi [46] developed a detection scheme based on training ransomware samples called HelDriod. Their approach detects whether a particular application will attempt to lock or encrypt a mobile device without the user's approval. It can also detect ransom requests from within the text of the application itself.

Stokkel, M. [47] proposed a code using an open-source intrusion detection system called Bro to detect many samples. Alfredo Cuzzocrea, Fabio Martinelli, and Francesco Mercaldo [48] presented a fuzzy logic classification method to identify whether a mobile application exhibits Ransomware behavior; they performed their evaluation based on a dataset containing 10,052 legitimate and illegitimate android mobile applications.

The work presented in [49] proposed a detection method leveraging a Support Vector Machine (SVM). This, inherently, is considered one of a group of supervised algorithms for machine learning. By using this approach, they can identify the API calls logs of Ransomware samples based on their features. These authors evaluated this scheme using 276 real Ransomware samples and they concluded that their technique indeed increases the predictive accuracy and the correct Ransomware detection rate. Ref. [50] conducted a survey on Ransomware Detection Using the Dynamic Analysis and Machine Learning from 2019 to 2021.

3.3. Recovery from Ransomware

This section provides an overview of the literature for recovery from ransomware attacks, the proposed schemes to counter them, and the efficiencies involved. Zimba A, Wang Z, and Simukonda in [51] examined samples from crypto Ransomware through reverse engineering and dynamic analysis to evaluate a Ransomware's underlying attack structures and deletion techniques. They conclude that no matter how disruptive a crypto Ransomware attack is, the key to data recovery is the underlying attack structure and the deletion technique applied. They show that data recovery based on the structure of the attack is possible. The work presented in [52] studies the recovery of lost files due to ransomware attacks in a network-shared volume scenario. It presents a software tool that monitors the traffic and records all user actions on the file. The authors demonstrate that their proposed tool can recover the file from previous and subsequent operations without taking the encrypted content as valid data. This tool, which could recover files successfully, is evaluated based on test-traffic records of 18 different families. The work presented in [53] presents a tool to perform evaluations for Ransomware backup systems during security-risk assessment; this study would make auditors analyze backup systems effectively and improve organizational abilities to detect and recover from Ransomware attacks.

RDS3 is a novel Ransomware Defense Strategy in which it stealthily backs up data in the spare space of a computing device so that the data encrypted by ransomware can be restored [54,55]. Kim et al. [56] proposed a method to decrypt Hive ransomware and recover infected data. Continella et al. [36] described a self-healing, ransomware-aware file system by monitoring low-level filesystem activity. If a process violates a previously trained model, its operations are deemed malicious, and the side-effects on the filesystem are transparently rolled back. The work carried out by Ye et al. [57] suggests monitoring and analyzing operating systems events to ensure that a back up is created whenever a suspicious event is detected. In case the misgiving comes true, it can be rolled back.

4. Proposed Version-Aware Ransomware Recovery Framework

In this section, we describe the proposed framework for Self-Healing Version-Aware Ransomware Recovery (SH-VARR). The main goal of the proposed framework is to serve as a version-control system and assist in recovery against ransomware attacks targeting XML-based documents. To achieve this goal, we implemented a distributed version-control system by adding the absolute URL path of the original file to keep track of file versions. Further, we employed access-control techniques to protect file versions from modification or deletion. These techniques ensure protection from ransomware attacks while allowing users to keep track of older versions of their files. Here, we point out that the novelty of our proposed framework relies on the way we combine well-known techniques from access-control theory and version-control mechanisms to achieve the desired Self-Healing Version-Aware Ransomware Recovery of XM-based documents.

Figure 1 depicts the overall framework architecture. In this framework, all XML-based documents in a predefined directory go through the version-control module at the time of file closing to maintain the latest version of each document. The access-control module is activated by invoking the root daemon service to perform write protection for the snapshot version, which would be already pointing to the original file.

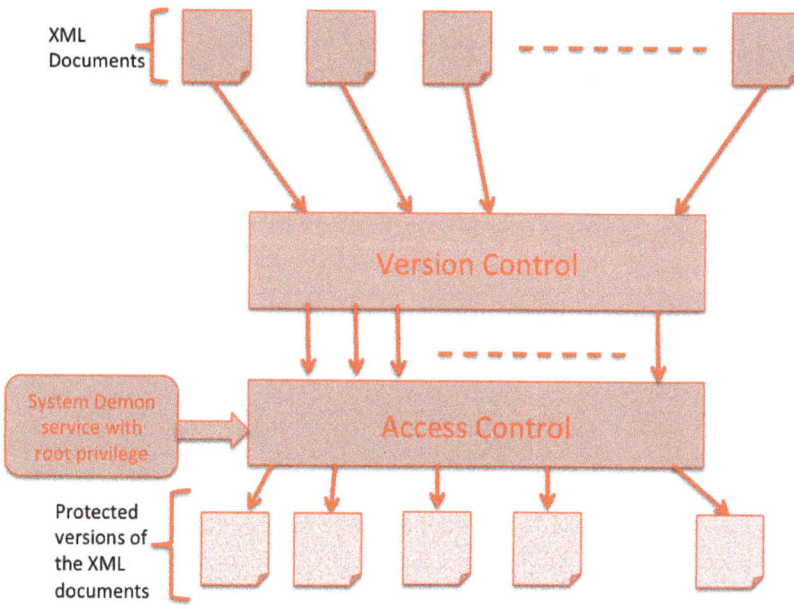

Figure 1. The overall architecture of SH-VARR framework.

4.1. Details of the Proposed SH-VARR Framework

We first describe the version-control module, illustrating the importance of using absolute URL links to keep track of old versions of a file. This is followed by a detailed description of the access-control module.

4.1.1. Version-Control Module

The version-control module is designed to maintain a copy of the XML-based file at the time of file closure so that the latest version can be retrieved in case of any corruption or system failure. We use the term *snapshot* to refer to the resulting file version. This can be achieved by adding a special plugin for Microsoft Word or LibreOffice. As part of this work, we have implemented a custom plugin for Microsoft Word 2013.

Our framework is specifically designed to recover XML-based documents in a predefined folder/directory in case of a ransomware attack. Microsoft documents and LibreOffice documents are XML-based documents that are originally compressed using the `zip` compression algorithm. To create a snapshot of a `.odt` or `.docx` file, the plugin performs the following steps:

- Step 1: Changing the `.odt`/`.docx` extension of the file to `.zip`.
- Step 2: Extracting the document archive. By unzipping the resulting `.zip` file, we obtain the document structure containing XML-based files and directories generated originally by Microsoft Word or LibreOffice. This includes configurations, meta information, content, settings, etc.

- Step 3: Adding a new XML file (`link.XML`) to the file archive that contains an absolute URL (i.e., a link) of the file version to be created in step 5.
- Step 4: Compressing the resulting ZIP archive, including the `link.XML` file.
- Step 5: Copying the resulting .zip file to a predefined directory that stores the protected versions. Access control permissions are added by the access control module as discussed in Section 4.1.2.
- Step 6: Changing the `.zip` extension of the file to `.odt`.

As an illustrative example, consider Figure 2, which shows the main steps performed by our distributed version-control module to obtain a new version for an XML-based file `abc.odt`. In this example, we assume that the file is in the user directory `/home/user/documents`. The version (i.e., a file snapshot) is created by renaming the file to `abc.zip` and then unzipping the resulting file to obtain the XML file archive. The main reason for performing this step is to add an absolute path (i.e., a link) to the location of the newly introduced version. Assuming that the file version will be stored in: `/home/user/versions` with the name `abc-version1.zip`, then the absolute path `/home/user/versions/abc-version1.zip` will be saved in the `link.XML` file that is added to the document archive in step 3. In step 4, the XML-based document archive is compressed back to obtain `abc.zip`. At this point, the file is copied to the predefined protected versions directory `/home/user/versions`. Finally, the file extension is changed to `.odt`.

Here, note that the version-control module is invoked at the time of closing the document. This ensures that a new snapshot of the XML-based document is saved each time the user closes the file. Here, we emphasize that keeping track of document history (i.e., versions) is achieved by following the absolute path stored in the link.XML file stored in each version. Figure 3 shows the approach used to retrieve older versions. Staring with the newest version (V_N), it is possible to retrieve the preceding version by following the link found in the link.XML file stored in the version itself. Older versions can be retrieved similarly. For recovery from a ransomware attack, it would be sufficient to keep the latest version only. However, suppose the objective was to retrieve older file versions while providing ransomware recovery capability. In that case, the system can be configured to store protected versions in precisely the same way as described in this section.

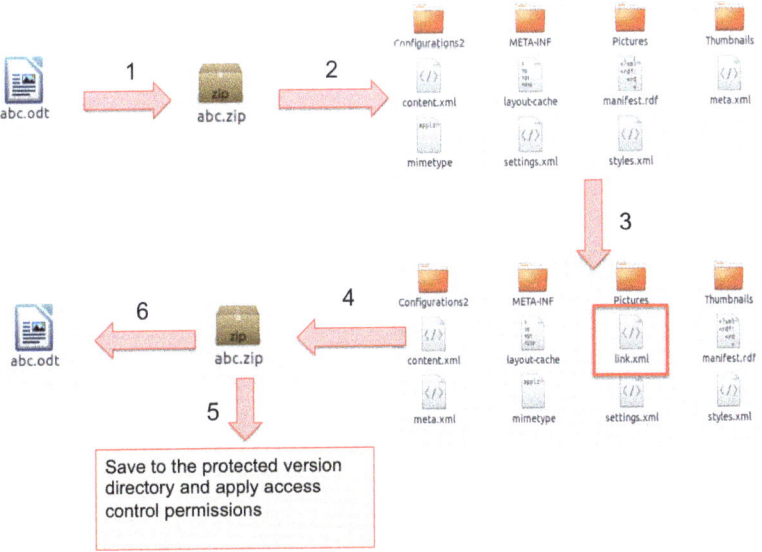

Figure 2. An illustrative example of the main steps of the version-control module.

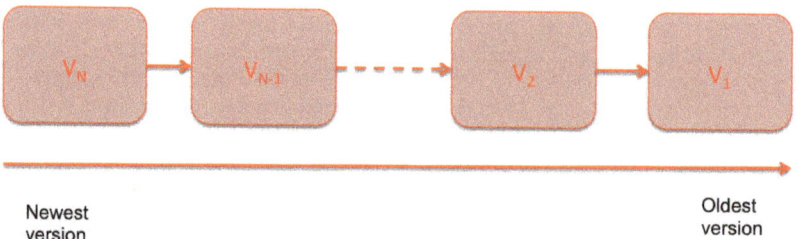

Figure 3. Keeping track of file version history based on link concept.

4.1.2. Access-Control Module

The access-control module is implemented as a root daemon that performs write/delete protection for the files produced by the version-control module each time a file version is created. This is achieved by running the chattr command (Change Attribute) with root privileges. chattr is a command line in Linux that is used to set/unset specific attributes to a file in a Linux environment to secure accidental deletion or modification of important files and folders, even by root users. Through this process, file snapshots are protected from corruption or deletion by using the change file attribute permissions with the immutable flag (i) under the Linux environment, preventing any user, including the root, from accidentally modifying and/or deleting files. An example using this command is shown in Figure 4.

```
justcb@justcb-virtual-machine:~/Desktop$ sudo chattr +i version1.zip
justcb@justcb-virtual-machine:~/Desktop$ lsattr
-------------e-- ./testnew
-------------e-- ./mate-terminal.desktop
-------------e-- ./DigOutv2.txt
-------------e-- ./file.zip
-------------e-- ./testnew.zip
-------------e-- ./in.txt
-------------e-- ./P.class
----i--------e-- ./version1.zip
```

Figure 4. An example using chattr command to perform file write/delete protection.

It is important to note that the default setting for standard users is assumed to be non-admins, with the access-control module configured as a system daemon with root access privileges executing the chattr command; this would inherently ensure the protection of newly created versions in the version-control directory. Any attempt to modify or delete a protected file will not be permitted, as shown in the example in Figure 5. This is considered a valid setting for two reasons: (i) users usually do not log into their systems as admins. In fact, one of the best practices of computer usage emphasizes that users never log in as admins. (ii) A recent report showed that 90% of ransomware instances in the wild could infect systems and encrypt files without administrative privileges [58]. This indicates that while users log in as non-admins, there is still a high possibility that Ransomware may encrypt their files. In our proposed solution, ensuring a specific access control process with administrative privileges will protect files created/edited by non-admin users.

```
justcb@justcb-virtual-machine:~/Desktop$ rm version1.zip
rm: remove write-protected regular file 'version1.zip'? y
rm: cannot remove 'version1.zip': Operation not permitted
justcb@justcb-virtual-machine:~/Desktop$
```

Figure 5. The file is immutable when trying to write or delete.

4.2. *Recovery from Ransomware Attack*

The focus of our framework for ransomware recovery is all about maintaining control of the latest possible versions of the files. As the proposed framework preserves protected versions of the files, we can gain access to the files in case of a ransomware attack. The result of the attack will corrupt the original file or even delete it. However, self-healing is achieved using the proposed SH-VARR framework by retrieving the protected version for each file stored in the version-control directory. In case the original file is deleted or encrypted by Ransomware, our SH-VARR framework allows immediate recovery of the last protected version of the file(s) involved, fulfilling the self-healing property. Based on the proposed framework, the protected snapshots will not be affected and can be recovered under root privileges assumed to be protected. The recovery process is performed by removing the sticky bit attribute to ensure that the file extension is .odt. Recovering a file from the protected versions directory is performed as follows:

- Removing the immutable flag (i) attribute. This is achieved by performing the command with root privileges only:
  $chattr -i file.dot.
- Changing the file name extension from .zip to .odt for Linux or .docx for a Windows environment.

4.3. Implementation Challenges and Limitations

Throughout this work, we conducted several experiments to ascertain that our goal of keeping a protected version of our XML-based files was achieved. Having set out to build a distributed version-aware control system for XML-based documents that ensures portability that would not depend on a centralized repository, the implemented approach was indeed found to warrant portability as it keeps a link to the original file as described above. During the implementation phase, the system was found to experience certain limitations, which can be summarized as follows:

- The proposed approach assumes a daemon is running with root privileges to keep versions protected.
- Under the Windows environment, and to ensure that our framework was well in place, we implemented a Microsoft office plugin working as a version-control system by keeping a complete snapshot of the active Word document inside the document itself upon document closure. A background process goes through iterations to span all files inside a directory or folder by calling this function. The main challenge here deals primarily with applying the permissions to the created version of each file; this is so because, under a Windows operating system, the read–write operation does not fall under permissions, but file attributes, which will be readily lost after compressing the file archive.

5. Performance Evaluation

In this section, we evaluate the proposed approach in terms of several performance metrics. To conduct our experiments, we use a repository of 500 .odt files collected from different sources, with different sizes ranging from 10 KB to 30 MB. All experiments were conducted on a Ubuntu 18.0 machine with a Core i5-1.8 GHz Intel processor and 4 GB RAM. Creating a protected version of each file was achieved by running a shell script that included all the steps outlined in the proposed framework discussed in Section 4. We performed multiple experiments to measure the performance of the proposed SH-VARR framework. SH-VARR uses zip/unzip for file compression/decompression as it is the default compression/decompression algorithm used in connection with XML documents. Meanwhile, SH-VARR still has the flexibility of operating with any other compression algorithm. Therefore, different compression algorithms were investigated investigated (zip, gzip, and bzip2) under our experimental set up. In this effort, we evaluate our proposed SH-VARR framework opposite storage overhead, time requirement, CPU utilization, and memory usage.

Creating a protected version of a file (i.e., a snapshot) represents a major step in our framework which results in extra storage requirements. Hence, our objective is to quantify the amount of the resulting storage. This overhead depends mainly on the compression algorithm used to create the snapshot. Figure 6a–c show how the storage overhead increases with the original file size for the cases when using the zip, gzip, and bzip2 compression algorithms. Figure 6d illustrates all cases together for the purpose of comparison. Generally, by increasing the file size, the size of the resulting snapshot increases proportionately. With that said, the size of the resulting file remains smaller than that of the original file. It is quite evident from the comparison that the bzip2-based SH-VARR slightly outperforms the other two versions. However, it consumes more time, as we will discuss next. This would also imply that there is a trade-off between time and storage overhead. Meanwhile, given the lower storage costs involved in today's technologies, the time required to create a protected snapshot may play out as a more pronounced factor.

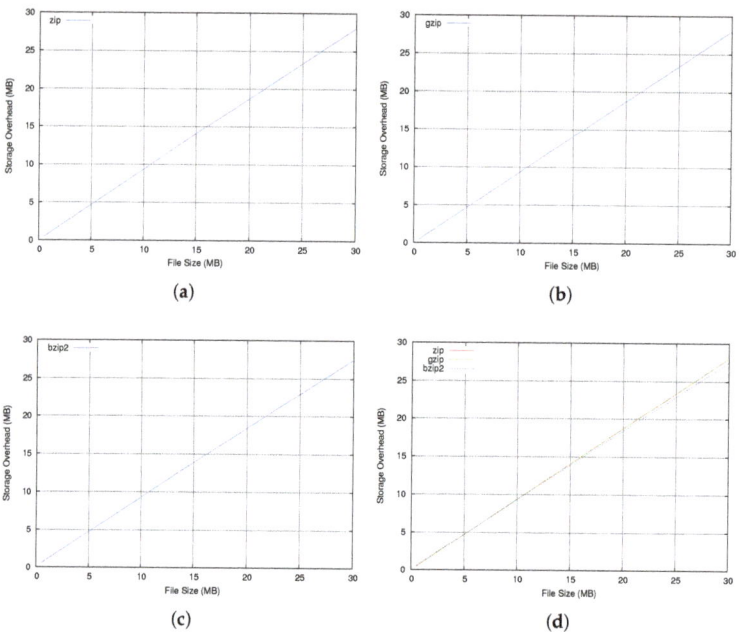

Figure 6. Storage overhead by SH-VARR snapshot based on three compression algorithms. (**a**) Using zip algorithm; (**b**) Using gzip algorithm; (**c**) Using bzip2 algorithm; (**d**) All algorithms.

The proposed SH-VARR framework involves several steps to create a protected snapshot for each file version. Therefore, it is important to measure the amount of time required to perform such an operation. Figure 7a–c show how the time requirement increases with the original file size for creating the snapshot in the proposed SH-VARR approach when leveraging the zip, gzip, and bzip2 compression algorithms, respectively. Figure 7d illustrates all cases together for comparison purposes. Creating a protected version for small files (e.g., less than 1 MB) takes a negligible amount of time that would, on average, not exceed 120 ms. However, for larger file sizes exceeding 10 MBs, more time is required to create the protected version. It can be observed that the amount of time varies as file compression depends on the amount of redundancy in each file and the type of content (e.g., text, images, etc.) contained in each file. It is evident from the outcomes of using both the zip and the gzip algorithms that the results are fairly comparable and they are seen to offer much better results than when using the bzip2 algorithm. In fact, the bzip2 is observed to consume considerable amounts of time to create the protected version, especially when the file sizes involved are quite large.

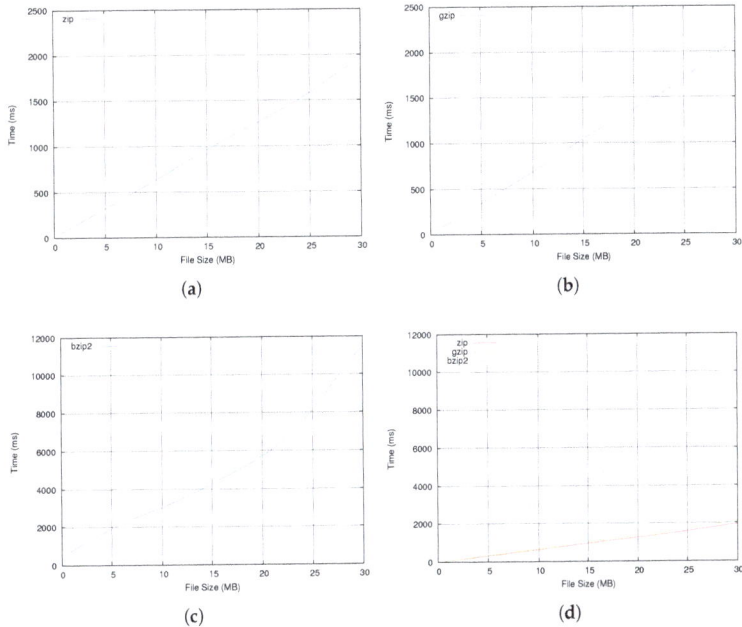

Figure 7. Time requirement for SH-VARR snapshot based on three compression algorithms. (**a**) Using zip algorithm; (**b**) Using gzip algorithm; (**c**) Using bzip2 algorithm; (**d**) All algorithms.

Figure 8a–c show how the CPU utilization varies against the original file size for creating the snapshot in the proposed SH-VARR schema when leveraging the zip, gzip, and bzip2 compression algorithms, respectively. Figure 8d illustrates all cases together for the purpose of comparison. Here, CPU utilization is the amount of work handled by the CPU while creating a protected version for each file. Generally, for small files, CPU utilization increases with increasing file size. However, for larger file sizes, it levels off to some decent value. By monitoring the CPU utilization for each job executed when creating a protected version, we observed that when the bzip2 compression algorithm was used the CPU utilization was evidently the highest.

Figure 9a–c show how the memory usage changes against the original file size to create the snapshot in the proposed SH-VARR schema when leveraging the zip, gzip, and bzip2 compression algorithms. Figure 9d illustrates all cases together for comparison purposes. It is readily seen that the memory usage, for the cases when the zip and gzip compression algorithms are used, is almost fixed (around 6.8 KBs) where it does not show any dependence on file size. Meanwhile, memory usage for the case involving the bzip2 compression algorithm is seen to increase with increasing file size, then it remains constant (around 28 KBs) for files with large sizes. This is because all the compression algorithms (zip, gzip, and bzip2) involved in our assessment of the proposed framework do not capture the entire file into the memory. Instead, they acquire it as a stream requiring a specific amount of memory each time (i.e., takes a chunk of data of a specific size each time), and the amount needed depends on the compression method used and the file size involved.

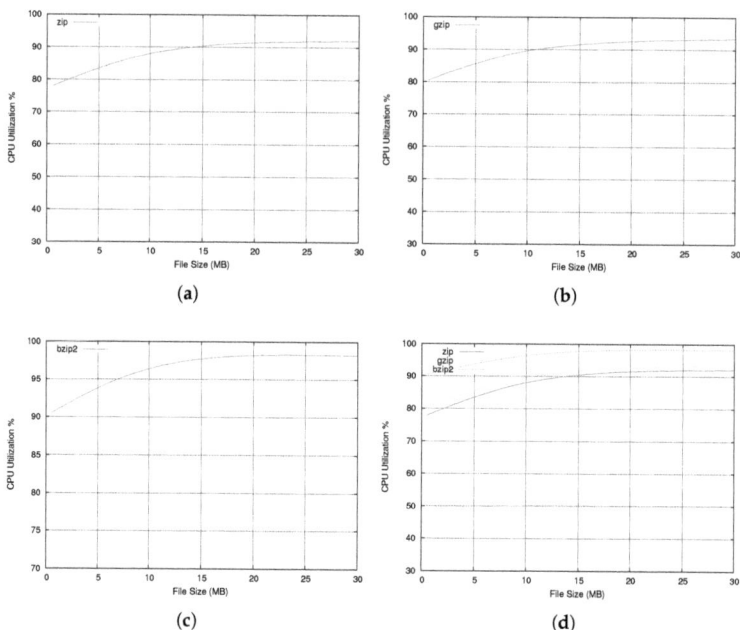

Figure 8. CPU utilization by SH-VARR snapshot based on three compression algorithms. (**a**) Using zip algorithm; (**b**) Using gzip algorithm; (**c**) Using bzip2 algorithm; (**d**) All algorithms.

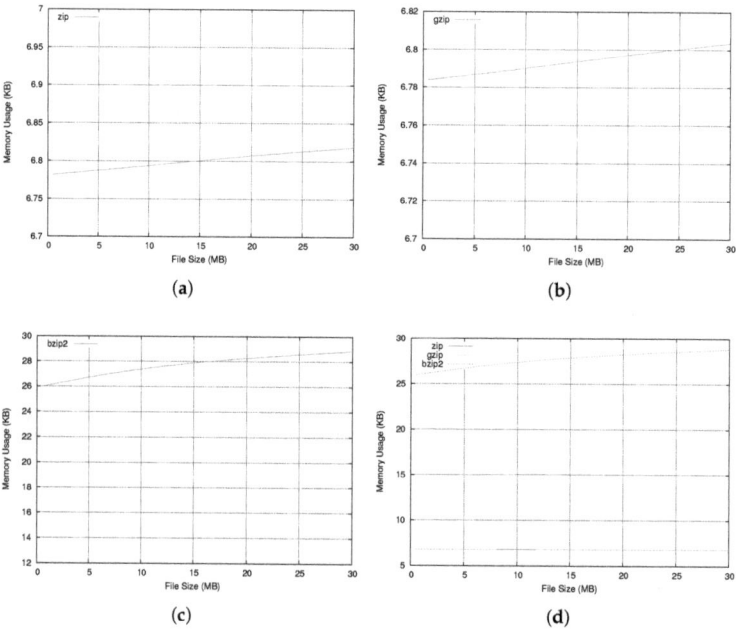

Figure 9. Memory usage by SH-VARR snapshot based on three compression algorithms. (**a**) Using zip algorithm; (**b**) Using gzip algorithm; (**c**) Using bzip2 algorithm; (**d**) All algorithms.

Finally, we compare the proposed mechanism with the work presented in [54,55]. In [55], the authors presented a Ransomware protection framework that depends on a network connection to backup files on a local or a remote server. However, they did not provide any performance evaluation of their framework in terms of time and storage requirements. In [54], the authors proposed backing up critical data in a fully isolated spare space that is not reachable by Ransomware, regardless of what privilege it can obtain. The authors assumed that the computing device has a particular portion of extra space, which can be utilized to create the backup volume to store encoded files with reverse deltas. This is different than the proposed work, where we can hold both reverse deltas and complete snapshots of files. We also used compression techniques to utilize the storage better. Moreover, our proposed work is portable because it can be shipped as a plugin that can be attached to documents; a feature that is not supported by [55] or [54].

6. Conclusions

In this paper, we introduced a Self-Healing Version-Aware Ransomware Recovery Approach (SH-VARR) of XML-based documents. This proposed system consists mainly of two modules. The first is a decentralized version-aware control system that periodically takes a backup version for each file and keeps the latest one. The second is the access-control module that executes special commands to protect the resulting versions from corruption or deletion caused by ransomware attacks; something that is carried out under administrator privileges.

The conducted set of experiments to assess the system focused on measuring the system performance in terms of the performance metrics: time, storage overhead, memory usage, and CPU utilization. Since compression is one of the main steps in the version-control system module, we evaluated these metrics by considering two commonly used compression algorithms: `bzip2` and `gzip`. Our technique (SH-VARR), introduced in this paper, uses the default zip algorithm. Comparisons show that the zip algorithm has the minimum time, size, utilization, and memory usage requirements. We conclude that this solution would protect XML-based files such as .docx and .odt files from ransomware attacks. The user can recover from such attacks even when the original files are deleted or encrypted. This is based on the assumption that these file types are compressed structures. In addition, we used a distributed version-aware control system to acquire a backup and keep track of each version. We observed access-control rules on these versions to achieve the core pillars of information security: Confidentiality, Integrity, and Availability.

Author Contributions: Conceptualization, A.S.S., O.A.-K. and B.A.-D.; methodology, M.A.-D.; software, M.A.-D. and A.S.S.; validation, M.A.-D.; writing—original draft preparation, M.A.-D.; writing—review and editing, M.A.-D., A.S.S., O.A.-K. and B.A.-D. All authors have read and agreed to the published version of the manuscript.

Funding: This research received no external funding.

Conflicts of Interest: The authors declare no conflict of interest.

References

1. Mashtalyar, N.; Ntaganzwa, U.N.; Santos, T.; Hakak, S.; Ray, S. *Social Engineering Attacks: Recent Advances and Challenges*, HCI for Cybersecurity, Privacy and Trust; Springer: New York, NY, USA, 2021; pp. 417–431
2. Mukhopadhyay, I. Cyber Threats Landscape Overview Under the New Normal. In *ICT Analysis and Applications*; Fong, S., Dey, N., Joshi, A., Eds.; Lecture Notes in Networks and Systems; Springer: Singapore, 2022; Volume 314. [CrossRef]
3. Djenna, A.; Harous, S.; Saidouni, D.E. Internet of Things Meet Internet of Threats: New Concern Cyber Security Issues of Critical Cyber Infrastructure. *Appl. Sci.* **2021**, *11*, 4580. [CrossRef]
4. Jang-Jaccard, J.; Nepal, S. A survey of emerging threats in cybersecurity. *J. Comput. Syst. Sci.* **2014**, *80*, 973–993. [CrossRef]
5. Zong, S.; Ritter, A.; Mueller, G.; Wright, E. Analyzing the Perceived Severity of Cybersecurity Threats Reported on Social Media. *arXiv* **2019**, arXiv:1902.10680.
6. Rudd, E.; Rozsa, A.; Günther, M.; Boult, T. A Survey of Stealth Malware Attacks, Mitigation Measures, and Steps Toward Autonomous Open World Solutions. *IEEE Commun. Surv. Tutor.* **2017**, *19*, 1145–1172. [CrossRef]

7. Nakashima, E. U.S. Aims to Thwart Ransomware Attacks by Cracking Down on Crypto Payments. *The Washington Post*. 2021. Available online: https://www.washingtonpost.com/business/2021/09/17/biden-sanctions-ransomware-crypto (accessed on 19 October 2021).
8. Kumar, M.; Ben-Othman, J.; Srinivasagan, K. An Investigation on Wannacry Ransomware and its Detection. In Proceedings of the 2018 IEEE Symposium on Computers and Communications (ISCC), Natal, Brazil, 25–28 June 2018; pp. 1–6.
9. Stallings, W. *Network Security Essentials: Applications and Standards*; Pearson: London, UK, 2016.
10. Peter, A.; Peter, S.; Van Ekert, L. An ontology for network security attacks. In *Proceedings of the 2nd Asian Applied Computing Conference (AACC'04), LNCS 3285*; Springer: Berlin/Heidelberg, Germany, 2004.
11. Richardson, R.; North, M. Ransomware: Evolution, mitigation and prevention. *Int. Manag. Rev.* **2017**, *13*, 10.
12. Everett, C. Ransomware: To pay or not to pay? *Comput. Fraud Secur.* **2016**, *2016*, 8–12. [CrossRef]
13. Yaqoob, I.; Ahmed, E.; Rehman, M.; Ahmed, A.; Al-garadi, M.; Imran, M.; Guizani, M. The rise of ransomware and emerging security challenges in the Internet of Things. *Comput. Netw.* **2017**, *129*, 444–458. [CrossRef]
14. Shashank, M.; Agrawal, A.K. Multi Pronged Approach for Ransomware Analysis. Available online: https://deliverypdf.ssrn.com/delivery.php?ID=529106093087077008125066087007008126061069029053059024023024048119007044109100058011016111014009004006028061086001098107006013106127099006095000116044119113035023073115003083030043113078009059098044124031019004068007115065011000084085080125073117006075066113004076094086068087090001095082&EXT=pdf&INDEX=TRUE (accessed on 10 March 2022).
15. What You Need to Know about the WannaCry Ransomware. Available online: https://symantec-enterprise-blogs.security.com/blogs/threat-intelligence/wannacry-ransomware-attack (accessed on 10 March 2022).
16. Leong, R.; Beek, C.; Cochin, C.; Cowie, N.; Schmugar, C. Understanding Ransomware and Strategies to Defeat It. 2016. Available online: https://www.mcafee.com/enterprise/en-us/assets/white-papers/wp-understanding-ransomware-strategies-defeat.pdf (accessed on 10 March 2022).
17. Al-rimy, B.; Maarof, M.; Shaid, S. Ransomware threat success factors, taxonomy, and countermeasures: A survey and research directions. *Comput. Secur.* **2018**, *74*, 144–166. [CrossRef]
18. Young, A.; Yung, M. Cryptovirology: The birth, neglect, and explosion of ransomware. *Commun. ACM* **2017**, *60*, 24–26. [CrossRef]
19. Young, A.; Yung, M. Cryptovirology: Extortion-based security threats and countermeasures. In Proceedings of the 1996 IEEE Symposium on Security and Privacy, Oakland, CA, USA, 6–8 May 1996; pp. 129–140.
20. Luo, X.; Liao, Q. Awareness education as the key to ransomware prevention. *Inf. Syst. Secur.* **2007**, *16*, 195–202. [CrossRef]
21. Gostev, A.; Unuchek, R.; Garnaeva, M.; Makrushin, D.; Ivanov, A. IT Threat Evolution in Q1 2016. Kapersky 2015 Report, Kapersky L. 2016. Available online: https://media.kasperskycontenthub.com/wp-content/uploads/sites/43/2018/03/07192617/Q1_2016_MW_report_FINAL_eng.pdf (accessed on 10 March 2022).
22. Thao, C.; Munson, E. Version-aware XML documents. In Proceedings of the 11th ACM Symposium on Document Engineering, Mountain View, CA, USA, 19–22 September 2011; pp. 97–100.
23. Coakley, S.; Mischka, J.; Thao, C. Version-Aware Word Documents. In Proceedings of the 2nd International Workshop on (Document) Changes: Modeling, Detection, Storage and Visualization, Fort Collins, CO, USA, 16 September 2014; p. 2.
24. Shatnawi, A.; Ethan, V.M.; Cheng, T. Maintaining integrity and non-repudiation in secure offline documents. In Proceedings of the 2017 ACM Symposium on Document Engineering, Valletta, Malta, 4–7 September 2017; pp. 59–62.
25. Shatnawi, A.S.; Ethan, V.M. Enhanced Automated Policy Enforcement eXchange framework (eAPEX). In Proceedings of the ACM Symposium on Document Engineering 2019, Berlin, Germany, 23–26 September 2019; pp. 1–4.
26. Gazet, A. Comparative analysis of various ransomware virii. *J. Comput. Virol.* **2010**, *6*, 77–90. [CrossRef]
27. Kharraz, A.; Kirda, E. Redemption: Real-time protection against ransomware at end-hosts. In *International Symposium on Research in Attacks, Intrusions, and Defenses*; Springer: Cham, Switzerland, 2017; pp. 98–119.
28. Bayer, U.; Kruegel, C.; Kirda, E. TTAnalyze: A Tool for Analyzing Malware. 2006. Available online: https://citeseerx.ist.psu.edu/viewdoc/download?doi=10.1.1.60.7584&rep=rep1&type=pdf (accessed on 10 March 2022).
29. Cabaj, K.; Mazurczyk, W. Using software-defined networking for ransomware mitigation: The case of cryptowall. *IEEE Netw.* **2016**, *30*, 14–20. [CrossRef]
30. Yen, T.; Heorhiadi, V.; Oprea, A.; Reiter, M.; Juels, A. An epidemiological study of malware encounters in a large enterprise. In Proceedings of the 2014 ACM SIGSAC Conference on Computer and Communications Security, Scottsdale, AZ, USA, 3–7 November 2014; pp. 1117–1130.
31. Zhang, T.; Antunes, H.; Aggarwal, S. Defending connected vehicles against malware: Challenges and a solution framework. *IEEE Internet Things J.* **2014**, *1*, 10–21. [CrossRef]
32. Hampton, N.; Baig, Z.; Zeadally, S. Ransomware behavioural analysis on windows platforms. *J. Inf. Secur. Appl.* **2018**, *40*, 44–51. [CrossRef]
33. Subedi, K.; Budhathoki, D.; Dasgupta, D. Forensic analysis of ransomware families using static and dynamic analysis. In Proceedings of the 2018 IEEE Security And Privacy Workshops (SPW), San Francisco, CA, USA, 24 May 2018; pp. 180–185.
34. Mansfield-Devine, S. Leaks and ransoms–the key threats to healthcare organisations. *Netw. Secur.* **2017**, *2017*, 14–19. [CrossRef]
35. Kolodenker, E.; Koch, W.; Stringhini, G.; Egele, M. PayBreak: Defense against cryptographic ransomware. In Proceedings of the 2017 ACM on Asia Conference on Computer And Communications Security, Abu Dhabi, United Arab Emirates, 2–6 April 2017; pp. 599–611.

36. Continella, A.; Guagnelli, A.; Zingaro, G.; De Pasquale, G.; Barenghi, A.; Zanero, S.; Maggi, F. ShieldFS: A self-healing, ransomware-aware filesystem. In Proceedings of the 32nd Annual Conference on Computer Security Applications, Los Angeles, CA, USA, 5–8 December 2016; pp. 336–347.
37. Gomez-Hernandez, J.; Gonzalez, L.; Garcia-Teodoro, P. R-Locker: Thwarting ransomware action through a honeyfile-based approach. *Comput. Secur.* **2018**, *73*, 389–398. [CrossRef]
38. Sathyanarayan, V.; Kohli, P.; Bruhadeshwar, B. Signature generation and detection of malware families. In *Australasian Conference on Information Security And Privacy*; Springer: Berlin/Heidelberg, Germany, 2008; pp. 336–349.
39. Scaife, N.; Carter, H.; Traynor, P.; Butler, K. Cryptolock (and drop it): Stopping ransomware attacks on user data. In Proceedings of the 2016 IEEE 36th International Conference On Distributed Computing Systems (ICDCS), Nara, Japan, 27–30 June 2016; pp. 303–312.
40. Shafiq, M.; Khayam, S.; Farooq, M. Improving accuracy of immune-inspired malware detectors by using intelligent features. In Proceedings of the 10th Annual Conference On Genetic And Evolutionary Computation, Atlanta, GA, USA, 12–16 July 2008; pp. 119–126.
41. Yüksel, Ö.; Hartog, J.; Etalle, S. Towards useful anomaly detection for back office networks. In *International Conference on Information Systems Security*; Springer: Cham, Switzerland, 2016; pp. 509–520.
42. Hurtuk, J.; Chovanec, M.; Kičina, M.; Billik, R. Case Study of Ransomware Malware Hiding Using Obfuscation Methods. In Proceedings of the 2018 16th International Conference on Emerging ELearning Technologies and Applications (ICETA), Stary Smokovec, Slovakia, 15–16 November 2018; pp. 215–220.
43. Lim, C.; Ramli, K. Mal-ONE: A unified framework for fast and efficient malware detection. In Proceedings of the 2014 2nd International Conference on Technology, Informatics, Management, Engineering & Environment, Bandung, Indonesia, 19–21 August 2014; pp. 1–6.
44. Poudyal, S.; Subedi, K.; Dasgupta, D. A Framework for Analyzing Ransomware using Machine Learning. In Proceedings of the 2018 IEEE Symposium Series on Computational Intelligence (SSCI), Bangalore, India, 18–21 November 2018; pp. 1692–1699.
45. Cusack, G.; Michel, O.; Keller, E. Machine learning-based detection of ransomware using sdn. In Proceedings of the 2018 ACM International Workshop on Security In Software Defined Networks & Network Function Virtualization, Tempe, AZ, USA, 21 March 2018; pp. 1–6.
46. Andronio, N.; Zanero, S.; Maggi, F. Heldroid: Dissecting and detecting mobile ransomware. In *International Symposium On Recent Advances in Intrusion Detection*; Springer: Cham, Switzerland, 2015; pp. 382–404.
47. Stokkel, M. Ransomware Detection with bro. Talk at BroCon '16. Available online: https://old.zeek.org/brocon2016/brocon2016_abstracts.html#toc-top (accessed on 20 January 2020).
48. Cuzzocrea, A.; Martinelli, F.; Mercaldo, F. A Novel Structural-Entropy-based Classification Technique for Supporting Android Ransomware Detection and Analysis. In Proceedings of the 2018 IEEE International Conference On Fuzzy Systems (FUZZ-IEEE), Rio de Janeiro, Brazil, 8–13 July 2018; pp. 1–7.
49. Takeuchi, Y.; Sakai, K.; Fukumoto, S. Detecting ransomware using support vector machines. In Proceedings of the 47th International Conference on Parallel Processing Companion, Eugene, OR, USA, 13–16 August 2018; p. 1.
50. Urooj, U.; Al-rimy, B.A.S.; Zainal, A.; Ghaleb, F.A.; Rassam, M.A. Ransomware Detection Using the Dynamic Analysis and Machine Learning: A Survey and Research Directions. *Appl. Sci.* **2022**, *12*, 172. [CrossRef]
51. Zimba, A.; Wang, Z.; Simukonda, L. Towards data resilience: The analytical case of crypto ransomware data recovery techniques. *Int. J. Inf. Technol. Comput. Sci.* **2018**, *10*, 40–51. [CrossRef]
52. Berrueta Irigoyen, E.; Morató Osés, D.; Magaña Lizarrondo, E.; Izal Azcárate, M. Ransomware encrypted your files but you restored them from network traffic. In Proceedings of the 2018 2nd Cyber Security in Networking Conference, CSnet 2018, Paris, France, 24–26 October 2018.
53. Thomas, J.; Galligher, G. Improving backup system evaluations in information security risk assessments to combat ransomware. *Comput. Inf. Sci.* **2018**, *11*. [CrossRef]
54. Subedi, K.P.; Budhathoki, D.R.; Chen, B.; Dasgupta, D. RDS3: Ransomware defense strategy by using stealthily spare space. In Proceedings of the 2017 IEEE Symposium Series on Computational Intelligence (SSCI), Honolulu, HI, USA, 27 November–1 December 2017; pp. 1–8.
55. Martínez-García, H.A. Facing ransomware: An approach with private cloud and sentinel software. *Comput. Fraud. Secur.* **2020**, *2020*, 16–19. [CrossRef]
56. Kim, G.; Kim, S.; Kang, S.; Kim, J. A Method for Decrypting Data Infected with Hive Ransomware. *arXiv* **2022**, arXiv:2202.08477.
57. Ye, H.; Dai, W.; Huang, X. File Backup to Combat Ransomware. U.S. Patent 9,317,686, 19 April 2016.
58. 90 Percent of Ransomware Can Execute without Administrator Rights-Business Reporter. Available online: https://engageemployee.com/90-per-cent-ransomware-can-execute-without-administrator-rights/ (accessed on 30 December 2019).

MDPI
St. Alban-Anlage 66
4052 Basel
Switzerland
Tel. +41 61 683 77 34
Fax +41 61 302 89 18
www.mdpi.com

MDPI Books Editorial Office
E-mail: books@mdpi.com
www.mdpi.com/books